PROPYLENE
AND ITS
INDUSTRIAL DERIVATIVES

CONTRIBUTORS

G. A. M. Diepen, dr ir, Professor of General and Inorganic Chemistry, The Technical University, Delft.

E. M. Evans, BSc, PhD, ARCS, DIC, FRIC, Research Associate, BP Chemicals International.

J. Falbe, Dr Rer Nat, Diplomchemiker, Direktor Research and Development, Ruhrchemie A. G.

J. C. Fielding, BSc, DipChemEng, CEng, MIChemE, Senior Technologist, Shell Chemicals UK Ltd.

A. J. Gait, MSc, CEng, FRIC, AMIChemE, Technical Expert, United Nations Industrial Development Organisation.

R. W. Gallant, BSChemE, Registered Professional Engineer, Production Superintendent, Dow Chemical Company.

J. Habeshaw, BSc, ARIC, BP Chemicals International Ltd.

D. J. Hadley, OBE, BSc, Research Liaison, BP Chemicals International Ltd.

P. Longi, Dr Ind Chem, Professor in Macromolecular Chemistry, Director of the Ferrara Research Centre, Montedison.

P. T. Mapp, AIRI, ANCRT, Technical Manager, General Marketing Division, Revertex Ltd.

P. Parrini, Dr Phys, Professor in Chemistry and Technology of Macromolecules ('70). Leader of the Fibres and Characterization Section of the Ferrara Research Centre, Montedison.

R. E. Sanders, BSc.(ChemE), MIChemE, Senior Process Engineer, CJB (Projects) Ltd.

G. Stahl, Ingénieur Diplomé, MSc.(Eng), Group Leader, Centre de Recherches d'Aubervilliers, Société Rhône-Progil.

H. M. Stanley, PhD, FRIC, FRS, Chemicals Adviser to BP Chemicals International Ltd.

J. C. Strini, Ingénieur Chimiste Diplomé, DrSc, Group Leader, Pilot Plant, Société Rhône-Progil.

A. Valvassori, Dr Chem, Professor in Macromolecular Chemistry, Director of Research Dept, Montedison SpA.

G. J. Woodhouse, BSc.(Tech), CEng, MIChemE, Manager, Process Design, Stone & Webster Engineering Ltd.

M. A. de Zeeuw, ir, Scientific Assistant, Institute of General and Inorganic Chemistry, The Technical University, Delft.

PROPYLENE

AND ITS
INDUSTRIAL DERIVATIVES

Edited by

E. G. HANCOCK

A HALSTED PRESS BOOK

JOHN WILEY & SONS
NEW YORK

Published in the U.S.A. by Halsted Press
A Division of John Wiley & Sons, Inc., New York

Library of Congress Cataloging in Publication Data
Main entry under title:
Propylene and its Industrial Derivatives
 "A Halsted Press Book"
 Includes bibliographical references.
 1. Propene. I. Hancock, E. G., ed.

 TP248.P75P76   661'.814   72–8512
 ISBN 0–470–34790–2

 © Ernest Benn Limited 1973

 Printed in Great Britain

PREFACE

THIS WORK WAS INTENDED to follow closely the lines of the companion volume on Ethylene published in 1969. Most unfortunately Dr. S. A. Miller, the editor of and a substantial contributor to the earlier book, died suddenly, shortly after he had started work on the present volume. The writer was asked to try and pick up the threads and carry on with the work Dr. Miller had started. This has proved difficult, not because of any difficulty in ascertaining the position, Dr. Miller had always kept meticulous records, but merely because of the very high standard that he had set and which we have tried, not always successfully, to maintain. Many of the contributors had been chosen by Dr. Miller and had already started work. We have therefore felt it right to continue along the lines Dr. Miller had begun and to make as few changes as possible.

An attempt has been made to keep a fair balance between purely scientific and technological information. As in the book on Ethylene, authors have been asked to give copious references and while there has inevitably been some variations in the interpretation of this request we feel confident that the book will prove of value not only to workers in fundamental research but to the complete spectrum of people interested in petrochemicals, right through to those concerned solely with technology and to technical salesmen.

We would like to thank all authors for the trouble they have taken over their contributions and to apologise for any complications that may have arisen from the change of editorship.

Compared with the book on Ethylene, although the number of chapters is only slightly reduced, these have in general been substantially shortened in length. We hope that this pruning will have produced a volume which is easier to handle and yet contains all the essential information.

We should perhaps add that authors have been encouraged to use metric units and the word 'tons' throughout the book refers to 'metric tons'.

E. G. HANCOCK

CONTENTS

CHAPTER 4
Oligomers and Co-oligomers of Propylene
J. Habeshaw

CHAPTER 5
Polypropylene
Prof. A. Valvassori, P. Longi and P. Parrini

CHAPTER 6
Isopropyl Alcohol and Acetone
J. C. Fielding

CHAPTER 7
Propylene Oxide
A. J. Gait

ACKNOWLEDGEMENTS

The editor, contributors and publishers wish to record their grateful thanks to the many companies and research associations who have helped with material for this book:

Stone and Webster Ltd.
BP Chemicals International
Sun Oil Co.
Sinclair Refining Co.
British Petroleum Co.
British Hydrocarbon Chemicals Ltd.
Phillips Petroleum Co.
Continental Oil Co.
Petrotex Chemical Corp.
Standard Oil of Ohio
Distillers Co. Ltd.
Farbenfabrik Bayer
Atlantic Refining
Kurashiki Rayon
Halcon International
Shell International Chemical Co. Ltd.
Goodyear Tyre and Rubber Co.
Pechiney St. Gobain
Scientific Design Co. Ltd.
Celanese Corp. of America
Union Carbide Corp.
Ethyl Corp.
California Research Corp.
Imperial Chemical Industries Ltd.
Shell Oil
Eastman Kodak
Montedison SpA for figs 2.26, 5.11, 5.27, 5.28
Dow Chemical Co. (London and Midland, Michigan USA)
Shell Chemicals (UK) Ltd.
I.G. Farbenindustrie AG.
Stauffer Chemical Co. (USA)
Solvay & Cie.
Chemische Werke Hüls
Esso Chemical Co. Inc.
Esso Research and Engineering Co.
Gulf Oil Corp.
Anglo-Iranian Oil Co. Ltd.

Farbwerke Hoechst AG.
Hoffman La Roche & Co.
Mitsubishi Corp.
Deutsche Hydrierwerke AG.
Du Pont de Nemours & Co.
BASF AG.
Acrolein Corp.
Asahi Chemical Ind.
Mitsubishi Rayon
American Cyanamid Co.
Nippon Kagaku
Daicel
Dynamit Nobel AG.
Ugine Kuhlmann
Rohm and Haas Co. (USA)
General Electric Co. Ltd.
Toa Gosei
BP Chemicals (UK) Ltd.
Allied Chemicals (UK) Ltd.
Allied Chemical Corp.
Erdölchemie
OSW
Wyandotte Chemical Corp.
Donau Chemie
Nitto
Monsanto Co.
Lummus
Fleissner for figs 11.13, 11.14 and 11.16b
Proctor Dalglish for 11.15a and 11.15b
Engelhard Sales Ltd., Special Products Division for 11.12a and 11.12b
Ruhrchemie AG.

CHAPTER I

HISTORY OF PROPYLENE AND ITS INDUSTRIAL DERIVATIVES

By H. M. Stanley

1.1. THE EARLY HISTORY OF PROPYLENE

1.1.1. Ethylene and its Homologues

ALTHOUGH ethylene, the first member of the homologous series of monolefinic hydrocarbons, was well known in the late 18th century and its preparation in pure form and characterisation by reaction with chlorine to form an oily dichloride described by the famous four Dutch chemists in 1795,[1] several decades elapsed before evidence accumulated that there existed a series of hydrocarbons of the empirical formula C_nH_{2n} with the common property of ready reaction with chlorine to yield oily dichlorides. Furthermore the discovery of the immediate homologue of ethylene which we now call propylene did not take place until about 1850, being anticipated by the preparation of several higher members of the series. Thus Faraday[2] in 1825 described the isolation of a hydrocarbon having the composition C_4H_8 from the pressure condensate derived from oil gas, i.e. the gas obtained by the thermal decomposition of vegetable oils. This material which was almost certainly a mixture of butenes was found to react readily with chlorine to furnish a colourless, heavy liquid of the composition $C_4H_8Cl_2$ and resembling closely the substance obtained in a similar way from ethylene. Later, Balard[3] dehydrated amyl alcohol, derived from wine fusel oil, by means of zinc chloride and obtained a hydrocarbon of the composition C_5H_{10} and boiling at 39°C which was evidently mainly 2-methyl-butene-2. Some years previously, Cahours[4] had decomposed the same alcohol with phosphoric acid and obtained an amylene dimer boiling at 160°C.

A much higher member of the series, namely cetene ($C_{16}H_{32}$) was obtained by Dumas and Péligot[5] by the action of phosphoric acid on cetyl alcohol, the latter being derived from spermaceti.

1.1.2. Discovery of Propylene

To Captain J. W. Reynolds, working in the laboratory of the Royal College of Chemistry under the famous Hofmann, belongs the credit of discovering the imme- diate homologue of ethylene, a hydrocarbon of the empirical formula C_3H_6 which he called 'propylene'. This discovery was announced in a paper of high historic interest, published in 1851.[6] He obtained this hydrocarbon, in admixture with other products, by passing the vapour of amyl alcohol (ex fusel oil) through a red hot tube of hard German glass followed by cooling to condense liquid products and separa- tion of the resulting gas. Under the best conditions of operation, about half of this gas consisted of propylene. To separate propylene from the gas mixture, Reynolds reacted the gas with bromine and fractionated the resulting mixture of brominated products. In this way he obtained a major fraction boiling at $143\,^{\circ}$C with a specific gravity of $1\cdot7$. It had the composition $C_3H_6Br_2$ and was decomposed by alcoholic potash to yield a mixture of compounds of the formula C_3H_5Br (bromopropenes). Similarly Reynolds prepared the corresponding chlorine derivative, the dichloride, $C_3H_6Cl_2$, an oily liquid boiling at $100\,^{\circ}$–$103\,^{\circ}$C.

It should be noted that Reynolds did not succeed in preparing pure propylene.

1.1.3. Nomenclature

Reynolds gave the newly discovered hydrocarbon the name of Propylene and remarked that 'this name is, like those of the rest of the series, derived from the corresponding alcohol, in this case still unknown, and for which the appellation of Propylic alcohol has been suggested by Dr. Hofmann'. This in turn is based on the name Propionic acid given by Dumas, Malaguti and Leblanc[7] to the immediate homologue of acetic acid and derived from two Greek words meaning 'the first fat' since it is the simplest acid, the salts of which have a soapy feel.

The generic name of 'olefin' hydrocarbons for ethylene and its homologues arose from the name 'gaz oléfiant' given originally by de Fourcroy to ethylene in reference to its reaction with chlorine to produce an oily product,[8] a property which was subsequently found to be common to this series of hydrocarbons.

Though the name propene was adopted for this hydrocarbon under the Geneva system of nomenclature, this has not been universally adopted to date, the name propylene being generally used in all but purely scientific publications.

1.2. EARLY CHEMISTRY OF PROPYLENE

The development of the chemistry of propylene in the two or three decades after its discovery followed a similar pattern to that which obtained in the case of ethylene, the main reactions studied being those with halogens, hydrogen halides, sulphuric acid and hypochlorous acid.

1.2.1. Reaction with Sulphuric Acid

The reaction of propylene with concentrated sulphuric acid was first studied by Berthelot[9] using propylene prepared from glycerol via allyl iodide. Berthelot found that propylene is rapidly absorbed by concentrated sulphuric acid and that the resulting solution on dilution with water, yields a propyl alcohol (now known as isopropyl alcohol) boiling at 81–$82\,^{\circ}$C and having a density of $0\cdot817$ at $17\,^{\circ}$C. This alcohol could be converted into esters by reaction with acetic and butyric acids. Treatment of this alcohol with concentrated sulphuric acid followed by neutralisation with barium hydroxide gave a barium propylsulphate salt which was identical with the salt produced by reacting propylene with concentrated acid followed by neutralisation with barium carbonate. The reaction of propylene with fuming sulphuric acid was found to yield a homologue of isethionic acid.

It should be emphasised that at this stage the true nature and structure of the propyl alcohol produced from propylene was not recognised. However in 1862,

Kolbe[10] deduced the structure of the alcohol produced by Friedel[11] by the reduction of acetone with sodium amalgam as:

and noted that its properties were similar to those recorded by Berthelot for the alcohol made from propylene. This conclusion was confirmed by Erlenmeyer[12] who was the first to put forward the structure of propylene as a methyl-substituted ethylene, namely:

$$CH_3.CH = CH_2$$

Many years after his pioneering work on the reaction of propylene with sulphuric acid Berthelot[13] re-examined the reaction and showed that when pure sulphuric acid was saturated with propylene at 18°C, addition of water precipitated a heavy oil which he identified as di-isopropyl sulphate. This reaction was later confirmed and further studied by Ormandy and Craven.[14]

1.2.2. Reaction with Halogens

The original work of Reynolds on the reaction of propylene with chlorine and bromine to yield the dihalides was confirmed by Cahours[15] and later by Würtz[16] on the basis of propylene produced in other ways. Würtz showed that in the presence of excess bromine, the addition reaction could be followed by bromine substitution to yield compounds such as $C_3H_5Br_3$. The reaction of propylene, produced from isopropanol and zinc chloride, with iodine chloride was shown by Friedel and Silva[17] to yield propylene chloroiodide, C_3H_6ClI.

Propylene di-iodide, a somewhat unstable substance was prepared by Berthelot and Luca[18] by reacting propylene with elemental iodine.[19]

1.2.3. Reaction with Hydrogen Halides

Berthelot[9] noted that propylene could be reacted with hydrogen chloride, bromide and iodide in aqueous solution to yield propyl halides of the formula C_3H_7X, where X is halogen. Thus propylene reacted with aqueous fuming hydrochloric acid slowly at room temperature but completely at 100°C (30 hours) to yield a volatile compound boiling at 40°C and of the formula $C_3H_6.HCl$. At this time the structure of the propyl chloride so produced was unknown.

An important advance was made by Erlenmeyer[12] in 1866. He reacted propylene with a concentrated aqueous hydriodic acid solution at 0°C to yield an organic iodide identified by him as isopropyl iodide (Erlenmeyer refers to it as pseudopropyl iodide) and related to the alcohol produced from acetone by Friedel.[11] These and other observations led Markovnikov[20] to enunciate his well-known Rule which states that in the addition of hydrogen halides to olefins, the halogen atom tends to become united to the carbon atom which has the least number of hydrogen atoms linked to it. It is of interest to note that more than sixty years were to pass before an exception to this rule was discovered in the so called 'abnormal' addition of hydrogen bromide to olefins in the presence of peroxides.[21] Thus, under such conditions, propylene reacts with hydrogen bromide to yield n-propyl bromide.[22] Of the hydrogen halides, only hydrogen bromide shows this effect though the abnormal mode of addition to yield n-propyl derivatives is observed with certain other reactants such as hydrogen sulphide,[23] mercaptans,[24] etc.

1.2.4. Propylene Chlorohydrin and Propylene Oxide

Following the discovery by Würtz[25] in the late 1850's that ethylene oxide could be obtained by the action of aqueous alkalies on ethylene chlorohydrin, similar reactions were applied to propylene a year or so after by Oser[26] working in the Würtz laboratory. Oser reacted propylene glycol, which he prepared from propylene dibromide, with dry hydrogen chloride to furnish the chlorohydrin (probably a mixture of the two isomers) which he then treated with aqueous potassium hydroxide. The resulting propylene oxide was described by Oser as a water-soluble, ethereal liquid of b.p. $35°C$ at atmospheric pressure.

A year or so later, Carius[27] discovered that ethylene chlorohydrin could be produced more directly by reacting ethylene with aqueous hypochlorous acid. However, it was not until about 1870 that this reaction was applied to propylene by Markovnikov[28] who described the production of the chlorohydrin, b.p. $127°C$, and its conversion into propylene oxide as well as the reaction of propylene oxide with hydrogen chloride, bromide and iodide to furnish the corresponding halohydrins. For some time there was some controversy as to the structure of the chlorohydrin formed from hypochlorous acid and propylene, Markovnikov believing it to be $CH_3.CHOH.CH_2Cl$ on the basis of its yielding some chloracetone on oxidation, while Henry[29] considered it had the structure $CH_3.CHCl.CH_2OH$ since oxidation with nitric acid yielded some α-chloropropionic acid. Later it was recognised[30] that propylene chlorohydrin is a mixture of these isomers, the proportions being about 95 per cent of the 1-chlor-compound and 5 per cent of the 2-chloro isomer.[31]

1.2.5. Alkylation

Though Friedel and Crafts[32] discovered that aromatic compounds could be alkylated by reaction with alkyl chlorides and bromides in the presence of anhydrous aluminium chloride, it was Balsohn[33] who first showed that ethylene could replace ethyl chloride in this reaction. Cumene, first prepared in 1841 from p-cuminic acid by Gerhardt and Cahours,[34] was made from benzene and isopropyl chloride or bromide by a number of workers[35] and later by Berry and Reid[36] from propylene, benzene and aluminium chloride as catalyst.

1.3. PREPARATION OF PROPYLENE ON THE SMALL SCALE

For the preparation of propylene on a laboratory or small production scale, the most convenient method is that based on the catalytic dehydration of isopropanol. Large scale industrial production, which is described in detail later, is of course based wholly on the cracking of petroleum hydrocarbons either as part of refinery operations for the production of gasoline or under conditions specifically designed to produce mixtures of propylene and its immediate homologues from selected petroleum fractions.

1.3.1. Methods Based on Thermal Decomposition of Organic Compounds

Following the discovery of Reynolds[6] that propylene, mixed with other gases, could be produced by the decomposition of amyl alcohol vapour in a red hot tube, several other investigators found that other organic raw materials could also be used to furnish propylene-containing gases by thermal treatment. Thus propylene was prepared in the form of its dibromide or dichloride, from fatty acids such as pelargonic acid, or its salts, or from the corresponding alcohols by Cahours[37] while Berthelot[38] used the dry distillation of salts of acetic acid and Dusart[39] the dry distillation of a mixture of calcium acetate and potassium oxalate for the same purpose. Of outstanding historic interest, however is the work of Prunier[40] who obtained propylene in admixture with other hydrocarbons by passing the vapours of a petroleum fraction b.p. 60°–90° through a red hot tube, thus anticipating by many decades one of the most important industrial routes employed today.

It should be emphasised that these pyrolytic methods all gave propylene in low yields and mixed with other gases so that the propylene was not recovered as a pure material but converted into its dichloride or dibromide which were thereafter separated from other olefin dichlorides.

1.3.2. Methods Based on Glycerol or Allyl Iodide

The discovery of a convenient laboratory method for preparing propylene in relatively pure form – a discovery which did much to stimulate research in propylene chemistry – was due to Berthelot and de Luca[41] who found that glycerol could be reacted with phosphorus iodide to yield allyl iodide, from which propylene could be produced by reaction with mercury and hydrochloric acid. The method was subsequently improved by substituting zinc and acetic acid[42] or, better still, granulated zinc in absolute alcohol[43, 19] in place of mercury and hydrochloric acid. It was shown by Erlenmeyer[12] that allyl iodide could also be converted into propylene by reaction with hydriodic acid. Allyl iodide was found by Simpson[44] to be converted into isopropyl iodide by reaction with hydriodic acid and Erlenmeyer[12] showed that isopropyl iodide is a convenient and efficient source of propylene by reaction with alcoholic potash.

1.3.3. Methods Based on Saturated Halogenated Compounds

Erlenmeyer's preparative method based on isopropyl iodide and alcoholic potash proved popular with succeeding workers including Markovnikov,[45] Birnbaum[46] and Tollens and Henninger.[48] Freund[49] made the important observation that propylene is produced by the action of alcoholic potash on both n-propyl and isopropyl iodides.

The corresponding chloro-compounds have also been found to be readily convertible to propylene. Thus, Mouneyrat[49] obtained propylene by the decomposition of n-propyl chloride in the presence of aluminium chloride as catalyst. Somewhat later the vapour phase dehydrochlorination of isopropyl chloride over solid catalysts, such as barium chloride and alumina was studied by Mailhe and Senderens.[50]

Propylene chloride may be converted into propylene by treatment with zinc and acetic acid although much polymeric material is simultaneously formed.[51] The reaction of propylene dibromide with potassium iodide is said to yield some propylene.[52] Berthelot[53] obtained propylene by heating propylene dibromide with copper and aqueous potassium iodide at 275°C.

1.3.4. Dehydration of n-Propanol and Isopropanol

In his classic paper published in 1855, Berthelot[9] showed that the propyl alcohol (later identified as isopropanol) prepared from propylene could be re-converted into propylene by heating with concentrated sulphuric acid. Many years later, he showed that n-propanol could be dehydrated in a similar way to give mainly propylene.[54] The addition of aluminium sulphate was found by Senderens[55] to catalyse the dehydration of these alcohols. Thus, heating 100 cc of n-propanol with 75 cc of concentrated sulphuric acid with addition of 10 g of anhydrous aluminium sulphate produced a rapid liberation of propylene at 100°–110°C, the yield being 95 per cent.

Other dehydrating agents have been used successfully for the decomposition of the propanols in the liquid phase. Thus, zinc chloride was employed by Friedel and Silva[56] for the preparation of propylene from isopropanol while Le Bel and Greene[57] similarly dehydrated n-propanol by allowing this alcohol to drop on to fused zinc chloride. While phosphorus pentoxide was found by Beilstein and Wiegand[58] to convert n-propanol into propylene, a more convenient reagent is syrupy phosphoric acid which was found by Newth[59] to decompose n-propanol rapidly at 250° and isopropanol at about 210°C to yield propylene with some liquid hydrocarbon by-products. In a modification of Newth's method, a slow stream of alcohol was introduced through a fine glass spiral dipping below a pool of syrupy phosphoric acid at about 240°C so that the alcohol was vaporised before contacting the acid medium.[60] An alternative, truly catalytic system was developed by White[61] who passed the vapour of isopropanol over a heated contact consisting of pumice impregnated with concentrated phosphoric acid, a method also employed by Ormandy and Craven.[62]

Since the beginning of this century there arose a great interest in the potentialities of heterogeneous catalysts and many different catalysts were found for the vapour phase dehydration of n- and iso-propanols to propylene. Thus Ipatiev[63] found that alumina would catalyse the dehydration of isopropanol even at 360°C (n-propanol required higher temperatures) to yield a gas containing 96 per cent propylene and 4 per cent hydrogen. Senderens[64] working with what was evidently a more active alumina catalyst, found that it catalysed the dehydration of isopropanol at temperatures of 235°C and above to yield a propylene of 99 per cent purity; n-propanol required a somewhat higher temperature. Goudet and Schenker[65] also studied this reaction using as catalyst alumina which had previously been heat treated. Over this catalyst, isopropanol reacted rapidly at 440°–450° to give a 98 per cent yield of propylene while n-propanol at 460°–470°C gave propylene in a yield of 89 per cent. Another active oxide catalyst consisting of zinc oxide, prepared from zinc oxalate and promoted with tungstic oxide has been described by Adkins and Millington.[66]

Various salts of phosphoric acid have been found to be active catalysts in this reaction. In particular aluminium phosphate was found by Senderens[67] to promote a rapid dehydration of isopropanol at 250–300°C to give an excellent yield of pure propylene while n-propanol required a higher temperature, e.g. about 340°C for comparable decomposition rate. Manganese orthophosphate and manganese pyrophosphate were found by Williamson and Taylor[68] to effect rapid dehydration of isopropanol at 425°C to give very pure propylene in high yield.

The dehydration of n- or iso-propanols is, of course, of little value in the large scale production of propylene. It is interesting to note, however, that limited supplies of propylene have been so prepared for use in small production or pilot plants designed to make propylene derivatives e.g. cumene. In Germany during the last War, propylene was produced by the catalytic dehydration of n-propanol at 350°C over an alumina catalyst on a scale sufficient to supply a synthetic glycerol plant with an output of 25 tons/week.[69] The n-propanol was produced as a by-product of a very large synthetic methanol plant.

1.4. BEGINNING OF THE PETROCHEMICAL INDUSTRY – PERIOD UP TO 1940

Up to the year 1920, the organic chemical industry of the world, including the USA, was mainly based on coal and its carbonisation products, including also calcium carbide, although in many countries substantial quantities of organic chemicals were produced by the fermentation of carbohydrate materials, by the carbonisation of wood and from the splitting of natural fats. It was quite natural that the first steps in the switch to petroleum as the primary raw material for organic chemical manufacture should have taken place in the USA with its immense petroleum production and refining industry and its highly developed market for chemical products of all kinds.

The fact that propylene became the first petroleum-derived hydrocarbon intermediate for chemical manufacture can be ascribed to two main reasons. In the first place, propylene was a cheap and abundant by-product of the cracking of petroleum fractions for gasoline production which was inititated by Burton in 1913 and rapidly expanded during the years of World War I. Secondly this War had created a great demand for acetone which was required as a solvent for aeroplane dopes based on cellulose acetate and propylene-derived isopropanol was seen to be the most likely raw material for the manufacture of additional acetone. In fact, the development of a large scale manufacture of isopropanol from propylene came too late to meet the wartime needs for acetone.

Unlike propylene, ethylene was not present in appreciable amounts in the by-product gases from cracking for gasoline production and its use as a chemical raw material depended on the development of a satisfactory means of production. Thanks to the work of Curme[70] and his colleagues of the Union Carbide Corporation, methods of producing ethylene by the cracking of ethane/propane mixtures were established as well as the development, and later large-scale manufacture, of a range of derivatives, including ethylene oxide, ethylene oxide derivatives, ethanol, etc. These important developments based on ethylene, which were taking place only a few years after the initiation of the manufacture of isopropanol from propylene, soon established ethylene as the most important olefin hydrocarbon intermediate for organic chemical manufacture, a position it has held ever since. Thus by 1940, the annual usage of ethylene for chemical manufacture in the USA was approximately 120,000 tons, the corresponding figure for propylene being about 91,000 tons.[71]

1.4.1. Isopropanol

The credit of having created the first petrochemical manufacture belongs to Carleton Ellis,* a consultant chemist and prolific inventor with his own laboratories in Montclair, New Jersey. After preliminary work in these laboratories, he founded the Melco Chemical Company to exploit this work and a small plant was erected in 1917 and operated at Bayonne, New Jersey, using gas from the adjoining Burton cracking plant of the Tidewater Oil Company. After preliminary purification, the gases were contacted with 87 per cent sulphuric acid at 15°–20°C until the specific gravity of the medium had fallen to 1·4. Thereafter the acid liquor was diluted with four volumes of water and distilled to furnish isopropanol as well as quantities of higher alcohols and small amounts of ethers.†

At this time, the U.S. Bureau of Aircraft Production took over the plant to develop a method, which proved successful, for the conversion of isopropanol to acetone by vapour phase oxidation over copper catalysts.

In 1920, Ellis sold his installation and patent rights to the Standard Oil Company of New Jersey which industrialised the process at their Bayway, New Jersey, refinery in 1920 and proceeded to develop markets for the alcohol under the trade-name of 'Petrohol'. At the same time important process improvements were incorporated including the use of a concentrated propylene as feedstock. By 1923, this plant was producing at the rate of 100,000 gallons per annum.

At this time a number of other companies became interested in the production of isopropanol from propylene. As early as 1921, the Carbide and Carbon Chemicals Corporation (now the Union Carbide Corporation) operated a small production plant[72] at Glendenin, West Virginia. Later in the decade, the same Company built a larger plant for isopropanol and acetone production at their new site at South Charleston, West Virginia but this was operated intermittently for some time. However the project gained momentum when the Company developed, and thereafter industrialised in 1933, a process for producing ketene, and thence acetic anhydride, by the cracking of acetone.

---

* Carleton Ellis was born in 1876, graduated at the Massachusetts Institute of Technology and founded the Ellis Laboratories in 1908. He died in 1941.

† Details of the process were first given at a meeting of the New Jersey Chemical Society (of which Ellis was President) on 13 December 1920. See *Chem. and Met. Eng.*, 1920, **23** (*23*), 1230.

Two oil companies, namely the Cities Service Oil Company and the Barnsdall Corporation also carried out development work on the isopropanol process but with little commercial success. The efforts of the Barnsdall Corporation resulted in the formation of the Petroleum Chemical Corporation which in 1932 was merged with the Standard Oil Company of New Jersey in the formation of the Standard Alcohol Company.[73]

A little later, the Shell Oil Company industrialised its own processes for the production of isopropanol and acetone at Dominguez, California, and in 1937–8 commenced the manufacture of a number of acetone derivatives, including diacetone alcohol, mesityl oxide and methylisobutyl ketone on that site.

1.4.2. Propylene Oxide and its Derivatives

Though of minor industrial significance during the period 1930–40, propylene oxide deserves mention since it was the only propylene derivative other than isopropanol (and the chemicals made from isopropanol) which was produced commerically during this period. Its development was a natural extension of the work of Curme and his collaborators of the Carbide and Carbon Chemicals Corporation which resulted during the 1925–30 period in the commercialisation of methods of producing ethylene chlorohydrin,[74] ethylene oxide and a range of ethylene oxide derivatives especially ethylene glycol. Similar reactions were applied on a modest commercial scale to propylene for the production of propylene oxide and propylene glycol by that Company at South Charleston, West Virginia during the decade 1930–40. However the market for these products was quite small at that time. Thus even in 1940 the total production of propylene oxide and its derivatives (mainly propylene glycol) did not amount to more than a few hundred tons per annum.[71] The great expansion in the usage of propylene oxide and its derivatives did not take place until after World War II.

1.5. PROPYLENE USAGE DURING THE PERIOD 1940–50

1.5.1. The entry of the United States into World War II in 1941 greatly stimulated the growth, and diversification, of the petrochemical industry in that country. A major factor in this greatly accelerated growth of the industry was the urgent need for the USA, faced with an interruption in supplies of natural rubber from the Far East, to build up as rapidly as possible a large synthetic rubber industry. Fortunately much of the know-how necessary for this purpose was available either as a result of development work by US oil and chemical companies or from German experience as a result of technical exchange agreements in operation before 1940. As a result of a truly gigantic technical effort, a vast synthetic rubber industry was established in an incredibly short time and the production of synthetic elastomers, mainly GR–S (a copolymer of butadiene and styrene), in the USA increased from about 8,000 tons in 1941 to over 800,000 tons in 1945. The major raw materials needed for this purpose were ethylbenzene, based on ethylene and benzene, and butadiene produced mainly from n-butenes but also to a substantial extent from ethanol.

In addition, the War also greatly stimulated the requirements for a wide range of synthetic organic chemicals including such petrochemical products as ethanol, ethylene oxide, styrene, ethylene dichloride and ethyl chloride as well as isopropanol and acetone. At the same time there was a great expansion in petroleum production and refining operations to meet the greatly increased needs caused by the War. New refining processes were put into operation to increase the supply and the octane rating of gasoline and aviation fuels. Particularly important from this point of view was catalytic cracking, first introduced by Houdry in 1935, and catalytic reforming over a fixed bed molybdenum oxide catalyst. As a result of this expansion of cracking facilities and the new techniques available, the availability of propylene and butenes was greatly enhanced. Two processes which made use of these lower olefins in the petroleum industry were the polymer gasoline process developed by the Universal Oil Products Company before the War and the alkylation process in which isobutane was reacted with propylene or propylene-butene mixtures either in the presence of sulphuric acid (initiated in 1938) or of liquid hydrofluoric acid (1942) to yield high octane gasoline.

Ethylene consumption for chemical purposes in the USA increased from 120,000 tons in 1940 to about 200,000 tons in 1944, the corresponding figures for propylene being 90,000 tons in 1940 to about 200,000 tons in 1944; however a further 230,000 tons of propylene were used in 1944 in the manufacture of cumene, a valuable additive to aviation gasoline.

After the War and a short period of adjustment, chemical manufacture based on ethylene and propylene continued to grow rapidly in the USA. Thus the production of isopropanol increased from an equivalent of 200,000 tons/annum of propylene in 1945 to about 360,000 tons/annum in 1950 while propylene oxide and its derivatives, mainly propylene glycol, also grew significantly in this period, reaching a propylene equivalent figure of 30,000 tons/annum in 1950.[71] However, the period 1945–50 witnessed the commercial début of a number of new propylene derivatives.

1.5.2. New Developments in Propylene Derivatives in the Period 1945–50

1.5.2.1. Allyl chloride and glycerol   An important new area in propylene chemistry was opened up in 1935 by the research workers of the Shell Development Company in Emeryville, California, who found that chlorination of propylene at elevated temperatures (the so-called 'hot' chlorination) could be used to produce allyl chloride in high yield. Using an excess of propylene at temperatures in excess of 500°C, yields of monochloropropylenes (over 96 per cent of which are allyl chloride) of 85 per cent could be achieved on a laboratory scale.[75] Furthermore it was shown in the same laboratories[76] that allyl chloride was a satisfactory material for the commercial synthesis of glycerol, based on the following reactions:

$$CH_3.CH = CH_2 \xrightarrow{Cl_2} CH_2 = CH.CH_2Cl \xrightarrow{HOCl} CH_2OH.CHCl.CH_2Cl$$
$$\text{Allyl chloride} \qquad\qquad \text{Dichlorohydrin}$$

$$CH_2OH.CHOH.CH_2OH \longrightarrow CH_2\!\!-\!\!CH.CH_2Cl$$
$$\text{Glycerol} \qquad\qquad \diagdown O \diagup$$
$$\text{Epichlorohydrin}$$

This process was carried through a pilot plant development in 1937 but was not commercialised until after the War. The Shell Chemical Company began producing allyl chloride in 1946–7 and in 1948 commissioned the first large scale (30,000 tons/annum) plant for the synthesis of glycerol at Houston, Texas.[77] Since that time several other plants have been operated commercially on this, or similar processes.

Shell's work on this glycerol synthesis, led the Company into the field of epoxy resins which are based on the intermediate epichlorohydrin. The production of these resins has grown rapidly in the last decade, production in the USA alone in 1965 being at the rate of 65,000 tons/annum.[78]

1.5.2.2. Propylene tetramer The great importance of the phosphoric acid-catalysed dimerisation and co-dimerisation of propylene and butenes in the production of polymer gasoline during World War II has already been mentioned. An extension of this technique for the production of the so-called propylene tetramer (a mixture of highly branched dodecenes) and also, to a limited extent, the trimer of propylene was applied in Germany during the war while a similar development took place in the USA[79] particularly by the California Research Corporation. The tetramer was employed to alkylate benzene (usually in the presence of hydrofluoric acid), the resulting dodecylbenzene being sulphonated to yield a surface-active agent which, because of its cheapness and good detergent properties, grew rapidly in importance in the USA and in many other countries, e.g. in Britain in 1954. Production of tetramer in the USA grew from an equivalent of 4,000 tons/annum of propylene in 1945 to nearly 90,000 tons/annum in 1950. Thereafter it continued to increase in importance until 1962–3 after which date its production has declined in most advanced countries because of the poor biodegradability of the derived detergent and consequent pollution problems.

1.5.2.3. Cumene This hydrocarbon was first produced on a commercial scale in the USA by the Shell Oil Company in 1942 by the vapour phase alkylation of benzene by propylene over a phosphoric acid catalyst.[80] It was used extensively during the last years of the War as an additive in aeroplane fuels. Because it could be made readily in existing polymer gasoline plants, its production in the USA grew rapidly from the equivalent of 60,000 tons of propylene in 1942 to a propylene equivalent of 230,000 tons in 1944.[71] At the end of the War, production of cumene virtually ceased until 1954 after which date it has become increasingly important as an intermediate in the production of phenol and acetone so that its production in the USA in 1969 was equivalent to 350,000 tons of propylene. It is of interest to note that an alternative method of making cumene based on liquid phase alkylation in the presence of 88–90 per cent sulphuric acid has also been used industrially, e.g. by the Petroleum Industrie Maatschappij in Curaçao.[81]

1.5.3. Developments in Germany During World War II

During the period 1939–44, several chemical products derived from propylene were manufactured on a modest scale in Germany, the propylene raw material being obtained ultimately from coal rather than from petroleum. These developments were terminated at the end of the War and it was not until 1954[82] that a substantial petrochemical industry was initiated in Germany.

1.5.3.1. Isopropanol   A plant for the simultaneous production of isopropanol and 2-butanol by sulphation of propylene and n-butenes with 75 per cent sulphuric acid was commissioned in 1939 by Rheinpreussen A.G. at Moers near Duisberg. The raw material used, consisting of a mixture of propylene, n-butenes (mainly butene-2) and the corresponding paraffin hydrocarbons with a total olefin content of around 35 per cent v/v, was a by-product of the production of liquid hydrocarbons in a large adjoining Fischer-Tropsch plant.[83] This plant produced isopropanol and 2-butanol at the rate of 1,000 tons/annum and 1,250 tons/annum respectively in 1943.[84]

An important development project in Germany at this time was aimed at establishing a process for the direct hydration of propylene to isopropanol.[85] This process was investigated on a pilot scale at both Ludwigshafen and Gendorf with a reactor volume of 1 litre and operating at around 200 atm at 230°C under mixed phase conditions using a catalyst of tungstic oxide promoted with zinc oxide. With a molar ratio of water to propylene in the feed of 10:1, conversions of propylene to isopropanol of up to 50 per cent per pass were achieved. This interesting development was not followed up in Germany at the time and the first large scale direct hydration plant for the production of isopropanol was commissioned by Imperial Chemical Industries Ltd at Wilton in 1951.

1.5.3.2. Glycerol   A plant for the production of glycerol on a scale of 1,500–2,000 tons/annum was erected by the I.G. at Oppau.[69] This plant was in operation in 1943 but due to bombing the highest production rate achieved was 25 tons per week. The propylene raw material was prepared by vapour phase dehydration over an alumina catalyst of n-propanol obtained as a by-product of a large methanol plant. By the high temperature chlorination of propylene, allyl chloride was produced but, unlike the method developed in the USA, this was first hydrolysed to allyl alcohol which was then chlorohydrinated to yield glycerol monochlorohydrin. Hydrolysis then yielded glycerol.

1.5.3.3. Propylene trimer and tetramer   Production of propylene oligomers, namely the trimer and tetramer, took place in Germany during the War but on a small scale, i.e. a few hundred tons per annum. At Uerdingen, propylene derived from the fractionation of coke oven gas[86] was polymerised by the use of a liquid catalyst system, consisting of a mixture of two parts of 85 per cent phosphoric acid and one part of phosphorus pentoxide, at 200°C and 40–50 atm pressure to yield a mixture of dimers, trimers and tetramers. The trimer fraction was polymerised by aluminium chloride to lubricating oils while the tetramer fraction was used to make isododecylphenol, the latter being applied as a plasticiser for polyamides.[87]

At Hüls, propylene recovered as a by-product in the manufacture of butadiene from 1:4-butandiol was polymerised over a supported phosphoric acid catalyst at 220°C and 50 atm, the scale of operations being about 300 tons/annum of propylene.

The resulting trimer and tetramer fractions were used in the manufacture of detergents.[88]

1.5.3.4. Oxo reaction   A discovery which was in due course to have an important influence on the growth of the petrochemical industry was made by Roelen of the Ruhrchemie AG in the middle of 1938.[89] It was found that ethylene could be reacted with carbon monoxide and hydrogen in the presence of a cobalt-thoria Fischer-Tropsch catalyst to yield propionaldehyde and soon this was shown to be a fairly general reaction yielding saturated aldehydes (and hence primary alcohols) from olefins in accordance with the overall equation:

$$C_nH_{2n} + CO + H_2 \longrightarrow C_nH_{2n+1}.CHO$$

The reaction was named the 'Oxo' reaction though 'hydroformylation' is also used and is more appropriate scientifically.

The importance of this discovery was quickly recognised and a great deal of research and development, including the operation of several pilot plants,[86] was carried out in Germany during the War. A plant at Holten for the production of 12,000 tons/annum of higher primary alcohols from C_{11}–C_{17} Fischer-Tropsch olefins was completed in 1944, but, because of bomb damage, never operated.[90] This plant was designed on the basis of batch operation using the standard cobalt-containing Fischer-Tropsch catalyst.

The subsequent development of this process involved continuous operation and the use of oil-soluble cobalt salts as catalyst source and is discussed later in this Chapter.

1.5.4. Petrochemical Developments in Great Britain in the Period 1940–50

The War years witnessed the founding of a petrochemical industry in Britain when British Celanese Ltd commissioned in 1942 at Spondon, near Derby, a plant for the production of ethanol from ethylene on a scale of 30 tons/day. The ethanol so produced was thereafter used as a source of relatively cheap acetic acid needed in the manufacture of cellulose acetate.[8] Gas oil cracking at 800–850°C under reduced pressure was employed as a source of ethylene which was then concentrated to 70–75 per cent concentration before conversion to ethanol by the usual sulphation process. Later, the plant was modified to use the economically more suitable naphtha as feedstock and was expanded to produce 120 tons/day of ethanol as well as substantial amounts of other chemicals including isopropanol.

In the immediate post-war period the infant British petrochemical industry grew rapidly. Thus, in 1947–8, the Shell Chemical Company commissioned plants for the manufacture of isopropanol (and also 2-butanol) from propylene derived initially from gas oil cracking and later from refinery gas.[91] A year or so later Petrochemicals Ltd began to operate at Carrington, near Manchester, the so-called Catarole process which had been developed in Britain during the War years. This process involved the high temperature pyrolysis of petroleum fractions under conditions giving rise to the production of much aromatic oils as well as ethylene and its immediate homologues.[92] The ethylene was used in the manufacture of ethylene oxide by the chlorohydrin process and, later, propylene oxide and its derivatives were also made. Since 1955 when Shell Chemical Company acquired the assets of Petrochemicals Ltd operations at the Carrington site have been based on the conventional steam cracking of naphtha.

Finally in 1950-1, two other companies, namely Imperial Chemical Industries Ltd and British Hydrocarbon Chemicals Ltd (now BP Chemicals International Ltd) entered the field with large scale plants for the production of a wide range of petrochemicals on the basis of naphtha cracking.

1.6. LARGE SCALE PRODUCTION OF PROPYLENE

Apart from a small production in Germany during the War years, 1939-44, all propylene since the beginning of the petrochemical industry has been made by the thermal decomposition of petroleum hydrocarbons in one or other of the following ways:

a. As a by-product in the manufacture of motor spirit by the cracking of higher hydrocarbons and to some extent of thermal reforming for gasoline improvement.

b. As a co-product with ethylene in the cracking of propane and butanes or mixtures of these with ethane.

c. As a co-product with ethylene (as well as higher hydrocarbons) in the high temperature pyrolysis of normally liquid petroleum fractions, especially naphtha.

1.6.1. Propylene as a By-product of the Petroleum Industry

Unlike ethylene, propylene is an important by-product of the cracking of petroleum fractions for gasoline manufacture as well as one of the key raw materials for the production of alkylate and polymer gasolines. The rapid growth of a vast cracking industry in the USA since 1913 was the result of an ever-increasing demand for gasoline arising from the development in that country of a tremendous automobile industry. Moreover the developments in automobile engines necessitated the production of gasolines of improved characteristics particularly higher octane number and this requirement very largely dictated the technological evolution of the cracking processes employed. Outside the USA, major refineries were mostly located at the oilfields as in Curaçao, Iran, etc., and a large cracking industry did not arise in Europe and Japan until after World War II.

The modern cracking industry had its beginnings in 1913 when the Burton process of pressure distillation was operated by the Standard Oil Co. of Indiana. This batch process operated at 700-750°F and a pressure of 100 p.s.i.g. in shell-type stills. This type of process was improved by Cross and Dubbs particularly in respect of the use of higher temperatures and pressures. In 1921 the Dubbs process achieved the first truly continuous operation and permitted the exploitation of heavy residues and asphaltic crude oils as feedstocks while in the mid 1920's the Tube and Tank process of the Standard Oil Co. of New Jersey was evolved and eventually operated under conditions as severe as 900-975°F and 250-1000 p.s.i.g.[93]

These processes usually referred to as liquid phase, or mixed phase, processes yielded a by-product gas containing very little ethylene and around 10 per cent by volume of propylene. Propylene production was usually somewhat less than 1 per cent by weight of feedstock used with about twice that quantity of propane.[94]

The demand for even higher octane number gasolines stimulated the development of vapour phase cracking processes which became popular in the USA in the early 1930's. These operated at low pressures and high temperatures and produced large quantities of olefin-rich gas (up to about 25 per cent by weight of feedstock) with propylene yields as high as 7 per cent by weight of feedstock as well as much ethylene.[95] However, these processes were generally not economically attractive for gasoline manufacture, process economics usually favouring the use of mixed phase units.[96]

A revolution in the cracking industry was initiated in the USA in 1936 when Houdry introduced the first commercial catalytic cracking process based on a silica-alumina catalyst. This process, though initially employing a fixed bed of catalyst, was developed to operate first with a moving catalyst bed and, later (in 1942) with a fluidised catalyst system. This type of process gives a good yield of high quality gasoline and has largely supplanted the older types of cracking process for converting heavy oils into gasoline. Thus, today, the installation of new mixed phase thermal cracking facilities for gasoline production cannot be justified economically because of the market demands for high anti-knock gasoline products and existing thermal cracking units are being employed for cracking heavy catalytic cycle stocks.[96]

Catalytic cracking is a major source of propylene and butenes, particularly in the USA, though very little ethylene is formed in this way. Total gas production (including C_4 hydrocarbons) is about 8–12 per cent by weight of feedstock (e.g. gas oil), the propylene yield being in the range of 3–5 per cent by weight of feedstock with much less propane.[94] However, the use of molecular sieve-type catalysts in place of the older silica-alumina catalysts is now widespread and is likely to increase. With these newer catalysts, gasoline yield is increased but propylene production is reduced,[97] possibly to below 3 per cent by weight of feedstock.

Another process which contributes substantially to by-product propylene availability is the thermal reforming process used to upgrade in anti-knock value the heavy portion of naphtha formed by catalytic cracking. This reforming process, first introduced into the USA by the Sun Oil Co. in 1955, yields propylene in amounts of 3–4 per cent by weight of feedstock together with similar quantities of propane and lesser amounts of ethylene.[98] The corresponding catalytic reforming process, introduced into the USA in about 1950, gives a gaseous by-product almost free from olefins.[99]

Finally mention should be made of the thermal coking process used to convert heavy, low-grade petroleum oils (on a scale of over 20 millions tons/annum of feedstock in the USA) into gas oil and smaller quantities of gasoline, gas and coke. This process yields appreciable quantities of a gas rich in propylene and its homologues.[100]

According to Curtiss,[101] a typical refinery gas might contain 9 per cent by volume of propylene with about 7 per cent of ethylene, the remainder being paraffin hydrocarbons and hydrogen. However it should be pointed out that refinery gas is a mixture of several different gas streams from various refinery operations and its composition will vary considerably, depending not only on the refining processes practised and conditions of operation but also on the type of crude oil employed.

1.6.2. Propylene and Ethylene from Cracking of Ethane, Propane and Butanes

The co-production of ethylene and propylene by the high temperature pyrolysis of ethane-propane mixtures was originally used by the Union Carbide Corpn. in their petrochemical developments in the 1920's. Since that time the cracking of propane, either alone or in admixture with ethane and butane, and also of the butanes has become an increasingly important source of propylene and ethylene in the USA. The cracking of isobutane to yield mainly isobutene and propylene is also practised on a limited scale in Texas and Oklahoma.[71] Outside the USA, relatively little propylene is produced from these raw materials.

Whereas the cracking of ethane gives very little propylene, propane cracking yields both ethylene and propylene in proportions depending upon the conditions, particularly the cracking severity. Typically, ethylene and propylene can be produced in a weight ratio of 1:0·3–0·6, a ratio of 0·4 parts by weight of propylene to 1 part of ethylene being about average.[71]

1.6.3. Propylene and Ethylene from the Crackers of Naphtha and Higher Petroleum Fractions

The co-production of ethylene and propylene (as well as higher unsaturated and saturated hydrocarbons) by the high temperature pyrolysis of naphtha and higher petroleum fractions has been known since the work of Prunier[40] but it is only in the last three decades that this method has become of major industrial significance. The cracking of gas oil for the production of ethylene and propylene has been practised in the USA on a limited scale for some time and the first cracking installation for this purpose in Europe, namely the plant of British Celanese Ltd. commissioned at Spondon (England) in 1942 was based on the pyrolysis of a paraffinic gas oil. However, it soon became apparent that the preferred raw material for gaseous olefin production in Europe (and also Japan) was naphtha, the reasons being as follows:[102]
a. The establishment of a substantial petroleum refining industry in Europe, which took place in the 1950's, was, and still is, based on Middle East or North African crudes with a relatively high content of light fractions.
b. The demand for gasoline in Europe is low compared with the USA while fuel oil requirements are relatively high.
c. It is thus possible in Europe to meet all motor gasoline requirements from straight run products, suitably upgraded by catalytic reforming and very little catalytic cracking of heavy fractions is necessary.
d. By-product propylene supplies in Europe are limited by the relatively modest catalytic cracking capacity and also by the fact that, unlike the USA, no abundant and cheap supplies of propane and butane are available.

It thus came about that the co-production of ethylene and propylene (together with butenes and butadiene as well as pyrolysis gasoline) in Europe and Japan is based largely on naphtha cracking though limited amounts are made from higher fractions. Because of their high content of paraffin hydrocarbons, the naphtha fractions of the crude oils normally available in Europe and Japan are very suitable for conversion in the steam cracking process.

In the early naphtha crackers such as those installed by Montecatini in Italy and by Imperial Chemical Industries in England (Wilton) in 1950–1, most of the heat of cracking was supplied by addition to the prevaporised and preheated naphtha of a relatively large quantity of steam previously heated to about 925°C. Subsequent technological improvements have made it possible to operate steam cracking of naphtha at much lower steam ratios in tubular reactions, the heat of cracking being supplied by external heating. Since about 1960 this mode of operation has been practised without appreciable interruptions due to coke formation even at very high conversions. Nowadays, propylene yields may be 15 per cent or more by weight on naphtha fed together with over 30 per cent by weight of ethylene.

1.6.4. Comparison Between USA and Europe and Japan

Whereas in Europe and Japan, a major portion of the total propylene used in chemical manufacture is supplied by the steam cracking of naphtha, the position in the USA is quite different. Of the 3·8 million tons of propylene converted into chemicals in the USA in 1969, more than half was derived from refinery streams with propane cracking next in importance followed by naphtha and gas oil cracking; the cracking of the butanes contributed only a minor proportion.

It should be noted that in addition to the 3·8 million tons of propylene mentioned above, about 5 million tons of propylene were consumed in the USA in 1969 for the production of gasoline components, namely alkylate and polymer gasoline.[71]

1.7. DEVELOPMENT OF PROPYLENE DERIVATIVES SINCE 1950

During the last two decades, the chemical utilisation of propylene has expanded very rapidly in many countries of the world. Thus, in the USA propylene consumption in chemical manufacture has increased from about 0·5 million tons/annum in 1950 to around 4·3 million tons/annum in 1970 and growth has been particularly rapid in the last five years during which it has practically doubled.[71] In Western Europe (comprising West Germany, Britain, Italy, France and the Benelux countries in that order of significance from the viewpoint of propylene usage), consumption of propylene has increased by a factor of about 2·5 during the past five years i.e., from 1·25 million tons in 1965 to an estimated 3·1 million tons in 1970. An even greater growth rate has been experienced in Japan where propylene usage on chemical manufacture has increased from nearly 0·5 million tons in 1965 to a figure (estimated) of over 1·5 million tons in 1970.[103]

The growth of the petrochemical industry generally is, of course, linked with the requirements of an expanding world industry for all types of chemical intermediates and finished products but by far the most important factor favouring growth has been the phenomenal increase in the size and scope of the industry of synthetic high polymers, including plastics, fibre-forming polymers and synthetic elastomers.

In the case of propylene, the following factors have contributed specifically to its rapid growth as a chemical raw material:

a. There has been a steady expansion in the demand for the older industrial derivatives of propylene especially for isopropanol and acetone, propylene oxide and, to a lesser extent, for glycerol.

b. The development of new processes has rendered possible the production of several important and rapidly growing intermediates more cheaply from propylene than from the older raw materials used, especially acetylene, with the result that propylene has become the preferred raw material. An outstanding example is that of acrylonitrile which is now produced exclusively from propylene.

c. As a result of new discoveries, propylene has become the raw material for several entirely new products (i.e., products new to world industry) one group of which, namely the high polymers and copolymers of propylene, has shown outstanding growth potential.

In the following brief historical survey of these developments details of processes and product uses have been omitted since they are discussed fully elsewhere in this volume.

1.7.1. Isopropanol and Acetone

1.7.1.1. Isopropanol   Though the production of isopropanol in the USA has more than doubled in the last twenty years and it is still the largest consumer of propylene for chemical purposes, its rate of growth has been much lower than that of a number of relatively new industrial derivatives such as acrylonitrile, polypropylene and oxo alcohols. In Western Europe, the amount of propylene used for making isopropanol is now much less than that consumed in the production of acrylonitrile and oxo alcohols and about the same as for polypropylene.[103]

One important factor which has slowed the growth of isopropanol is that more and more acetone, one of the major outlets for isopropanol, is being produced by the cumene-based process for the co-production of acetone and phenol. Thus, while 80 per cent of all the acetone produced in the USA in 1960 was made from iso-propanol, the corresponding figure for 1969 is about 50 per cent.

Up to the moment virtually all isopropanol has been produced by the sulphation process, the exception being a direct hydration process operated by ICI in Britain since 1951. However, there are signs that the direct hydration method may become increasingly important in the future and several new plants of this type have been recently announced. VEBA-Chemie AG (formerly Hibernia-Chemie) have developed and commercialised a direct hydration process for propylene based on the use of a supported phosphoric acid catalyst.[104]

1.7.1.2. Acetone   The supply position of acetone has been greatly influenced by the tremendous growth of the cumene-based process for the co-production of phenol and acetone which now provides a major part of all phenol made in the world. Thus in the USA in 1969 more than 70 per cent of all phenol and 50 per cent of all acetone was produced by the cumene route.[103] The process involves the alkylation of benzene with propylene to cumene, the oxidation of cumene to its hydroperoxide and the acid decomposition of the latter to yield phenol and acetone, a sequence of reactions that effectively results in the conversion of propylene to acetone simultaneously with the oxidation of benzene to phenol. The basic oxidation and cleavage reactions were first observed by Hock and Lang[105] in Germany during the War and the first commercial plant to operate these processes was commissioned by B A Shawinigan Co. at Montreal (Canada) in 1953. Since that time a very large number of plants have been operated in all the important industrial countries.

Another and, at the moment, less important route to acetone is the direct oxidation process discovered by Smidt[106] and his colleagues of the Wacker-Chemie GmbH of Munich in the late 1950's. This reaction involves oxidation of propylene by air in the presence of an aqueous catalytic medium of palladium chloride and copper chlorides, the overall reaction being expressed by the equation:

$$C_3H_6 + \tfrac{1}{2}O_2 \longrightarrow CH_3.CO.CH_3$$

The process has been developed and has been industrialised recently in several commercial plants in Japan.

1.7.2. Propylene Oxide

Although the consumption of propylene in the manufacture of propylene oxide in the USA has increased from about 30,000 tons/annum in 1950 to nearly 500,000 tons/annum in 1969,[71] nevertheless the growth rate of this important intermediate has been considerably less than that of some other propylene derivatives so that it now occupies fourth place in the list of propylene users in that country whereas it ranked second only to isopropanol a few years ago. In Western Europe, usage of propylene for propylene oxide production is estimated at nearly 300,000 tons per annum in 1970 and it is likely to be doubled in the next five to six years;[107] even so it represents only about 10 per cent of the propylene used for all chemical purposes.

Major uses for propylene oxide include propylene glycol and the condensation products of propylene oxide with polyhydric alcohols which find application in polyurethane foams, the growth rate of the latter being particularly marked.

Up until quite recently, practically all propylene oxide has been produced by the chlorohydrin route and all attempts to develop a commercially viable direct oxidation route (similar to that so successfully accomplished in the case of ethylene) have been abortive. However, in the last few years, a process not involving chlorine has been developed by Halcon International Inc.[108] in which the hydroperoxides of such hydrocarbon as isobutane and ethylbenzene (prepared from these hydrocarbons and oxygen (air) in a separate operation) are reacted with propylene in the liquid phase in the presence of certain dissolved heavy metal catalysts in accordance with the overall equation:

$$ROOH + CH_3.CH = CH_2 \longrightarrow CH_3.CH{-}CH_2 + ROH$$
$$\diagdown O \diagup$$

The alcohols simultaneously produced in this reaction (t-butanol or methylphenyl carbinol) may be dehydrated to the corresponding olefins, namely isobutene or styrene so that this process can be used for the simultaneuos production of propylene oxide and styrene (or isobutene). Since 1969, this process has been operated on a large commercial scale (70,000 tons/annum of propylene and 150,000 tons/annum of t-butanol, the latter being probably dehydrated to isobutene, at least in part) at

Bayport, Texas, by the Oxirane Chemical Co., a joint venture of Halcon International and Atlantic Richfield Corp.[109] A second facility with a capacity of 150,000 tons/annum of propylene oxide is under construction at the same site and is due to be commissioned in late 1971. Another plant, designed to produce propylene oxide and styrene at the rate of 32,000 tons/annum and 80,000 tons/annum, respectively is being built at Puertollano, Spain, and is due on stream in early 1971[110] while other plants utilising this process are being erected in other countries, e.g. the Netherlands, in this case using isobutane.

1.7.3. Oxo Alcohols from Propylene

Reference has already been made to the important discovery made in Germany in 1938 by Roelen of the Ruhrchemie AG and its subsequent development during the War years by Ruhrchemie and by the IG at Ludwigshafen, Leuna and Oppau. A commercial plant for the production of 12,000 tons/annum of primary alcohols from C_{11}–C_{17}. Fischer-Tropsch olefins were erected by the Oxo-Gesellschaft at Holten but was never commissioned because of War damage in 1944 and several years elapsed before this process was applied industrially in the USA, Britain and Germany. However, in 1948 Esso Standard in the USA industrialised a continuous process for converting isoheptene (ex polymer gasoline) into iso-octanol at Baton Rouge, La.[111]

The application of the Oxo reaction (sometimes called 'hydroformylation') to propylene furnishes a mixture of n- and iso-butyraldehydes from which the corresponding primary alcohols may be produced. Furthermore other propylene-derived hydrocarbons can be used as feedstocks to this reaction to give useful primary alcohols, e.g. isooctanol from isoheptene from polymer gasoline, iso-decanol from propylene trimer and iso-tridecanol from propylene tetramer. Even more important, however, is the fact that n-butyraldehyde may be converted into the highly desirable primary alcohol, 2-ethylhexanol which is a key material for the manufacture of plasticiser esters, mainly phthalates.

Because of its relative cheapness compared with the alternative multistage synthetic routes for the production of n-butanol and 2-ethylhexanol from acetaldehyde, the Oxo process for the manufacture of these important alcohols is tending to displace the alternative in most countries.

The first plant for the conversion of propylene into mixtures of the two butyraldehydes was commissioned in 1952 by Tennessee Eastman Corp. at Kingsport, Tenn.[112] Since that time the production of these aldehydes and the corresponding alcohols as well as 2-ethylhexanol from propylene by this method has greatly expanded throughout the world but particularly in the USA, and Germany. In the latter country, the BASF Company has the largest propylene-based Oxo operation in the world, producing 360,000 tons/annum of C_4 products, at Ludwigshafen and this is now being expanded to give a capacity of 500,000 tons/annum of these products by 1971.[113] In this plant a ratio of approximately 3 parts by weight of n-butyraldehyde to 1 part of isobutyraldehyde is achieved.

The world capacity for propylene-based Oxo alcohols is now in the region of $2\frac{1}{2}$ million tons/annum[114] divided roughly equally between the two butanols, 2-ethylhexanol, iso-octanol and other alcohols from higher olefins.

In the USA about one half of all the n-butanol and one third of all the 2-ethyl-hexanol made in 1969 was derived from propylene by this reaction. The following quantities of propylene were used in the USA in 1969 in the manufacture of C_4 and higher primary alcohols:[71]

Alcohol	Propylene Usage tons/annum
Isobutanol	45,000
n-Butanol	85,000
2-Ethylhexanol	75,000
Iso-octanol	95,000
Iso-decanol	90,000
Iso-tridecanol	10,000
Total	400,000

Whereas in 1969 the quantities of n-butanol and 2-ethylhexanol were about the same in the USA, in Western Europe 2-ethylhexanol output is 50 per cent higher than that of n-butanol and is expected to be twice as large by 1975.[115]

Since isobutyraldehyde has only limited application (almost entirely as a source of isobutanol), a high ratio of n- to iso-butyraldehyde in the Oxo reaction product is obviously highly desirable. Some success has been achieved in raising this ratio by modification of process conditions and, for example, the Mitsubishi Chemical Industries Ltd. claim a ratio of 4 to 1 (n- to iso) in their process which is operated industrially in Japan and Czechoslovakia.[116] However, recent discoveries[117] have opened up the possibilities of achieving much higher ratios, e.g. 10 to 1 (n- to iso) by modifying the cobalt carbonyl catalyst with tertiary organophosphines and commercial plants are now being built on this basis.

A modification of the Oxo reaction involving the use of a mixed cobalt and zinc catalyst has been developed by Esso Research and Engineering Co. in which 2-ethylhexanol as well as a mixture of butanols are produced. It is known as the Aldox process[118] and has been commercialised at Kawasaki (Japan) on a scale of 10,000 tons/annum of 2-ethylhexanol and 5,400 tons/annum of butanols.

Finally mention may be made of the Reppe process in which a mixture of approximately 84 per cent of n-butanol and 16 per cent of isobutanol is produced directly by reaction of propylene with water and carbon monoxide in the presence of an iron pentacarboxyl catalyst in accordance with the equation:

$$C_3H_6 + 3CO + 2H_2O \longrightarrow C_4H_9OH + 2CO_2$$

This process has been industrialised on a scale of 15,000 tons/annum of butanols by the Japan Butanol Co. at Yokkaichi, Japan.[119]

1.7.4. Acrylonitrile

It is remarkable that within a year or so of Natta's discovery of the stereospecific polymerisation of propylene, another discovery in propylene chemistry of comparably great technical importance was made and equally rapidly applied industrially.

This was the reaction of propylene, oxygen and ammonia in the vapour phase over specific solid catalysts to yield acrylonitrile in accordance with the overall equation:

$$C_3H_6 + NH_6 + 1\tfrac{1}{2}O_2 \longrightarrow CH_2 = CH.CN + 3H_2O$$

Until the development of this process, usually referred to as the 'ammoxidation' process,* acrylonitrile was produced mainly from acetylene and hydrocyanic acid although some was made from ethylene oxide and that acid. The switch from these rather expensive raw materials to the much cheaper propylene and ammonia combined with a single stage process operated on very large scales of production has led to a rapid reduction in acrylonitrile production cost so that its price has been almost halved since the introduction of this process. This has in turn greatly stimulated the utilisation of acrylonitrile in the production of fibre-forming polymers and also of oil-resisting rubbers, latices and ABS- and SAN-type moulding materials; the ABS-type plastic materials represent a very rapidly expanding outlet for acrylonitrile. As a result, acrylonitrile production from propylene in the USA has increased from around 10,000 tons in 1960 (the first year in which it was produced by the 'ammoxidation' route) to 460,000 tons in 1969 and is expected to reach an output rate of around 650,000 tons/annum by 1975.[120]

In Western Europe, propylene usage for acrylonitrile production is estimated at around 0·5 million tons in 1970, a nearly ten-fold increase in the last five years.[103] Acrylonitrile is thus one of the more rapidly expanding derivatives of propylene, the others being polypropylene and propylene-derived Oxo alcohols.

The first indication that acrylonitrile could be obtained, though in low yield, by the catalytic reaction of propylene, ammonia and oxygen was given in a patent granted to the Allied Chemical and Dye Corporation[121] in 1949. A few years later the research workers of the Distillers Co. Ltd. (Edinburgh)[122] showed that acrolein could be converted to acrylonitrile in good yield over molybdenum oxide catalysts and envisaged a two stage process for the oxidation of propylene, acrolein being produced in the first stage and reacted with ammonia and air in the second stage to furnish acrylonitrile.[123] However, the credit of first discovering a commercially viable, single stage process for converting propylene into acrylonitrile must be accorded to the workers of the Standard Oil Co. of Ohio ('Sohio') who in 1957 found that a catalyst of bismuth molybdate or phosphomolybdate was effective for this purpose.[124] A commercial plant with a capacity of 18,000 tons/annum embodying this process in a fluidised bed system was commissioned by Sohio at Lima, Ohio, in 1960 and this was the first of a very large number of plants which have subsequently been operated all over the world on the basis of this process which has undergone a number of improvements especially in respect of catalyst composition and performance.

About this time the Distillers Co. discovered other catalyst compositions capable of efficiently catalysing the ammoxidation of propylene and developed in collaboration with the French Company,[125] Société d'Electrochemie d'Ugine, a fixed bed process which was industrialised in France on a scale of 30,000 tons/annum and also in 1965 at Grangemouth (Scotland) on a scale of 45,000 tons/annum of acrylonitrile.[126] Since that time, capacity has been substantially increased.

* A term first used by Hadley, *Chem. and Ind.*, 1961, 238.

During the last few years, several other companies have been active in developing new catalytic systems for this reaction including a catalyst containing cerium, molybdenum and tellurium oxides developed by Montecatini Edison[127] and now industrialised in Italy on a scale of 40,000 tons/annum. However, the bulk of acrylonitrile production from propylene is produced by the Sohio process.

An alternative to the direct ammoxidation process was developed by du Pont and industrialised at Beaumont, Texas. This was based on the vapour phase reaction of propylene with nitric oxide over a silver catalyst in accordance with the equation:

$$4C_3H_6 + 6NO \longrightarrow 4CH_2 = CH.CN + N_2 + 6H_2O$$

However, it is believed that this process is no longer operated.

1.7.5. Acrolein and Acrylic Acid

The development of processes for the conversion of propylene into acrolein and acrylic acid by catalytic vapour phase oxidation has opened up opportunities for the future use of acrolein as a relatively cheap and highly reactive intermediate and, in the case of acrylic acid, has provided an alternative route for the manufacture of the acrylic esters which have become well established as valuable intermediates for film- and fibre-forming polymers.

1.7.5.1. Acrolein   Although the oxidation of propylene by aqueous acid solutions of mercuric sulphate to yield acrolein was observed as early as 1898,[128] attempts to use this reaction as the basis of a commercial process have been unsuccessful. Until the appearance of the propylene oxidation route, acrolein was made on a modest scale in Germany since 1942, and later in the USA, by the catalytic vapour phase condensation of acetaldehyde and formaldehyde, most of the acrolein so produced being used in the manufacture of methionine.

However, in 1946, workers of the Shell Development Co. discovered the vapour phase oxidation of propylene to acrolein in the presence of a supported cuprous oxide catalyst[129] and this process was industrialised on a scale of 15,000–20,000 tons/annum in 1959 at Norco, La., a large part of the product being subsequently utilised as an intermediate in the large scale synthesis of glycerol. About this time, several other companies developed processes based on the use of other catalysts. Thus, the workers of the Standard Oil Co. of Ohio[130] developed the use of bismuth molybdate or phosphomolybdate for this reaction and this process was applied commercially in the early 1960's at Lima, Ohio. Other active catalytic systems were developed by the Distillers Co. Ltd. whose process has been operated commercially in France since about 1965.

The consumption of acrolein, apart from that used in the manufacture of glycerol and methionine, is relatively small and this in spite of the large amount of research and development effort which has been deployed in attempts to develop industrial applications of numerous acrolein derivatives. In the USA, propylene usage in acrolein production has increased from about 3,000 tons in 1962 to barely 20,000 tons in 1969.[71] It is believed that the considerably smaller European production is used mainly for methionine manufacture.

1.7.5.2. Acrylic acid   Acrylic acid has become important industrially in the last few decades as an intermediate in the production of film-forming polymers (and to some extent also, fibre-forming polymers) based on the polymerisation of its various esters. Until recently these materials have been made for the most part from acetylene and carbon monoxide (modified Reppe process) but interest on the propylene-based route is likely to increase and at least one large acrylate plant based on propylene oxidation is now in operation.

The production of acrylic acid from propylene can be effected either in a single catalytic reaction stage or in two stages in series, the first of which produces mainly acrolein and the second one (usually employing a different catalyst and reaction conditions) converts this acrolein into acrylic acid. Catalytic systems for these reactions have been developed by a number of companies notably the Standard Oil Co. of Ohio and the Distillers Co. Ltd. and in the last few years a somewhat bewildering number of different catalysts have been proposed for this purpose. The first commercial plant, based on two-stage catalysis and incorporating technology of Distillers Co. Ltd. was commissioned by the Union Carbide Corp. at Taft, La., in 1969 on a scale of 90,000 tons/annum.[131] Smaller plants employing both single and two stage operations have been announced in Japan[132] while a 45,000 tons/annum plant is to be built in Britain by Lennig, the UK subsidiary of Rohm and Haas Co.[133]

World demand for acrylic esters at the moment is roughly 200,000 tons/annum of which the USA accounts for 60 per cent, Western Europe 25 per cent and Japan less than 15 per cent. Use patterns differ considerably from country to country. Thus, in the USA surface coatings and latex paints, requiring mainly the ethyl and 2-ethylhexyl esters, represent the bulk of the demand which in Japan, methyl acrylate accounts for more than half of the demand because of its increasing consumption as a co-monomer in acrylic fibre-forming polymers.[134]

1.7.6. Glycerol

Relatively small quantities of propylene are consumed in the synthesis of glycerol which has a rather modest growth rate compared with several other propylene derivatives. Thus in the USA, where 56 per cent of all glycerol is made synthetically (the corresponding figure for Western Europe is 26 per cent) production from propylene has increased from about 70,000 tons/annum in 1960 to around 120,000 tons/annum in 1970; in that year about 50,000 tons of epichlorohydrin was made as a co-product of the allyl chloride-based route.

No less than three different processes are currently in use for the manufacture of glycerol from propylene. By far the most important in terms of quantity is the allyl chloride—glycerol dichlorohydrin route already referred to previously and used by Shell Chemical Co. since 1948 at their Beaumont, Texas plant and more recently at Pernis in Holland. Large plants operating this process also exist in the USSR and in Italy. An intermediate in this process is epichlorohydrin which has now become an important co-product of glycerol by this route.

Another route, also pioneered by Shell Chemical Co. and forming the basis of a 16,000 tons/annum plant commissioned in 1959 at Norco, La., involves the reduction of acrolein to allyl alcohol by catalytic reaction with isopropanol followed by hydroxylation of the allyl alcohol by means of hydrogen peroxide produced in a separate operation by the oxidation of isopropanol with air. Acetone is produced in two of these stages and the overall result can be summarised as follows:

$$CH_2 = CH.CHO + CH_3.CHOH.CH_3 \rightarrow CH_2 = CH.CH_2OH + CH_3.CO.CH_3$$
$$CH_3.CHOH.CH_3 + O_2 \longrightarrow H_2O_2 + CH_3.CO.CH_3$$
$$CH_2 = CH.CH_2OH + H_2O_2 \longrightarrow CH_2OH.CHOH.CH_2OH$$

Yet a third process is based on the vapour phase isomerisation of propylene oxide to allyl alcohol which is then epoxidised to glycidol by means of peracetic acid, the latter step being an adaptation of the Prileschajeff reaction[135] discovered in about 1909. Finally glycidol is hydrolysed to glycerol. This sequence of operations is the basis of a plant with a capacity of 18,500 tons/annum now operated by the Food Machinery and Chemical Corporation at Bayport, Texas.

It is of interest to note that another propylene oxide-based process was operated for some years by the Olin Mathieson Corporation.[136] In this process, propylene oxide was isomerised to allyl alcohol which was then (as in the process developed in Germany during World War II) converted into glycerol monochlorohydrin, the latter being then hydrolysed to glycerol.

1.7.7. Polymers of Propylene

1.7.7.1. High polymers   The highly important discoveries on the polymerisation of α-olefins and especially propylene, to yield high polymers of a stereo regular nature and with very desirable physical and mechanical properties, made by Natta and his colleagues of the Institute of Industrial Chemistry of the Milan Polytechnic were sparked off by the work of Ziegler and his associates of the Max Planck Institute in Mülheim (Ruhr) on the polymerisation of ethylene in the presence of certain organometallic catalysts. Natta has recorded[137] that his interest was first aroused by hearing a lecture by Ziegler at Frankfurt in 1952 when the possibilities of obtaining perfectly linear polymers of ethylene with molecular weights of several thousands was first disclosed.[138] The catalysts first described by Ziegler for polymerising ethylene either did not polymerise higher olefins at all or only converted them into low molecular weight products. However, Natta and his collaborators began working in 1953 and by early 1954 had obtained α-olefin high polymers which showed crystallinity when examined by X-ray diffraction.[139] When these polymers were extracted with various solvents, Natta and his colleagues were surprised to find that there were obtained amorphous, soluble fractions with relatively high molecular weight (20,000–30,000) and nearly insoluble fractions having similar molecular weights but exhibiting high crystallinity and a high melting point. Natta immediately thought that these differences in solubility and crystallinity were to be attributed not to slight differences in molecular weight but to an entirely different chain structure.[140] Detailed examination of these insoluble polymers convinced Natta that a highly stereospecific polymerisation had been effected to yield a polymer constituted by chain segments with a regular repetition of monomer units containing asymmetric carbon atoms with the same steric configuration. To this type of polymer, Natta gave the name 'isotactic' while the amorphous, soluble types of polymer were designated 'atactic' since they contained an irregular distribution of the steric configurations of the monomer units.[141] Later, another type of α-olefin stereoregular high polymer, also crystalline, was described[142] and termed 'syndiotactic'; these polymers differ from the isotactic polymers in that they contain a regular sequence of units each tertiary carbon of which has a configuration opposite to that of its neighbours.

The original identification of isotactic polypropylene resulted from the fractionation of a mixture of propylene polymers consisting of 40–60 per cent of insoluble, crystalline material, the remainder being amorphous polypropylene. Somewhat later, highly selective catalysts, especially those based on certain crystalline forms of titanium trichloride in the presence of aluminium trialkyls, were developed which polymerised propylene to a product containing more than 90 per cent of the desirable, isotactic material.

These discoveries were quickly recognised as being not only of the greatest theoretical interest but also of great practical and industrial importance. Natta himself recognised this[143] and postulated a rapid development of the industrial production of these new isotactic polypropylenes on account of their excellent mechanical properties, low raw material cost and high yield in a polymerisation process capable of being carried out under mild conditions of temperature and pressure. In fact, the speed with which the stereospecific polymerisation of propylene was industrialised on a world wide basis is quite remarkable and only paralleled in propylene chemistry by the rapid commercial application of the 'ammoxidation' process for acrylonitrile, discovered and developed at nearly the same time.

The first production plant was commissioned in 1957 at Ferrara, Italy, by the Montecatini Co. which had collaborated with Natta and his co-workers on this project and now fostered the industrialisation of the process by suitable arrangements with Ziegler in regard to the patent situation. Soon afterwards the Hercules Company started making isotactic polypropylene in the USA while manufacture was also established by Farbwerke Hoechst in Germany and Shell Chemical Co. in Britain. In the USA, the first year for which statistics are available is 1958 when about 7,500 tons of propylene were consumed in the production of polypropylene; by 1969 this figure had risen to an annual consumption of about 560,000 tons.[71]

A detailed account of the technology of production and of the properties and applications of polypropylene is to be found in Chapter 5. However it may be mentioned that industrially important copolymers of propylene with small amounts of other olefins, particularly ethylene, have been developed and these have grown rapidly in significance because of their improved impact strength at room temperature and below as well as their good melt flow properties.[144] It is anticipated that this type of polymer will shortly amount to about one half of all polypropylene produced.[145] In the UK, polymers of this type have been produced since about 1960.

Another type of copolymer of propylene which, though small at the moment, may ultimately become of major significance is the ethylene-propylene-diene terpolymer rubber, production of which in 1969 was around 120,000 tons on a world basis.[146] The diene component of these materials, usually a cyclic diene, is incorporated to facilitate vulcanisation.[147]

1.7.7.2. Propylene dimers While propylene trimer and tetramer as well as propylene-butene co-dimer are substantial intermediates for the chemical industry (propylene usage in their manufacture in the USA is now of the order of 600,000 tons/annum),[71] it is only within recent years that propylene dimers have found application in that industry. The selective dimerisation of propylene to one or other of its various dimers has been the subject of much study[148] from which have emerged two dimerisation processes of some industrial significance.

a. The dimerisation of propylene by the Ziegler[149] method, using aluminium trialkyl as catalyst, furnishes mainly 2-methylpentene-1. This is the raw material used by the Goodyear Tire and Rubber Co. in a plant operated in Beaumont, Texas since about 1961 for the large scale (originally 20,000 tons/annum but since enlarged) production of isoprene. The process employed involves the isomerisation of 2-methylpentene-1 to 2-methylpentene-2 which is thereafter pyrolysed in the presence of small amounts of hydrogen bromide to give isoprene and other hydrocarbons.

b. Dimerisation of propylene under the influence of alkali metals such as potassium was first observed by Schramm.[150] More recently, the British Petroleum Co. has developed a continuous dimerisation process in the presence of a supported alkali-metal catalyst[151] which gives high yields of 4-methylpentene-1, a useful monomer in the production of thermoplastic high polymers of good physical properties and high optical clarity. The monomer is now produced by the British Petroleum Co. in a plant at Grangemouth (Scotland) while the polymer is manufactured by Imperial Chemical Industries Ltd. and sold under the designation TPX.

1.8. PRESENT POSITION AND FUTURE TRENDS IN PROPYLENE-BASED CHEMICALS

1.8.1. Growth of Propylene Usage in Chemical Manufacture

In the USA where the chemical utilisation of propylene was first established in the early 1920's, propylene consumption for chemical purposes has grown steadily and, except for the two or three years following the end of World War II, uninterruptedly. This is shown graphically in Fig. 1.1 which also shows the growth of the main chemical derivatives of propylene in the USA since 1940 in terms of the quantities of propylene consumed annually in their manufacture.[71] In the three decades 1940–50, 1950–60 and 1960–70 propylene usage for chemicals in the USA has increased five times, 2·5 times and 3·3 times respectively.

In other countries where the petrochemical industry only became substantial after 1950, the growth of propylene consumption for chemical purposes has been much higher, particularly in West Germany and Japan. Detailed figures are not readily available for many countries but it has been estimated[152] that in Western Europe, propylene consumption for chemicals has increased from 0·5 million tons in 1960 to 3·6 million tons in 1970 while in Japan it has grown from 0·1 million tons/annum to 1·6 million tons/annum during the same period. For the non-communist world, Collingwood[152] has estimated a propylene consumption in chemical manufacture in 1970 of 9·3 million tons of which the shares of the USA

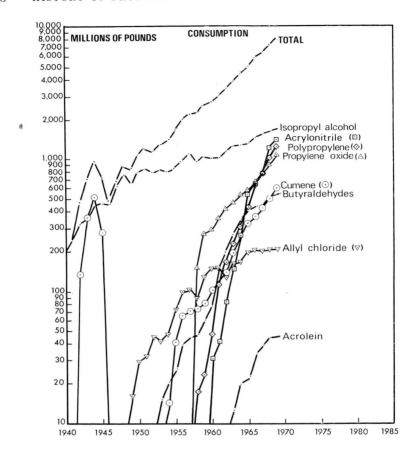

FIG. 1.1   *Growth of Propylene Usage in Chemical Manufacture from 1940.*

and Canada, Western Europe and Japan are respectively 39 per cent, 39 per cent and 20 per cent, with only 8 per cent in other non-communist countries. The corresponding estimate of ethylene usage is 16·6 million tons, a weight ratio of propylene to ethylene of 0·56. For the whole world, Imhausen[153] has estimated a petrochemical usage of propylene for 1970 of 12·4 million tons and an ethylene consumption of 25 million tons in this year, a weight ratio of propylene to ethylene of nearly 0·5.

It is noteworthy that in spite of its steady growth rate, petrochemical propylene usage is tending to fall in relation to the corresponding ethylene consumption. Thus the weight ratio of propylene to ethylene in the USA has fallen from about 0·9 in the early 1940's to around 0·5 in 1970. In Western Europe this ratio varies considerably from country to country, being rather low (0·4) for Western Germany and high (0·65) in France and Italy with Britain occupying an intermediate position.[154]

In Japan a high ratio (0·7) reflects the proportionately greater development of poly-propylene and acrylonitrile (for fibres) in that country.

1.8.2. Propylene Use Pattern

The most important chemical derivatives of propylene are the following, in order of their world significance as propylene consumers:

a. Polypropylene including high molecular weight co-polymers and also elastomers containing propylene and ethylene.
b. Acrylonitrile.
c. Oxo alcohols including n- and iso- butanols, 2-ethylhexanol-1 and branched primary alcohols based on propylene-butene codimer (isoheptene) and propylene trimer and tetramer.
d. Isopropanol.
e. Propylene oxide and its derivatives.
f. Cumene (and thence acetone).
g. Oligomers of propylene especially the tetramer and trimer but including the dimers, 2-methylpentene-1 (an intermediate for isoprene) and, more recently, 4-methylpentene-1.
h. Allyl chloride, and intermediate for epichlorohydrin and glycerol.
i. Acrolein and its derivatives, including allyl alcohol and glycerol.
j. Acrylic acid and its esters.

Other derivatives of less significance at the moment are:

k. Chlorinated solvents, perchlorethylene and carbon tetrachloride.[155]
l. Carbon disulphide.
m. Lactic and oxalic acids.[156]
n. Ethylene and butene-2 by the so-called Triolefin dismutation process (operated in Canada).

TABLE I.1

Propylene Usage in the USA in Thousands of Tons per Annum

Derivative	1940	1945	1950	1955	1960	1965	1969
Acrolein						10	22
Acrylonitrile					14	238	694
Allyl Chloride			12	30	63	82	93
Butyraldehydes				11	33	179	224
Cumene*				21	51	163	343
Dodecene			87	194	320	162	282
Heptenes			2	27	40	162	147
Isopropanol	91	204	359	384	448	581	812
Nonene				38	78	132	148
Polypropylenes					21	192	555
Propylene Oxide	1†	4†	28†	25†	130	254	478
Total	92	208	488	730	1198	2155	3798

* For chemical purposes only.
† Propylene glycol only.

1.8.2.1. Use pattern in the USA   Table I.1 shows the quantities of propylene (in thousands of long tons) consumed annually in the USA in the manufacture of various chemical derivatives during the period 1940–69. Propylene used in the manufacture of a number of miscellaneous products, including isoprene, ethylene-propylene rubbers and acrylic acid are not included. The variation of propylene usage with time for some of the major products is shown graphically in Fig. 1.1.

1.8.2.2. Use pattern in Western Europe and Japan   Estimates of propylene usage in the production of various derivatives in Western Europe and Japan for 1965 and 1970 have been put forward by Ockerbloom, Brownstein and Root[103] and are tabulated in Table I.2. Their figures for the USA, which are reasonably in line with the Stanford Research Institute data,[71] are included for comparison.

TABLE I.2

Estimate of Propylene Usage in Western Europe, Japan and USA (in thousands of metric tons per annum)

	1965			1970		
	W. Europe	Japan	USA	W. Europe	Japan	USA
Polypropylene	90	70	190	410	565	710
Acrylonitrile	55	185	195	510	470	590
Oxo Alcohols	215	25	550	655	155	680
Isopropanol	300	30*	595	435	85*	680
Propylene Gxide	125	40	260	295	75	455
Cumene	205	20	105	340	30	240
Others	275	110	445	465	145	745
Total	1265	480	2340	3110	1525	4100

* Acetone.

In terms of propylene usage in chemical manufacture, West Germany had in 1969 a substantial (more than 50 per cent) lead over its nearest competitors, France and Britain, while usage in Italy was appreciably less than that for these two countries.[154] However the use pattern was different somewhat in these countries, West Germany having a relatively large propylene-based Oxo alcohols production and Britain a relatively large isopropanol manufacture.[157]

1.8.3. Future of Propylene-based Chemicals

The rapid growth of the petrochemical industry during the last two decades has been dependent on a vast expansion of the consuming industries and, in particular, of those based on synthetic high polymers and including plastics, surface coating materials, synthetic rubbers and synthetic fibre-forming polymers as well as the associated auxiliary materials such as solvents and plasticisers. In turn, these consuming industries have been greatly stimulated by the fact that the raw materials and intermediates necessary for their expansion have been produced on a petro-chemical basis in adequate quantities and considerably more cheaply than was possible from other carbon sources.

In the case of propylene, the discovery and industrial development of poly-propylene and of elastomers co-polymers directly from propylene has played a major role in promoting the expansion of propylene usage throughout the world. The future is certain to witness a rapid growth in these materials which will doubtless be increased in range and scope by appropriate modification.

Another major development in propylene usage which has had an important impact on the industry of high polymers is the ammoxidation process for the direct conversion of propylene into acrylonitrile. This development, in which propylene has completely displaced acetylene as the raw material for acrylonitrile manufacture, is certain to become increasingly important with time. More recently, the manu-facture of acrylic acid and its esters (valuable in the production of emulsion paints, etc.) has been initiated on a large scale on the basis of the direct oxidation of propy-lene and it may be that this route will ultimately displace the alternative synthetic method based on acetylene.

TABLE I.3

Future Propylene Usage in USA, Western Europe and Japan
(thousands of metric tons per annum)

		1975			1980	
	USA	W. Europe	Japan	USA	W. Europe	Japan
Polypropylene	1320	890	975	2280	1415	1515
Acrylonitrile	935	895	640	1100	1230	950
Propylene Oxide	705	565	115	1100	810	175
Isopropanol	835	530	120*	1050	685	180*
Oxo Alcohols	730	980	250	870	1255	410
Cumene	545	495	80	690	630	125
Others	830	505	220	2110	509	345
Total	5900	4860	2400	9200	6534	3700

* Acetone only.

It is likely that some propylene derivatives which at the moment are produced on a relatively modest scale will grow in importance in the future. Thus, acrolein which is a highly reactive material readily and cheaply producible by the direct oxidation of propylene has so far had a disappointing growth record in spite of a vast amount of research and development effort directed towards its industrial utilisation. This situation could change radically during the next decade.

Various projections have been made of the likely growth of propylene usage dur-ing the next decade. Imhausen[153] suggests that propylene demand for chemical manufacture throughout the entire world will increase from 12·4 million tons in

1970 to 18·5–22·1 million tons in 1975 and 24·6–28·0 million tons in 1980, with the demand ratio (weight) of propylene to ethylene declining from about 0·5 to 0·4 in the period 1970–80.

The estimates put forward by Ockerbloom, Brownstein and Root[103] on the likely usage of propylene in the manufacture of various derivatives in the USA, Western Europe and Japan are given in Table I.3.

1.8.4. Future Raw Material Situation

Even if the estimates of growth of propylene-based chemicals during the next decade prove to be over optimistic, it is clear that a large expansion of facilities for producing propylene (and also, of course, ethylene) will be necessary to meet the demand for these raw materials. There is no technical difficulty in producing mixtures of propylene and ethylene from a wide range of petroleum fractions and even from crude oil. However, certain petroleum feedstocks, which may differ in different countries, are likely to be preferred for economic reasons.

A complicating factor in any discussion of the future raw material situation is the probability that in the next few years the use of tetraethyl lead (TEL) as a gasoline additive will be severely reduced or even entirely eliminated in the interests of reducing atmospheric pollution. The result of such a restriction would almost certainly be to intensify the demand by refineries for high octane blending materials which could be used to up-grade the anti-knock value of gasoline. One such material is the alkylate obtainable by reacting the lower olefins with isoparaffins such as isobutane. Although the alkylate from ethylene is appreciably better in octane rating than those from butenes and propylene, it seems likely, at any rate in the USA, that all these olefins will be used for this purpose. If the increased demand for alkylate does materialise, then refineries will compete with petrochemical plants for the available supplies of olefins, including propylene.[158]

In the USA, where a substantial proportion of petrochemical propylene is derived from refinery gases, this competition could result in the installation of very large new capacity for producing ethylene and propylene by cracking liquid petroleum fractions. Naphtha for olefins production is generally imported into the USA but the quantities so imported are restricted by the Government.[159] If, for this and other reasons, adequate supplies of naphtha do not become available for this purpose, the US producers will be forced to switch to other petroleum fractions, such as gas oils, or even to crude oil. An alternative feedstock for propylene is isobutane which is currently in long supply in the USA although a greatly increased alkylate demand may change this situation. Finally, a promising method of producing propylene in the future is the catalytic dehydrogenation of propane though this process has not yet been applied industrially.[97]

In Western Europe and Japan where the production of ethylene and propylene is, at the moment, based almost entirely on naphtha steam cracking, it seems likely that increased production over the next few years will be achieved by expanding very largely on the same raw material basis. The naphtha supply position in Western Europe in relation to expected petrochemical and refinery requirements has been critically surveyed by Docksey[160] who has concluded that there should be sufficient naphtha available to meet these requirements at least until 1975. However, several methods of augmenting supply will in the interim be actively considered, such as the diversion of naphtha now used for ammonia production, the recovery and import of field gases and the utilisation of the C_2–C_4 paraffins in imported crude. In the long run, cracking of heavier feedstocks is likely to become necessary.

1.9. PROPYLENE PRODUCTION CAPACITIES

Table I.4 contains a list of propylene-producing plants, either existing or planned, in the non-communist world with present capacities and authorised expansions. Where the information is available, an indication is given of the raw materials used (or to be used) and the approximate date of commissioning of the additional capacity. It must be stressed that only petrochemical propylene is dealt with and propylene solely used in refinery operations is not included.

While every attempt has been made to render this list complete and accurate, the rapidly changing situation makes certain errors almost inevitable. One uncertainty arises from the fact that when new, and usually larger, plants are installed, the older units are sometimes maintained as stand-by facilities or, in some cases, scrapped. It is thus not always certain, though it will often be the case, that total capacity on completion of expansion is equal to present capacity plus authorised expansions.

In some cases where naphtha cracking is being used, only the ethylene capacity is quoted. To obtain an approximate indication of the potential propylene production (quoted in brackets), the ethylene capacity has been multiplied by the factor 0·6.

The abbreviation 'NA' indicates that the necessary information is not available. 'RG' refers to refinery gas and 'NG' to natural gas and liquefied petroleum gases.

TABLE I.4

Petrochemical Propylene Capacities in Non-Communist World
(thousands of tons per annum)

Company	Location	Present capacity	Authorised expansion Additional capacity	Commissioning date	Feedstock	Ref.
Argentina						
Pasa	San Lorenzo	NA	110	1972/3	NA	ooo
Australia						
Altona Petrochemical	Altona	30			Naphtha	161, 162, 164
Esso/BHP	Westernport	—	NA		NG	162
ICIANZ	Botany	(40)			Naphtha	162
Paccal	Clyde	NA			RG heavy oil	162
Austria						
Österr. Mineralöl Verwaltung	Schwechat	50	40	1972	Naphtha and RG	ooo

Company	Location	Present capacity	Authorised expansion Additional capacity	Commissioning date	Feedstock	Ref.
Belgium						
Petrochim	Antwerp	230			Naphtha	161, 165
Brazil						
Petrobras	Capuava	—	155	1972	Naphtha	
Cia Petroquimica						
Unico	Capuava	—	93	1972/3	Naphtha	
Canada						
British American						
Oil	Montreal, Que.	NA			RG	164
Dow Chemical	Sarnia, Ont.	NA			RG	164
Imperial Oil	Sarnia, Ont.	56			Naphtha	164
Shawinigan						
Chemicals	Varennes, Que.	120			Naphtha	164
Shell Canada	Montreal, Que.	NA			RG	164
Union Carbide	Montreal, Que.		45	1971	RG	173
Colombia						
Empresa Colomb.						
de Petroleos	Barrancabermeja		10	1971	Naphtha	
Denmark						
A. P. Møller	Maersk	45			Naphtha	165
Egypt						
State Authority	Alexandria	20			Naphtha	164
Finland						
Neste O.Y.	Porvoo		(70)	1972	Naphtha	
France						
Antar	Donges	12			RG	
Soc. Chim. des						
Charbonnages	Carling	120			Naphtha	163
California-						
Atlantique	Gonfreville	60	150	1972	RG and Naphtha	163
Shell Chimie	Berre	60			Naphtha	
Rhône-Alpes	Feyzin	160	135	1973	Naphtha	165
Naphthachimie	Lavera e la Mede	150	320	1972	Naphtha	165
Esso Chimie	Port-Jérôme	150			Naphtha	
Rhône-Progil	Gonfreville		(200)	end 1971	Naphtha	161
German Federal Republic (Western Germany)						
BASF	Ludwigshafen	120	240	1972	Crude oil and Naphtha	
Caltex-Deutschland	Raunheim	160			Naphtha	165
Chemische Werke						
Hüls	Marl	40			RG	165
Chemische Werke						
Hüls/VEBA	Gelsenkirchen	100	100	NA	Naphtha	165
Deutsche Erdöl	Heide	58			Naphtha	161, 165
Dow Chemical	Stade		(240)	1975	Naphtha	165
Erdölchemie	Köln-Worringen	380			Naphtha	165, 172
Esso Chemie	Köln	80			Naphtha	161, 165
Gelsenberg AG	Münchmünster		130	1972	Naphtha	165, 166

Company	Location	Present capacity	Authorised expansion Additional capacity	Commissioning date	Feedstock	Ref).
Rheinische Olefinwerke	Wesseling	285	300 +250	1972 1974	Naphtha	
Union Rheinische Braunkohlen Kraftstoff	Wesseling	50	135	1972	Naphtha	165
VEBA-Chemie	Gelsenkirchen	117	200	1973	Naphtha	
Holland						
DSM	Geleen	100	200	1972	Naphtha, gas oil	
Dow Chemical	Terneuzen	200	200	1972	Naphtha	165
Gulf Oil	Europoort	150			Naphtha	161, 165
Shell Nederland	Pernis	170			Naphtha and RG	
	Moerdijk	—	(150)	1973	Naphtha	
India						
NOCIL	Thana, Bombay	35			Naphtha	161
State Authority	Koyali-Gujarat	60			Naphtha	161
Union Carbide	Bombay	6			Naphtha	164
Esso Standard	Trombay	10			RG	164
Iran						
State Authority	—	4			Propane	164
Israel						
Israel Petrochemical Enterprises	Haifa	10			RG	164
Italy						
ANIC	Gela	40	80	1972	Naphtha	161, 165
Montedison	Brindisi	140	90	1973/4	Naphtha	
	Ferrara	40			Naphtha	165
	Mantova	50			Naphtha	165
	Porto Marghera	150	150	1972	Naphtha	165
Rumianca	Cagliari	37	33	1973	Naphtha	165
Sincat	Priolo	160			Naphtha	165
SIR	Porto Torres	135	35	1972/3	Naphtha	
Japan						
Daikyowa Sekiyu K.	Yokkaichi	21	190	1972	Naphtha	
Idemitsu Petrochemical	Tokuyama	175			Naphtha	161
Mazuren Petrochemical	Chiba	78			Naphtha	161
Mitsubishi Yuka	Yokkaichi	(220)			Naphtha	161
	Kashima		(180)	1970/1	Naphtha	161
Mitsui Sekiyu K.	Chiba	60			Naphtha	161
	Iwakuni	126			Naphtha	161
Mitzushima Ethylene	Mitsushima	210			Naphtha	161
Nippon Petrochemicals	Kawasaki	120			Naphtha	
Sanyo Ethylene	Mijushima		(180)	end 1971	Naphtha	161

Company	Location	Present capacity	Authorised expansion Additional capacity	Commissioning date	Feedstock	Ref.
Sumitomo K.K.	Ohe	75			Naphtha	164
	Anegasaki	(250)			Naphtha	161
	Niihama ⎫ Kikumoto ⎭	(65)			Naphtha	161
Tonen Sekiyu	Kawasaki	175			Naphtha	161
Tsurusaki Yuka	Tsurusaki	70			Naphtha	161
Ukishima Petrochemicals	Ukischima	160			Naphtha	
Mexico						
Petroleos Mexicanos (PEMEX)	Atzcapotzaica	NA ⎫			RG	
	Cuidad Madero	NA ⎬	110	1972/3	RG	
	Minatitlan	NA ⎭			RG	
	○	—	110		Propane	166
Puerto Rico						
Commonwealth Oil	Penuelas	NA	300		Naphtha	161, 168
Puerto Rico Olefins	Penuelas	NA	270	1971	Naphtha and gas oil	161
Union Carbide	Penuelas	NA	225	1971	Naphtha	161
South Africa						
African Explosives & Chemical Ind.	Sasolburg	NA			Propane	164
Sasol	Sasolburg	70			Naphtha	164
South Korea						
State Authority	Ulsan	—	44	1972	Naphtha	169
Spain						
Calvo Sotelo	Puertollano	35	90		Naphtha	
Industrias Quimicas Associadas	Tarragona	32			Naphtha	
Sweden						
Esso Chemical	Stenungsund	150			Naphtha	
Taiwan						
Chinese Petroleum	Kaohsiung	(30)	32	1973/4	Naphtha	
Turkey						
Petkim Petrokimya	Izmit	20	20	1972	Naphtha	161, 163
	Aliaga		146	1974	Naphtha	165
United Kingdom						
BP Chemicals Int.	Baglan Bay	40	200	1972	Naphtha	
	Grangemouth	275			Naphtha	165
British Celanese	Spondon	20			Naphtha	165
Esso Chemical	Fawley	80			Naphtha	165
I C I	Wilton	400			Naphtha	165
Shell Chemicals	Carrington	130	270	1973	Naphtha	165, 161
	Stanlow	RG 40	300	1976	Naphtha	171, 177
United States of America						
Allied Chemical Corp. & Wyandotte Chemicals	Geismar, La.	40			NG	164

Company	Location	Present capacity	Authorised expansion Additional capacity	Commissioning date	Feedstock	Ref.
Amerada Hess	Port Reading, NJ	60			RG	174
Ashland Oil	Ashland, Ky. Canton, Ohio Louisville, Ky. Tonawarda, NY	60			RG	174, 71
Atlantic Richfield	Channelview, Tex.				By-product butadiene	174, 71
	East Chicago, Ill.	100			RG	
	Wilmington, Cal.				RG	
Chemplex Co.	Clinton, La	60			NG	174
Chevron Chemical	Pascagonia Miss.	90			RG	
Cities Service Co.	Lake Charles, La.	150			RG, NG	174, 71
Clark Oil & Refining Co.	Blue Island, Ill.	30			RG	71
Coastal States Petrochem. Co.	Corpus Christi, Tex.	NA			RG	71
Continental Oil Co.	Lake Charles, La.	NA			RG	71
Dart Industries	Odessa, Tex.	68			NG	
Dow Chemical Co.	Bay City, Mich.				NG	
	Freeport, Tex.	250			NG	71
	Plaquemine, La.				NG	
Du Pont	Orange, Tex.	90			NG	174
Eastman Kodak	Longview, Tex.	100			Propane	71
Enjay Chemical Co.	Baytown, Tex.		235	1972	RG	174
	Bayway, NJ	855			RG	71
	Baton Rouge, La.				RG	
BF Goodrich	Calvert City, Ky.	45			Propane	71
Gulf Oil	Cedar Bayou, Tex.	270	320		RG	
	Port Authur, Tex.				RG	
	Philadelphia, Pa.				RG, NG	
Jefferson Chemical	Port Neches, Tex.		90	NA	NG	
Marathon Oil	Detroit, Mich.		68	NA	RG	
Mobil Oil	Beaumont, Tex.	112	112	1972	Naphtha	
Monsanto	Alvin, Tex.		158	NA	NG	71
	Texas City, Tex.				NG	71
Northern Petrochemical	Joliet, Ill.	—	90	1971	NG	
Novamont Corp.	Neal, W. Va.	45			RG	71
Phillips Petroleum	Sweeney, Tex.	65			NG	71, 174
Phillips Petroleum/ Houston Natural Gas Products	Sweeney, Tex.	55			NG	71
Shell Oil Co.	Deer Park, Tex.	210			RG	71
	Norco, La.	95	90	1972	NG	71
	Dominguez, Cal.	NA			RG	71
Signal Oil and Gas	Houston, Tex.	35			RG	71, 174

Company	Location	Present capacity	Authorised expansion Additional capacity	Commissioning date	Feedstock	Ref.
Sinclair-Koppers	Houston, Tex.		45	NA	NG	
Skelly Oil	El Dorado, Ka.	40			RG	71, 174
Standard Oil Co. of Indiana (including Amoco)	Texas City, Tex.					71
	Whiting, Ind.					
	Yorktown, Va.	450			RG	174
	El Dorado, Ark.					
Sun Oil	Marcus Hook, Pa.	158	80	1973	RG (later Naphtha)	71
	Corpus Christi, Tex.	35			RG	71
	Duncan, Okla.	NA			NG	71
Texaco	Port Arthur, Tex.	NA	} 90		RG	
	Westville, NJ	NA			RG	
Texas City Refining	Texas City, Tex.	45			RG	71
Tidewater (Getty)		112				
Union Carbide	Institute, W. Va.	90			NG	71
	Seadrift, Tex.	110			NG	71
	Texas City, Tex.	110			NG	71
	Whiting, Ind.	135			NG, RG	71
	Taft, La.	110			Naphtha, NG	176
Union Oil of California	Nederland, Tex.	NA			RG	71
Vistron Corp.	Lima, Ohio	80			RG	71
Venezuela						
Inst. Venez. de Petroquimica	El Tablazo	—	85	1971/2		161

REFERENCES

[1] Deimann, Van Troostwyk, Bondt and Louwrenburgh, *Crell's Annalen*, 1795, **195** (*2*), 310, 430.

[2] Faraday, *Phil. Trans.*, 1825, 440.

[3] Balard, *Ann. Chim. Phys.*, 1844, **12** (*3*), 294.

[4] Cahours, *Ann. Chim. Phys.*, **70** (*2*), 81.

[5] Dumas and Péligot, *Ann. Chim. Phys.*, **62** (*2*), 1.

[6] Reynolds, *J. Chem. Soc.*, 1851, **3**, 111; also *Ann.*, 1851, **77**, 114.

[7] Dumas, Malaguti and Leblanc, *Compt. rend.*, 1847, **25**, 783; *Ann.*, 1847, **64**, 329.

[8] '*Ethylene and its Industrial Derivatives*' edited by S. A. Miller (Ernest Benn, London, 1969), 4.

[9] Berthelot, *Compt. rend.*, 1855, **40**, 102; *Jahresber. Fortschritte Chem.*, 1855, 611; *Ann. Chim. Phys.*, 1855, **43** (*3*), 399.

[10] Kolbe, *Zeit. Chem. u. Pharm.*, 1862, **5**, 687.

[11] Friedel, *Compt. rend.*, 1862, **55**, 53; *Zeit. Chem. u. Pharm.*, 1862, **5**, 460.

[12] Erlenmeyer, *Ann.*, 1866, **139**, 214.

[13] Berthelot, *Ann. Chim. Phys.*, 1895, **4** (7), 100; *Compt. rend.*, 1894, **118**, 1009.

[14] Ormandy and Craven, *J. Soc. Chem. Ind.*, 1930, **49**, 362T.

[15] Cahours, *Ann.*, 1850, **76**, 283; *Compt. rend.*, 1850, **31**, 142, 291.

[16] Würtz, *Ann.*, 1857, **104**, 242.

[17] Friedel and Silva, *Compt. rend.*, 1873, **76**, 1594; *Jahresber. Fortschritte Chem.*, 1873, 322.

[18] Berthelot and de Luca, *Ann. Chim. Phys.*, 1856, **47**, 279.

[19] Malbot, *Ann. Chim. Phys.*, 1890, **19**, 345.

[20] Markovnikov, *Ann.*, 1870, **153**, 251; *Compt. rend.*, 1875, **81**, 668.

[21] Kharasch and Reinmuth, *J. Chem. Educat.*, 1931, **8**, 1703, 1725; Mayo and Walling, *Chem. Reviews*, 1940, **27**, 351.

[22] Kharasch and McNab, *J. Am. Chem. Soc.*, 1934, **56**, 1425.

[23] Vaughan and Rust, *J. Org. Chem.*, 1942, **7**, 474.

[24] Jones and Reid, *J. Am. Chem. Soc.*, 1938, **60**, 2453.

[25] Würtz, *Ann.*, 1859, **110**, 125.

[26] Oser, *Ann.*, 1861, Suppt. Bd. I, 253.

[27] Carius, *Ann.*, 1863, **126**, 197; Butlerov, *Ann.*, 1866, **144**, 40.

[28] Markovnikov, *Ann.*, 1870, **153**, 251.

[29] Henry, *Compt. rend.*, 1874, **79**, 1203.

[30] Henry, *Compt. rend.*, 1902, **134**, 1071; Michael, *J. Prakt. Chem.*, 1899, **60**, 463.

[31] Smith and Style, *Acta Chem. Scand.*, 1951, **5**, 1415.

[32] Friedel and Crafts, *Compt. rend.*, 1877, **84**, 1392.

[33] Balsohn, *Bull. Soc. Chim.*, 1879, **31**, 539.

[34] Gerhardt and Cahours, *Ann.*, 1841, **38**, 88.

[35] Gustavson, *Ber.*, 1878, **11**, 1251; Konovalov, *Bull. Soc. Chim.*, 1895, **16**, 864; Radziewonowsky, *Ber.*, 1895, **28**, 1135.

[36] Berry and Reid, *J. Am. Chem.*, 1927, **49**, 3148.

[37] Cahours, *Compt. rend.*, 1850, **31**, 142, 291.

[38] Berthelot, *Ann.*, 1858, **108**, 188; *Ann. Chim. Phys.*, 1858, **53**, 69.

[39] Dusart, *Ann.*, 1856, **97**, 122; *Ann. Chim. Phys.*, 1855, **45**, 339.

[40] Prunier, *Jahresber. Forschritte Chem.*, 1873, 347.

[41] Berthelot and de Luca, *Ann.*, 1854, **92**, 306; *Ann. Chim. Phys.*, 1855, **43**, 251.

[42] Linnemann, *Ann.*, 1872, **161**, 54.

[43] Gladstone and Tribe, *J. Chem. Soc.*, 1874, **27**, 211; also Niederist, *Ann.*, 1879, **196**, 358.

[44] Simpson, *Proc. Roy. Soc.*, 1863, **12**, 533.

[45] Markovnikov, *Ann.*, 1866, **138**, 364; *Ann.*, 1870, **153**, 251.

[46] Birnbaum, *Ann.*, 1868, **145**, 67.

[47] Tollens and Henninger, *Ann.*, 1870, *156*.

[48] Freund, *Monatsh. Chem.*, 1882, **3**, 625.

[49] Mouneyrat, *Bull. Soc. chim.*, 1899, **21**, 616.

[50] Mailhe, *Chem. Ztg.*, 1921, **111**, 467; Senderens, *Compt. rend.*, 1935, **200**, 614, 2138.

[51] Prunier, *Compt. rend.*, 1873, **76**, 98.

[52] Patterson and Robertson, *J. Chem. Soc.*, 1924, **125**, 1527.

[53] Berthelot, *Ann. Chim. Phys.*, 1857, **51**, 561.

[54] Berthelot, *Compt. rend.*, 1899, **129**, 487.

[55] Senderens, *Bull. Soc. chim.*, 1911, **9**, 373; *Compt. rend.*, 1910, **151**, 394.

[56] Friedel and Silva, *Compt. rend.*, 1873, **76**, 1594.

[57] Le Bel and Greene, *Am. Chem. J.*, 1880-1, **2**, 23.

[58] Beilstein and Wiegand, *Ber.*, 1882, **15**, 1498.

[59] Newth, *J. Chem. Soc.*, 1901, **79**, 915.

[60] Clutterbuck and Cohen, *J. Chem. Soc.*, 1922, **121**, also Frick, *J. Am. Chem. Soc.*, 1923, **45**, 1796.

[61] White, *J. Chem. Soc.*, 1924, **125**, 2395.

[62] Ormandy and Craven, *J. Soc. Chem. Ind.*, 1928, **47**, 317T.

[63] Ipatiev, *Ber.*, 1903, **36**, 1997.

[64] Senderens, *Bull. Soc. chim.*, 1908, **3**, 201.

[65] Goudet and Schenker, *Helv. chim. Acta.*, 1927, **10**, 132.

[66] Adkins and Millington, *J. Am. Chem. Soc.*, 1929, **51**, 2449.

[67] Senderens, *Bull. Soc. chim.*, 1907, **1**, 687; *Compt. rend.*, 1907, **144**, 382.

[68] Williamson and Taylor, *J. Am. Chem. Soc.*, 1931, **53**, 3272.

[69] BIOS Miscellaneous Report no. 24.

[70] For a description of the work of Curme and his associates, see '*Ethylene and its Industrial Derivatives*', edited by S. A. Miller (Ernest Benn Ltd., London, 1969), 21–3.

[71] Erskine, Chemical Economics Handbook, 'Propylene', March 1971, Stanford Research Institute.

[72] Curme and Reid, *Chem. and Met. Eng.*, 1921, **23**, 1049.

[73] Hayes, W., 'American Chemical Industry', Vol. IV, pp. 193–4 (D. Van Nostrand Co., New York, 1945).

[74] The semi-technical production of ethylene and propylene chlorohydrins by reaction of oil gas (containing 20–25 per cent each of ethylene and propylene) was described in 1920 by Brooks in *Chem. and Met. Eng.*, 1920, **22**, 629.

[75] Groll and Hearne, *Ind. Eng. Chem.*, 1939, **31**, 1530.

[76] Williams, *Ind. Eng. Chem. News Edit.*, 1938, **16**, 630; *Chem. and Met. Eng.*, 1940, **47**, 834; *ibid*, 1941, **48**, 87; *Trans. Amer. Inst. Chem. Eng.*, 1941, **37**, 157.

[77] McCaslin Jr., *Oil Gas J.*, 1948, 16 September, 88.

[78] Anon., *Chem. Ind.*, 1964, **16**, 484.

[79] Jones, *Petroleum Refiner*, 1954, **33**, 186; Resen, *Oil Gas J.*, 1954, **52**, 203, 209.

[80] *Trans. Amer. Inst. Chem. Eng.*, 1945, **41**, 463.

[81] McAllister, Anderson and Bullard, *Chem. Eng. Prog*), 1947, **43**, 189 (*Trans. Am. Inst. Chem. Eng.*).

[82] Horn, *Brennstoff-Chem.*, 1963, **44**, 363.

[83] BIOS Final Report no. 131.

[84] BIOS Overall Report no. 1.

[85] FIAT Final Report no. 968.

[86] BIOS Final Report no. 1646.

[87] BIOS Final Report no. 1154.

[88] BIOS Final Report no. 1060.

[89] Stanley, *Chem. and Ind.*, 1965, 1196–7.

[90] BIOS Miscellaneous Report no. 113.

[91] Holroyd, *Chem. and Ind.*, 1958, 906.

[92] Steiner, *J. Inst. Petrol.*, 1947, **33**, 410.

[93] Gohr, *Ind. Eng. Chem.*, 1958, **50**, 52A.

[94] Murphree, *Compt. rend. Semaine Techn. de l'AFTP*, June 1949, p. 300.

[95] Wagner, *Science of Petroleum*, 1930, **3**, 2112 (Oxford University Press).

[96] Petroleum Processing Handbook edited by Bland and Davidson (McGraw-Hill Inc., 1967).

[97] Weiss, *Hydro. Proc.*, 1969, **48** (*10*), 125.

[98] Pollock, *Petroleum Refiner*, 1955, **34** (*2*), 127.

[99] Whalley, *Chem. and Ind.*, 1953 August, p. 3.

[100] Nelson, *Oil Gas J.*, 1956, 3 September, p. 129.

[101] Curtiss, *Chem. and Ind.*, 1953 August, p. 18.

[102] Howes, *Chem. and Ind.*, 1968, 30 November, p. 1671.

[103] Ockerbloom, Brownstein and Root, *Oil Gas Intern.*, 1969, **9**, 79; *Chim. et Ind. Génie chimique*, 1970, **103**, 937.

[104] Anon, *Europ. Chem. News*, 24 July 1970, p. 32.

[105] Hock and Lang, *Ber.*, 1943, **76**, 1130; 1944, **77**, 257.

[106] Smidt, *Chem. and Ind.*, 1962, 54.

[107] Piombino, *Chem. Week*, 1970, 20 May, 71.

[108] Landau, Brown, Russell and Kollar, Proc. 7th World Petroleum Congress (Mexico), 1967, **5**, 67.

[109] Anon, *Europ. Chem. News*, 1969, 21 February, p. 22.

[110] Anon, *Europ. Chem. News*, 1969, 11 April, p. 14.

[111] Anon, *Chem. Eng. News*, 1948, **26**, 3695.

[112] Warren, *Chem. Eng.*, 1951, **58**, 117; Bernard, *Chim. et Ind.*, 1954, **72**, 919.
[113] Dümbgen and Neubauer, *Chem. Ing. Techn.*, 1969, **41**, 974.
[114] Murfitt, *Europ. Chem. News Trade Report Polymer Intermediates*, 30 October 1970, p. 14.
[115] Anon, *Europ. Chem. News*, 8 December 1970, 50.
[116] Ishikawa, *Chem. Eng. Econ. Rev. (Japan)*, 1970, January, p. 31.
[117] (Shell Oil Co.), USP's 3,274,263 and 3,239,569.
[118] Hargis and Young, *Ind. Eng. Chem. Prod. Res. Develop.*, 1966, **5**, 72.
[119] Kindler and Eisfeld, *Ind. chim. Belge*, 1967, **32**, Spec. No. 650–3.
[120] Anon, *Chem. Ind.*, 1968, **20**, 537.
[121] Cosby, USP 2,481,826 (1949).
[122] Bellringer, Bewley and Stanley, BP 709,337 (1954).
[123] Hadley, *Chem. and Ind.*, 1961, 238; Stanley, *ibid*, 1965, 1198.
[124] Veatch, Callahan, Idol and Milberger, *Hydroc. Proc.*, 1962, **41**, 187; Hughes, *Chem. and Ind.*, 1969, 1568.
[125] Lichtenberger, *Rev. Ass. France Techn. Petr.*, 1967, no. 183, 29.
[126] Anon, *Ind. Chem.*, 1963, May, 242.
[127] Cevidalli, Nenz and Caporali, *Chim. et Ind.*, 1967, **49**, 809.
[128] Denigès, *Compt. rend.*, 1898, **126**, 1147; *Ann. Chim. Phys.*, 1899, **18**, 389.
[129] Shell Development Co., USP's 2,451,485 and 2,486,842.
[130] Veatch, Callahan, Milberger and Foreman, *Actes. Congr. Intern. Catatyse, Paris*, 1960, **2**, 2647.
[131] Andreas and Gröbe, *Propylene Chemie, Akademie Verlag Berlin*, 1969, p. 192.
[132] Nakatani, *Japan Chem. Quarterly*, 1968, **4**, 50; *Hydroc. Proc.*, 1969, **48**, 152.
[133] Anon, *Europ. Chem. News*, 1 August 1969, 19.
[134] Kijima, *Japan Chem. Quarterly*, 1969, **5**, (3), 12.
[135] Prileschajeff, *Ber.*, 1909, **42**, 4811.
[136] Thwaites, *Chem. and Ind.*, 1969, 1001.
[137] *Stereoregular Polymers and Stereospecific Polymerisations* edited by G. Natta and F. Danussi, Vol. I (Pergamon Press, 1967).
[138] Ziegler, *Brennstoff-Chem.*, 1952, **33**, 193.
[139] Natta, Pino and Mazzanti, Ital. P. 535,712 (1954) and 537,425 (1954).
[140] Natta, *Experientia*, 1957, Suppl. 8, 21; *Materie Plastiche*, 1958, **24**, 3.
[141] Natta, *Atti. Accad. Naz. Lincei, Memorie*, 1958, **4**, 61.
[142] Natta, *Angew Chem.*, 1956, **68**, 393; Natta and Porri, Ital. P. 538,453 (1955).
[143] Natta, *Makromol. Chem.*, 1955, **16**, 213.
[144] Horner, *Trans. Plastics. Inst.*, 1967, **35**, 415.
[145] Anon, *Chemische Ind.*, 1968, **20**, 81.
[146] Anon, *Chemische Ind.*, 1967, **19**, 190.
[147] Brennan, *Chem. Eng.*, 1965, **72**, 94.
[148] Bogdanovic and Wilke, *Brennstoff-Chem.*, 1968, **49**, 323.
[149] Zeigler *et al.*, *Brennstoff. Chem.*, 1954, **35**, 321.
[150] USP 2,986,588 to California Research Corp. Group.
[151] Hambling, *Chemistry in Britain*, 1969, **5**, 354.
[152] Collingwood, *World Development in Olefin Supplies* presented at ACS/CIC Meeting in Toronto, on 26 May 1970, *Chem. Eng.*, 13 July 1970, p. 34.
[153] Imhausen, *Chem. and Ind.*, 5 December 1970, p. 1559.
[154] Anon, *Europ. Chem. News*, 3 July 1970, p. 10; *ibid.*, 17 July 1970, 10.
[155] Miller, *Chem. Process. Eng.*, 1967, **48**, 79.
[156] Anon, *Europ. Chem. News*, 30 January 1970, p. 16.
[157] Anon, *Chem. Ind.*, 1968, **20**, 539.
[158] Sherwood, *Erdöl u. Kohle.*, 1970, **23**, 590.
[159] Strelzoff, *Chem. Eng.*, 24 August 1970, p. 75.
[160] Docksey, *Chem. and Ind.*, 10 August 1968, p. 1078.
[161] Anon, *Oil and Gas J.*, 7 September 1970, 122.
[162] Anon, *Chem. Ind. Int.*, 1970, **2**, 55.
[163] Anon, *Europ. Chem. News Supplement*, 'new plants', 13 March 1970, 84.

[164] Anon, *Europ. Chem. News Supplement*, 'Propylene', 25 February 1966.

[165] Anon, *Europ. Chem. News Supplement*, 'Olefins and their Derivatives', 25 September 1970, 103.

[166] Anon, *Europ. Chem. News*, 26 January 1970, 28.

[167] Anon, *Chem. Age*, 13 November 1970, 23.

[168] Anon, *Europ. Chem. News*, 4 September 1970, 28.

[169] Anon, *Chem. Week*, 16 December 1970, 38.

[170] Anon, *Chem. Ind. Int.*, 1970, **1**, 24.

[171] Anon, *Chem. Age*, 30 October 1970, 21.

[172] Anon, *Chem. Age*, 27 November 1970, 13.

[173] Anon, *Europ. Chem. News*, 2 October 1970, 18.

[174] Piombino, *Chem. Week*, 20 May 1970, 73.

[175] Anon, *Chem. Week*, 27 May 1970, 20.

[176] Private Sources.

CHAPTER 2

ISOLATION OF PROPYLENE

By **G. J. Woodhouse**

2.1. INTRODUCTION

No significant quantities of propylene occur naturally in oil or gas deposits and the availability of propylene stems from various manufacturing processes. Regardless of the source, propylene is nearly always produced as the co-product in several processes where the prime product is either gasoline or ethylene.

In 1970 in the United States, 85 per cent of the propylene produced originated from refinery processes with the remaining 15 per cent coming from plants producing ethylene as the prime product.[1] In Europe and Japan, however, the reverse is true with the majority of the propylene being produced from ethylene plants. The reason for this lies in the fact that demand for gasoline in Europe and Japan is much less than in the United States keeping Catalytic Cracking capacity at a relatively low level. At the same time most ethylene originates from the pyrolysis of liquid feedstocks yielding much larger quantities of propylene than does the pyrolysis of ethane and propane, recovered from natural gas, which are the main feedstocks for ethylene plants in the United States.

2.2. ISOLATION OF PROPYLENE FROM REFINERY OPERATIONS

Propylene is produced as a by-product from several refinery processes, the main objective of which is to produce either high quality gasoline or coke. These processes all involve cracking of a saturated hydrocarbon feedstock and those yielding significant quantities of propylene are Catalytic Cracking, Thermal Cracking and Coking of which Catalytic Cracking produces more than all other processes combined.

Recovery of about 80 per cent of the propylene contained in the gas streams from these processes is economic, the main objective being to provide feedstock for other refinery processes, principally catalytic polymerisation and alkylation, both of which produce high octane gasoline. Although some propylene is recovered for sale as liquified petroleum gas (LPG) this is not usual since it is much more valuable as a feedstock for other processes. LPG is normally derived from natural gas and from refinery operations yielding saturated hydrocarbons. However, where natural gas is not available as a source for LPG, recovery of propane and propylene from refinery gas does fulfil this need. The specification for propane/propylene LPG is given in Table II.1.

TABLE II.1

Propane-Propylene LPG Specification[2]

Vapour pressure at 37·8°C	bars gauge, max.	14·0
Temperature at 95% evaporation	°C, max.	−38·3
or		
Butane and heavier	% vol, max.	2·5
Residue on evaporation of 100 ml	ml, max.	0·05
Corrosion	copper strip, max.	No. 1
Volatile sulphur	g per m³, max.	0·34

The compositions of typical feedstocks for catalytic polymerisation and alkylation plants are given in Table II.2, and a typical composition of a propane-propylene product meeting the LPG specification is given in Table II.3.

TABLE II.2

Composition of Typical Feeds for Cat. Poly. and Alkylation Plants

	HF[3] Alkylation % vol (liq)	H_2SO_4[4] Alkylation % vol (liq)	Catalytic Polymerisation % vol (liq)
Methane		0·3	3·7
Ethylene		1·2	10·2
Ethane		5·0	7·0
Propylene	16·5	50·0	52·1
Propane	12·9	35·0	25·9
Butylenes	28·0	2·3	0·5
Iso-Butane	25·9	5·9	0·5
n-Butane	9·2	0·3	0·1
Pentane and heavier	7·5		
	100·0	100·0	100·0

TABLE II.3

Composition of Typical C₃ Stream Meeting LPG Specifications

	% mol
Ethylene	0·3
Ethane	1·0
Propylene	59·7
Propane	37·0
Butanes	2·0
	100·0

It can be seen from Table II.2 that the composition of feedstock suitable for alkylation can vary over a wide range. In some refineries a mixed propylene-butylene stream is used, whilst in others a fairly pure propane-propylene stream is isolated depending on the refinery balance and the other uses to which the propylene and butylene are put.

The recovery of propylene on a refinery is either carried out on the individual plants producing the material or in a centralised gas recovery plant, into which is fed the combined gas flows from all units. Although the composition of the gas varies, depending on the particular refinery arrangement, the processing scheme for the recovery of the propylene is similar.

Refinery gas recovery plants are usually based on oil absorption and are designed to recover between 70 and 90 per cent of the propane-propylene in the gas stream. Processes have been developed based on adsorption using a moving bed of activated charcoal[5] but these have not proved commercially attractive. Straightforward fractionation at low temperatures using refrigeration could also be used and this route is the one normally followed in the recovery sections of olefin production plants, where a high recovery of ethylene and propylene is sought and separation of hydrogen and methane is also required. On refinery processes the concentration of valuable ethylene and propylene in the gas streams is much lower than in the ethylene plant streams and the extra cost and complication of this scheme is not justified over the simpler oil absorption process. Whilst the gas recovery sections of refinery units and olefin plants can usually be considered quite separately, in special circumstances, where the olefin plant forms part of an oil refinery complex, the recovery of ethylene, propylene and other light products is sometimes combined in one common unit.

A straightforward absorption process using a heavy hydrocarbon liquid stream to absorb the propylene and heavier components from the gas stream followed by fractionation of the rich oil to separate the propylene, provides the simplest recovery scheme.

In an absorber, the recovery of any component from the gas feed can be defined in terms of the absorption factor A, which is related to the other variables in the system by the following expression:

$$A = \frac{L}{KV}$$

where A = absorption factor
L = mols per hour of liquid downflow[*]
V = mols per hour of vapour
K = equilibrium constant of the desired component at the absorber conditions.

Furthermore the fraction of any component in the absorber feed which is absorbed by the lean oil can be determined by the following expression:

$$\frac{\text{mols of component absorbed}}{\text{mols of component in feed}} = \frac{A^{n+1}-A}{A^{n+1}-1}$$

where n = the number of theoretical contacting stages in the absorber.

It can be seen from these expressions that high absorption factors and hence high levels of recovery can be achieved in three main ways:

1. By maintaining a high absorbing oil (lean oil) flow to the absorber.
2. By using a lean oil of low molecular weight in order to keep the molal flow of liquid high.
3. By operating the absorber at low temperature and high pressure in order to keep the equilibrium constant of the desired component low.

Since the absorption factor for all components is directly proportional to L/V and inversely proportional to the equilibrium constant of the component, it follows that increasing the recovery of the desired component, in this case propylene, by any of the above three measures will also increase the recovery of components lighter than propylene, which is undesirable. Also steps 1 and 3 increase the operating costs of the process whilst step 2 increases the loss of lean oil in the absorber tail gas, since this gas must be saturated with lean oil as it leaves the absorber. The factors affecting the selectivity of absorption between the desired component and lighter components, are the relative volatilities of the components and the number of trays in the absorber. The greater the relative volatility between propylene and ethane the more selective will be the absorption and there are advantages in this respect by designing the absorber to operate at as low a temperature as is practical. The effect of increasing the number of trays in the absorber becomes insignificant above about 20 theoretical trays.

From this brief discussion of the principles involved in a simple absorption operation it will be appreciated that, at best, only a crude separation of propylene from the lighter components in the gas can be realised.

Modern recovery plants using oil absorption normally employ a fractionating absorber, or absorber-stripper as it is sometimes called, rather than a simple absorber. This technique uses a combination of fractionation and oil absorption and gives a much improved separation by elimating to a large extent the undesirable light components from the rich oil.

Normally the fractionating absorber is designed to make a good separation between the ethane and lighter components and the propylene and heavier components. The sharpness of separation required is governed by the propane-propylene product specification required and also by the effect the presence of lighter components have on the operating conditions of downstream processing equipment. By the proper design and operation of the de-ethanising fractionating absorber it is possible to arrange the downstream fractionators, i.e. depropaniser and debutaniser so that total condensation of their overhead product is possible using normally available cooling water and reasonable operating pressures. The amount of ethane and ethylene which can be tolerated in the propane-propylene stream from a product quality point of view depends on what use the stream is to be put. Although, as discussed earlier, a refinery gas plant is normally designed to provide feed for alkylation or catalytic conversion processes rather than to produce LPG, if the feed to the conversion process is not adequately de-ethanised the propane effluent from that process will need de-ethanising to make it suitable for sale as for LPG. It is clear, therefore,

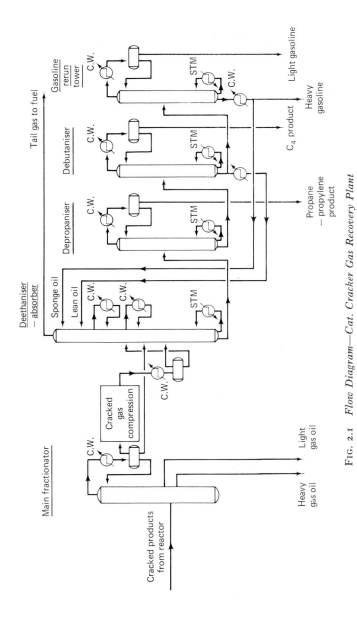

FIG. 2.1   *Flow Diagram—Cat. Cracker Gas Recovery Plant*

that a well designed and efficient fractionating absorber is needed if a high recovery of a good quality propane-propylene product is to be obtained.

Although the basic principles regarding the effect of operating variables on recovery in a simple absorber still apply, design of fractionating absorbers is quite complex and has been fully discussed in the literature.[6, 7]

2.2.1. Propylene Recovery from a Catalytic Cracking Unit

A flow diagram for the gas recovery section of a typical catalytic cracking unit is shown in Fig. 2.1. The products from the catalytic reactor are first of all passed to the main fractionator where they are quenched by circulating oil streams. This fractionator also serves to separate the heavy components from the cracked gas, i.e. the light and heavy gas oils, some of these being recycled back to the reactor for further cracking. The overhead vapour product from the main fractionator which operates at substantially atmospheric pressure is cooled to condense reflux for the tower. The overhead vapour product is then compressed to about 11 bars gauge and further cooled by cooling water and the vapour and liquid separated. Both vapour and liquid streams are fed to the de-ethanising fractionating absorber, the liquid entering the tower at a higher point than the gas since as well as being stripped of light components the heavier components increase the flow of lean oil down the tower.

Debutanised gasoline is used as absorber lean oil and this is fed to the tower a few trays from the top. Since the vapour leaving the lean oil feed tray must be in equilibrium with the liquid on the tray it follows that the vapour will be saturated with lean oil, i.e. gasoline, at the operating temperature and pressure of the tray. This would lead to considerable losses of valuable gasoline and to avoid such losses, a 'sponge oil' section consisting of about five trays is incorporated at the top of the tower. Sponge oil consisting of heavy gasoline from the base of the gasoline rerun tower is fed to the top of this section where it contacts the vapour rising up the tower. The sponge oil absorbs most of the lean oil from the gas and since the sponge oil is less volatile than the lean oil the amount of heavy gasoline lost in the 'dry gas' leaving the absorber is very much reduced. The heavy gasoline leaving the sponge oil section mixes with the lean oil in the lower part of the tower.

Ethane and lighter components are stripped from the oil in the stripping section of the tower by the vapour generated in the reboiler. In order to increase the absorption and keep lean oil circulation rates to a minimum, the temperature in the absorption section of the tower is kept as low as possible by removing the heat of absorption in two intercooling circuits. Liquid is removed from the tower and after passing through water cooled heat exchangers is returned to the tower immediately below the withdrawal point. On some installations striving for very high recovery of propylene, these intercoolers are cooled by refrigerant to achieve lower temperatures than can be obtained with ordinary cooling water.

The rich oil from the bottom of the fractionating absorber is pumped to the depropaniser where the propylene-propane product is separated as overhead product. The bottoms is fractionated in the debutaniser to give a C_4 overhead product. The debutanised gasoline from the bottom of the tower is split into light and heavy gasoline products in the rerun tower. Debutanised gasoline from the base of the debutaniser is returned to the absorber as lean oil and as discussed earlier, a small quantity of heavy gasoline is fed as sponge oil to the top section of the tower.

A typical material balance for such a plant is shown in Table II.4 which indicates a 90 per cent recovery of propylene and the production of a propane propylene stream suitable for feedstock for catalytic polymerisation and most alkylation processes or for sale as LPG. At higher recoveries, the ethane and ethylene content of the product would increase, making downstream de-ethanisation necessary to meet some product specifications.

TABLE II.4

Material Balance for Cat. Cracker Gas Recovery Plant

	Main Fractionator Overhead liquid % mol	Main Fractionator Overhead vapour % mol	Lean oil % mol	Sponge oil % mol	Absorber tail gas % mol	Absorber bottoms rich oil % mol	Propane–Propylene product % mol	C₄ product % mol
Nitrogen		4·6			6·3			
Carbon monoxide		1·0			1·3			
Carbon dioxide		2·1			2·9			
Hydrogen		14·6			20·0			
Methane		25·7			35·3			
Ethylene	2·6	5·0			8·9			
Ethane	3·3	11·5			18·0	0·1	0·9	
Hydrogen sulphide		0·5			0·3	0·1	0·7	
Propylene	21·0	12·1			3·3	8·0	67·3	
Propane	7·0	7·3			1·6	3·7	28·9	2·4
Butylenes	25·8	5·8			1·3	7·2	2·2	52·3
iso-Butane	15·8	4·0	0·9		0·5	5·3		35·3
n-Butane	3·3	1·0	2·2		0·3	2·5		8·0
Pentane and heavier	21·2	4·8	96·9	100·0		73·1		2·0
	100·0	100·0	100·0	100·0	100·0	100·0	100·0	100·0
Molecular weight	54	30	125	150	21	107	43	57
kg/hr	8,280	7,764	61,875	15,000	4,023	88,896	3,636	5,401

On some refineries a mixed C_3, C_4 stream is separated for use as polymerisation or alkylation unit feed and in this case the rich oil from the base of the absorber is fed direct to the debutaniser where the C_4 and lighter material is taken off as an overhead product.

The propane-propylene stream separated from refinery gases normally contains hydrogen sulphide and mercaptan sulphur and these must be removed to make the stream suitable for feeding to catalytic polymerisation or alkylation plants. This treatment is normally carried out on the consumer unit by washing with caustic soda followed by a water wash to remove any traces of caustic. The water wash also serves to remove nitrogen compounds which are also undesirable in feedstocks for catalytic polymerisation units since they deactivate the catalyst.

2.3. ISOLATION OF PROPYLENE AS A BY-PRODUCT IN THE MANUFACTURE OF ETHYLENE

In the United States ethylene is produced mainly from the pyrolysis of ethane and propane recovered from natural gas although there are several plants based on naphtha and heavier gas oil feedstocks. Pyrolysis of ethane yields very little propylene and the main source of propylene in the United States is from the refinery processes discussed earlier. In Europe and Japan almost all ethylene produced is from the pyrolysis of liquid feedstocks which yield significant quantities of propylene.

Until around 1968 ethylene plants were generally being built and operated to achieve the highest possible yield of ethylene with little interest being shown in the propylene by-product which was often an embarrassment. A large part of the propylene product was frequently burnt as fuel. However, in Europe since 1968 there has been a growing demand for propylene, and ethylene plants are now being planned and built with almost as much emphasis being placed on propylene as ethylene.

The propylene is recovered mainly to provide feedstock for the manufacture of chemical derivatives such as cumene, acrylonitrile, isopropanol and propylene oxide or as polymer grade material for the manufacture of polypropylene. Again, as was the case with propylene recovered from refinery processes, some propylene is produced for sale as LPG in cases where the ethylene plant is located in a country with no other source for this material.

TABLE II.5

Yields of Major Products from Medium Severity Cracking of Hydrocarbons

	Feedstock			
	Ethane % wt	Propane % wt	Naphtha % wt	Gas oil % wt
Hydrogen	3·5	1·6	1·0	0·5
Methane	3·6	20·6	14·3	8·8
Ethylene and acetylene	50·3	34·3	27·0	20·7
Ethane	38·2	2·9	3·6	4·2
Propylene and C_3 acetylenes	2·4	17·0	15·2	14·6
Propane	0·2	15·0	0·5	0·6
C_4 hydrocarbons	0·6	4·0	10·0	9·6
Gasoline	1·2	4·6	25·9	18·0
Fuel oil			2·5	23·0
	100·0	100·0	100·0	100·0

2.3.1. Yields

The yields of propylene and ethylene which can be obtained from the pyrolysis of hydrocarbons, depends on the composition of the hydrocarbon feedstock and on the cracking conditions employed. This subject has been covered comprehensively by Zdonik, Green and Hallee.[8] Table II.5 gives typical yields obtainable from four widely different feedstocks and Table II.6 shows the effect on the yield structure of varying the cracking severity for a naphtha feedstock.

TABLE II.6

Effect of Severity on Yields of Major Products from Naphtha Cracking

	Low % wt	Medium % wt	High % wt
Hydrogen	0·8	1·0	1·2
Methane	12·0	14·3	16·5
Ethylene and acetylene	24·0	27·0	29·5
Ethane	3·5	3·6	3·4
Propylene and C_3 acetylenes	17·3	15·2	13·0
Propane	0·8	0·5	0·3
C_4 hydrocarbons	12·0	10·0	8·0
Gasoline	28·1	25·9	23·1
Fuel oil	1·5	2·5	5·0
Total	100·0	100·0	100·0

In areas where the demand for propylene is increasing the trend is moving away from high severity cracking operations where the ratio of propylene to ethylene yield is in the region of 0·5, to less severe conditions where this ratio is nearer to 0·7. At the same time the cracking process should be highly selective in producing high yields of both propylene and ethylene so that feedstock consumption is kept to a minimum. In this respect there are advantages in operating at low residence times in the cracking furnaces,[9] one example of a process using this principle being the Stone and Webster USC (Ultra Selective Conversion) cracking process.[10]

Ethylene plants are now in operation and many more are being built to produce as much as 450,000 tons per year of ethylene and at the same time over 300,000 tons per year of propylene.

2.3.2. Propylene Recovery

Until around 1960 two processing schemes were in general use to recover the propylene and ethylene from the products of the pyrolysis section of the plant, one involving compression of the gas followed by absorption and fractionation and the other compression followed by low temperature fractionation. The first scheme used a fractionating absorber in a similar way to that described for the recovery of propylene from refinery gases.[11] This method has now been replaced almost entirely by the scheme employing fractionation at low temperatures.[12]

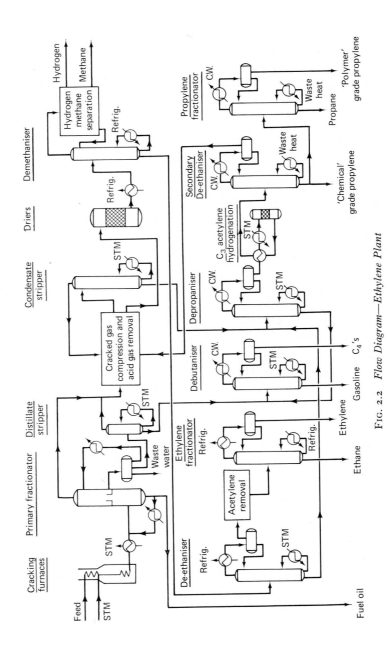

FIG. 2.2   *Flow Diagram—Ethylene Plant*

The flow diagram shown in Fig. 2.2 represents an ethylene manufacturing process of Stone and Webster. The cracked gas from the pyrolysis furnaces is cooled and quenched and then passed to the primary fractionator, where the fuel oil and some of the gasoline fraction is separated. The cooled gas is then compressed from the low pressure (about 0·3 bar gauge) of the primary fractionator to a pressure of about 35 bars gauge in four stages of compression. With the large plants being built in the 1960's and 70's great emphasis has been placed on low operating costs and thus highly integrated designs with high utilisation of waste heat have been developed. High pressure steam generated in the pyrolysis section of the plant is used in the steam turbines driving centrifugal compressors used for both compression of the cracked gas and for providing the refrigeration used in the low temperature sections of the plant. Hydrocarbons condensed by the interstage cooling in the cracked gas compression system are flashed back to the lower stages and finally are fed to the distillate stripper where debutanised gasoline is produced as a bottoms product. Gas from the third stage of the compression, after cooling, is scrubbed with caustic soda to remove hydrogen sulphide and low boiling mercaptans. Hydrocarbons condensed after cooling the gas at the final discharge pressure are fed to the condensate stripper which functions as a de-ethaniser yielding a bottoms product rich in propylene. The uncondensed cracked gas, containing the rest of the propylene, from the final stage of compression is dried over activated alumina or molecular sieves and then enters the demethanisation system where methane and hydrogen are removed. The extent to which further separation into methane and hydrogen products is carried out depends on product requirements but some production of high purity hydrogen is necessary for use in downstream hydrogenation processes. The demethanised stream is fed to the de-ethaniser where a mixed stream of ethylene, ethane and acetylene is taken overhead. This stream is treated in a catalytic process where the acetylene is hydrogenated before it is fractionated to yield ethylene and ethane products. In many plants the ethane is recycled and cracked in the pyrolysis furnaces to yield more ethylene.

The bottoms from the de-ethaniser containing propylene and heavier material, together with the bottoms from the condensate stripper, are passed to the depropaniser for separation of the propane and propylene. The depropaniser bottoms are further processed in the debutaniser for separation of the C_4 product and gasoline.

Table II.7 gives the composition and quantities of the main propylene bearing streams in an ethylene plant producing 450,000 tons per year of ethylene and 300,000 tons per year of propylene from mild severity cracking of a naphtha feedstock. Essentially all of the propylene available in the cracked products is recovered as depropaniser overheads, since usually the quality specification of the ethylene, ethane and C_4 products prohibit the presence of any appreciable amounts of propylene.

The design and operation of the condensate stripper and depropaniser systems require great care to avoid fouling of the equipment by polymerisation of the highly unsaturated hydrocarbons contained in the streams being processed, such as butadiene, pentadiene, vinyl acetylene, etc. Operating conditions in these towers are set so that temperatures are not so high as to cause excessive fouling. Maximum operating pressures are thus fixed which establish the location of the condensate stripper in the cracked gas compression system and set the cooling medium necessary for condensing the overheads of the depropaniser. With cooling water available at temperatures in the region of 25 °C the depropaniser can be operated at pressures in the order of 14 bars gauge. At such conditions fouling can normally be kept to a reasonable level although this cannot be avoided completely and spare reboilers are generally needed in order to allow cleaning without shutting down the tower. Commonly a spare stripping section is provided for the depropaniser so that this section of the tower can also be cleaned whilst maintaining continuous production of propylene product. Anti-fouling inhibitors are often used in an attempt to reduce the effects of the fouling. On some plants where cooling water temperatures are high or where maintenance costs associated with cleaning the equipment are very high, refrigeration is used to lower the operating pressure and temperature of the depropaniser to a level where fouling is so slight as not to be a problem.

The recovery section of an ethylene/propylene plant built for BP Chemical International Ltd. at Grangemouth by Stone and Webster is shown in Fig. 2.3.

TABLE II.7

Composition and Qualities of Main Propylene Bearing Streams in Ethylene Plant

	Total cracked gas % mol	Condensate stripper bottoms % mol	De-ethaniser bottoms % mol	Depropaniser overhead % mol
Ethane and lighter	66·3		0·1	0·2
Methyl acetylene and propadiene	0·5	1·2	2·4	3·1
Propylene	13·3	29·6	73·4	92·7
Propane	0·5	1·5	2·8	3·8
Butylenes and heavier	19·4	67·7	21·3	0·2
	100·0	100·0	100·0	100·0
Molecular weight	35·0	55·7	45·3	42·1
Flow kg/hr	241,630	39,150	43,610	41,560

2.3.3. Propylene Purification

Further treatment and purification of the propane-propylene stream from the depropaniser depends on the final product purity required.

As mentioned earlier propylene is usually recovered from ethylene plants as 'chemical grade' to provide feedstock for the manufacture of derivatives such as cumene and acrylonitrile or as 'polymer grade' for the production of polypropylene. The propane/propylene stream from the depropaniser would be suitable for sale as LPG without further treatment.

Typical specifications for the two product grades are given in Table II.8.

FIG. 2.3   *Recovery Section of an Ethylene Plant built for*
BP Chemical International Limited at Grangemouth by Stone & Webster

TABLE II.8

Typical Propylene Product Specifications

		'Chemical' grade	'Polymer' grade
Propylene	% mol min.	93·00	99·5
Hydrogen	ppm mol max.	300	300
Methane	ppm mol max.	300	300
Ethylene	ppm mol max.	300	300
Ethane	ppm mol max.	1000	1000
Propane	% mol	balance	balance
Methyl acetylene	ppm mol max.	10	10
Propadiene	ppm mol max.	30	30
Butadiene	ppm mol max.	10	10
Sulphur	ppm mol max.	5	5
Water	ppm mol max.	5	5

2.3.3.1. Production of 'chemical grade' propylene. Methyl acetylene and propadiene removal Both product specifications shown in Table II.8 require the removal of methyl acetylene and propadiene down to a level of 40 ppm and sometimes even lower levels are required.

For 'chemical' grade propylene where no separation of propane from propylene is needed, the methyl acetylene and propadiene are usually removed by hydrogenation in either the vapour or liquid phase. In the vapour phase a high purity hydrogen stream produced within the ethylene plant is mixed with the propylene stream and the mixture passed over a palladium based catalyst under carefully controlled temperature conditions. The reaction is exothermic and it is common to use at least two catalyst beds with intercooling to avoid excessively high temperatures within the reactors. Under well controlled conditions the hydrogenation process is selective, the methyl acetylene and propadiene being hydrogenated mainly to propane, and losses of propylene due to hydrogenation should be insignificant.

Processes offered by Farbenfabriken Bayer AG and Institut Française du Petrole, hydrogenate the methyl acetylene and propadiene in the propylene stream in the liquid phase. The Bayer process uses a tubular reactor cooled by refrigerant whilst the IFP process uses two fixed bed reactors arranged in series with refrigerated precooling and intercooling. Higher selectivity in hydrogenating the methyl acetylene and propadiene to propylene is claimed than for the vapour phase processes. Thus some propylene gain should result.

With the development of high severity cracking, the yield of methyl acetylene and propadiene has increased and in some locations the recovery of these compounds becomes attractive for two reasons.

Firstly, hydrogenation of the acetylenes becomes more complicated due to the high concentrations in the propylene stream, and also when hydrogenated, the propylene concentration in the resulting propylene-propane stream can be lower than desirable due to the hydrogenation of the acetylenes to propane.

Secondly, the methyl acetylene-propadiene mixture can be marketed as a welding gas having properties in many ways superior to acetylene.[13] Recovery of the methyl acetylene and propadiene from the depropaniser overhead stream can be achieved either by fractionation or by absorption although care has to be exercised in keeping the concentration of the methyl acetylene plus propadiene below about 60 per cent to avoid the risk of detonation and explosion. If a chemical grade propylene product with a very low permissible content of methyl acetylene and propadiene is required a system involving two fractionators with intermediate hydrogenation may be necessary.

After hydrogenation, the propane-propylene stream contains small concentrations of hydrogen and methane originating from the hydrogen added to the hydrogenation reactors. Also the stream is saturated with water since only part of the feed to the depropaniser has passed through the cracked gas driers. Both the light ends and the water must be removed in order to meet the 'chemical' grade specification, and this can normally be done in a simple stripping tower. Where very low water contents are required it may be necessary to pass the product through a bed of molecular sieves or activated alumina. The hydrogen sulphide, mercaptan sulphur and carbonyl sulphide of the propylene stream must also be reduced to a low level in order to meet the sulphur specification of the product. Removal of hydrogen sulphide and mercaptans presents no problems since the processing sequence in the cracked gas compression section of the plant can generally be arranged so that all of the propylene and propane contained in the cracked gas passes through the caustic scrubber and hence is freed from these compounds. Where this is not possible a simple caustic wash system followed by water washing of the depropaniser overhead stream is sufficient. Where very low sulphur specifications (1 ppm) have to be met, removal of carbonyl sulphide may be necessary and schemes involving solid absorbents such as soda ash have been suggested.[14]

2.3.3.2. Production of 'polymer grade' propylene. Propylene-propane separation To produce the high purity 'polymer grade' propylene required for polypropylene manufacture it is necessary to separate the propane and propylene. This is a difficult separation since the relative volatility between propylene and propane is very low and the separation falls into the category of superfractionation. The average relative volatility between propylene and propane is 1·12 at 21 bars abs, 1·15 at 15·8 bars abs and 1·21 at 7 bars absolute.[15] Even though the higher relative volatility at the lower pressure would result in significant reductions in the reflux ratio and the number of theoretical plates needed for the separation, the extra costs and complications of operating at low pressures cannot usually be justified. Normally, therefore, the operating pressure of the fractionator depends on the temperature of the cooling water available for condensing the column overheads. With cooling water available at 25 °C a column overhead pressure of 17 bars gauge would be possible. At lower operating pressures the overhead vapour could be compressed and condensed in the reboiler using a heat pump principle to save operating costs. However, due to the very large quantities of vapour to be handled in a modern day ethylene plant this is not an economic solution. Also, although the cooling water load on the overhead condenser is very big, the cost of heating the reboiler can usually be made very small since waste heat from the pyrolysis section of the plant can be used.

A plant having a capacity corresponding to the flow quantities shown in Table II.7

would require a propylene-propane fractionator some 6–8 m in diameter and 60–70 m in height. Since the height of the fractionating column is so great, the column is often split into two sections operating in series. This is very common in plants located in countries subject to earth tremors where the problems of designing and building very tall columns are exaggerated.

Where all the propylene by-product from the plant is required as high purity 'polymer grade' it is sometimes possible to eliminate the methyl acetylene-propadiene hydrogenation system since in carrying out the fractionation of propylene from propane the methyl acetylene and propadiene will be removed in the propane bottoms product. The feasibility of this separation is evident from examination of the vapour-equilibrium data.[16] Care has to be exercised however in keeping the concentration of the acetylenes in the bottoms stream to the safe limits discussed earlier and this inevitably results in losses of propylene in this stream.

Since the equipment cost and utilities consumptions involved in separating the propane and propylene are so high various ways of using extractive distillation to make the separation easier have been suggested. In broad terms, extractive distillation is distillation in which the relative volatilities of the key components are altered by the presence of a third component. For the propylene-propane separation possible extractive agents which increase the relative volatility include acrylonitrile and benzonitrile.[17, 18] Although extractive distillation would appear to have some merit it has not been used on ethylene plants as a means of separating the high purity polymer grade propylene product.

2.4. STORAGE

For the storage of LPG both above ground and underground methods are used. In the United States where large quantities of propane are used to supplement the supply of natural gas at times of peak demand, the use of underground storage in salt formations and in mined caverns has been found generally to be the cheapest form of storage where geological formations exist to make it possible.

However, for propylene destined for use as feedstock in other manufacturing processes, above ground storage is normally used. Three main methods are used:

Pressure storage at ambient temperature,
Refrigerated storage at atmospheric pressure,
Refrigerated storage at an intermediate pressure.

2.4.1. Pressure Storage

For many years pressure vessels were used almost exclusively for storing liquified gases such as propylene.

Because of the high vapour pressure of propylene at ambient temperature i.e. 9·1 bars absolute at 15°C, the liquid must be stored in vessels designed to withstand the vapour pressure at the maximum temperature which can be expected at the particular location of the storage vessel; this is normally the temperature which could exist in the storage vessel when exposed to the sun. In temperate climates such as the United Kingdom this is considered to be 35°C which corresponds to a maximum operating pressure of 14·5 bars gauge and results in a design pressure of about 16·5 bars gauge. In tropical climates designs are based on a temperature of 40°C. Some companies installing pressure storage specify the use of the maximum ambient temperature as a more reasonable basis for establishing design pressures, and where necessary install shading or cooling sprays to prevent loss of gas through the safety valves when temperatures exceed those chosen for design. In order to keep down the design pressure of the storage vessel without suffering losses of vapour at high ambient temperatures, it is not uncommon to connect the vapour space of the storage vessel back to the production plant for recovery of vapour. In the case of storage located close to an ethylene plant the vent line from the storage vessel is often connected to the depropaniser or propylene splitter tower so that vapours released from the storage vessel are condensed in the tower overhead condenser and eventually returned to storage.

Propylene is stored under pressure in either spherical or cylindrical vessels designed to one of the design codes, e.g. ASME Section VIII for unfired pressure vessels or British Standard 1500. In laying out storage vessels in a tank farm, the minimum spacing distances are recommended for both storage at refineries or at bulk storage installations by such authorities as the National Fire Protection Association[19] (NFPA) of American and the British Institute of Petroleum.[20]

The volume of liquid which can be stored in a vessel must be limited to allow sufficient room for thermal expansion. Ways of calculating this volume are given by NFPA.[19]

As with any pressure vessel, the storage sphere or cylinder must be protected by safety valves. These are designed to prevent the pressure in the vessel exceeding the design pressure in the event of overfilling or for the extreme emergency when the vessel may be subjected to external fire. For the latter case the capacity of the safety valves depends on the surface area of the vessel which might be exposed to the fire and the quantity of heat which can be transferred to the contents of the vessel. Recommendations for calculating relief valve capacities are given by the American Petroleum Institute.[21] Safety valves must also be installed in the piping leading to and from storage vessels in the sections of pipe between valves which if closed create a liquid filled system. Without such safety valves serious overpressuring of the piping could result due to increase in pressure caused by increase in ambient temperature. A storage sphere with propylene recovery plant in background is shown in Fig. 2.4.

2.4.2. Refrigerated Storage

For the storage of large quantities of liquid propylene, in excess of about 400 tons there are usually economic advantages in using refrigerated storage. With this method, the liquid propylene is stored at a pressure only slightly above atmospheric —about 0·07 bars gauge and at the corresponding boiling point—42·8°C. Although spherical tanks covered with insulating material can be used, the more common method is to use tanks which are similar to conventional flat-bottomed oil storage tanks except that the roofs are of the dome type. These tanks can be either of single or double wall construction. With the single wall, the tank is insulated with foamed glass blocks or polyurethane foam with an outer jacket—usually aluminium—for weather protection. With the double wall method of construction the inner tank is surrounded by an outer shell and the annular space is filled with an insulating material such as perlite. In both cases the tank is fabricated from a grade of carbon steel which retains its ductility at the storage temperature. The annular space is kept under a slight positive pressure of dry nitrogen or inert gas to prevent ingress of moisture. To prevent the ground freezing and lifting, the bottom of the tank is laid on an insulating concrete slab supported on piles with an air space between the slab and the ground.

FIG. 2.4   *View of the same plant shown in Fig. 2.3 with Propylene Storage Sphere in the foreground*

In a typical refrigerated storage system, at low pressure, vapour is generated in the storage tank in two ways. Firstly, the propylene being fed to the tank will normally be at a higher pressure and temperature than the tank itself and vapour will therefore be released as the pressure of the propylene is reduced to the tank pressure. Secondly, however good the insulation, there will be some heat leak into the tank which will vaporise the contents. The vapour from these two sources is usually compressed to a pressure at which the propylene can be condensed either using air or water-cooled heat exchangers and the condensed liquid then returned to the tank. A flow diagram for this system is shown in Fig. 2.5.

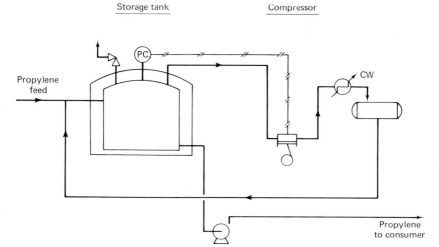

Fɪɢ. 2.5   *Flow Diagram—Refrigerated Storage System*

The pressure in the tank is controlled automatically by adjusting the capacity of the compressor either by unloading the compression cylinders or by stopping and starting the machine so that the capacity of the compressor corresponds to the volume of the gas generated in the tank. Vent gas from refrigerated storage tanks associated with an ethylene plant can be fed into the propylene refrigeration system within the ethylene plant thus obviating the need for a separate system in the storage area. The quantity of vapour produced as the propylene fed to the tank expands to the tank pressure is usually far greater than that generated by heat leak and this wide variation in vapour flow can complicate the design and control of the compression system. To minimise this, it is preferable to cool the propylene to as low a temperature as possible before it is fed to the tank.

Safety valves located on the tank roof vent vapour to atmosphere at a safe location should the compression system fail or should the amount of vapour generated in the tank exceed the capacity of the compression system.

2.4.3. Refrigerated Storage at Intermediate Pressure

This system uses refrigeration in a similar way to that described above so that storage is possible at temperatures intermediate between ambient and the atmospheric boiling point of $-42.8°C$ and at the corresponding pressure. This method has the advantage that the storage vessel—usually a sphere—designed for a pressure of 4·2 bars gauge can be used either for propylene using refrigeration or for butane at ambient temperature.

Since the vapour leaves the storage sphere at elevated pressure, the vent compressor can be located away from the storage area since there is ample pressure drop available for the interconnecting piping.

On L P G storage installations and on any installation involving loading or unload-ing facilities (e.g. road car loading) devices to limit the escape of the propylene in the event of damage to a pipeline or flexible hose are recommended. These devices include excess flow valves, back pressure check valves and remotely controlled stop valves. An excess flow valve is designed to close should the flow through it be 50 per cent greater than the maximum operating flow rate, or when a preset pressure dif-ferential across the valve is exceeded. These valves are installed in the outlet piping from storage vessels so that should there be a rupture in the piping, the valve will close preventing the escape of large quantities of liquid. A partial breakage of a pipe line or hose can result in the escape of liquid which although presenting a serious hazard may not be large enough to close the excess flow valve. To cover this eventuality, remotely operated valves or check valves can be installed to isolate the storage vessel and thus provide additional security. Check valves (non-return valves) installed in the pipe line feeding the storage vessel prevent liquid escaping from the vessel should there be a break in the feed line. On pressure storage vessels, connec-tions located below the liquid level are kept to a minimum and a single line is used for both filling and emptying.

In handling liquified gas such as propylene it has always to be borne in mind that since the gas is heavier than air, any gas or liquid escaping through leaking valves or equipment will tend to sink to ground level and accumulate at low points. On wind-less days the gas can travel considerable distances and may reach a source of ignition. Should this happen, the fire can 'flash' back to the source of the leak and result in a serious accident in the storage area. The procedures adopted by operating companies for draining, venting, sampling, etc. are, in consequence, extremely strict.

2.5. TRANSPORT

Liquified propylene is transported by road, rail and sea and to a lesser extent by pipeline.

Road tankers are usually designed with the tanks capable of withstanding the vapour pressure of the propylene at atmospheric temperature, i.e. about 15 bars gauge. The tanks are cylindrical pressure vessels fabricated from carbon steel, fitted with internal baffles to form approximately equal compartments so that the move-ment of liquid inside the tank is restricted. The tanks are fitted with pressure, temperature and flow gauges and safety valves. These fittings and also the inlet and outlet connections are protected from accidental damage by locating them in recesses in the tank shell or dished ends. Further protection is given by hinged, lockable covers over the fittings.

The tank can be mounted directly on a lorry chassis or alternatively on a trailer for use as an articulated vehicle. Most road tankers are fitted with a comprehensive pumping, metering and gauging system to deliver the propylene directly to the receiving storage vessel without the need for external pumps or power supply. Railway wagons are also designed to carry liquified gases such as propylene in uninsulated pressure vessels of similar construction to those used for road tankers.

Ships built to transport liquified petroleum gases have conventionally used cylindrical tanks designed to withstand the vapour pressure of the material at ambient temperature. Recently, ships have also been built with spherical tanks for semi-refrigerated transport.[22]

Pipelines exist in the United States to transfer propane recovered from natural gas from producing areas to consumers. In Europe as well as the United States pipelines are in use and more are being built to transport ethylene large distances, but at the present time no equivalent pipeline network exists for propylene.

2.6. ANALYSIS

It will already have become apparent from the previous sections that the specifications involved in producing 'chemical' and 'polymer' grade propylene from an ethylene plant are much more stringent than those for the product from a refinery unit producing feedstock for other refinery processes.

Whilst the detailed analysis of the propylene products necessary to ensure that the product meets specification are carried out in the laboratory, the use of on-stream analysers to monitor the more critical impurities in the propylene product is becoming commonplace on ethylene plants. These analysers usually take the form of gas chromatographs which are fed continuously with a sample of the product. The analysers are usually housed in a building located near the plant battery limits and are linked to recording instruments on the panel of the main control room. The method of sampling is extremely important in order to reduce the time interval between the sample being taken and the results of the analysis appearing on the recorder. Since there can be considerable distances between the sampling point and the analyser, a continuous flow of sample is fed through the analyser house and back to a convenient low pressure point in the plant. From this loop, a short sample line feeds the analyser and to keep time lags to a minimum the ratio of sample flow by-passing the analyser to that portion entering the analyser is kept quite high. All samples must either be taken in the vapour phase or means must be provided in the sampling system to vaporise liquid samples.

It is also common to analyse continuously, in the same way, samples drawn from intermediate points in the recovery of purification plant so that the operator can follow the operation of discrete sections of the plant and take corrective action before the final product goes off specification. In this way the production of specification product can be maintained at a much higher level than was the case when the operation had to rely on samples taken for laboratory analysis. The propylene streams which are typically analysed continuously on a modern ethylene plant are shown in Table II.9. Detailed recommendations covering the installation of gas chromatographs for on-line analysis are given by the American Petroleum Institute.[23]

Detailed analyses of the product in the laboratory also rely very much now a days on gas chromatography, whilst analyses which cannot be carried out in this way follow standard methods of test such as those of the American Society for Testing and Materials (ASTM),[24] the Institute of Petroleum,[25] and Universal Oil Products Company.[26]

In refineries where propylene product specifications are usually less severe than in ethylene plants, gas chromatography is used to determine the composition of the product and routine laboratory analysis is normally adequate for quality control purposes. When the propylene is being produced for sale as LPG then the physical properties of the material, e.g. vapour pressure, are more important than the actual composition and standard methods of test are employed. Where refinery gas recovery units are designed for very high recovery of propylene then an on-stream analyser monitoring the propylene content of the de-ethaniser absorber tail gas can be easily justified since the absorber operating conditions can then be adjusted more easily to keep losses of propylene to a minimum.

Some standard test methods commonly used for propylene analysis are listed in Table II.10.

TABLE II.9

Location of Continuous Analysers for Analysing Propylene Streams on Ethylene Plants

Location	Components analysed
De-ethaniser bottoms	Ethylene, Ethane
Depropaniser overheads	C_4's
Outlet of hydrogenation reactors	Methyl Acetylene, Propadiene
High purity propylene product	Ethylene, Ethane, Methyl Acetylene, Propadiene

TABLE II.10

Some Methods Used for Propylene Analysis

Analysis	Test method
Vapour pressure	ASTM D1267–67
Specific gravity	ASTM D1657–64
Volatility	ASTM D1837–64
Copper strip corrosion	ASTM D1838–64
Residue	ASTM D2158–65
Methane, ethylene, ethane ⎫	ASTM D2163–66 or
Propylene, propane, C_4's ⎭	UOP 373–59
Butadiene	ASTM D1096–54
Methyl acetylene	UOP 373–59
Propadiene	UOP 373–59
Sulphur	ASTM D1266–627
Water	UOP 312–59

TABLE II.II

Analysis	Test method
Vapour pressure	ASTM D1267–67
Specific gravity	ASTM D1657–64
Volatility	ASTM D1837–64
Copper strip corrosion	ASTM D1838–64
Residue	ASTM D2158–65
Methane, ethylene, ethane, propylene, propane, butadiene, C₄'s, methyl acetylene, propadiene	Methods using gas chromatography such as ASTM D2163–70, ASTM D2712–70 and UOP 373–59
Butadiene	ASTM D1096–54
Sulphur	ASTM D1266–627
Water	UOP 312–59

REFERENCES

[1] Ockerbloom, *Pet. Chem. Eng.*, April 1970, 21.
[2] Natural Gas Processors Association, Publication 2140–70.
[3] *Hydroc. Proc.*, 1966, **45**, 209.
[4] *Hydroc. Proc.*, 1950, **29**, 103.
[5] *Trans. Amer. Inst. Chem. Eng.*, 1946, **42**, 665.
[6] Edmister, *Pet. Eng.*, 1948, **19**, 68.
[7] Hannah, *Pet. Ref.*, 1949, **28**, 135.
[8] Zdonik, Green and Hallee, *Oil and Gas J.*, 1967, **65**, 26.
[9] Zdonik, Green and Hallee, *Oil and Gas J.*, 1967, **67**, 19.
[10] *Europ. Chem. News*, 19 April 1968.
[11] Kniel and Slager, *Chem. Eng. Prog.*, 1947, **43**, 335.
[12] Pratt and Foskett, *Trans. Amer. Inst. Chem. Eng.*, 1946, **42**, 149.
[13] Hembree, *Welding Journal*, 1963, **42**, 395.
[14] Khelghatian (Sun Oil Co.), USP 3,315,003.
[15] Zdonik, *Advances in Petroleum Chemistry and Refining*, Vol. I, 226.
[16] Hill, McCormick, Barton and Fenske, *Amer. Inst. Chem. J.*, 1962, **8**, 681.
[17] Hafslund, *Chem. Eng. Prog.*, 1969, **65**, No. 9.
[18] Teter and Shand (Sinclair Refining Co.), USP 2,588,056.
[19] National Fire Protection Association, NFPA Standard No. 58.
[20] Liquefied Petroleum Gases Safety Code, Institute of Petroleum.
[21] American Petroleum Institute, API RP 520 and 521.
[22] Oil and Gas International, January 1962.
[23] American Petroleum Institute, API RP 550 Pt II.
[24] ASTM Standard, Part 18.
[25] Institute of Petroleum Standards for Petroleum and its Products.
[26] UOP Laboratory Test Methods for Petroleum and its Products.

CHAPTER 3

THE PHYSICAL PROPERTIES OF PROPYLENE

3.1. THERMODYNAMIC AND TRANSPORT PROPERTIES
By R. W. Gallant

3.1.1. Introduction

PROPYLENE is encountered in distillation and heat exchange activities in light hydrocarbon plants, mixed in with propane, ethane, ethylene, and other low molecular weight hydrocarbons. Propylene and propane are very similar in physical properties in all aspects. Compared with other olefins, propylene's transport properties have lower values than ethylene at the same temperature; higher values than butene. Thus as a liquid, propylene has higher viscosity, surface tension, and densities at both the same temperature and at their respective boiling points. Propylene has a much higher enthalpy and entropy value than butene on a gram basis; and about half the ethylene enthalpy value.

Accurate thermodynamic and transport properties of propylene are essential for the proper design of equipment in light hydrocarbon plants and many petrochemical plants using propylene as a raw material. This chapter presents the important physical properties of propylene.

3.1.2. Thermodynamic Properties of Propylene Gas

3.1.2.1 Molecular weight. The molecular weight of propylene is 42·078.

3.1.2.2. Enthalpy, entropy, and volume. Table III.1 presents the enthalpy, entropy, and volume for super-heated propylene gas. The thermodynamic data were taken from the extensive investigations reported by Canjar.[1] The volumetric data are based on three studies.[1, 2, 3] The data were converted to metric units.

The saturated liquid and gas properties are tabulated in Table III.2.[1]

TABLE III.1

Thermodynamic Properties of Propylene

Atmospheres		Temperature. °C						
		−40	−20	0	20	40	60	80
0	h	10,769	11,042	11,328	11,622	11,933	12,258	12,602
1	v	18,501	20,253	21,971	23,691	25,412	27,114	28,808
	h	10,729	11,005	11,293	11,589	11,905	12,237	12,583
	s	60·13	61·31	62·40	63·37	64·30	65·39	66·44
2	v	—	9734·9	10,665	11,610	12,511	13,376	14,227
	h	—	10,961	11,253	11,557	11,877	12,214	12,562
	s	—	59·83	60·93	61·98	63·07	64·13	65·14

Atmospheres		−40	−20	0	20	40	60	80
					Temperature. °C			
4	v	—	—	5143·3	5605·6	6091·5	6567·0	7008·3
	h	—	—	11,171	11,484	11,819	12,165	12,520
	s	—	—	59·37	60·51	61·56	62·65	63·66
10	v	—	—	—	—	2211·8	2427·2	2632·1
	h	—	—	—	—	11,615	11,996	12,377
	s	—	—	—	—	59·25	60·42	61·56
20	v	—	—	—	—	—	1029·7	1166·3
	h	—	—	—	—	—	11,641	12,094
	s	—	—	—	—	—	58·19	59·58

		100	120	140	160	180	200	220	240
0	h	12,952	13,324	13,705	14,105	14,517	14,942	15,382	15,835
1	v	30,484	32,155	33,815	35,475	37,132	38,787	40,442	42,097
	h	12,938	13,310	13,694	14,093	14,505	14,935	15,372	15,826
	s	67·45	68·46	69·43	70·40	71·32	72·25	73·17	74·06
2	v	15,078	15,924	16,775	17,615	18,456	19,289	20,121	20,954
	h	12,922	13,294	13,680	14,079	14,493	14,923	15,363	15,817
	s	66·10	67·11	68·04	69·00	69·98	70·86	71·74	72·67
4	v	7433·8	7872·5	8300·7	8731·5	9154·4	9582·6	10,000	10,426
	h	12,882	13,261	13,652	14,054	14,472	14,900	15,342	15,798
	s	64·67	65·68	66·65	67·58	68·55	69·47	70·40	71·28
10	v	2831·7	2981·4	3202·1	3388·6	3575·1	3758·9	3929·7	4108·3
	h	12,761	13,158	13,565	13,977	14,400	14,837	15,288	15,749
	s	62·61	63·66	64·63	65·64	66·61	67·54	68·50	69·39
20	v	1287·1	1402·7	1510·4	1615·5	1717·9	1815·1	1912·3	2004·2
	h	12,529	12,971	13,404	13,841	14,283	14,732	15,190	15,658
	s	60·76	61·94	62·99	64·04	65·05	65·98	66·99	67·87
40	v	475·4	575·3	654·1	722·4	785·4	843·2	901·0	950·9
	h	11,910	12,511	13,041	13,535	14,019	14,505	14,994	15,485
	s	58·07	59·62	60·97	62·11	63·24	64·25	65·31	66·23
60	v		265·3	357·2	420·3	472·8	520·1	562·1	604·2
	h		11,779	12,564	13,172	13,724	14,255	14,776	15,293
	s		57·27	59·25	60·68	61·94	63·07	64·17	65·18
80	v		144·5	212·8	270·6	320·5	362·5	399·3	430·8
	h		10,972	11,964	12,754	13,404	13,995	14,554	15,103
	s		55·00	57·52	59·33	60·80	62·11	63·24	64·34
100	v		123·5	149·7	194·4	233·8	270·6	302·1	331·0
	h		10,734	11,550	12,361	13,072	13,729	14,325	14,905
	s		54·24	56·22	58·19	59·79	61·18	62·44	63·58
120	v		115·6	131·3	157·6	186·5	215·4	241·7	267·9
	h		10,619	11,349	12,085	12,805	13,479	14,112	14,713
	s		53·78	55·59	57·35	58·95	60·42	61·73	62·95
150	v		105·1	118·2	131·3	149·7	168·1	191·8	210·1
	h		10,530	11,187	11,861	12,534	13,198	13,841	14,465
	s		53·35	55·00	56·59	58·11	59·54	60·89	62·11
170	v		102·4	110·3	123·5	136·6	152·4	170·7	186·5
	h		10,495	11,124	11,777	12,410	13,060	13,705	14,330
	s		63·14	54·74	56·22	57·65	59·08	60·42	61·64
200	v		—	105·1	115·6	126·1	136·6	152·4	162·9
	h		—	11,035	11,676	12,293	12,917	13,544	14,161
	s		—	54·41	55·80	57·18	58·57	59·88	61·10

v = volume, cc/mole   h = enthalpy, cal/mol   s = entropy, cal/mole-°C.

TABLE III.2

Properties of Liquid and Vapour in Equilibrium

Temperature °C	Pressure atm	Volume cc/mole		Enthalpy cal/mole			Entropy cal/mole-°C		
		Liquid	Gas	Liquid	ΔH	Gas	Liquid	ΔS	Gas
−47·8	1·000	68·56	17,794	6,215	4,400	10,615	40·16	19·51	59·67
−40	1·401	69·85	12,966	6,394	4,315	10,709	40·91	19·50	59·41
−30	2·097	71·37	9,010	6,613	4,198	10,811	41·88	17·28	59·16
−20	3·023	73·29	6,404	6,835	4,072	10,907	42·62	16·25	58·87
−10	4·257	74·97	4,639	7,066	3,937	11,003	43·69	14·97	58·66
0	5·772	76·83	3,423	7,302	3,794	11,096	44·55	13·85	58·40
10	7·685	79·09	2,569	7,547	3,651	11,198	45·39	12·80	58·19
20	10·046	81·80	1,957	7,790	3,467	11,257	46·21	11·77	57·98
30	12·911	84·32	1,510	8,043	3,249	11,292	47·06	10·71	57·77
40	16·307	87·52	1,177	8,303	3,033	11,336	47·88	9·68	57·56
50	20·299	91·47	922·0	8,565	2,800	11,365	48·68	8·67	57·35
60	24·978	97·19	719·7	8,835	2,542	11,377	49·47	7·63	57·10
70	30·435	104·7	551·6	9,161	2,184	11,345	50·42	6·34	56·76
80	36·639	114·4	396·6	9,600	1,662	11,262	51·59	4·71	56·30
90	44·160	157·7	223·1	10,282	558	10,840	53·60	1·58	55·18
91·8	45·609	191·0	191·0	10,705	0	10,705	54·54	0	54·54

Temperature °C	Density g/cc	
	Liquid	Gas
−47·7	0·6137	0·002365
−40·0	0·6024	0·003245
−28·9	0·5883	0·004878
−17·8	0·5715	0·007104
− 6·6	0·5556	0·01010
+ 4·5	0·5383	0·01403
15·6	0·5209	0·01912
26·7	0·5050	0·02567
37·8	0·4854	0·03394
48·9	0·4619	0·04450
60·0	0·4329	0·05846
71·1	0·3983	0·07891
82·2	0·3503	0·1161
87·8	0·3215	0·1511
91·76	0·2203	

3.1.2.3. *Ideal gas properties.* Kobe has calculated the ideal gas enthalpy and heat capacity (Table III.3).[4] The equation for the heat capacity is the following:

$$Cp = 14.240 + .04233t - 2.430t^2 \times 10^{-5} + 5.88t^3 \times 10^{-9}$$

where heat capacity is in calories/mole-°C and temperature is in °C.

TABLE III.3

Ideal Gas Properties

Temperature °C	Heat capacity cal/mole-°C	Enthalpy cal/mole
0	14·32	0
18	15·01	264
25	15·27	370
100	18·11	1,623
200	21·71	3,617
300	24·92	5,945
400	27·69	8,579
500	30·09	11,470
600	32·19	14,580
700	34·01	17,900
800	35·60	21,380
900	36·98	25,020
1000	38·19	28,780
1100	39·25	32,650
1200	40·17	36,620

3.1.2.4. Constant volume heat capacity. Nevers and Martin have studied the constant volume heat capacity (Table III.4).[5]

TABLE III.4

Constant Volume Heat Capacity
cal/mole-°C

Temperature, °C	20·75	13·86	10·84	9·07	7·13	6·22
70	16·45	—	—	—	—	—
80	16·59	17·54	—	—	—	—
90	16·81	17·63	18·24	18·86	—	—
100	17·09	18·71	18·28	18·71	19·81	20·33
110	17·44	17·94	18·40	18·70	19·55	19·73
120	17·70	18·18	18·60	—	—	—
130	18·44	18·46	18·83	—	—	—
140	18·71	18·76	—	—	—	—

3.1.2.5. Fugacity. The fugacity-pressure ratios are shown in Table III.5.[6]

3.1.2.6. Compressibility. The compressibility factor is a measure of how a gas deviates from an ideal gas in its pressure-volume-temperature-relationship. The compressibility factor is defined as the pressure times the volume divided by the gas constant times the temperature. Thus, it is dimensionless. The compressibility factor of the vapour increases as the temperature increases, and decreases as the pressure increases. saturated vapour compressibility factor drops as it approaches the critical point. The Above the critical point, the factor increases both with pressure and temperature.

The data in Table III.6 were calculated from the experimental data of three investigators.[1, 2, 3]

TABLE III.5

Fugacity—Pressure Ratios

Atmospheres	Temperature, °C								
	−47·7	−40	−10	0	20	40	60	80	91·8
0·075	0·997	0·998	0·998	0·999	0·999	0·999	1·000	1·000	1·000
1·0	0·983	0·984	0·989	0·990	0·993	0·996	0·997	0·997	0·998
2·0	—	—	0·959	0·962	0·969	0·975	0·980	0·983	0·989
5·0	—	—	—	—	0·925	0·940	0·950	0·958	0·968
10	—	—	—	—	0·846	0·876	0·899	0·916	0·930
20	—	—	—	—	—	—	0·798	0·833	0·852
Saturation, atm	1·00	1·401	4·257	5·772	10·046	16·307	24·978	36·639	45·609
Ratio	0·983	0·970	0·914	0·891	0·847	0·802	0·750	0·695	0·645

TABLE III.6

Compressibility Factors for Propylene

Pressure atm	Temperature, °C									
	−40	−20	0	50	91·4	100	150	200	250	300
1·0	0·966	0·974	0·980	0·982	0·984	0·985	0·996	0·998	0·999	0·999
2·3	—	0·937	0·954	0·974	0·982	0·984	0·990	0·994	0·996	0·997
9·1	—	—	0·028	0·890	0·930	0·936	0·960	0·974	0·983	0·986
18·2	—	—	0·055	0·744	0·848	0·861	0·911	0·942	0·961	0·972
27·2	—	—	0·083	0·090	0·748	0·774	0·864	0·912	0·939	0·957
36·3	—	—	0·109	0·119	0·625	0·673	0·814	0·882	0·920	0·942
49·9	—	—	0·151	0·161	0·202	0·418	0·735	0·838	0·893	0·924
54·5	—	—	0·165	0·175	0·209	0·268	0·707	0·823	0·885	0·918
63·6	—	—	—	—	0·230	0·256	0·652	0·795	0·867	0·908
81·7	—	—	—	—	0·278	0·288	0·543	0·736	0·838	0·890

Temperature °C	Saturated liquid	Saturated vapour
−47·8	—	0·966
−40	—	0·949
−20	0·010	0·931
0	0·018	0·875
25	0·036	0·798
50	0·067	0·703
75	0·124	0·555
90	0·212	0·359
91·8	0·273	0·273

TABLE III.7

Compressibility Factor Propylene-Propane Mixture

Pressure atm	Temperature, °C				
	$-12 \cdot 2°$	$37 \cdot 8°$	$104 \cdot 4°$	$137 \cdot 8°$	$204 \cdot 4°$
Mole fraction propylene = 0·2411					
1·00	—	0·9855	0·9922	0·9938	0·9968
2·72	—	0·9598	0·9782	0·9843	0·9908
6·80	0·0252	0·8955	0·9446	0·9593	0·9763
20·4	0·0753	0·0734	0·8300	0·8761	0·9278
68·0	0·2476	0·2364	0·2774	0·4995	0·7656
204·4	0·7210	0·6682	0·6484	0·6604	0·7307
680·0	2·2425	2·0118	1·8043	1·7332	1·6307
Mole fraction propylene = 0·6289					
1·00	—	0·9870	0·9938	0·9952	0·9972
2·72	—	0·9634	0·9817	0·9854	0·9913
6·80	0·0245	0·9050	0·9518	0·9628	0·9777
20·4	0·0731	0·0715	0·8403	0·8840	0·9325
68·0	0·2406	0·2300	0·2777	0·5337	0·7788
204·4	0·6989	0·6490	0·6334	0·6467	0·7244
680·0	2·1631	1·9415	1·7537	1·6827	1·5792

3.1.2.7. Vapour-liquid and vapour-solid equilibria. Vapour-liquid equilibria of propylene have been measured by numerous investigators.[2, 6, 6a, 7–11, 23, 24] For pressures

below 10 mm $\log P \text{ mm} = -\dfrac{1151}{T} + 8 \cdot 154$

10 mm–760 mm $\log P \text{ mm} = -\dfrac{1074}{T} + 7 \cdot 677$

1–40 atm $\log P \text{ mm} = -\dfrac{975 \cdot 1}{T} + 7 \cdot 205$

Critical constants[9] are $t = 91 \cdot 76 \pm 0 \cdot 015°C$
$p = 45 \cdot 61 \pm 0 \cdot 02 \text{ atm}$
$V = 0 \cdot 191 \text{ l/mole}$ (see also refs. [10] and [12]).

Densities of liquid and gaseous propylene in equilibrium are shown in Table III.2.[10, 13]

The triple point is at $87 \cdot 85°K$,[8, 11, 14] heat of fusion 717·6 cal/mole.[11] The pressure at the triple point is $1 \cdot 13 \times 10^{-5} \text{mm}$ and the heat of vaporisation of the liquid is 5266 cal/mole so that the heat of vaporisation of the solid propylene has to be 5984 cal/mole. The sublimation equilibrium can be calculated as

$$\log P \text{ mm} = -\frac{1308}{T} + 9 \cdot 941.$$

3.1.3. Transport Properties of Propylene

3.1.3.1. Viscosity. Lambert has measured the vapour viscosity of propylene from 35°C to 91°C (Table III.8).[16] Other data is shown in Table III.9.[10, 17]

TABLE III.8

Vapour Viscosity Data of Lambert[8]

°C	Micropoise
35	92·2
50	95·4
60	99·1
71	102·4
81	105·6
91	108

TABLE III.9

Vapour Viscosity Data of Sherwood and Reid[9]

°C	Micropoise
20	84·3
50	93·3
150	121·0
250	146·7

Thodos and Flynn have developed an equation for calculating the viscosity of hydrocarbons at normal temperature.[18] This method has been used to calculate the viscosities from −50 to 500° (Table III.10).

TABLE III.10

Calculated Vapour Viscosities

°C	Micropoise
−50	64
0	78
50	93
100	106
150	121
200	133
250	147
300	159
400	183
500	205

The liquid viscosity of propylene is available from -115 to $-95°$ (Table III.11).[19] A plot of the log of the viscosity versus the reciprocal of the absolute temperature has been used to extrapolate the daat to $0°C$ (Table III.12).

TABLE III.11

Liquid Viscosity Data of Dreisbach[11]

°C	Centipoise
-115	0·32
-105	0·266
-95	0·23

TABLE III.12

Calculated Liquid Viscosity

°C	Centipoise
-80	0·194
-60	0·156
-40	0·131
-20	0·113
0	0·100

3.1.3.2. Thermal conductivity. Lambert has measured the vapour thermal conductivity and found this to be $52·4$ cal/cm-sec-°K at 66°C.[8] Table III.13 was calculated by the method of Misic and Thodos.[20]

TABLE III.13

Calculated Vapour Thermal Conductivity

°C	cal/cm-sec-°K
100	59×10^{-6}
200	90
300	123
400	157
500	191
600	225
700	256
800	292
900	322
1000	344

The calculated liquid thermal conductivities are presented in Table III.14.[21]

TABLE III.14

Liquid Thermal Conductivity[13]

°C	cal/cm-sec-°K
−70	3.83×10^{-4}
−60	3.62
−50	3.44
−40	3.28
−30	3.12
−20	2.97
−10	2.84
0	2.71
10	2.59
20	2.46
30	2.35

3.1.3.3. Diffusivity. The data of Unver and Himmelblau are tabulated in Table III.15.[22] The equation for the coefficient of diffusion is shown below:

$$D = (0.29773 + 0.008086t + 0.00024837t^2) \times 10^{-5}$$

TABLE III.15

Diffusivity of Propylene in Water[14]

°C	*Coefficient of diffusion, cm²/sec*
7	0.376×10^{-5}
25	0.681
35	0.895
53	1.373
65	1.868

3.1.4. Properties of liquid propylene

The important properties of liquid propylene are presented in Table III.16. The vapour pressure data are based on the measurements of four different investigators.[1, 3, 23, 24] The liquid densities are taken from the data of Maas[2] and Canjar.[1] Powell and Giaque have determined the liquid heat capacities from the melting point to the boiling point (−47.7°C).[26] Above that temperature, the heat capacities have been estimated. Maas[8] reports the surface tension of propylene from −62 to −22°C.[25] The surface tension at other temperatures was estimated from the Parachor and the volumetric data of Canjar.[1]

TABLE III.16

Properties of Liquid Propylene

Temperature °C	Pressure atm	Density g/cc	Heat capacity cal/mole-°C	Surface tension dyn/cm
−120	0·0076		20·82	
−110	0·0205		20·84	
−100	0·0430		20·91	
−90	0·0950		21·03	
−80	0·1755	0·6545	21·19	21·60
−70	0·2990	0·6423	21·38	20·14
−60	0·544	0·6301	21·61	18·64
−50	0·905	0·6180	21·83	17·16
−40	1·401	0·6045	22·10	15·67
−30	2·097	0·5900	22·75	14·21
−20	3·023	0·5757	23·25	12·72
−10	4·257	0·5614	23·88	11·31
0	5·772	0·5471	24·50	9·90
10	7·685	0·5322	25·21	8·49
20	10·046	0·517	25·95	7·18
30	12·911	0·5011	26·24	5·97
40	16·307	0·4822		4·78
50	20·299	0·4613		3·61
60	24·978	0·4353		2·44
70	30·435	0·4032		1·41
80	36·639	0·3665		0·55
90	44·160	0·2705		0·05
91·8	45·609	0·2210		0·00

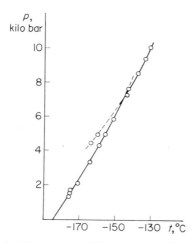

Fig. 3.1 *Melting Temperature of Propylene as a Function of Pressure*

3.1.5. Properties of solid propylene

The heat capacity of solid propylene is shown in Table III.17.[11] In measuring the melting curve up to 10,000 bars a transition point was found at 7,150 bars[15] at 129°K where 'solid propylene I', 'solid propylene II' and 'liquid' co-exist. Above this pressure the 'solid propylene II' is the stable one.

The melting point of propylene at various pressures is shown in Fig. 3.1.

TABLE III.17

The Properties of Solid Propylene

°K	Heat capacity cal/mole-°C
15	1·27
20	2·29
25	3·59
30	4·78
35	5·92
40	6·92
45	7·78
50	8·56
60	10·05
70	11·45
80	12·79
87·85 (solid)	13·83
87·85 (liquid)	22·33

Melting point = 87·85°K = −185·25°C
Heat of fusion = 717·6 cal/mole

N.B.—The information on vapour-liquid and liquid-solid equilibria in this section were contributed by Prof. Diepen and Mr. de Zeeuw.

3.2. EQUILIBRIA DATA OF PROPYLENE MIXTURES WITH OTHER CHEMICALS

By **G. A. M. Diepen and M. A. de Zeeuw**

3.2.1. Binary Systems

3.2.1.1. Hydrogen—Propylene Vapour-liquid equilibria in this system were measured[25] at several temperatures between 23·9 and −156·7°C and pressures up to 550 atm. The results are given in Fig. 3.2.

From this plot it is clear that the critical locus in this system moves from the critical point of propylene (92·3°C, 45 atm) very fast to high pressures. Therefore it might be possible that this locus passes through a temperature minimum, thereafter rising to higher temperatures and pressures thus giving immiscibility in the gas phase at temperatures above the critical temperature of propylene at very high pressures.

These measurements are used for calculation of solubility of hydrogen in cryogenic solvents.[26]

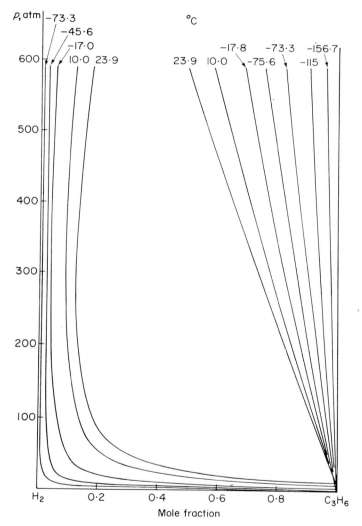

Fig. 3.2   *Vapour Pressure Equilibria, Hydrogen—Propylene*

3.2.1.2. Oxygen—Propylene   The solubility of solid propylene in liquid oxygen was measured.[27, 28] Whereas the first gives a solubility of 38·5, 7·93, 2·82, 0·771 and

0·202 mole per cent at 86·5, 83, 80·6, 75·0 and 67·0°K respectively, the second gives a solubility of 0·66, 0·62, 0·23 and 0·14 mole per cent at 87·4, 86·8, 83·3 and 83°K respectively. The great differences are possible due to the fact that there is immiscibility in the liquid phase.[28] Tsins method depends on filtering a mixture of the saturated solution and excess solute followed by analysis of the filtrate. The presence of two liquid layers would give erroneous results in such a procedure.

Cox determined the solubilities by measurement of the lowering of the vapour pressure of the solvent in a saturated solution. This method is based upon the assumption which was experimentally verified, that the solution obeys Raoults law and that the vapour pressure of the solute is negligible.

TABLE III.18

Vapour-Liquid Equilibria in the Propylene-Water System

Temp. °C	Pressure atm	L_2 $X_1 \times 10^4$	G $X_2 \times 10^4$	Temp. °C	Pressure atm	L_2 $X_1 \times 10^4$	G $X_2 \times 10^4$
37·81	2·52	1·5236	269·6	104·56	104·38	13·2022	111·3
37·83	4·51	2·9767	126·1	104·56	124·67	13·5276	97·1
37·78	6·33	4·1805	90·0	104·58	138·66	13·7631	
37·81	8·85	5·5750	63·9	104·56	144·46		89·8
37·83	11·41	7·0699	46·5	104·56	166·71	14·2283	84·7
37·83	13·95	7·6421	39·2	104·56	207·27	14·9856	84·3
37·78	14·67	7·9896	36·5	104·58	252·31	15·5092	81·4
71·25	5·25	2·0360	619·2	104·58	291·48	16·0112	77·3
71·22	9·30	4·2861	344·9	104·61	319·88		78·0
71·25	12·72	5·5390	234·6	137·79	14·24		2585·8
71·22	14·90	5·3539	183·7	137·79	15·60	4·6776	
71·25	19·70	6·9974	139·9	137·79	21·06	6·0831	1809·5
71·25	22·32	8·1204	125·5	137·81	27·76	8·0909	1186·8
71·11	25·21	8·4676	104·0	137·81	34·50	9·5986	968·6
71·25	29·40	9·1853	96·4	137·73	40·56	12·4658	785·4
104·50	7·97	1·9709	1629·3	137·79	48·31		692·5
104·47	12·46	3·0152	926·0	137·79	57·70	13·9960	554·4
104·56	16·81	4·9579	757·4	137·69	76·42	16·3154	446·1
104·56	23·46	6·2384	510·0	137·69	90·09	17·1891	356·1
104·56	26·78	6·8842	440·9	137·69	101·32	17·1560	341·7
104·56	33·21	8·4674	328·8	137·69	129·97	18·1504	276·6
104·56	39·87	9·7630	283·8	137·69	150·38	18·7990	278·4
104·61	47·19	11·1659	248·1	137·69	191·41	19·5650	253·0
104·56	55·53	12·1382	192·1	137·69	222·78	20·3464	248·6
104·56	70·67	12·8830	161·6	137·69	255·51	21·0183	234·3
104·58	85·23	13·3389	131·6	137·69	317·77	22·6356	207·3

L_2 = water-rich liquid.
G = vapour.
X_1 = mole fraction of propylene.
X_2 = mole fraction of water.

3.2.1.3. Water—Propylene The water—propylene system affords a clathrate compound, propylene hydrate. The three phase lines: solid hydrate, aqueous liquid, gas; solid hydrate, aqueous liquid, hydrocarbon liquid; and aqueous liquid, hydrocarbon liquid, gas, were measured.[29] The p-t projection of these lines is shown in Fig. 3.3.

TABLE III.19

Co-existing Phases, Liquid-Liquid and Vapour-Liquid-Liquid on the System Propylene-Water

Temp. °C	Pressure atm	L_2 $X_1 \times 10^4$	L_1 $X_2 \times 10^4$	Temp. °C	Pressure atm	L_2 $X_1 \times 10^4$	L_1 $X_2 \times 10^4$	G $X_2 \times 10^4$
37·81	31·98	8·0866	9·012	37·74	15·94	8·932	8·537	38·463
37·78	74·39	9·8059	8·331	49·72	20·78	8·777	18·053	51·644
37·78	119·20	10·2631	7·699	60·72	25·86	8·913	27·224	72·900
37·89	156·62	11·2611	7·328	71·86	31·83	9·458	40·956	86·660
37·81	198·22		7·470	83·39	39·34	10·005	53·174	104·142
37·92	216·93	12·2040	7·001	90·39	44·98	10·478	66·935	104·336
37·94	239·08	12·9589	7·022	92·11	45·96		96·542	95·126
37·94	280·53	13·1481	6·877					
37·92	325·33	13·4538	6·402					
71·26	52·39	9·5567	36·135					
71·22	63·21	9·6743						
71·22	86·89		35·630					
71·28	119·55	11·2017						
71·28	152·63	11·4832						
71·19	154·34		32·790					
71·11	187·02		30·180					
71·28	208·44	12·3313	30·890					
71·22	257·19		30·650					
71·28	264·90	13·9303						
71·28	276·16		30·110					
71·28	317·62	14·0519						
71·24	326·06		28·310					

L_1 = propylene-rich liquid.
L_2 = water-rich liquid.
G = vapour.
X_1 = mole fraction of propylene
X_2 = mole fraction of water.

The mutual solibilities of propylene and water in the two-phase region of water-rich liquid and vapour, water-rich liquid and propylene-rich liquid and in the three-phase region of water-rich liquid, propylene-rich liquid and gas were measured.[30] The results are collected in Tables III.18 and III.19.

3.2.1.4. Nitrogen—Propylene The solubility of solid propylene in liquid nitrogen was measured.[27] At 67·0, 75·0 and 83·0°K the solubility was found to be 0·17, 0·70 and 7·2 mole per cent respectively. The last figure is possibly too high because there is immiscibility in the liquid phase. (See also the system Oxygen—Propylene.)

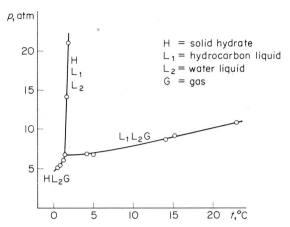

FIG. 3.3   p-t *Projection 3-Phase Lines of the Propylene-Water System*

The vapour pressure of propylene-rich liquid, the two immiscible liquids and the nitrogen content of the propylene-rich liquid were measured at some different temperatures.[31] The results are shown in Fig. 3.4. The presence of the two liquids was also observed visually.

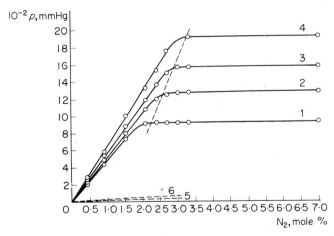

FIG. 3.4   *Equilibrium Isotherms for Propylene– Nitrogen Solutions in the Region of Low Nitrogen Concentrations; temperature (°K): (1 and 5) 79; (2) 82; (3) 84; (4 and 6) 86*

3.2.1.5. Ammonia—Propylene   Liquid ammonia and liquid propylene are not completely miscible below the critical temperature −8·1 °C.[32] The composition of the coexisting phases is given in Table III.20 and shown in Fig. 3.5.

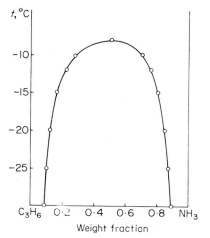

FIG. 3.5 *Composition of Co-existing Phases of Liquid Ammonia and Liquid Propylene*

TABLE III.20

Compositions of Co-existing Liquid Phases in the System Propylene-Ammonia

Temperature °C	L_1 Weight fraction NH_3	L_2 Weight fraction NH_3
−30·0	0·0846	0·8939
−25·0	0·1011	0·8762
−20·0	0·1264	0·8505
−15·0	0·1673	0·8100
−12·0	0·2210	0·7605
−10·0	0·2795	0·7076
−8·1	0·5050	

3.2.1.6. Carbon Dioxide—Propylene The melting points of solid carbon dioxide in various mixtures with propylene were measured.[33, 34] The results are collected in Table III.21. Vapour-liquid equilibria were measured by several investigators.[34, 35, 36, 37] The low temperature dew and bubble points[34] are given in Table III.22., p-t measurements for five different compositions[35] together with the boiling lines of carbon dioxide and propylene and the critical locus are shown in Fig. 3.6. The dew and bubble points at higher temperatures[36] are given together with the molar volumes of the saturated liquid and gas phases in Table III.23 for five different mixtures. Finally a temperature–composition diagram for three different pressures[38] is shown in Fig. 3.7.

8

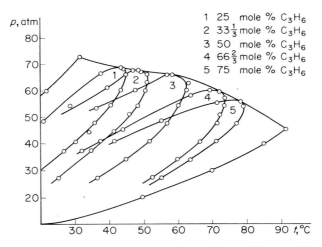

FIG. 3.6   p-t *Measurements for Five Different Compositions, together with Boiling Lines of Carbon Dioxide and Propylene and Critical Locus*

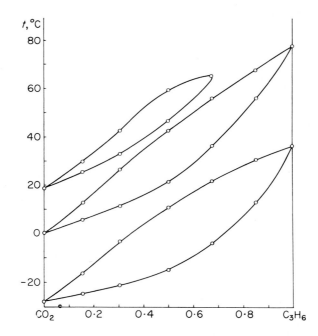

FIG. 3.7   *Temperature-Composition Diagram for Three Different Pressures in the System Carbon Dioxide-Propylene*

TABLE III.21

Melting Points of Solid Carbon Dioxide in Mixtures with Propylene

Mole % CO_2	Melting point °K
0	87·9
0·68	120·4
2·1	140·0
2·4	140·0
3·8	142·5
9·2	161·2
19·7	181·8
21·9	183·9
33·1	192·3
45·3	197·8
100	215·7

TABLE III.22

Dew and Bubble Points in the System Carbon Dioxide-Propylene

Mole fraction CO_2	Dew points pressure atm	Temperature °K	Mole fraction CO_2	Bubble points pressure atm	Temperature °K
0·2116	0·2486	192·7	0·089	1·9979	225·1
	0·4484	202·9		1·5340	218·3
	0·9925	218·9		1·0019	208·4
	1·4433	226·8		0·6946	200·6
	1·9438	234·4		0·4926	193·8
				0·3053	185·3
				0·2096	179·1
				0·1361	172·5
				0·0919	166·4
0·3976	0·2467	188·9	0·197	2·0796	212·3
	0·4866	200·3		1·833	208·9
	1·0131	213·9		1·5050	203·4
	1·5607	223·3		1·3629	200·8
	2·0052	229·1		1·3145	200·3
				1·0788	195·6
				0·9215	191·7
				0·7473	186·8
0·6001	0·2291	182·4	0·331	2·1490	205·7
	0·5103	194·7		1·8701	202·0
	1·0192	206·4		1·7397	200·4
	1·5363	214·6		1·4702	196·0
	1·9530	220·1		1·3412	194·1
0·7992	0·2553	173·2	0·453	2·1222	202·5
	0·5054	183·4		1·9788	200·8
	1·0167	195·3		1·8034	198·7
	1·5249	203·2			
	1·9245	207·4			

TABLE III.23

Dew Points, Bubble Points and Volumes of Saturated Liquid and Vapour Phases in the System Carbon Dioxide-Propylene

$t°C$	15·81 mole % C_3H_6				30·80 mole % C_3H_6				50·28 mole % C_3H_6				68·63 mole % C_3H_6				85·33 mole % C_3H_6			
	Dew bubble point atm		Molar volume l/mol		Dew bubble point atm		Molar volume l/mol		Dew bubble point atm		Molar volume l/mol		Dew bubble point atm		Molar volume l/mol		Dew bubble point atm		Molar volume l/mol	
			Gas	Liquid			Gas	Liquid			Gas	Liquid			Gas	Liquid			Gas	Liquid
−30·0	9·39	12·72	1·868	0·0471	6·01	11·43	3·043	0·0520	3·76	9·56	4·891	0·0584	2·85	7·34	6·472	0·0633	2·39	4·89	7·538	0·0685
−15·0	15·73	20·26	1·109	0·0498	10·32	18·01	1·796	0·0549	6·52	14·81	2·932	0·0611	5·07	11·38	3·818	0·0662	4·15	7·24	4·675	0·0711
0·0	24·79	30·55	0·6832	0·0538	16·67	26·80	1·109	0·0588	10·59	21·79	1·812	0·0645	8·23	16·61	2·395	0·0692	6·63	11·18	3·011	0·0742
15·0	37·38	43·84	0·4232	0·0596	25·56	38·03	0·7009	0·0640	17·02	30·54	1·117	0·0689	12·53	23·19	1·520	0·0730	10·17	15·86	1·927	0·0778
30·0	55·38	60·82	0·2371	0·0723	38·91	51·92	0·4215	0·0725	25·62	41·26	0·7184	0·0752	18·64	31·27	1·015	0·0787	14·94	21·74	1·297	0·0822
35·3	64·08	67·48	0·1707	0·0871																
40·0					51·24	62·43	0·2848	0·0862	32·88	49·40	0·5286	0·0820	24·06	37·55	0·7654	0·0834	19·09	26·38	1·001	0·0862
44·4					58·41	66·21	0·2201	0·1059												
50·0									42·23	58·98	0·3795	0·0820	30·46	44·39	0·5816	0·0895	24·01	31·72	0·7708	0·0908
60·0									57·18	62·53	0·2206	0·1525	38·73	51·59	0·4247	0·0990	30·06	37·61	0·5775	0·0970
70·0													49·66	57·93	0·2799	0·1279	37·13	44·11	0·4286	0·1080
80·0																	46·16	50·76	0·2872	0·1332
82·0																	48·55	51·53	0·2481	0·1523

3.2.1.7. Methane—Propylene   The total vapour pressure of solutions containing 2·6; 8·2; 13·5; 22·2; 34·5; 42·4; 50·0; 62·3; 72·5; 85·6; 95·1 and 96·2 mole per cent methane in the liquid phase were measured[39] at various temperatures. The results are shown in Fig. 3.8. From these results it is clear that at low concentration of propylene the mixtures are nearly ideal, whilst at higher concentrations of propylene there is a positive deviation from ideality.

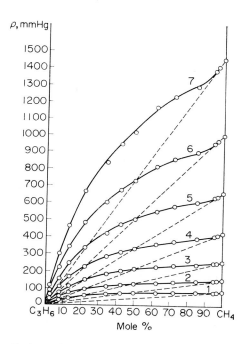

FIG. 3.8   *Equilibrium Isotherms for Propylene—Methane Solutions; temperature (°K): (1) 90; (2) 95; (3) 100; (4) 105; (5) 110; (6) 115; (7) 120; the dashed lines represent the variation of the vapour pressure with concentration in the liquid phase for an ideal solution*

3.2.1.8. Ethane—Propylene   The liquid-vapour equilibrium of this binary system had been determined[40] at a number of temperatures between −30 and 90°C over a wide pressure range. Dew points, bubble points and the volumes of the pure hydrocarbons and of nine mixtures have been determined. The p-t relations are shown in Fig. 3.9.

Phase compositions of vapour-liquid equilibria at four temperatures and various pressures together with the molar volumes of these phases[41] are collected in Table III.24. Critical points of mixtures are given in Table III.25. Comparison of these critical points with Fig. 3.9 shows that the critical points determined by McKay *et al.* are somewhat lower in pressure than those determined by Lu *et al.*, possibly due to impurities.

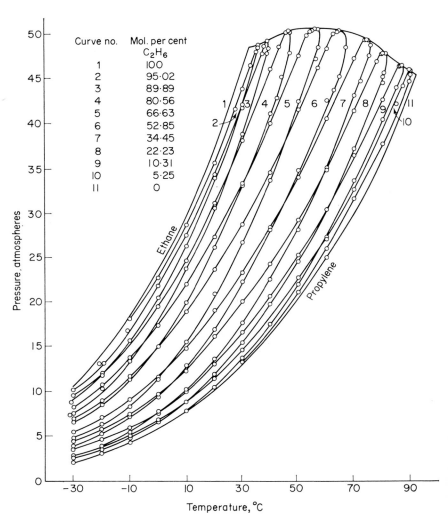

Curve no.	Mol. per cent C_2H_6
1	100
2	95·02
3	89·89
4	80·56
5	66·63
6	52·85
7	34·45
8	22·23
9	10·31
10	5·25
11	0

FIG. 3.9   *Dew Points and Bubble Points of the Pure Hydrocarbons, and of Nine Mixtures.*
p-t Relations Shown with Critical Locus

TABLE III.24

Phase Composition and Molar Volumes of Saturated Liquid and Gas Phases in the System Ethane-Propylene

Temperature °C	p atm	Mole fraction propylene		Volume cm³/mole	
		Liquid	Gas	Liquid	Gas
−12·2	3·95	1·000	1·000	74·79	4876
	6·80	0·772	0·487	73·29	2709
	10·21	0·506	0·239	71·61	1719
	13·61	0·255	0·106	70·11	1231
	17·01	0·023	0·009	68·80	943·3
	17·3	0·000	0·000	68·67	921·4
4·4	6·56	1·000	1·000	77·72	2984
	6·80	0·986	0·955	77·66	2872
	10·21	0·791	0·548	76·97	1832
	13·61	0·610	0·354	76·35	1319
	17·01	0·432	0·225	75·85	1013
	20·41	0·261	0·127	75·41	807·8
	23·82	0·104	0·049	75·04	659·2
	26·20	0·000	0·000	74·91	575·0
37·8	15·47	1·000	1·000	86·96	1240
	17·01	0·952	0·882	87·09	1108
	20·41	0·843	0·683	87·46	890·2
	23·82	0·740	0·553	87·96	735·4
	27·22	0·639	0·457	88·52	619·3
	30·62	0·539	0·374	89·33	526·3
	34·02	0·446	0·303	90·33	448·2
	37·43	0·357	0·241	91·83	381·4
	40·83	0·273	0·187	94·39	322·1
	44·23	0·191	0·137	99·26	264·7
	47·63	0·106	0·088	115·49	199·1
	49·13	0·070	0·070	142·96	142·96
71·1	30·98	1·000	1·000	104·44	528·8
	34·02	0·938	0·889	104·88	460·1
	37·43	0·870	0·795	106·50	393·9
	40·83	0·801	0·723	109·87	334·0
	44·23	0·731	0·670	116·30	273·4
	47·63	0·662	0·643	143·27	186·2
	47·97	0·650	0·650	160·63	160·63

TABLE III.25

Critical Data in the System Ethane-Propylene

$t°C$	p atm	Critical mole fraction Propylene	Critical volume cm³/mole
37·8	49·13	0·070	143·0
43·3	49·27	0·153	144·2
48·9	49·27	0·247	146·7
54·4	49·13	0·346	149·2
60·0	48·86	0·449	152·3
65·6	48·45	0·549	156·7
71·1	47·97	0·650	160·4
76·7	47·36	0·747	165·4
82·2	46·68	0·844	171·1
87·8	45·86	0·938	177·9

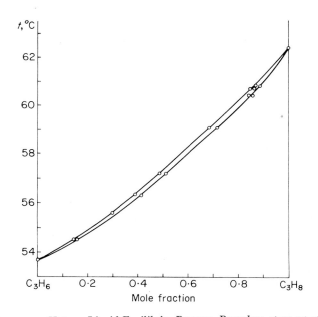

FIG. 3.10   *Vapour-Liquid Equilibria, Propane–Propylene at 21.94 atm*

3.2.1.9. Propane—Propylene   Vapour-liquid equilibria were measured at −12·2, 4·4, 37·8, 71·1, 87·8°C[42] and at −3·6, 28·2 and 57·2°C.[43]

The results are given in Table III.26. At 21·94 atm the temperature-composition relation was measured.[43] The results are shown in Fig. 3.10.

p-x diagrams for low temperatures[44] are shown in Fig. 3.11.

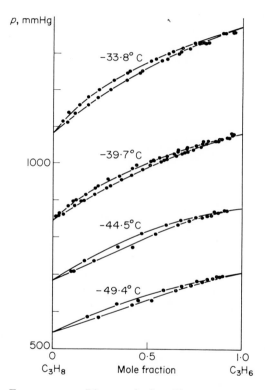

FIG. 3.11   p-x *Diagrams for Low Temperatures*

TABLE III.26

Vapour-Liquid Compositions at various Temperatures and Pressures in the System Propane-Propylene

| Temp. °C | Pressure atm | Mole fraction propane | | Temp. °C | Pressure atm | Mole fraction propane | |
		Liquid	Gas			Liquid	Gas
−12·2	3·95	0·0000	0·0000	37·8	15·47	0·0000	0·0000
	3·84	0·1532	0·1358		15·04	0·2096	0·1892
	3·83	0·2076	0·1950		14·51	0·4440	0·4077
	3·73	0·3520	0·3055		14·02	0·6230	0·5872
	3·68	0·4443	0·3952		13·43	0·8206	0·7949
	3·55	0·5864	0·5347		12·86	1·0000	1·0000
	3·54	0·6232	0·5656	57·2	23·58	0·000	0·000
	3·40	0·7947	0·7497		23·29	0·126	0·120
	3·37	0·8182	0·7832		23·15	0·158	0·150
	3·20	1·0000	1·0000		22·89	0·244	0·240
−3·6	5·20	0·000	0·000		21·94	0·522	0·497
	5·19	0·037	0·034		20·48	0·862	0·849
	5·18	0·034	0·031		19·74	1·000	1·000
	5·12	0·143	0·135	71·1	30·98	0·0000	0·0000
	4·76	0·534	0·485		30·17	0·2099	0·1972
	4·76	0·528	0·484		29·14	0·4422	0·4208
	4·33	0·902	0·878		28·20	0·6251	0·6024
	4·19	1·000	1·000		27·13	0·8231	0·8050
4·4	6·57	0·0000	0·0000		26·12	1·0000	1·0000
	6·45	0·1546	0·1315	87·8	42·27	0·0000	0·0000
	6·24	0·3446	0·3128		41·07	0·2105	0·2041
	6·14	0·4396	0·3989		39·71	0·4312	0·4420
	5·97	0·5868	0·5386		39·65	0·4480	0·4372
	5·93	0·6214	0·5734		38·48	0·6249	0·6124
	5·67	0·7942	0·7536		37·01	0·8228	0·8127
	5·64	0·8217	0·7812		35·72	1·0000	1·0000
	5·38	1·0000	1·0000				
28·2	12·30	0·000	0·000				
	12·19	0·100	0·093				
	12·01	0·201	0·182				
	11·45	0·486	0·454				
	10·49	0·893	0·876				
	10·20	1·000	1·000				

3.2.1.10. Isobutane—Propylene Vapour-liquid equilibria were measured at five pressures.[45] The results are collected in Table III.27 and shown in Fig. 3.12.

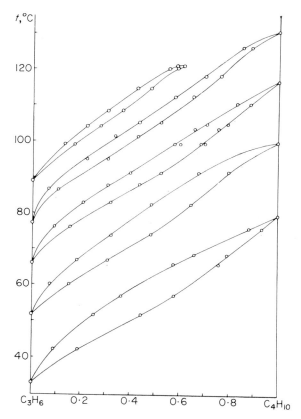

FIG. 3.12 *Vapour-Liquid Equilibria, Isobutane–Propylene*

3.2.1.11. Ethylene—Propylene From phase equilibria of a 1 : 1 ethylene-propylene mixture in the critical region,[46] the densities and phase compositions of the two phases are derived, as shown in Figs. 3.13 and 3.14. The critical density of the 1 : 1 mixture is 0·230 g/cm³, and the critical temperature 58·30±0·05°.

Bubble and dew points and molar volumes of the phases have been measured[35] for the mixture 79·42 per cent C_2H_4, 20·58 per cent C_3H_6. They are given in Table III. 28.

Critical data[35] for various compositions are given in Table III.29.

TABLE III.27

Vapour-Liquid Compositions at various Temperatures and Pressures in the System Isobutane-Propylene

		Mole fraction isobutane	
p atm	$t°C$	Liquid	Gas
13·61	42·2	0·190	0·084
	51·1	0·445	0·253
	56·7	0·581	0·364
	65·6	0·762	0·577
	68·3	0·794	0·663
	75·6	0·935	0·882
20·41	60·0	0·152	0·073
	66·7	0·309	0·184
	73·9	0·485	0·323
	82·2	0·647	0·485
	91·1	0·800	0·679
27·22	76·1	0·154	0·091
	82·8	0·319	0·206
	87·8	0·438	0·309
	91·1	0·525	0·400
	99·4	0·687	0·581
	99·4	0·703	0·602
	103·3	0·758	0·662
	104·4	0·796	0·711
	110·0	0·887	0·832
34·02	86·7	0·101	0·065
	95·0	0·306	0·221
	101·1	0·430	0·335
	105·0	0·524	0·434
	111·7	0·652	0·581
	117·8	0·766	0·702
	124·4	0·889	0·854
40·83	98·9	0·172	0·132
	103·9	0·277	0·223
	108·3	0·367	0·308
	114·4	0·485	0·428
	120·0	0·590	0·554
	120·6	0·615	0·588

TABLE III.28

Bubble and Dew Points and Molar Volumes for Propylene—Ethylene

Temperature, °C	−30·0	−15·0	0·0	15·0
Dew point pressure, atm	7·72	12·82	20·27	30·84
Bubble point pressure, atm	15·02	21·98	30·82	41·78
Molar volume of vapour, l/mole	2·300	1·385	0·8517	0·5180
Molar volume of liquid, l/mole	0·0644	0·0688	0·0749	

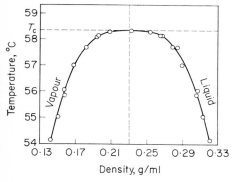

FIG. 3.13   Density of the Co-existing Vapour and Liquid Phases of a 1:1 mixture of Propylene and Ethylene in the Critical Temperature Region

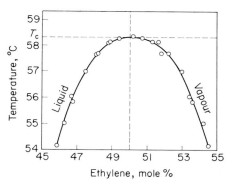

FIG. 3.14   t-x Curve of the Co-existing Vapour and Liquid Phases of a 1:1 mixture of Propylene and Ethylene in the Critical Region

TABLE III.29

Critical Data for Propylene—Ethylene Mixtures

Ethylene, mol-%	91·4	75·9	62·5	49·4	38·4	25·6	11·7
Critical temperature, °C	19·8	37·1	49·6	59·6	68·0	76·5	84·8
Critical pressure, atm	52·0	53·7	54·0	52·9	51·6	49·7	47·7
Critical density, mole/l	6·20	5·38	5·13	4·86	4·68	5·15	5·35

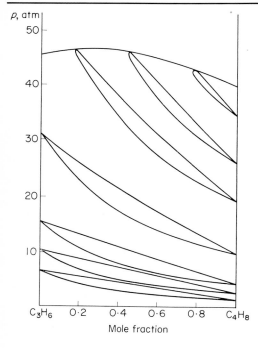

FIG. 3.15   Vapour-Liquid Equilibria, Butylene–Propylene

3.2.1.12. Butylene—Propylene Phase compositions of vapour-liquid equilibria in this system at seven temperatures and various pressures, together with the molar volumes of these phases and some critical data[47] are collected in Table III.30, the pressure-composition diagrams for these temperatures are shown in Fig. 3.15.

TABLE III.30

Mole Fractions and Molar Volumes of Co-existing Liquid and Vapour Phases at various Temperatures and Pressures in the System Butylene-Propylene

Temperature °C	p atm	Mole fraction n-butylene		Volume cm³/mole	
		Liquid	Gas	Liquid	Gas
4·4	1·490	1·000	1·000	91·40	14521
	2·041	0·887	0·674	88·84	10401
	2·722	0·743	0·436	87·84	7835
	3·402	0·600	0·291	85·78	6330
	4·083	0·467	0·199	83·78	5349
	4·763	0·336	0·126	82·16	4511
	5·444	0·207	0·071	80·28	3823
	6·124	0·086	0·027	78·85	3281
	6·634	0·000	0·000	77·97	2950
21·1	2·599	1·000	1·000	95·27	8565
	2·722	0·983	0·943	94·89	8128
	3·402	0·888	0·698	94·08	6411
	4·083	0·793	0·531	91·77	5296
	5·444	0·606	0·319	89·15	3891
	6·805	0·429	0·188	87·09	3054
	8·507	0·216	0·080	84·34	2372
	10·21	0·017	0·006	82·09	1925
	10·34	0·000	0·000	81·96	1893
37·8	4·253	1·000	1·000	98·82	5339
	5·444	0·880	0·702	96·83	4091
	6·805	0·746	0·491	95·14	3225
	8·506	0·587	0·329	92·96	2542
	10·21	0·440	0·213	91·71	2052
	11·91	0·284	0·129	89·71	1715
	13·61	0·144	0·062	88·15	1463
	15·47	0·000	0·000	87·14	1239
71·1	9·724	1·000	1·000	108·3	2359
	10·21	0·970	0·930	107·7	2220
	13·61	0·781	0·608	105·2	1470
	17·01	0·604	0·410	103·5	1227
	20·41	0·440	0·272	102·4	812·8
	23·82	0·284	0·166	102·4	667·4
	27·22	0·142	0·082	103·0	632·4
	31·07	0·000	0·000	104·6	525·9
104·4	19·23	1·000	1·000	121·6	1128
	20·41	0·954	0·917	121·2	1034
	23·82	0·834	0·735	121·0	839·7
	27·22	0·720	0·600	121·7	686·7
	30·62	0·610	0·490	122·2	566·8
	34·02	0·508	0·399	123·2	485·1
	37·42	0·411	0·324	126·1	422·0

Temperature °C	p atm	Mole fraction n-butylene		Volume cm³/mole	
		Liquid	Gas	Liquid	Gas
	40·83	0·320	0·255	136·3	370·5
	42·53	0·276	0·224	143·5	324·6
	44·23	0·232	0·196	158·5	296·7
	45·93	0·188	0·172	181·4	261·1
	46·27	0·180	0·172	183·8	222·2
	46·37	0·176	0·176	194·1	194·1
121·1	25·99	1·000	1·000	133·1	777·9
	27·22	0·961	0·938	123·3	716·0
	30·62	0·860	0·796	134·6	591·1
	34·02	0·766	0·686	135·8	500·2
	37·42	0·674	0·593	142·8	421·6
	40·83	0·585	0·519	141·3	391·3
	42·19	0·550	0·491	146·0	376·0
	43·55	0·515	0·467	161·6	361·6
	44·91	0·479	0·447	182·8	338·0
	45·93	0·453	0·446	205·6	226·3
	46·01	0·449	0·449	210·8	210·8
137·8	34·43	1·000	1·000	157·5	464·8
	35·72	0·966	0·952	152·8	439·5
	37·42	0·920	0·893	147·6	395·4
	39·47	0·871	0·834	152·5	355·8
	40·83	0·835	0·797	178·0	322·5
	42·19	0·801	0·771	208·0	281·9
	42·53	0·793	0·770	217·6	270·7
	42·87	0·783	0·783	228·9	228·9

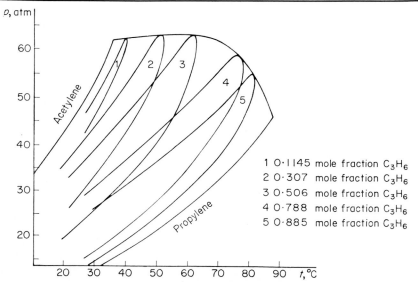

FIG. 3.16   *Bubble Points, Dew Points and the Border Curves of Five Mixtures of Propylene and Acetylene together with Boiling Lines of Pure Acetylene and Propylene*

3.2.1.13. Acetylene—Propylene Bubble points and dew points have been determined for five mixtures of propylene and acetylene.[48] The measurements are collected in Table III.31 whilst the border curves together with the boiling lines of pure acetylene and propylene are given in Fig. 3.16.

TABLE III.31

Dew and Bubble Points in the System Acetylene-Propylene

Mole fraction propylene	Temperature °C	Bubble point atm	Dew point atm
0·1145	28·0	47·5	43·7
	37·7	58·5	55·8
	39·9	61·2	58·3
	43·2	single phase	
0·307	24·2	38·8	28·2
	38·0	50·0	39·5
	48·7	59·9	50·5
	50·2	61·4	53·3
	52·7	62·6	57·3
	54·1	single phase	
0·506	22·4	33·3	21·1
	33·3	40·5	27·6
	43·7	48·3	33·9
	55·2	56·6	43·3
	58·8	59·5	47·6
	61·7	61·9	53·3
	64·4	62·3	58·2
0·788	27·3	29·9	15·9
	37·3	33·7	19·7
	48·8	40·3	24·8
	59·1	47·3	32·3
	69·3	52·3	40·5
	76·8	57·6	47·6
	79·2	58·2	50·0
	80·1	58·3	52·7
	80·7	fogg	54·2
0·885	28·7	26·0	14·3
	43·8	31·6	20·1
	53·6	36·9	25·5
	68·1	45·0	33·9
	74·3	48·4	38·4
	79·4	51·6	44·1
	81·1	52·5	45·5
	83·2	54·4	47·8

3.2.2. Ternary Systems

3.2.2.1. Hydrogen—Propane—Propylene Determinations of the coexisting liquid and vapour phases at −73·3°C in this ternary system[49] are given in Table III.32.

TABLE III.32

Compositions of some Co-existing Liquid and Vapour Phases in the System Hydrogen-Ethylene-Propylene

Pressure, atm	68		544	
Phase	Liquid	Vapour	Liquid	Vapour
Hydrogen	2·8	99·26	18·1	99·07
Propane	42·2	0·30	36·6	0·41
Propylene	55·0	0·44	45·3	0·52

3.2.2.2. Hydrogen—Ethylene—Propylene Determinations of the coexisting liquid and vapour phase at $-73\cdot3\,°C$ in this ternary system[49] are given in Table III.33.

TABLE III.33

Compositions of some Co-existing Liquid and Vapour Phases in the System Hydrogen-Ethylene-Propylene

Pressure, atm	68		544	
Phase	Liquid	Vapour	Liquid	Vapour
Hydrogen	3·1	93·41	21·7	95·01
Ethylene	55·1	6·22	37·9	4·30
Propylene	41·8	0·37	40·4	0·65

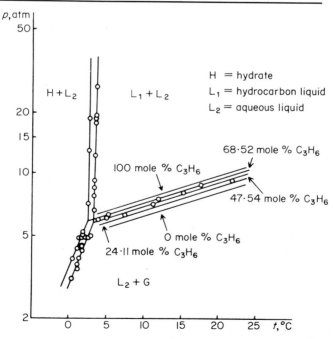

FIG. 3.17   *Condition for Hydrate Forming in the System Water–Propane–Propylene*

9

3.2.2.3. Water—Propane—Propylene The conditions for formation of mixed hydrates in this system have been studied.[50] The pressure-temperature relation is shown in Fig. 3.17 for some different propane-propylene ratios.

3.2.2.4. Water—Ammonia—Propylene Liquid-liquid equilibria in this system[32] were measured at 20°C. The results are given in Table III.34. At 25°C[51] the system can be represented by Fig. 3.18.

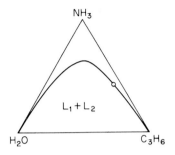

FIG. 3.18 *The Ternary System, Water–Ammonia–Propylene at 25°C.*
o = plait point

TABLE III.34

Liquid Phase Compositions in the System Water-Ammonia-Propylene

L_1 Weight fraction		L_2 Weight fraction	
Ammonia	Propylene	Ammonia	Propylene
0·000 00	0·999 69	0·000 000	0·002 286
0·011	0·989	0·263	0·008
0·012	0·987	0·279	0·008
0·023	0·976	0·403	0·012
0·045	0·953	0·547	0·020
0·091	0·906	0·674	0·034
0·110	0·886	0·701	0·039
0·128	0·868	0·717	0·044
0·224	0·768	0·764	0·097
0·323	0·664	0·751	0·158
0·410	0·572	0·709	0·228
0·543	0·427	0·543	0·427

3.2.2.5. Ammonia—Propane—Propylene In this system propane and ammonia are not completely miscible in the liquid state whereas ammonia and propylene as well as propylene and propane are completely miscible in the temperature area examined.

Equilibria of co-existing hydrocarbon-rich (L_1) and ammonia-rich (L_2) liquids[32] at 0°C and 20°C are collected in Table III.35. Fig. 3.19 shows the equilibria at 0°C.

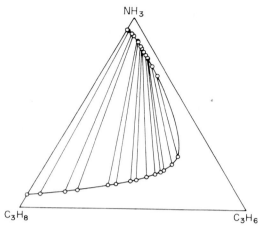

FIG. 3.19   *The Ternary System, Ammonia–Propane–Propylene at 0°C*

3.2.2.6. Benzene Sulphonic Acid—Isopropyl Benzene Sulphonate—Propylene At 25°C[52] solid benzene sulphonic acid can co-exist with an isopropyl benzene sulphonate-rich liquid, a propylene-rich liquid and with both liquids. The unsaturated liquids have a critical point. See Fig. 3.20.

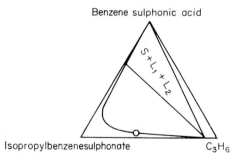

FIG. 3.20   *The Ternary System, Benzene Sulphonic Acid–Isopropyl Benzene Sulphonate–Propylene at 25° C.* *o = plait point*

3.2.2.7. Hydrogen Cyanide—Propane—Propylene The behaviour of this system at 25°C[53] is shown in Fig. 3.21.

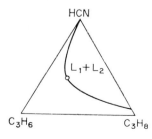

FIG. 3.21   *The Ternary System, Hydrogen Cyanide–Propane–Propylene at 25°C.* *o = plait point*

TABLE III.35

Mutal Solubility of Hydrocarbon-rich Liquid and Ammonia-rich Liquid in the System Ammonia-Propane-Propylene

t°C	L_1 Weight fraction		L_2 Weight fraction	
	Ammonia	Propylene	Ammonia	Propylene
0	0·067	0·000	0·935	0·000
	0·070	0·054	0·930	0·012
	0·076	0·078	0·926	0·016
	0·087	0·153	0·915	0·031
	0·093	0·194	0·911	0·038
	0·094	0·205	0·908	0·041
	0·122	0·329	0·882	0·071
	0·130	0·359	0·874	0·078
	0·143	0·417	0·853	0·100
	0·152	0·438	0·843	0·109
	0·169	0·475	0·825	0·128
	0·183	0·504	0·807	0·146
	0·197	0·521	0·793	0·161
	0·204	0·532	0·782	0·171
	0·239	0·550	0·749	0·204
	0·280	0·557	0·699	0·252
	0·560	0·375	0·560	0·375
20	0·162	0·000	0·864	0·000
	0·177	0·058	0·835	0·020
	0·179	0·062	0·834	0·022
	0·222	0·136	0·808	0·050
	0·274	0·188	0·764	0·081
	0·331	0·213	0·694	0·118
	0·559	0·177	0·559	0·177

3.2.3. Multicomponent Systems

3.2.3.1. Hydrogen—Methane—Ethane—Propane—Ethylene—Propylene Phase compositions have been determined in this six-component[54] system at two temperatures and two pressures, and are given in Table III.36.

TABLE III.36

Phase Compositions, mol-per cent, in a Six-Component System

| Temp., °C | −17·8 | | −73·3 | | −73·3 | | −17·8 | | −73·3 | |
| Pressure, atm | 34 | | 34 | | 34 | | 68 | | 68 | |
Phase	Liquid	Vapour	Liquid	Vapour	Liquid	Vapour	Liquid	Vapour	Liquid	Vapour
H_2	1·04	38·28	1·34	65·79	1·48	68·60	2·36	46·93	1·97	77·26
CH_4	7·67	31·37	14·84	30·45	12·77	25·45	16·45	35·18	17·82	19·66
C_2H_4	9·42	8·56	9·06	2·00	18·23	4·06	17·32	9·64	17·97	2·09
C_2H_6	5·07	3·44	4·66	0·65	4·21	0·60	4·02	1·73	3·74	0·38
C_3H_6	8·39	2·63	7·77	0·26	7·07	0·24	6·67	0·86	6·00	0·10
C_3H_8	68·41	15·72	62·33	0·85	56·24	1·05	53·18	5·66	52·50	0·51

3.2.3.2. Water—Ammonia—Propane—Propylene Liquid-liquid equilibria were measured[32] at 20 °C at two different water-ammonia ratios. The results are given in Table III.37.

TABLE III.37

Liquid-Phase Composition in the System Water-Ammonia-Propane-Propylene

	L_1 Weight fraction				L_2 Weight fraction			
	H_2O	NH_3	C_3H_6	C_3H_8	H_2O	NH_3	C_3H_6	C_3H_8
Propane—Propylene	0·0008	0·102	0·000	0·897	0·051	0·880	0·000	0·069
95% NH_3—5% H_2O	0·0008	0·118	0·080	0·797	0·051	0·866	0·018	0·065
	0·0011	0·145	0·184	0·670	0·053	0·852	0·038	0·057
	0·0009	0·157	0·212	0·630	0·051	0·846	0·046	0·057
	0·0012	0·201	0·359	0·439	0·055	0·821	0·080	0·044
	0·0017	0·223	0·435	0·340	0·058	0·803	0·101	0·038
	0·0023	0·242	0·479	0·277	0·065	0·803	0·102	0·030
	0·0030	0·270	0·470	0·257	0·052	0·793	0·121	0·034
	0·0050	0·300	0·499	0·196	0·055	0·781	0·137	0·027
	0·0069	0·346	0·499	0·148	0·050	0·751	0·171	0·028
	0·0046	0·288	0·553	0·154	0·066	0·770	0·144	0·020
	0·0109	0·379	0·536	0·074	0·057	0·745	0·184	0·014
	0·0114	0·392	0·531	0·066	0·055	0·740	0·191	0·014
	0·0143	0·399	0·546	0·041	0·059	0·726	0·206	0·009
	0·0192	0·434	0·547	0·000	0·058	0·693	0·249	0·000
Propane—Propylene	0·0014	0·094	0·000	0·905	0·104	0·867	0·000	0·029
90% NH_3—10% H_2O	0·0016	0·102	0·074	0·823	0·109	0·858	0·006	0·027
	0·0017	0·121	0·186	0·691	0·106	0·850	0·018	0·026
	0·0019	0·136	0·282	0·580	0·106	0·842	0·029	0·023
	0·0023	0·157	0·378	0·462	0·105	0·834	0·041	0·020
	0·0029	0·183	0·485	0·329	0·106	0·819	0·059	0·016
	0·0030	0·187	0·500	0·310	0·106	0·817	0·061	0·016
	0·0045	0·222	0·603	0·171	0·104	0·802	0·084	0·010
	0·0065	0·244	0·673	0·076	0·105	0·788	0·102	0·005
	0·0094	0·262	0·729	0·000	0·115	0·765	0·120	0·000

3.2.3.3. Water—Methane—Ethylene—Propylene The conditions for the formation of a hydrate in the quaternary system of water, methane, ethylene and propylene, have been studied.[55] Fig. 3.22 gives, for various pressures and two gas compositions the temperatures at which the hydrate decomposes into a water-rich liquid and a gas.

3.2.3.4. Water—Propane—Isobutane—Isobutylene—Propylene The conditions for hydrate forming in this system were determined.[56] The results are collected in Table III.38.

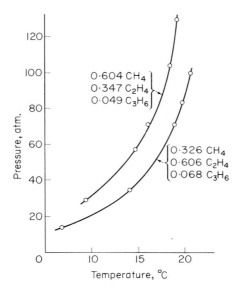

0·604 CH$_4$
0·347 C$_2$H$_4$
0·049 C$_3$H$_6$

0·326 CH$_4$
0·606 C$_2$H$_4$
0·068 C$_3$H$_6$

FIG. 3.22   *Hydrate Forming Conditions in the System Water—Methane—Ethylene —Propylene*

TABLE III.38

**Conditions for Hydrate Formation in the System
Water-Propane-Isobutane-Isobutylene-Propylene**

Temperature °C	0·65	0·65	0·65
Pressure, atm	1·44	1·77	1·9
Composition mole %			
Gas above hydrate			
C$_3$H$_8$	25·86	45·6	31·6
C$_3$H$_6$	8·62	22·62	45·1
iso C$_4$H$_{10}$		3·73	5·3
iso C$_4$H$_8$	65·52	28·05	18
Hydrate			
C$_3$H$_8$	13·69	35·31	30
C$_3$H$_6$	4·31	14·15	30·5
iso C$_4$H$_{10}$		5·34	8·5
iso C$_4$H$_8$	82	45·2	31

3.2.3.5. Propane—Propadiene—Propyne—Propylene   At 10 and 32·2 °C some liquid-vapour equilibria were measured.[57] The results are collected in Table III.39.

TABLE III.39

Composition Data of Liquid-Vapour Equilibria in the System Propane-Prodadiene-Propyne-Propylene

Temp. °C	Pressure atm	Liquid composition mole %			Relative volatility component/propane		
		Propane	Propylene	Propadiene	Propylene	Propadiene	Propyne
10·0	7·49	4·0	93·5	1·3	1·105	0·839	0·819
	7·49	4·1	94·1	0·9	1·100	0·828	0·822
	7·08	49·2	47·0	1·9	1·159	0·929	1·005
	6·80	47·7	48·8	1·8	1·156	0·905	0·984
	6·46	82·2	12·5	2·8	1·162	1·000	1·137
		83·4	12·0	2·4	1·189	1·004	1·167
32·2	13·13	9·3	86·5	2·0	1·083	0·896	0·912
	13·13	9·5	86·2	2·0	1·097	0·884	0·900
	12·59	46·5	53·5		1·136		
	12·45	51·6	47·0	0·7	1·127	0·948	1·025
	12·45	48·2	49·4	1·2	1·129	0·966	1·051
	12·25	54·0	40·0	2·8	1·130	0·945	1·015
	11·50	91·2	5·8	1·3	1·171	1·025	1·181
	11·43	92·4	4·7	1·4	1·152	1·028	1·148
	11·57	87·1	8·4	2·1	1·179	1·022	1·161
	9·59	2·6	2·7	93·5	0·969	0·694	0·605
	9·59	1·2	3·7	93·3	0·963	0·683	0·654
	9·66	1·0	3·6	94·1	0·968	0·683	0·616

3.2.4. Solubility

In this Section we give the solubility of gaseous propylene in various liquids when only the composition of the liquid is known. It will be clear, that some solubilities also can be found in the sections on phase equilibria where the compositions of coexisting phases are given.

Solubilities are often given as solubility coefficients. We shall use the following symbols:

β = Bunsen absorption coefficient, i.e. the volume of the gas (reduced to 0° and 760 mm) absorbed by one volume of the liquid when the pressure of the gas itself excluding any liquid tension amounts to 760 mm.

λ = The volume of the gas reduced to 0° and 760 mm in cm³ which is absorbed by 1 g of the liquid per atm.

3.2.4.1. In water The oldest measurements[58] give for the temperature dependance of the solubility λ for pressures below 1 atm:

$$\lambda = 0·446506 - 0·022075t + 0·0005388t^2 \text{ (temp. °C)}$$

At higher pressures[59] the solubilities at six temperatures are given in Table III.40.

TABLE III.40

Solubility of Propylene in Water in Mole Fraction $\times 10^4$

Pressure, atm	21.1	37.8	54.4	71.1	87.8	104.4
1.0	1.36	0.76	0.42	0.27	0.13	
1.4	1.78	0.92	0.56	0.40	0.25	0.08
2.7	3.12	1.85	1.16	0.88	0.68	0.49
4.1	4.52	2.76	1.76	1.35	1.12	0.90
5.4	5.86	3.55	2.38	1.82	1.53	1.30
6.8	7.14	4.33	2.96	2.30	1.96	1.72
8.2	8.24	5.10	3.59	2.77	2.41	2.12
9.5	9.32	5.87	4.16	3.25	2.83	2.52
10.9		6.64	4.76	3.71	3.25	2.94
12.2		7.40	5.32	4.20	3.65	3.35
13.6		8.17	5.92	4.64	4.06	3.75
15.0		8.93	6.49	5.07	4.45	4.16
16.3			7.05	5.48	4.84	4.57
17.7			7.60	5.87	5.22	4.98
19.0			8.14	6.22	5.61	5.39
20.4			8.68	6.55	5.97	5.80
21.8			9.16	6.84	6.31	6.22
23.1				7.11	6.57	6.60
24.5				7.35	6.85	6.96
25.9				7.58	7.11	7.30
27.2				7.80	7.34	7.63
28.6				8.02	7.58	7.95
29.9				8.22	7.82	8.27
31.3				8.41	8.05	8.57
32.7					8.28	8.84
34.0					8.49	9.11

3.2.4.2. In aqueous solutions The Bunsen absorption coefficient β as function of temperature for two Cu^+ salt solutions[60][61] is given in Fig. 3.23.

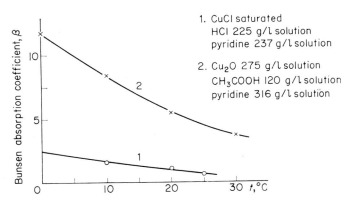

1. CuCl saturated
   HCl 225 g/l solution
   pyridine 237 g/l solution

2. Cu_2O 275 g/l solution
   CH_3COOH 120 g/l solution
   pyridine 316 g/l solution

FIG. 3.23 *Bunsen Absorption Coefficient of C_3H_6 in Aqueous Cu^+ Salt Solutions*
$$P_{C_3H_6} \approx 1 \ atm$$

The effect of pressure on the solubility of propylene in CuNO$_3$-Ethanolamine solution is shown in Fig. 3.24.[62] The solubility of propylene in Cu$^+$ solutions with HCl and Ethanolamine is also reported at 20°C and pressures up to 10 atm.[63] Aqueous AgNO$_3$ solutions dissolve also propylene.[64] A 71 per cent AgNO$_3$ solution dissolves at 25° 8 volumes of C$_3$H$_6$ at atmospheric pressure and 185 volumes at 11·5 atm. In mixtures with propane the solubility is given in Table III.41.

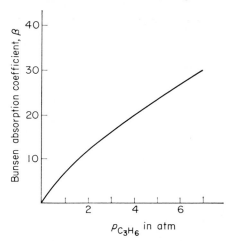

FIG. 3.24   *Solubility of Propylene in an Aqueous CuNO$_3$-ethanolamine Solution at Room Temperature, as a Function of the Propylene-pressure*

TABLE III.41

Extraction of Propylene from Mixtures with Propane by an Aqueous AgNO$_3$-Solution

Mole fraction C$_3$H$_6$ in the gas	Soly in 71% AgNO$_3$ wt %
0·945	18·7
0·78	14·1
0·405	10·4
0·33	4·7
0·22	4·4

Potassium oleate in aqueous solutions raises the solubility of propylene.[65] Fig. 3.25 shows the effect of the addition of soap to water. The final equilibrium pressure was 400 mm.

The effect of typical detergents on the solubility of propylene in water[66] is given in Table III.42.

TABLE III.42

Effect of Detergents on the Solubility of Propylene in Water at 25°C for the Pressure Range 500–700 mm Hg

Solution*	Solubility in grams $\times$ 10^{-5} per g of water per mm pressure	Mean	Effect
Water only	3·5, 3·4, 3·4, 3·3	3·4	standard
15% potassium oleate		18·7	great
12% potassium oleate	14·3, 15·6, 14·8	14·9	great
9% potassium oleate		12·5	great
1% potassium oleate	3·4, 3·3, 3·5, 3·9, 3·9	3·7	slight
15% pure Tergitol 4		13·6	great
1·5% Aerosol OT		7·5	fair
1% Aerosol OT	3·9, 3·0, 3·3, 3·2, 3·4	3·4	none
15% Aerosol MA	11·2, 10·8	11·0	good
15% mixture; 1 pt. OT, 4 pts. MA	9·8, 10·4	10·1	good
15% Aerosol AY	7·7, 8·0, 10·9	8·9	good
1% Aerosol AY	2·8, 3·1	3·0	none?
15% Aerosol IB	5·5, 5·8	5·7	fair
10% Aerosol IB	4·9, 4·3	4·6	fair
15% Aerosol OS	7·0, 7·1	7·1	fair
15% sodium deoxycholate		6·1	fair
15% sodium dehydrocholate		0·8	salts out!
15% Aquasol AR (Turkey Red Oil)		14·4	great
15% Igepon A (thixotropic)	7·6, 7·8	7·7	fair
15% of 95% Triton NE		11·3	good
15% of 90% Alronal	11·5, 13·3	12·4	good
25% Triton K 60		32·6	great
15% potassium novenate		11·2	good
10% Nacconol NR		9·2	good
10% Sapamine KW		5·4	fair
15% diethylcarbitol		3·7	slight
0·4% Calgon	3·6, 3·6, 4·1	3·8	none?
2% Calgon		3·3	none
5% Calgon		3·8	none
12% KOl + 0·2% Calgon		14·6	Calgon none
12% KOl + 3% Calgon		15·2	Calgon none
5% K_2CO_3		3·0	negative
0·5% K_2CO_3		3·5	none?
12% KOl + 5% K_2CO_3		15·9	K_2CO_3 slight
12% KOl + 0·2% K_2CO_3		14·1	K_2CO_3 slight
12% KOl + 2% tetrasodium pyrophosphate		14·5	phosphate none
12% KOl + 5% tetrasodium pyrophosphate	15·0, 16·1	15·6	phosphate none

* For a description of the active agent in materials here designated only by trade names, see: *Ind. Eng. Chem.*, 1939, *31*, 66, 1941, *33*, 16, 740.

The distribution constant h for propylene between the vapour phase and aqueous perchloric acid solutions at 30°C[67] is given in units of kmole/l. atm in Table III.43.

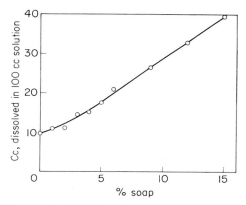

FIG. 3.25   *Effect of Soap on the Solubility of Propylene in Water at 25°C and 400 mm Hg*

TABLE III.43

Distribution Constant h for the Distribution of Propylene Between the Vapour Phase and Aqueous Solutions of $HClO_4$

$HClO_4$ wt %	h, kmole. l^{-1}. atm^{-1}
0·00	4·36 ± 0·04
9·75	4·28 ± 0·04
20·80	4·48 ± 0·04
32·01	4·40 ± 0·06
43·60	4·01 ± 0·04
53·73	3·11 ± 0·04

The solubility of mixtures of propane and propylene in aqueous solutions of acetic acid and acetic acid in mixtures with phenol at 30°C and ≈ 1 atm[68] are collected in Table III.44.

TABLE III.44

Solubility of Propane and Propylene from their Mixtures in Several Mixtures of Water, Acetic Acid and Phenol

Solvent mol %			Gas mol %			
H_2O	CH_3COOH	C_6H_5OH	C_3H_6	C_3H_8	$\lambda_{C_3H_6}$	$\lambda_{C_3H_8}$
	100		15·4	84·6	2·32	0·83
25	75		17·2	82·8	0·44	0·30
40	60		18·3	81·7	0·18	0·14
50	50		18·5	81·5	0·08	0·11
	50	50	16·5	83·5	2·15	0·70
20	40	40	17·5	82·5	0·78	0·35
33·3	33·3	33·3	17·4	82·6	0·42	0·28
50	25	25	17·2	82·8	0·15	0·26

3.2.4.3. In liquid ammonia   The results of solubility measurements over the temperature range -60 to $45°$ and concentrations of propylene up to 0·5 per cent[69] have been expressed in the relation:

$$\log K p v = A - \frac{\Delta H}{2 \cdot 3\, R.T}$$

where $K = y/x$.

y is the molar percentage of propylene in the gas phase
x is the molar percentage of propylene in the solution
p is the total pressure in atm
v is the fugacity coefficient of propylene
ΔH is the heat of solution in kcal/mole and found to be 1·215 and
A is a constant and found to be 2·88.

Since x is proportional to ypv by Henry's law Kpv is independent of the pressure and solely a function of temperature. This presentation of solubility data is somewhat unusual, a high value of Kpv meaning a low solubility and vice versa. An example is therefore worked out below of the solubility of propylene at 2 atm total pressure in liquid ammonia at $-20°$.

$$\log K p v = 2 \cdot 88 - \frac{1 \cdot 215}{2 \cdot 3 \times 1 \cdot 987 \times 10^{-3} \times 253 \cdot 2} = 1 \cdot 83$$
$$= 67 \cdot 6 \text{ atm.}$$

The vapour pressure of ammonia at $-20°$ is 1·88, therefore $y = 6$ per cent. Under these conditions is $v = 1$. The solubility becomes therefore:

$$x = \frac{6 \times 2}{67 \cdot 6} = 0 \cdot 18 \text{ mole per cent.}$$

3.2.4.4. In hydrocarbons   The solubility of propylene in some aliphatic and aromatic hydrocarbons was determined.[70] The results are given in Table III.45.

TABLE III.45
Solubility of Propylene in Several Hydrocarbons

trC	*n-hexane*		*n-heptane*		*n-nonane*		*benzene*		*toluene*		*xylene**	
	p atm	x mol %	p atm	x mol %	p atm	x mol %	p atm	x mol %	p atm	x mol %	p atm	x mol %
20	2·38	0·195	2·44	0·211	2·48	0·222	0·64	0·033	2·51	0·175	2·40	0·167
	4·64	0·420	4·88	0·434	4·78	0·432	4·68	0·356	4·66	0·362	4·84	0·377
	6·87	0·701	7·01	0·683	7·23	0·698	6·94	0·610	6·86	0·598	6·88	0·605
40	2·47	0·116	2·53	0·137	2·55	0·143	2·46	0·086	2·59	0·119	2·46	0·105
	4·88	0·271	5·10	0·288	4·96	0·292	4·90	0·213	5·01	0·234	5·03	0·244
	7·41	0·508	7·40	0·474	7·52	0·478	7·38	0·376	7·30	0·386	7·31	0·397
60	2·53	0·070	2·56	0·093	2·58	0·099	2·50	0·057	2·63	0·074	2·51	0·071
	5·00	0·177	5·21	0·202	5·05	0·205	5·01	0·142	5·01	0·156	5·08	0·170
	7·56	0·367	7·54	0·350	7·61	0·335	7·48	0·254	7·48	0·271	7·49	0·282

**Mixed isomers*

The solubility of propylene in n-hexane and 2,2-dimethylbutane is given at low pressures and $15\,^{\circ}C^{71}$ in Table III.46 as the distribution coefficient K, defined as:

$$K = \frac{\text{moles dissolved gas per volume liquid}}{\text{moles dissolved gas per volume of the gas phase}}$$

TABLE III.46

Distribution of Propylene Between the Gas and Liquid Phase for Solutions in n-hexane and 2,2′-dimethylbutane

p mm Hg	n-Hexane K	2,2′-dimethylbutane K
27·8	18·48	36·28
23·8	20·06	36·00

The solubility of propylene in gas oil for 0–5 atm and in heavy naphtha for 3–7 atm can be given[72] as $Y = 12\cdot1\ X$ for gas oil and $Y = 15\cdot1\ (X+0\cdot7)$ for heavy naphtha. Y = volume of gas at $25\,^{\circ}$ and 1 atm per volume of liquid and X is absolute pressure in atm. The solubility of propylene in some aromatic hydrocarbons was determined[73] at 1 atm and 0–80 °C. The results are given in Table III.47.

TABLE III.47

Solubility of Propylene in Benzene and Some Alkylbenzenes at 1 atm

Solubility in cm³ gas/cm³ solution. x calculated for the pure hydrocarbon

$t\,^{\circ}C$	Benzene	Benzene + 3 mol % nitromethane x	Toluene	Toluene + 10 mol % nitromethane x	Isopropylbenzene	Diisopropylbenzene	Triisopropylbenzene
0		57·20	15·50	26·54	19·87	16·65	20·34
10	20·98	21·10					
20	13·14	9·11	8·89	12·63	10·49	10·12	11·50
30	9·91	8·03	5·90	9·58			
40	6·28	5·10	4·52	7·57	5·81	6·47	9·19
50	4·25	3·54					
60	2·87	2·10	2·72	5·15	3·40	3·43	5·88
70	1·39	1·07	1·30				
80					1·72	2·59	4·84

The solubility of propylene in toluene[74] at propylene pressures ≈ 1 atm are given in Table III.48.

TABLE III.48

Absorption Coefficients for Propylene in Toluene

Temp. °C	β	λ
−60	220	226
−40	120	126
−20	53	57
0	27	30

The solubility of propylene in several hydrocarbon fractions[75] at different pressures and temperatures is given in Table III.49.

TABLE III.49

Solubility in Petroleum Fractions

Pressure mm	Solubility at NPT vol. gas/volume solvent									
	Baku Kerosene (boiling range 140–210°C, 737 mm)					Xylene				
	−21°	−10°	0°	20°	40°	−21°	−10°	0°	20°	40°
760	76·1	33·0	23·0	13·8	8·5	76·0	53·4	34·7	20·0	12·6
700	47·1	30·5	21·0	12·0	7·5	69·9	48·5	31·3	16·4	11·6
600	45·0	26·0	18·0	10·3	6·5	59·4	40·5	27·6	15·8	10·0
500	27·5	21·5	15·0	8·5	5·5	48·0	33·2	23·5	13·2	7·9
400	21·0	17·0	12·0	7·0	4·5	37·9	26·0	13·5	10·8	6·4
300	15·5	13·0	9·0	5·0	3·0	28·4	19·1	14·0	8·0	5·0
200	10·5	8·5	6·0	3·0	2·0	19·1	13·0	9·4	5·0	3·2
100	5·5	4·5	3·0	1·8	1·0	9·7	6·5	5·0	3·4	1·6
50	3·0	2·5	2·0	1·0	0·5	4·6	3·2	2·4	1·4	0·8

Pressure mm	Cracked gasoline 30% unsaturated (boiling range 61·8–200°C at 750 mm)				
	−21°	−10°	0°	20°	40°
760	72·8	47·4	33·5	17·2	10·0
700	67·5	44·0	31·3	16·0	9·0
600	58·0	38·5	26·5	14·0	7·5
500	48·0	32·6	22·5	12·0	6·0
400	38·0	26·0	18·0	10·0	5·0
300	28·0	19·0	14·0	8·0	4·0
200	19·0	13·0	9·0	5·0	2·0
100	10·0	7·0	4·5	3·0	0·9
50	5·0	4·0	2·5	1·5	0·2

Solubilities of propylene in some hydrocarbons have been expressed in terms[69] of the constants ΔH, A and $(Kpv)_{20}°$ which have been defined where the procedure for converting these factors in solubilities has been indicated. These 'Hannaert' constants are given in Table III.50.

TABLE III.50

Solubilities in Some Organic Solvents in Terms of the Hannaert Constants

Solvent	Temperature range	Conc. range mol %	A	ΔH kcal/mole	$(Kpv)_{20}°$ atm
Kerosene b.p. 150–280°	+60 to −40°	5–10	3·94	4·025	8·5
Kerosene b.p. 150–185°	+20 to −40°	5–10	3·92	3·95	9·2
Kerosene b.p. 207–255°	+20 to −30°	5–10	3·76	3·80	8·8
Kerosene b.p. 140–260°	−20 to −40°	10	4·25	4·34	(10)
Kerosene b.p. 200–285°	+30 to −40°	5–10	3·83	3·85	9·2
Petrol b.p. 94–168°	+20 to −30°	1–10	4·08	4·20	8·7
Toluene	+20 to −30°	1–10	3·90	3·80	11·6
Xylene	+20 to −30°	1–10	4·23	4·20	12·15

3.2.4.5. In acetone   The solubility of propylene in acetone has been determined,[70] the results being given in Table III.51.

TABLE III.51

Solubilities of Propylene in Acetone

$t°C$	Acetone	
	p atm	x mol %
20	3·17	0·183
	5·38	0·345
	6·67	0·424
40	3·28	0·090
	5·56	0·179
	7·06	0·246
60	3·36	0·056
	5·64	0·111
	7·16	0·154

3.2.4.6. Selectivity factor for solubility of propane-propylene mixtures in various solvents   A selectivity factor[68] for the solubility of mixtures of propane and propylene is

defined as $\delta = \dfrac{y_1 \cdot x_2}{x_1 \cdot y_2}$ where x_1 = mole fraction C_3H_8 in the liquid

$x_2 =$,,   ,, C_3H_6 in the liquid

$y_1 =$,,   ,, C_3H_8 in the gas

$y_2 =$,,   ,, C_3H_6 in the gas

δ for different solvents is given in Table III.52.

TABLE III.52

Selectivity Factor for the Solubility in Several Organic Agents

Solvent	δ
Carbon tetrachloride	0·83
Acetic acid	1·96
Amyl alcohol	1·49
Furfural	2·11
Phenol	1·95
Benzaldehyde	1·89
Pyridine	1·77
Aniline	2·98
Xylidine	2·00
Morpholine	2·30
Nitrobenzene	1·65

REFERENCES

[1] Canjar, L. N., Goodman, M. and Marchman, H., *Ind. Eng. Chem.*, 1951, **43**, 1186.

[2] Farrington, P. S. and Sage, B. H., *Ind. Eng. Chem.*, 1949, **41**, 1734.

[3] Vaughan, W. E. and Graves, N. R., *Ind. Eng. Chem.*, 1940, **32**, 1252.

[4] Kobe, K. A. and Long, E. G., *Pet. Refiner*, 1949, **28**, 125.

[5] Nevers, N. D. and Martin, J. J., *Aiche Journal*, 1960, **6**, 43.

[6] Hanson, G. H. *et al.*, *Ind. Eng. Chem.*, 1952, **44**, 604.

[6a] Reamer, H. H. and Sage, B. H., *Ind. Eng. Chem.*, 1951, **43**, 1628.

[7] Landolt-Börnstein, 'Zahlenwerte und Funktionen', 6th ed., II 2a (Berlin, 1960), 90, 151, 158, 173; Rossini, F. *et al.*, 'Selected Values of Physical and Thermodynamic Properties of Hydrocarbons and Related Compounds' (Amer. Petr. Inst., Carnegie Press, Pittsburgh, Penn., USA, 1953), 52, 332, 355, 358, 411 and 778; Burrell, G. A. and Robertson, T. W., *US Bur. Mines, Tech. Paper*, 1916, **142**; Stull, D. R., *Ind. Eng. Chem.*, 1947, **39**, 517; Delaplace, R., *Compt. rend.*, 1937, **204**, 493; Echols, L. S. and Gelus, E., *Anal. Chem.*, 1947, **19**, 670; Francis, A. W. and Robbins, G. W., *J. Am. Chem. Soc.*, 1933, **55**, 4339; Gilliland, E. R. and Scheeline, H. W., *Ind. Eng. Chem.*, 1940, **32**, 48; Griswold, J. and Brooks, W. B., *Pet. Refiner*, 1947, **26**, 818; Myers, H. S., *Ind. Eng. Chem.*, 1955, **47**, 1659; Negishi, R., *Bull. chem. soc. Japan*, 1942, **17**, 481; Seibert, F. M. and Burrell, G. A., *J. Am. Chem. Soc.*, 1915, **37**, 2683; Stull, D. R., *Ind. Eng. Chem.*, 1947, **39**, 520; Tickner, A. W. and Lossing, F. P., *J. Physic. & Colloid Chem.*, 1951, **55**, 733; Villard, P., *Ann. chim. phys.*, 1897, **10** (7), 387.

[8] Maass, O. and Wright, C. H., *J. Am. Chem. Soc.*, 1921, **43**, 1098.

[9] Marchman, H., Prengle, H. W. and Motard, R. L., *Ind. Eng. Chem.*, 1949, **41**, 2658.

[10] Vaughan, W. E. and Graves, N. R., *Ind. Eng. Chem.*, 1940, **32**, 1252.

[11] Powell, T. M. and Giauque, W. F., *J. Am. Chem. Soc.*, 1939, **61**, 2366.

[12] Lu, H., Newitt, D. M. and Ruhemann, M., *Proc. Roy. Soc.*, 1941, A**178**, 523.

[13] Landolt-Börnstein, 'Zahlenwerte u. Funktionen', 6th ed., II 2a (Berlin, 1960), p. 194; Lawrence, N. C., Goldman, M. and Marchman, H., *Ind. Eng. Chem., Ind. Ed.*, 1951, **43**, 1186; Winkler, C. A. and Maass, O., *Canad. J. Res.*, 1933, **9**, 612.

[14] Hoge, H. J. in: 'Temperature, its Measurement and Control in Science and Industry' (New York, 1941), p. 141.

[15] Reeves, L. E., Scott, G. J. and Babb, S. E., *J. Chem. Phys.*, 1964, **40**, 3662.

[16] Lambert, J. D. *et al.*, *Proc. Roy. Soc.*, 1955, A**231**, 280.

[17] Reid, R. C. and Sherwood, T. K., 'The Properties of Gases and Liquids' (McGraw-Hill Book Co., New York, 1958).

[18] Flynn, L. W. and Thodos, G., *J. of Chem. Eng. Data*, 1961, **6**, 457.
[19] Dreisbach, R. R., *'Physical Properties of Chemical Compounds'*, Vol. II. Advances in Chemistry Series No. 22 (American Chemical Society, 1959).
[20] Misic, D. and Thodos, G., *Aiche Journal*, 1961, **7**, 264.
[21] Gallant, R. W., *'Physical Properties of Hydrocarbons'*, Vol. I (Gulf Publishing Co., Houston, Texas, 1968).
[22] Unver, A. A. and Himmelblau, D. M., *J. of Chem. Eng. Data*, 1964, **9**, 428.
[23] Lamb, A. B. and Roper, E. E., *J. Am. Chem. Soc.*, 1940, **62**, 806.
[24] Burrell, G. A. and Robertson, I. W., *J. Am. Chem. Soc.*, 1915, **37**, 2188.
[25] Williams and Katz, *Ind. Eng. Chem.*, 1954, **46**, 2512.
[26] Orentlicher and Prausnitz, *Chem. Eng. Sci.*, 1964, **19**, 775.
[27] Tsin, *J. phys. chem. (USSR)*, 1940, **14**, 418.
[28] Cox and De Vries, *J. Phys. & Colloid Chem.*, 1950, **54**, 665.
[29] Reamer, Selleck and Sage, *J. Petroleum Technol.*, 1952, **4**, no. 8, 197.
[30] Li and McKetta, *J. Chem. Eng. Data.*, 1963, **8**, 271.
[31] Blagoi, Orobins'kii and Trofimow, *Zhur. Fiz. Khim.*, 1965, **39**, 2022, *Russ. Journ. Phys. Chem.*, 1965, **39**, 1073.
[32] Ishii, Hayami, Shirai and Ishida, *J. Chem. Eng. Data.*, 1966, **11**, 288.
[33] Cheung and Zander, *Chem. Eng. Prog. Symp. Ser.*, 1968, **64**, 34.
[34] Haselden and Snowden, *Trans. Faraday Soc.*, 1962, **58**, 1515.
[35] Haselden, Holland, King and Strickland-Constable, *Proc. Roy. Soc.*, 1957, **A240**, 1.
[36] Winkler and Maass, *Can. J. Res.*, 1932, **6**, 458.
[37] Yorizane, Yoshimura and Masuoka, *Kagaku Kogaku*, 1967, **5**, 175.
[38] Haselden, Newitt and Shah, *Proc. Roy. Soc.*, 1951, **A209**, 1.
[39] Blagoi and Orobins'kii, *Zh. Fiz. Khim.*, 1966, **40**, 3031; *Russ. Journ. Phys. Chem.*, 1966, **40**, 1625.
[40] Lu, Newitt and Ruhemann, *Proc. Roy. Soc.*, 1941, **A178**, 506.
[41] McKay, Reamer, Sage and Lacey, *Ind. Eng. Chem.*, 1951, **43**, 2112.
[42] Reamer and Sage, *Ind. Eng. Chem.*, 1957, **43**, 1628.
[43] Hanson, Hogan, Nelson and Cines, *Ind. Eng. Chem.*, 1952, **44**, 604.
[44] Hirata, Hakuta and Onoda, *Sekiyu Gakkai Shi*, 1967, **10**, 440. Translation in: *Int. Chem. Eng.*, 1968, **8**, 175.
[45] Gilliland and Scheeline, *Ind. Eng. Chem.*, 1940, **32**, 48.
[46] Schneider and Maass, *Can. J. Res.*, 1941, **B19**, 231.
[47] Goff, Farrington and Sage, *Ind. Eng. Chem.*, 1950, **42**, 735.
[48] McCurdy and Katz, *Oil and Gas J.*, 1945, **43**, 102.
[49] Williams and Katz, *Ind. Eng. Chem.*, 1954, **46**, 2512.
[50] Reamer, Selleck and Sage, *J. Petroleum Technol.*, 1952, **4**, 197.
[51] Francis, *J. Chem. Eng. Data*, 1965, **10**, 327.
[52] Francis, *J. Chem. Eng. Data.*, 1965, **10**, 45.
[53] Francis, *J. Phys. Chem.*, 1953, **63**, 753.
[54] Benham and Katz, *A. I. Ch. E. Journal*, 1957, **3**, 33.
[55] Snell, Otto and Robinson, *Am. Inst. Chem. Eng. Journal*, 1961, **7**, 482.
[56] Fomina and Byk, *Khim. Tekhnol. Topl. Masel* (Chem. and Technol. of Fuels and Oils), 1967, **12**, 683.
[57] Hill, McCormick, Barton and Fenske, *A.I.Ch.E. Journal*, 1962, **8**, 681.
[58] Than, *Ann.*, 1862, **123**, 187.
[59] Azarnoosh and McKetta, *J. Chem. Eng. Data*, 1959, **4**, 211.
[60] Robey, USP 2,245,719 (1938).
[61] Landolt-Börnstein, *'Zahlenwerte und Funktionen'*, 6th ed., II, 2b (Berlin, 1962), pp. 1–178.
[62] Sergeys, FIAT Final Report no. **1294** (1948), pp. 7, 8 and 18.
[63] Joshua and Stanley, USP 2,005,500 (1935), BP 428,108 (1935).
[64] Francis, USP 2,463,482 (1949).
[65] McBain and O'Connor, *J. Am. Chem. Soc.*, 1941, **63**, 875.
[66] McBain and Soldate, *J. Am. Chem. Soc.*, 1942, **64**, 1556.

[67] Lee Purlee and Taft, *J. Am. Chem. Soc.*, 1956, **78**, 5811.
[68] Jones and Shewell, USP 2,372,085 (1942).
[69] Hannaert, Haccuria and Mathieu, *Ind. Chim. Belg.*, 1967, **32**, 156.
[70] Konobeev and Ljapin, *Khim. Prom.*, 1967, **43**, no. 2, 114.
[71] Tilquin, Decannière, Fontaine and Claes, *Ann. Soc. Sci. Bruxelles*, Ser. I, 1967, **81**, 191.
[72] Frolich, Tauch, Hogan and Peer, *Ind. Eng. Chem.*, 1931, **23**, 548.
[73] Marshtupa, Babin, Kolesnikov, Maryshkina and Borodina, *Khim. Prom.*, 1965, **41**, 585.
[74] Klempt, *Ber. Ges. Kohlentechn.*, 1950, **5**, 460; Landolt-Börnstein, 6th ed., II, 2b, 1–108.
[75] Kirejev, Kaplan and Romanchuk, *Zhur. Obshch. Khim.*, 1935, **5**, 444 (English translation No. 61-16833, Europ. Transl. Centre, Doelenstraat 101, Delft, Netherlands).

CHAPTER 4

OLIGOMERS AND CO-OLIGOMERS OF PROPYLENE
By J. Habeshaw

4.1. INTRODUCTION

SINCE the first use of propylene oligomerisation to form liquid gasoline components from the by-products of heavy oil cracking, higher valued chemical outlets for these oligomers and co-oligomers have been developed. Catalytic cracking[1] in the USA and steam cracking in Europe and Japan, have become the major sources of propylene, superseding the older cracking processes. In all these areas, the chemical uses for propylene oligomers and co-oligomers have developed similarly, the main outlets being as olefinic intermediates for the manufacture of plasticisers for PVC, and of synthetic detergents, particularly the anionic alkylbenzene sulphonates. The requirements for these outlets have determined the direction of much of the applied research on oligomerisation reactions and processes. The following account summarises this research and describes the preparation and main applications of those oligomers which have become industrially important.

The dimerisation of propylene to form olefins of specific structure by recently discovered catalysts has made possible the production of pure monomers, particularly of isoprene precursors and of 4-methylpentene-1, but so far these applications have not achieved the importance of the older plasticiser and detergent intermediate processes. However, these new developments provide opportunities for the synthesis of olefins of defined structure in pure form, which may yet find large volume uses in new or improved polymers as well as in the applications already established, where the structural requirements are becoming more specific. In the detergent field, for example, the need for biodegradability necessitates the use of alkyl benzenes with linear, or at least near-linear, side chains. Similarly the need for plasticised polymers of specific properties has stimulated investigations on the variations possible by changing the structure of the plasticiser, particularly of the alcohol part of the esters used, and hence of the olefin from which the alcohol is produced by hydroformylation.

4.2. Chemistry of Propylene Oligomerisation and Co-Oligomerisation

4.2.1. Cationic oligomerisations Many years ago the formation of low polymers from propylene and butenes was commercialised to convert low olefins formed as a by-product of oil cracking operations to gasoline range hydrocarbons.[2] These processes for producing polymer gasoline used acidic catalysts (phosphoric acid on a solid support such as Kieselguhr for example), and gave complex mixtures of products, generally with a relatively low selectivity to dimers. The reactions proceed by a carbonium ion mechanism, the acid catalyst transferring a proton to the olefin. This mechanism is capable of explaining the products obtained[3] but not easy to use to predict the final product composition, because the carbonium ions readily rearrange and undergo further reaction. Evidence has been presented[4] from experiments at low temperatures (to avoid side reactions) that the primary product of propylene dimerisation on these acid catalysts is 4-methylpentene-2. Terres considered from his evidence that only intermediate olefins having terminal double bonds are capable of further polymerisation. However this may be, these acid catalysts under practically useful conditions are relatively unselective for dimer formation and give 'scrambled' products containing isomers produced by carbonium ion rearrangement and by isomerisation of the primary olefinic products as well as olefins and paraffins of intermediate carbon numbers produced by more complex side reactions. Ipatieff and his co-workers[5] showed that a phosphoric acid catalyst copolymerised propylene and isobutene at 135°C to give 40–45 per cent of heptenes, 10–15 per cent octenes, and 40–50 per cent of higher olefins, but that temperatures around 190°C were needed to give substantial reaction of the normal butenes with propylene. Despite the unselective nature of the reaction, these acid catalysts nevertheless still find application in the commercial production of heptenes, nonenes and dodecenes used as raw materials for the manufacture of plasticisers and detergents. Other catalytic mineral acids, notably sulphuric acid have been studied,[6] but only phosphoric acid and supported phosphoric acid have so far been used commercially with propylene as starting material.

The cationic polymerisations can be carried out with Lewis acids and mixed oxide catalysts, as well as with mineral acids. Regarding the catalyst as the ion pair $CAT^+ AN^-$; (CAT^+ will be H^+ for protonic acids), a formal scheme for propylene dimerisation can be written as:

$$CAT^+AN^- + \overset{\delta-}{CH_2} = \overset{\delta+}{CH_2} - CH_3 \longrightarrow CAT - CH_2 - \overset{\overset{\displaystyle AN^-}{+}}{CH} - CH_3$$

$$\downarrow C_3H_6$$

$$\longleftarrow CAT - CH_2 - CH - CH_2 - \overset{\overset{\displaystyle AN^-}{+}}{CH} - CH_3$$

$$CAT^+AN^- + \overset{CH_3}{\underset{CH_3}{\diagdown}} CH - CH = CH - CH_3 \qquad CH_3 \mid C_3H_6$$

$$\downarrow$$

Higher Oligomers

The Lewis acids (e.g. BF_3, $AlCl_3$, $AlRCl_2$, $TiCl_4$, $ZnCl_2$, $SnCl_4$) may catalyse oligomerisation alone in solution, but usually a co-catalyst such as water or hydrogen chloride is added to increase the activity by generating protonic acidity. In general these Lewis acids have not proved efficient for forming the lower oligomers of propylene, although transition metal modified chloro-aluminium catalysts such as $CoCl_2$—$AlCl_3$[7, 8] or $NiCl_2$—$AlCl_3$[9] are capable of quite high selectivities of dimer formation (ca. 60 per cent).

The mixed oxide catalysts owe their activity to acid sites on the catalyst surface.

Silica/alumina, the well-known cracking catalyst, is active for oligomerisation, but resembles the mineral acid catalysts in giving poor selectivity to dimer, and a 'scrambled' complex product.[11-14] Improved efficiency of dimer formation and a considerably simpler mixture of isomers is obtained by modifying the silica/alumina catalyst with nickel oxide.[15-19] The yields of heptenes from propylene and butenes on these modified oxide catalysts are quite high, and they give quite high yields (60–80 per cent) of dimer from propylene. The linearity of the dimer can reach 50 per cent or even more at higher temperatures (150°C). A considerable literature on these modified catalysts now exists, but no commercial use appears to have been made of them so far.

Aluminium ethyl dichloride (AlEtCl$_2$) and its co-catalysts with nickel and cobalt compounds give good yields of dimer from propylene, but aluminium diethyl chloride (AlEt$_2$Cl) is inactive for propylene oligomerisation under mild conditions (i.e. below about 100°C) although at 180–200°C this catalyst will form 2-methylpentene-1.[10]

Many other protonic and Lewis acid-type catalysts have been described in the literature, including zinc chloride[30-32] and the heteropolyacids of silicic and tungstic acid:[33] again none have found significant industrial applications as yet.

4.2.2. Anionic oligomerisations   Alkali metal alkyls and some alkali metals react with olefins to form carbanions,[22, 23] metallation occurring most easily in the allylic position.[20, 21] Generally potassium alkyls are more powerful metallating agents than sodium alkyls.[29] Vinyl metallation does not occur to any great extent under conditions where propylene dimerises efficiently. The allyl anion formed adds to another propylene molecule, giving the anion of 4-methylpentene-1. This is considerably less stable than the allyl anion and readily exchanges with another propylene molecule, yielding 4-methylpentene-1 together with another allyl anion to continue the dimerisation cycle:

$$CH_2=CH-CH_3 \xrightarrow{\text{Na}} (CH_2=CH=CH_2)^- \, Na^+$$

$$CH_2=CH-CH_2-\underset{\underset{CH_3}{|}}{\overset{\overset{CH_3}{|}}{C}}H + (CH_2=CH=CH_2)^- Na^+ \quad C_3H_6 \left\{ CH_2=CH=CH-\underset{\underset{CH_3}{|}}{\overset{\overset{CH_3}{|}}{C}}H \right\}^- Na^+$$

Because of the relative rapidity of the 4-methylpentene-1 anion exchange with propylene to reform the allyl anion, compared with further propylene addition the reaction is, under proper conditions, very selective to dimer formation. Addition of the allyl anion to the other side of the propylene double bond (the alternative primary reaction), finally gives hexene-1, and this is normally a minor co-product of the dimerisation.

Secondary reactions can also occur. Remetallation of the dimer can occur to some extent, and this can lead to the formation of isomers as follows:

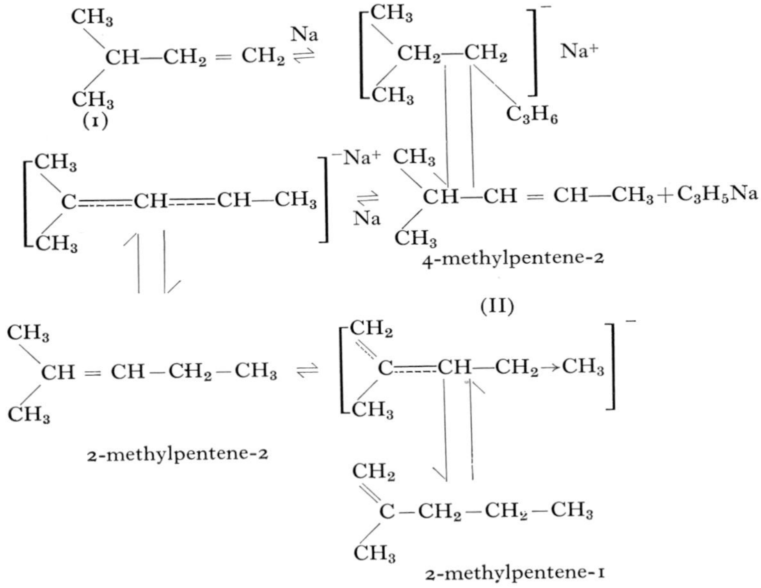

$$\begin{array}{c} CH_3 \\ \diagdown \\ CH-CH_2 = CH_2 \\ \diagup \\ CH_3 \\ (I) \end{array} \underset{}{\overset{Na}{\rightleftharpoons}} \left[\begin{array}{c} CH_3 \\ \diagdown \\ CH_2-CH_2 \\ \diagup \\ CH_3 \end{array} \right]^{-} Na^{+}$$

$$\downarrow\uparrow \; C_3H_6$$

$$\left[\begin{array}{c} CH_3 \\ \diagdown \\ C=\!=\!=CH=\!=\!=CH-CH_3 \\ \diagup \\ CH_3 \end{array} \right] \underset{Na}{\overset{{}^{-}Na^{+}}{\rightleftharpoons}} \begin{array}{c} CH_3 \\ \diagdown \\ CH-CH = CH-CH_3 + C_3H_5Na \\ \diagup \\ CH_3 \end{array}$$

4-methylpentene-2

(II)

$$\begin{array}{c} CH_3 \\ \diagdown \\ CH = CH-CH_2-CH_3 \\ \diagup \\ CH_3 \end{array} \rightleftharpoons \left[\begin{array}{c} CH_2 \\ \diagdown\!\!\diagdown \\ C=\!=\!=CH-CH_2\rightarrow CH_3 \\ \diagup \\ CH_3 \end{array} \right]^{-}$$

2-methylpentene-2

$$\downarrow\uparrow$$

$$\begin{array}{c} CH_2 \\ \diagdown\!\!\diagdown \\ C-CH_2-CH_2-CH_3 \\ \diagup \\ CH_3 \end{array}$$

2-methylpentene-1

The anions of (I) and (II) in the above scheme are particularly liable to diene formation through possible cyclic intermediates.[24] This type of reaction, which can lead to accumulation of polymers on the catalyst and consequent deactivation is obviously to be avoided, and is one of the reasons why practicable dimerisation reactions can be carried out efficiently only in a relatively narrow range of temperature conditions.

Pines and Mark[20] compared the effects of sodium-anthracene and sodium on alumina catalysts on the isomerisation of butene-1; their results show that the environment of the alkali metal has a large effect on the isomerisation activity. J. B. Wilkes and his co-workers[25, 26] believed that potassium allyl was the active intermediate in dimerising propylene. Potassium or potassium allyl dispersed on alumina are both good catalysts for double bond isomerisation, whereas potassium allyl dispersed on potassium carbonate shows very little isomerisation activity. These results emphasise the importance of the alkali metal environment on the catalytic behaviour. The development of catalysts reducing these undesirable side reactions to a minimum, and hence giving high efficiencies of reaction and a good catalyst life, has been the object of much of the applied research in this field (see also refs. 27 and 28). This has resulted in the development of efficient catalysts for dimerising propylene to 4-methylpentene-1 with minimum double bond isomerisation.

The conditions needed for high selectivity to 4-methylpentene-1 are critical, and the reaction temperature, feedstock contact time and alkali metal:olefin ratio are important parameters in avoiding unwanted side reactions. Freedom of the feed-stock from oxygen, oxygenated compounds, sulphur or sulphur compounds, amines and water (which all poison the catalysts) must be ensured for successful operation of the dimerisation.[27]

Similar mechanisms operate in forming co-oligomers with these catalyst systems. Ethylene alone does not metallate under dimerisation conditions, and does not there-fore polymerise with the heavier alkali metal catalysts. It is, however, the best acceptor of anions of all the simple monolefins, the order of receptivity being reported as:[23]

$$C_2H_4 > C_3H_6 > nC_4H_8 > iso\ C_4H_8$$

The rapid reaction with ethylene of the allylic anions generated from propylene thus leads to high yields of the expected pentene-1 codimer of ethylene and propy-lene, with smaller amounts of 4-methylpentene-1 and some heptenes, the latter derived most probably from secondary reaction of the pentene-1 allylic anion with ethylene. Reactions between ethylene and the butenes show some further properties of the dimerisation.[25-27, 34] Butene-1 and butene-2, as might be expected with a common allylic intermediate, yield with ethylene the same products, primarily 3-methylpentene-1 with some hexene-2, and have similar reaction rates. Hexene-3, which is thermodynamically more stable and the product expected from vinyl metallation of the normal butene is not found as a primary product, again indicating absence of vinyl metallation under the low temperature conditions effective for dimer formation. The results also indicate that isomerisation of 3-methylpentene-1 to 3-methylpentene-2 is much slower than the 4-methylpentene-1 to 4-methyl-pentene-2 conversion, probably because tertiary carbon atoms ($>CH$) metallate less readily than secondary ($-CH_2-$), and these again less readily than primary ($-CH_3$) carbon atoms. Isobutene and ethylene form 2-methylpentene-1. Results of codimerising ethylene with the butenes also indicate that the stability of the derived anions increases (i.e. the rate of reaction decreases) in the order:

$$n\text{-Butenes} < propylene < isobutene.$$

However, with butene mixtures the results can be complicated by transmetalla-tion reactions, which result in isobutene reducing the concentration of n-butene anions by the exchange reaction:

$$C\!-\!\overset{-}{C}\!-\!C = C\!+\!C = \overset{\overset{\displaystyle C}{|}}{C}\!-\!C \ \rightleftharpoons \ C\!-\!C\!-\!C = C\!+\!C = \overset{\overset{\displaystyle C}{|}}{\underset{-}{C}}\!-\!C$$

and giving a product by reaction of mixed butenes with ethylene containing a considerably higher proportion of the isobutene-ethylene co-dimer than would otherwise be expected. The rapid exchange of pentenyl anion with propylene similarly prevents further reaction in co-dimerisation of ethylene and propylene, so that large amounts of heptenes are not formed easily from ethylene and propylene. However, in the absence of propylene, ethylene readily reacts with normal pentenes to form 3-ethylpentene-1 as main product, with heptene-3 as by-product.

Reaction of propylene with a mixture of the butenes (bearing in mind the foregoing discussion) should give the primary products from reaction of the three allylic anions produced with the parent olefins, thus:

$$
\begin{array}{lll}
\overset{(-)}{C}-C-C+\text{propylene} & \longrightarrow & \text{hexenes} \\
\quad\quad +n\text{-butenes} & \longrightarrow & \text{4-methylhexene-1} \\
\quad\quad +\text{isobutene} & \longrightarrow & \text{4·4-dimethylpentene-1} \\
\end{array}
$$

$$
\overset{(-)}{C}-C-C = C+\text{propylene} \longrightarrow \text{3·4-dimethylpentene-1}
$$

$$
\searrow \quad\quad \text{3-methylhexene-1}
$$

$$
C-C = \underset{\underset{C}{|}}{C}-\overset{(-)}{C}+\text{propylene} \longrightarrow \text{5-methylhexene-2}
$$

$$
\overset{(-)}{C}-\underset{\underset{C}{|}}{C} = C+\text{propylene} \longrightarrow \text{2·4-dimethylpentene-1}
$$

The reaction can be conducted under conditions where octene production is small, and all these products have been found. The relative amounts of the individual products formed depend on the relative reactivities and stabilities of the various allylic intermediates, and on the rates of trans-metallation reactions between the anions and olefins when mixtures of olefins are reacted. This makes quantitative prediction of the product composition complex and difficult in such cases. However, under conditions giving efficient dimer and co-dimer formation (generally at temperatures below about 175 °C), there is again no indication that vinyl metallation occurs. The chemistry involved is essentially similar to that concerned with double-bond isomerisation and dehydrogenation reactions on this type of catalyst.[35-38]

A family of homogeneous anionic lithium catalysts capable of converting ethylene-propylene mixtures to linear oligomers has been described by Eberhardt and others.[39-41] These catalysts are made by reaction of a lithium alkyl (lithium butyl, for example) with some fully substituted diamines, tetraethylenediamine seeming to be the best. The propylene anion reacts with successive molecules of ethylene, giving a mixture of linear olefins containing odd numbers of carbon atoms:

$$
\begin{array}{c}
CH_3 \quad CH_2-CH_2 \quad CH_3 \\
\diagdown \quad \diagup \quad \diagdown \quad \diagup \\
N \quad\quad\quad N \quad\quad \text{Catalyst} \\
\diagup \quad \diagdown \quad\quad \diagdown \\
CH_3 \quad\quad\quad\quad CH_3 \\
\diagdown \quad\quad \diagup \\
\underset{+}{Li}
\end{array}
$$

$$
C = C-C \longrightarrow \overset{-}{C} = C-C\ \overset{+}{Li} \xrightarrow{+\ C_2H_4} C = C-\overset{-}{C}-C-C\ \overset{+}{Li} \longrightarrow \text{Pentene-1}
$$

$$
\Big\downarrow C_2H_4
$$

$$
C = C-C-\overset{-}{\underset{|}{C}}-C-C-C\ \overset{+}{Li} \longrightarrow \text{Heptene-1}
$$

$$
\Big\downarrow C_2H_4
$$

$$
C = C-C-C-C-\overset{-}{\underset{|}{C}}-C-C-C\ \overset{+}{Li} \longrightarrow \text{Nonene-1}
$$

The reaction also occurs with other olefins such as isobutene and butene-2 instead of propylene, but the products are not then linear alpha olefins. The degree to which product isomerisation and rearrangement to cyclic compounds occurs with these catalysts, and hence the selectivity with which linear alpha olefins can be formed is not yet clear. The catalysts are also said to be active for anionic alkylation of aromatic hydrocarbons. No industrial applications of these systems has so far been reported.

4.2.3. Insertion oligomerisations About twenty years ago Ziegler showed[42, 46] that long chain alkyl compounds are formed by reaction of aluminium alkyls with ethylene at about 100°C. This growth reaction can be succeeded at higher temperatures by a displacement reaction with ethylene resulting in the liberation of a mixture of linear alpha olefins and regeneration of aluminium trialkyl, the oligomerisation of ethylene being the overall result. A statistical distribution of molecular weights in the product is obtained. Growth and displacement can be conducted simultaneously at the higher temperatures (around 200°C), resulting in a rather flatter molecular weight distribution curve.[43] The displacement reaction can be catalysed by a number of transition metals, notably nickel, cobalt and platinum[44] as well as titanium and zirconium esters.[47] Lower temperatures in the range 50–120°C are quoted for the catalysed displacement reaction with ethylene, producing products ranging from butenes to dodecenes.[52] The growth reaction with ethylene followed by conversion of the aluminium alkyls formed with oxygen to alkoxides, from which alcohols are formed by hydrolysis has been applied commercially to produce linear primary alcohols.[45]

Higher olefins will also react with the aluminium alkyls, propylene giving a branched alkyl chain, which is considerably less stable than the straight chain alkyls. The displacement reaction with the derived aluminium alkyls therefore occurs much faster relative to growth than with ethylene, and high efficiencies of conversion of propylene to 2-methylpentene-1 can be obtained, thus:

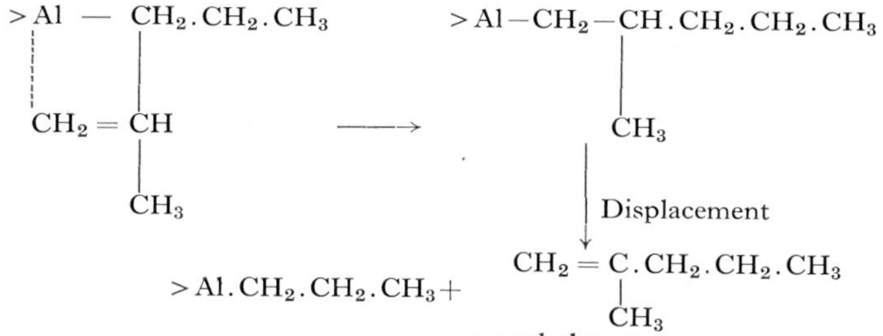

The mechanism of this propylene dimerisation has been discussed by Bass.[106] The insertion reaction may well proceed through a 4 centre strained transition state as suggested by Hay,[107, 108] of the type:

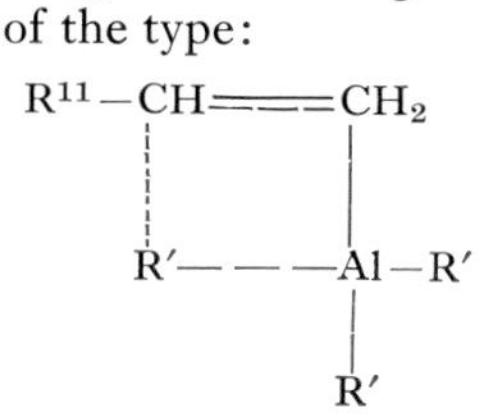

The overall general reaction is then formally:

$$AlR'_3 + R''CH-CH_2 \longrightarrow R''CH-CH_2-AlR'_2$$
$$\underset{R'}{\Big|}$$

$$R'CH_2CH_2AlR'_2 \longleftarrow \underset{R'CH=CH_2}{\overset{R''CH=CH_2}{\Big|}} AlHR'_2 + R''-CH=CH_2$$
$$\underset{R'}{\Big|}$$

The dimer normally contains around 10 per cent of hexene-2 formed by insertion of propylene the other way round. Thermal decomposition of the aluminium alkyl is reduced by operating under higher pressures, particularly at the higher temperatures used for the uncatalysed displacement reaction. The growth reaction does not easily proceed beyond the dimer stage with other higher olefins and these, like propylene, give good yields of dimers (see also refs. 48, 49). Selective dimerisation and codimerisation of C_2 to C_4 olefins at temperatures around 200–350°C and pressures in the range 100–200 atm. are described.[50, 51] With transition metal catalysed-displacement reactions, propylene dimerisation can be carried out at temperatures as low as 40°C[44] but selectivity to 2-methylpentene-1 is lower.

Reaction products of aluminium alkyls with transition metal compounds, particularly cobalt or nickel give catalysts active at low temperatures (e.g. 40°C) when Lewis acid components are formed. Some typical results are shown in Table IV.1 for the dimerisation of propylene (see also ref. 59).

TABLE IV.1

Dimerisation of Propylene

Catalyst	Temp. °C	Selectivity to dimer formation	Hexene composition, wt. %			Refs.
			n-hexenes	methylpentenes	2·3 dimethyl-butenes	
π-Allyl NiCl—AlCl$_3$/AlEtCl$_2$	30–50	—	22	74	4	53, 55
Ni-oleate—AlBu$_2$Cl	20	95	27	ca. 70	3	56
(π-Allyl)$_2$ Ni—AlBr$_3$	20	—	22	ca. 75	3	56
Ni π complexes—AlEtCl$_2$	20	95	22	70–75	0–5	57
Ni (acac)$_2$ etc.—AlEtCl$_2$	0–30	90	24–32	67–74	0–5	58

A wide range of nickel compounds has been used: all show high activity at low temperatures, high efficiency to dimer and give hexene products containing 20–30 per cent straight-chain hexenes (mainly hexene-2) and 70–80 per cent methylpentenes (mainly 4-methylpentene-2). Addition of phosphine ligands gives increasing proportions of 2·3-dimethylbutenes (2·3-dimethylbutene-1 appearing to be the primary product) as the basicity of the phosphine is increased.[53] However, the presence of a Lewis acid such as AlEtCl$_2$ appears to be necessary in all these systems and mixtures of a nickel salt with an aluminium trialkyl and triphenylphosphine are reported to be inactive.[58] This class of catalyst, containing a nickel or cobalt compound, a Lewis acid component which is usually an aluminium alkyl halide and a Lewis base which is usually a tertiary phosphine, is the subject of much recent literature on olefin dimerisation.[59] The similarity in composition of the propylene dimers given by these catalysts and by the nickel oxide-silica-alumina catalysts described earlier is noteworthy. This has been discussed recently by J. R. Jones,[10, 59] who considers that the formation of 4-methylpentene-2 and hexene-2 in the phosphine-free systems may be a cationic rather than an insertion mechanism.

A number of other catalysts have been reported which are claimed to function by an insertion mechanism. Nickelocene[60] has low activity and requires a high temperature. The nitrogen complex of cobalt $H(N_2)Co(PR_3)_3$ converts propylene to 2-methylpentene-1 at room temperature.[61] Rhodium chloride,[62] palladium chloride[64] and ruthenium chloride[63] all dimerise ethylene rapidly to an equilibrium mixture of normal butenes, but propylene and higher olefins react much less readily, although giving relatively high selectivity to linear dimers. The palladium catalysts are notable for the high selectivity of linear hexene formation from propylene.

These transition metal-containing catalysts produce a wide range of product structures and there is no clear boundary between those showing an insertion and those showing a cationic type of activity in which the dimer products approximate closely to those formed on alumino-silicate catalysts. Application of cobalt-charcoal catalysts[65] to form linear dimers of propylene at low reaction temperatures[66, 67] has been described. The mechanism of reaction over this catalyst is rather obscure, but may be related to that with a catalyst formed by reacting nickel acetylacetonate with an alkyl aluminium in the absence of either strong Lewis acids or strong Lewis bases. This gives a somewhat similar higher selectivity (about 65 per cent) to linear dimer formation, the predominant product being hexene-2.[10, 59]

4.2.4. Radical polymerisations Thermal polymerisation was an early but short-lived process for converting hydrocarbon gases to gasoline,[68] and is not a practically important method for making propylene oligomers. The thermal reaction is reported[69] to give 4-methylpentene-1 and hexene-1 as major products at 380–450°C, but about 50 per cent of the dimers are of other structures not explicable by any simple radical mechanism (see ref. 70). The thermal reaction with propylene is very unselective and cannot compete with the catalytic methods which have been reviewed above.

4.2.5. Synthetic possibilities for propylene oligomer production The enormous amount of experimental work carried out in recent years in both academic and industrial research laboratories and summarised above has produced catalyst systems capable of making propylene dimers and co-dimers of defined structure and in some cases of high purity. The possibilities for propylene dimerisation are summarised in the following Table IV.2. Although in the above discussion oligomerisations have been classified as anionic, cationic or insertion, there is no very clearly marked demarcation between some types of catalysts in this sense, and Table IV.2 indicates the products obtained using the principal types of catalyst which have been described in the literature.

TABLE IV.2

Reported Propylene Dimers

Catalyst system used	Olefins produced	References
Trialkyl aluminium	2-methylpentene-1	46, 71, 72
Dispersed alkali metals on supports	4-methylpentene-1	27
Cobalt on charcoal	Hexene-2 (Hexene-3 and 2-methylpentene-2 also formed, *ca.* 50% linear dimer)	65, 66, 67
PdCl₂ and complexes	Hexene-2 (65—*ca.* 90% linear dimer)	10, 59, 73
Nickel-β-diketone complexes with aluminium alkyls	Hexene-2 (*ca.* 65% linear dimer)	10, 59
Ni X₂/AlRCl₂/PR₃	2·3-dimethylbutene-1	54, 74

No dimerisation of propylene to 3·3-dimethylbutene-1 appears to have been reported. Although 2-methylpentene-2 is formed as a co-product in many dimerisations, as might be expected this being thermodynamically the most stable of the hexene isomers in the relevant temperature range, this olefin is probably best made by isomerisation of 2-methylpenetene-1, from which it is fairly easily separated. Similarly, 2·3-dimethylbutene-2 may be prepared by isomerising 2·3-dimethylbutene-1.

Most of the new catalysts are very selective for the formation of dimers and co-dimers, and are hence not easily applied to the production by single stage conversions of C_2–C_4 olefins of oligomers higher than C_8 suitable for use in detergent manufacture, for example.

The present status of industrial manufacture and applications of propylene oligomers and co-oligomers made on the newer catalyst systems, as well as by the older methods, is described in the following sections.

4.3. 4-Methylpentene-1 and Poly-4-methylpentene-1

4.3.1. Production of monomeric 4-methylpentene-1   This is the least stable thermodynamically of the propylene dimers, and production in good yield therefore requires a catalyst selective for dimer formation, giving 4-methylpentene-1 as primary product and at the same time substantially free of isomerising acitvity, whether for skeletal isomerisation or for double-bond movement. From the foregoing discussion the anionic alkali metal based catalysts will give 4-methylpentene-1 as primary product, but many of these (for example potassium on alumina)[75] have marked activity for double-bond isomerisation. However, the many studies carried out have demonstrated that the activity of these catalysts for side reactions is very dependent on the alkali metal environment, and catalysts capable of dimerising propylene to 4-methylpentene-1 have now been described in the literature (see for example refs. 76, 77 and 78). These are formed by dispersing an alkali metal (sodium or potassium) on supports such as alkali metal carbonates or talc.

Some information on the 2,000 tons per annum plant of British Petroleum has been published,[79] although precise details of the catalyst have not been disclosed. The general form of the process is shown in Fig. 4.1. Purified propylene is fed from

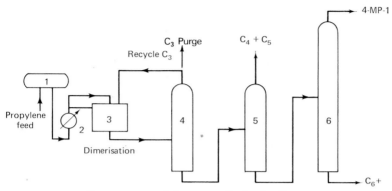

FIG. 4.1   *4-methylpentene-1 Production Process*

the tank 1 to the heater 2, and thence to the dimerisation reactor 3. This is operated to give more than 30 per cent conversion per pass, with efficiencies of 4-methylpentene-1 formation based on propylene converted in the range 80 to 90 per cent. The reaction product passes from the reactor to the depropaniser 4, where unreacted propylene is taken overhead and recycled to the reactor. A purge stream to remove propane introduced with the feed is withdrawn from the recycle stream. The depropanised product forming the residue from the column 4 is passed to the light ends column 5, where the small amount of C_4 and C_5 hydrocarbons formed is taken overhead, leaving residual hexenes and higher hydrocarbons which are passed to the final 4-methylpentene-1 column 6. Here 4-methylpentene-1 is removed as distillate, with the higher boiling hexenes and higher hydrocarbons left as residue. This final fractionation requires a high efficiency, since cis- and trans-4-methylpentene-2 are formed together with hexene-1 as by-products, and their boiling points are close to that of 4-methylpentene-1, actual boiling points (760 mm) being:

4-methylpentene-1	53·88°C
cis-4-methylpentene-2	56·30°C
trans-4-methylpentene-2	58·55°C
hexene-1	60·70°C

Hexene-1 is a primary but minor co-product: 4-methylpentene-2 is produced by isomerisation of the 4-methylpentene-1 first formed. A typical 4-methylpentene-1 product from the process is reported as having the following composition:

C_5	0·1 per cent
4-methylpentene-1	97·4 per cent
cis-4-methylpentene-2	0·9 per cent
trans-4-methylpentene-2	1·5 per cent
hexene-1	0·1 per cent

TABLE IV.3

Properties of Poly-Methylpentene-1 Compared with other Thermoplastics

	Poly-(4-MP-1) TPX	Polymethyl methacrylate	Transparent rigid PVC	Polystyrene GP	Polystyrene HI	Polypropylene	LD polyethylene
Relative density 20°C/20°C	0·83	1·19	1·4	1·06	1·05	0·90	0·92
Transparency	90%	92%					
Tensile strength (50% per mm) lb/m²	4,000	10,500	7,500	6,500	4,500	5,000	1,500
Elongation at break (%)	15						
Tensile modulus (0·2% strain, 100 sec value) lb/m²	$1·6 \times 10^5$	$4·5 \times 10^5$	$4·5 \times 10^5$	$4·6 \times 10^5$	$3·5 \times 10^5$	$1·7 \times 10^5$	$0·25 \times 10^5$
Rockwell hardness	L67–74	M103	R115	M68–80	M30–45	R95	D41–60
Vicat softening point (°C)	179	109–112	80–85	88–92	85	145–150	80–90
Volume resistivity, ohm/cm³	$<10^{16}$						
Permittivity at 20°C (10^2–10^{16} Hz)	2·12					2·25	2·28
Dielectric strength volts/m⁻³	700						
Izod impact, kg cm/cm	0·28–0·44	0·22			0·14–0·22		
Flammability cm/min	2·5						
Mould shrinkage cm/cm	0·15–0·3						
Thermal conductivity	4×10^{-4}						
Coefficient of expansion	$11·7 \times 10^{-5}$	7×10^{-5}	5×10^{-5}	7×10^{-5}	7×10^{-5}	10×10^{-5}	22×10^{-5}

Precise reaction conditions have not been disclosed, but from information on the reaction in the literature and patents,[20, 27, 76, 77] temperatures below about 200°C are used, high temperatures reducing both the efficiency of the reaction and the life of the catalyst.

A Russian process under development for producing 4-methylpentene-1 from propylene has been described:[80] this process appears to be similar to the one described above.

4.3.2. Applications of 4-methylpentene-1 and poly-4-methylpentene-1   Isomerisation of 4-methylpentene-1 to 4-methylpentene-2 and to 2-methylpentene-2 is easily brought about on a number of catalysts, including supported alkali metal catalysts.[81, 82] The 2-methyl-pentene-2 is a possible isoprene precursor, but is at present produced by propylene dimerisation using aluminium alkyl catalysts, forming 2-methylpentene-1 and subsequent isomerisation (see below) rather than from 4-methylpentene-1. Oxidation of 4-methylpentene-1 with palladium chloride systems to give methyl isobutyl ketone has been described,[83] and dimerisation with aluminium alkyl catalysts to give 2-isobutyl-6-methyl-heptene-1 has been reported:[84] neither of these reactions has so far achieved commercial use.

The most important application of 4-methylpentene-1 is in producing the high polymer. Little has so far been disclosed on the detailed polymerisation process used, but Ziegler/Natta type catalyst systems of a similar type to those used in producing polypropylene are employed.[79, 85, 86] Information on the polymers produced by I C I from 4-methylpentene-1 (their T P X range of polymers) has been published,[87] and some properties of the homopolymers and the copolymers with hexene-1 and decene-1 reported[79] by Hambling and Northcott, who indicate that the major part of the monomer produced in the British Petroleum plant is used by I C I in the manufacture of their T P X polymers. Some properties of the T P X polymer, compared with other thermoplastics, are given in Table IV.3.

The principal properties of 4-methylpentene-1 polymers significant in possible applications are as follows:

(1) The polymer has the lowest density (approximately 0·83) of any thermoplastic, giving advantage in volume cost at equal weight costs.

(2) The crystalline melting point is very high, figures up to 240°C being quoted for homopolymer, and dimensional stability is good at temperatures somewhat lower than the crystalline melting point. However, at temperatures below about 100°C, the polymer is slightly inferior to polypropylene in tensile modulus.

(3) The polymer has very high transparency, only very slightly inferior to polymethylmethacrylate.

(4) Very good electrical properties (resistivity, permissivity and loss factor) are shown, especially at very high frequencies.

(5) Although hydrocarbons and chlorinated solvents have some effect, the resistance to chemicals and solvents is otherwise good.

The polymer, as film, has fairly high permeability to oxygen and nitrogen. In common with other polyolefins, resistance to ultra-violet light and oxidation is not marked. Although the high moulding temperature and somewhat narrow softening range make moulding of homopolymer rather difficult, most copolymers have much better processing properties.[90]

Applications of the polymers have been described.[88, 89] These mostly follow from the high transparency, high softening point and good electrical properties of the polymer. A large number of applications, where these properties and the resistance to chemicals at fairly high temperatures give an advantage, have been mentioned.[91-94] In reported uses in disposable syringes, containers, packages and tubing for medical use and in laboratory ware, ability to withstand steam sterilisation is important. Moulded food containers to withstand hot filling, foil or film packaging to resist oven temperatures up to 220°C, and liquid dispensers in milking machines to resist disinfection by hot solutions are other uses where the polymer properties give it a premium value. Generally competition with other polyolefins, notably polypropylene, may be difficult except where the higher softening point and transparency of polymethylpentene compensates for its inevitably higher cost. The electrical properties in combination with the high softening point make the polymer attractive for coaxial cable insulation and for insulation of electronic assemblies where ability to withstand soldering and encapsulation temperatures is valuable. Other electrical applications are in motors where continuous exposure at up to 125°C is encountered, and in rigid bases for printed circuits. Although the formation of fibres from the polymer has been described[95, 96] the fibres have not so far found wide application, which also seems to be the case with fibres from polymers of other branched olefins such as 3-methylbutene-1.[97]

4.4. Production of Monomeric Isoprene

4.4.1. Propylene oligomers as isoprene precursors The fact that relatively high yields of isoprene could be obtained by pyrolysis of some methylbutenes has long been known, and as early as 1946 a process to produce isoprene by catalytically dimerising propylene and cracking the methylpentenes so made was described.[98] This process used acid centre catalysts such as silica-alumina for the dimerisation and uncatalysed pyrolysis to convert the methylpentenes of isoprene, and in such systems the overall efficiency to isoprene could not be made high. Moreover until the advent of stereo specific polymerisation catalysts made possible the production of high cis-polyisoprene polymers, isoprene outlets were limited to a small tonnage as comonomer in butyl rubber. The application of the Ziegler–Natta stereo specific aluminium alkyl based systems,[99, 105] and of lithium alkyl catalysts[100] to the production of high-cis polyisoprene made available a potential synthetic alternative to natural rubber with possibilities of large volume use.

4.4.2. Goodyear-Scientific Design process Application of the aluminium trialkyl catalysts to give very high yields of 2-methylpentene-1 from propylene, and the easy isomerisation of this to 2-methylpentene-2, which yields isoprene on pyrolysis, provided the basis for a technically and economically feasible process for the production of monomeric isoprene from propylene. This was commercialised in the Goodyear-Scientific Design process, of which general descriptions have been given.[101-104] A simplified flow diagram is shown in Fig. 4.2.

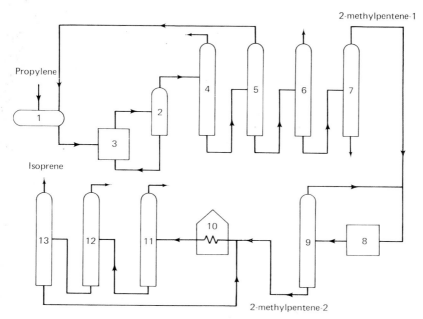

FIG. 4.2 *Isoprene From Propylene Process* (*Goodyear-Scientific Design*)

Purified fresh propylene feed, mixed with tripropyl aluminium catalyst (recycled from the flash tower 2) is fed from the feed tank 1 to the dimerisation reactor 3. The dimerisation is carried out in a shell and tube heat exchanger type of reactor, with propylene in the tubes and coolant in the shell to control the temperature at 150–250°C by removing the appreciable (*ca.* 23 kcals/mol methylpentene) heat of dimerisation. The dimerisation pressure is around 3000 psig. The reaction product passes to the flash tower 2, where the pentenes formed are flashed overhead, leaving the catalyst in a heavy hydrocarbon carrier as residue which is recycled to the reactor. The product then passes to the column 4, where gaseous by-products are removed, unreacted propylene then being recovered and recycled as overhead from the column 5. Small quantities of material boiling below 2-methylpentene-1 are taken overhead and discarded in column 6, the main 2-methylpentene-1 product being recovered as distillate from column 7, from which trimer and higher polymers are discarded as residue. The per-pass conversion in this dimerisation stage is probably around 70 per cent, and the efficiency of 2-methylpentene-1 formation about 80 per cent of the propylene converted. The dimer passes to the isomerisation reactor 8, where isomerisation to 2-methylpentene-2 is effected over a solid acid-centre type catalyst. The efficiency of this conversion is very high at the operating temperature which is in the range 150–300°C, and preferably about 200°C, but the per pass conversion is limited by the thermodynamic equilibrium between 2-methyl-pentene-1 and 2-methylpentene-2 to approximately 85 per cent. The isomerised product is fractionated in column 9. 2-Methylpentene-1 boils at 61·5°C, and is recovered as distillate from 2-methylpentene-2 boiling at 67·31 °C and recycled to the reactor 8: the 2-methylpentene-2 passes to the pyrolysis reactor via a surge-tank. The methylpentene is mixed in the surge tank with hydrogen bromide used as catalyst, and cracked in 10 at temperatures in the range 650–800°C, with contact times in the range 0·03 to 0·5 secs: steam is used to control the residence time. The product is fractionated to remove any material boiling below C_5 in columns 11 and 12, isoprene being recovered as distillate in the column 13. Catalyst is recirculated from this recovery system to the pyrolysis reactor.

Although other catalysts can be used, probably none will form 2-methylpentene-1 or 2-methylpentene-2 from propylene with an efficiency as high as aluminium trialkyls. Alternative possible routes to hexenes capable of pyrolysis to isoprene by, for example, catalytically co-dimerising ethylene and butene to 3-methylpentene-1 over lithium or sodium dispersed on an anhydrous potassium compound such as the carbonate,[108] or by isomerising 4-methylpentene-2 or 2-methylpentene-2[109] have been described in the patent literature. There are no reports of these methods being used commercially.

TABLE IV.4

Competitive Routes to Monomeric Isoprene

Feedstocks used	Type of process	References
C_5 fractions from steam crackers	Extraction of isoprene present in the steam cracked product using various solvents	124, 125, 126, 132, 133, 134
Isoamylenes	Catalytic dehydrogenation to isoprene	135, 136
Isoamylenes	Catalytic oxidative dehydrogenation to isoprene	130, 131
Mixed butenes (from butadiene extraction plants after removing butadiene) plus formaldehyde	Reaction of isobutene with formaldehyde to form 4·4-dimethyl-1·3-dioxane, followed by cleavage to isoprene and formaldehyde	137, 138
Isobutene plus formaldehyde	Reaction to form isopentenyl alcohol, followed by catalytic dehydration to isoprene	137
Acetone and acetylene	Methylbutynol formed, which is hydrogenated to methylbutenol, and this dehydrated to isoprene	127, 128, 129
Isopentane	Oxidised to t-amyl hydroperoxide, which is reacted with propylene to form propylene oxide and t-amyl alcohol, yielding isoamylenes on dehydration, these being dehydrogenated to isoprene	139

A number of catalysts capable of isomerising 2-methylpentene-1 to 2-methylpentene-2, notably phosphoric acid on Kieselguhr,[110] calcium aluminate molecular sieve,[111] potassium on graphite[112] and complexes of palladium chloride[113] are described in the patent literature. However, high conversions and efficiencies favour silica-alumina catalysts for the isomerisation at relatively low temperatures and pressures.[114-116] As an acid centre silica-alumina catalyst would inevitably give considerable polymer formation with tertiary olefins, some pretreatment with acid to neutralise these centres is required in order to obtain the high reaction efficiencies necessary for economic operation.

A number of patents describe the pyrolysis step without catalyst,[110, 117-119] but higher yields are obtained in the presence of hydrogen bromide as promoter,[115, 120] and this appears to have been the catalyst first used commercially. Addition of 2-methylbutene-2 together with hydrogen bromide has been claimed to give improvement,[121] but somewhat later patents describe the effect of added ethylene dichloride[122] and of n-propyl mercaptan:[123] these would have the advantage of reducing or avoiding the corrosion problems inseparable from the use of hydrogen bromide.

Details of the purification method used to make final product isoprene have not been reported, but as the final reaction is a pyrolysis, and some piperylenes and monolefins are almost certainly present the purification is unlikely to be simpler or cheaper than for the product from the direct catalytic dehydrogenation of isoamylenes.

4.4.2. Other routes to isoprene There are now a number of possible routes to synthetic isoprene monomer which will compete with the method described above using propylene dimers. These, with the raw materials used and references in which further details can be found, are shown in the following Table IV.4.

Commercial plants for extracting the isoprene present in the C_5 fraction separated from the liquid products of steam-cracking plants have been announced by Japanese Geon in Japan and by Erdölchemie in Germany[126, 141, 161] Both plants will use C_5 fractions collected from a number of large steam-crackers, since even with the very large ethylene plants now coming into use, the amount of isoprene available will not be sufficient from a single plant to provide feedstock for an extraction unit of viable commercial scale. Clearly, with the amount of isoprene formed in the steam cracking operation varying substantially with the type of naphtha feedstock cracked and with the cracking conditions used, but amounting to not more than about 5 per cent of the tonnage of ethylene produced, this source of isoprene is limited in ultimate tonnage and is dependent for any growth on the expansion of ethylene capacity based on the cracking of naphtha or heavier feedstocks.

The per-pass conversion in direct dehydrogenation of isoamylenes is limited by the thermodynamic equilibrium between mono- and di-olefin to around 40 per cent. Much work is in progress to overcome this limitation and hence reduce the overall costs[155] by using oxidative dehydrogenation. Although fully developed processes for producing butadiene from butenes by oxidative dehydrogenation have been reported,[156] the status of development of isoprene production by oxidative dehydrogenation is not yet clear. In particular, the published literature does not make clear whether the reduction in processing costs through operating the dehydrogenation in the presence of oxygen to give higher overall conversions is sufficient to counterbalance the somewhat lower efficiency of the oxidative reaction compared with the well established and commercially operated direct dehydrogenation process.

The direct catalytic dehydrogenation of isoamylenes requires very roughly 1·2 to 1·4 tons of isoamylenes per ton of isoprene made, and the route from propylene through 2-methylpentene-2 described above needs approximately 2 tons of propylene per ton of isoprene made.[14, 126] In both routes costs depend on the value and availability of feedstocks, but the propylene based synthesis must be more sensitive to raw material cost. In Europe and Japan propylene is derived mostly as co-product with ethylene from naphtha crackers, in the USA mostly from refinery (catalytic-cracking) operations. In none of these areas is any large surplus of propylene expected to develop, and in the long term propylene value in other uses is more likely to increase than to diminish.[142] Isoamylenes have so far been produced by extraction from the C_5 fraction of catalytically cracked gasoline, and this source is unlikely to provide sufficient material at low enough cost to provide for any very large expansion in isoprene monomer production.[126] However, recent reports and patents have disclosed possible alternative sources of isoamylene. Banks and co-workers[143] have described the production of 2-methylbutene-2 by reaction of isobutene with butene-2:

$$CH_3-\overset{\overset{\displaystyle CH_3}{|}}{C}=CH_2+CH_3-CH-CH-CH_3 \longrightarrow CH_3-\overset{\overset{\displaystyle CH_3}{|}}{C}=CH-CH_3+CH_3-CH=CH_2$$

or by the reaction of isobutene with propylene:

$$CH_3-\overset{\overset{\displaystyle CH_3}{|}}{C}-CH_2+CH_3-CH=CH_2 \longrightarrow CH_3-\overset{\overset{\displaystyle CH_3}{|}}{C}=CH-CH_3+CH_2=CH_2$$

The reaction of mixed butenes containing isobutene in which the butene-1 has been isomerised to butene-2, so that the isobutene reacts with butene-2 to give isoprene and propylene is described in recent patents,[144] and Bradshaw and co-workers have described the dismutation reaction with a number of olefins.[145] The reaction[143, 145] is a general one, following the course:

$$\begin{array}{c} \overset{\overset{\displaystyle R_2}{|}}{R_1-C}=\overset{\overset{\displaystyle R_3}{|}}{C}-R_4 \\[2mm] + \\[2mm] \overset{\displaystyle R_5-C}{\underset{\underset{\displaystyle R_6}{|}}{|}}=\overset{\displaystyle C-R_8}{\underset{\underset{\displaystyle R_7}{|}}{|}} \end{array} \longrightarrow \left[\begin{array}{c} \overset{\overset{\displaystyle R_2}{|}}{R_1-C}-\overset{\overset{\displaystyle R_3}{|}}{C}-R_1 \\[2mm] \vdots \quad \vdots \\[2mm] R_5-\underset{\underset{\displaystyle R_6}{|}}{C} \quad \underset{\underset{\displaystyle R_7}{|}}{C}-R_8 \end{array} \right] \longrightarrow \begin{array}{c} \overset{\overset{\displaystyle R_2}{|}}{R_1-C} \quad \overset{\overset{\displaystyle R_3}{|}}{C}-R_1 \\[2mm] \parallel \quad + \quad \parallel \\[2mm] R_5-\underset{\underset{\displaystyle R_6}{|}}{C} \quad \underset{\underset{\displaystyle R_7}{|}}{C}-R_8 \end{array}$$

with a four-centre transition state. This scheme readily predicts the products formed, but does not of course clarify the basic mechanism of the reaction. Practically, however, commercial development of this reaction would allow the production of isoamylene from butene mixtures, and particularly from the isobutene-normal butene mixtures remaining after extraction of butadiene from the C_4 fraction of ethylene plants using naphtha or heavier feedstocks. This extraction of butadiene is already practised on a very large scale in Europe and Japan, and such a process could, if its economic viability is proved, considerably increase the quantities of isoamylenes available for isoprene manufacture. The other possible source of isoamylenes is from dehydrogenation of isopentane. This is, however, a valuable lower boiling gasoline component[126] and its availability on a large scale at low cost cannot be taken for granted. The single stage dehydrogenation of isopentane to isoprene, analogous to the single stage production of butadiene from normal butane, requires roughly 2 tons isopentane per ton isoprene, and again availability of isopentane at low enough cost cannot be assumed.

Overall, although the tonnage of isoamylenes available from present commercially proved technology is not likely to provide for any very large expansion in monomeric isoprene capacity, new olefin chemistry which could be capable of greatly increasing isoamylene availability has been described, and if this is successfully applied commercially could make the routes to isoprene via isoamylenes relatively more attractive in the long run than the route based on propylene dimerisation.

The two-stage isoprene process developed by Institut Francais du Petrole (IFP) has been described in the literature.[138, 146-148] In this process, isobutene, present in a mixture of C_4 hydrocarbons (e.g. raffinates from butadiene extraction plants), is reacted with formaldehyde in sulphuric acid solution to form 4·4-dimethyl-1·3-dioxane.

$$CH_3-\underset{\underset{CH_3}{|}}{C}=CH_2+2H.CHO \longrightarrow \underset{\underset{CH_3}{/}}{\overset{\overset{CH_3}{\backslash}}{C}} \underset{\underset{O-\!\!-\!\!-CH_2}{\diagup}}{\overset{\overset{CH_2-CH_2}{\diagup \quad \backslash}}{\backslash \qquad O}}$$

The product is recovered from the reaction mixture (it is formed in about 90 per cent efficiency on both isobutene and formaldehyde). A complex mixture of higher boiling products and some tertiary butanol are formed as by-products. The dioxane is cracked over a granular quartz catalyst impregnated with phosphoric acid to form isoprene, formaldehyde and water:

$$\underset{\underset{CH_3}{/}}{\overset{\overset{CH_3}{\backslash}}{C}} \underset{\underset{O-\!\!-\!\!-CH_2}{\diagup}}{\overset{\overset{CH_2-CH_2}{\diagup \quad \backslash}}{\backslash \qquad O}} \longrightarrow H_2C=\underset{\underset{CH_2}{|}}{C}-CH_2=CH_2+H.CHO+H_2O$$

The isoprene so formed is recovered by fractionation: an advantage of the process is said to be that polymerisation grade isoprene can be produced using a relatively simple purification method.

A number of different methods for reacting isobutene and formaldehyde to isoprene have been reported. A single-step reaction is described,[149] but this requires a large excess of isobutene in the reaction. Hence a large recycle and a fairly pure isobutene feedstock is needed and operation with the raffinate from butadiene extraction plants is not possible. Alternative catalysts for the dioxane decomposition step are described in patents,[149, 150] and variations of the dioxane forming step have been investigated,[151, 152] cation exchange resins being claimed effective. The old reaction[137] between isobutene and formaldehyde to form the unsaturated C_5 alcohol, which is subsequently dehydrated to isoprene is also made the basis of isoprene production processes in several patents.[153] A variation of the isobutene-formaldehyde process by Marathon Oil Co.[159] uses the reaction sequence:

$$HCl + CH_3OH + H.CHO \rightleftharpoons H_2O + CH_3 - O - CH_2Cl$$

$$CH_2 = C(CH_3)_2 + CH_3 - O - CH_2Cl \longrightarrow CH_3 - \underset{\underset{Cl}{|}}{\overset{\overset{CH_3}{|}}{C}} - CH_2 - CH_2 - O - CH_3$$

$$\downarrow Heat$$

$$CH_3OH + HCl + CH_2 = \overset{\overset{CH_3}{|}}{C} - CH = CH_2$$

No plans for commercialisation of this process have yet been announced.

In the synthesis of isoprene from acetone and acetylene, the first reaction is the formation of methylbutynol:

$$\underset{CH_3}{\overset{CH_3}{\diagdown}} C = O + HC \equiv CH \longrightarrow CH_3 - \underset{\underset{OH}{|}}{\overset{\overset{CH_3}{|}}{C}} - C \equiv CH$$

This is hydrogenated to methylbutenol, which is then catalytically dehydrated to form isoprene. Few details of the commercial process have been reported, and although high efficiencies of reaction are obtained in each step the raw materials, particularly acetylene, are relatively costly.

4.4.3. Outlets for isoprene–polyisoprene rubbers　The earliest use of isoprene was as a comonomer in butyl rubber, added to introduce into this rubber a small number of unsaturated linkages for cross-linking.[156] Analogous applications in other largely saturated polymers, such as polyethylene and polypropylene, have been described.[157] The use in butyl rubber is the only commercially established such application, and such co-monomer outlets can never be more than relatively small tonnage-consumers of isoprene.

None of the chemical applications in current use require large volumes of isoprene, and have not yet shown signs of developing into large volume outlets. The major outlets are in stereospecific elastomeric polymers, notably cis-1·4 polyisoprene of basic structure:

$$\begin{bmatrix} \underset{CH_3}{\overset{CH_2}{\diagdown}} C = \underset{}{\overset{CH_2}{\diagup}} CH \end{bmatrix}_n$$

This is a tough elastomeric polymer, essentially similar to natural rubber. The trans-1·4-polyisoprene of basic structure:

$$\begin{bmatrix} \underset{CH_3}{\overset{CH_2}{\diagdown}} C = CH \underset{CH_2}{\diagdown} \end{bmatrix}_n$$

resembles balata or gutta-percha and is used in golf-ball covers. The large volume polymer is clearly the cis-1·4-polymer, which is produced using either Ziegler catalyst systems[158] or a lithium alkyl catalyst, the Ziegler systems yielding a polymer of higher cis-content.[159] The close similarity of high cis-polyisoprene to natural rubber makes it technically competitive with the natural product in important applications, notably in tyres, and particularly large commercial tyres.[159] The main factors affecting the isoprene expansion rate[126, 154, 159] are the cost of polyisoprene rubber relative to natural rubber, the availability of synthetic monomer and of the raw materials from which it is derived, and the uncertain political situation in South-East Asia where the greater part of natural rubber supplies are produced. The last factor may persuade rubber users to accept synthetic polyisoprene for a part of their needs at a price somewhat higher than that prevailing for natural rubber. Other secondary factors are the fluctuations in price and variability in quality of natural rubber,[160] although developments are in progress to improve the quality standardisation of the natural product. One estimate of the natural rubber consumption growth rate is shown in Table IV.5 following:[160]

TABLE IV.5

World Production of Natural Rubber ('000 Long Tons)

	Production	*% of total rubber*
1960	1,990	49
1969	2,780	34
1970	2,910 (Est.)	—
1975	3,550 (Est.)	—

TABLE IV.6

World Synthetic Rubber Capacities 1966–72 ('000 Long Tons)
(Excludes Soviet Bloc and Communist China)

	1966		*1969*		*1972*	
	USA	*Free world total*	*USA*	*Free world total*	*USA*	*Free world total*
SBR	1,587	2,721	1,624	3,332	1,674	3,876
Poly-butadiene	206	461	396	788	496	1,059
Poly-isoprene	90	135	150	212	142	445
EPDM	42	47	107	129	205	360
Butyl rubber	190	303	160	330	160	370
Neoprene	145	235	164	341	219	411
Nitrile	90	182	112	247	115	277
Others	60	100	80	130	100	140
Total	2,410	4,184	2,793	5,509	3,111	6,938

This may well rise to around 4·5 million tons by 1980, giving an annual growth rate of around 3 per cent for 1975–80, and about 4 per cent for 1970 to 1975. Consumption seems likely to follow a similar increase, with a continuation of the trend reducing the proportion of natural rubber in the total consumed. One estimate of synthetic rubber capacities in the period 1966 to 1972 is shown in Table IV.6 above.[159]

In addition to the polyisoprene capacities in the above table, it is reported that two 50,000 metric ton/annum plants are operating in the USSR.[159] All these figures confirm that there may be room for large expansions of polyisoprene capacity in the next ten years, but costs will have to be competitive with natural rubber, assuming that no serious disturbances in South-East Asian production occur. The three main competitive routes to isoprene monomer start from propylene, isobutene and isoamylenes. There are, of course, large tonnages of propylene available, but also a large number of competing outlets (including gasoline production by alkylation), so that the propylene based route to isoprene, requiring about 2 tons propylene per ton isoprene, will be vulnerable to rises in value of propylene by pressure of these competing outlets. The extraction route from steam-cracked C_5 fractions will be limited to roughly 5 per cent of the ethylene capacity in naphtha cracking plants, and moreover requires collection from several plants to provide enough feedstock for a viable isoprene extraction plant. Nevertheless, this is likely to provide the next stage of expansion in isoprene monomer production. Present evidence points to competition between the isobutene/formaldehyde routes and isoamylene dehydrogenation for subsequent isoprene expansions. If such expansions become large, new sources of isoamylenes to supplement material at present extracted from catalytically cracked gasoline may be needed. Application of the new dismutation technology to the butene raffinate from butadiene extraction plants might be one such source but will compete with the use of the same feedstock to provide isobutene for the isobutene/formaldehyde synthesis. Present evidence does not allow of a clear demonstration of significant cost differences between these possible isoprene processes. It is however quite clear that the propylene dimer route will not remain the only synthetic route to isoprene competing with isoamylene dehydrogenation in the next ten years.

4.5. Propylene Oligomers and Detergent Intermediates

The general situation showing the growth of soaps, detergents and cleaning agents in the world as a whole is illustrated by the figures in Table IV.7 given by Henkel and Cie GmbH and reported recently.[162]

TABLE IV.7

World Production of Soaps, Detergents and Cleaning Agents

Product	1960 '000 tons	1960 %	1966 '000 tons	1966 %	1968 '000 tons	1968 %	1969 '000 tons	1969 %
Soaps	6,885·6	63·1	6,762·5	47·4	6,492·9	42·0	6,440	40·0
Synthetic detergents	3,433·3	31·4	5,957·1	41·8	7,369·5	47·7	8,050	50·0
Scouring agents	461·4	4·2	653·6	4·6	584·5	3·8	560	3·4
Others	138·9	1·3	887·5	6·2	1,014·8	6·5	1,080	6·6
Total	10,919·2	100·0	14,260·7	100·0	15,461·7	100·0	16,130	100·0

Much the most important of the synthetic materials are the anionic active agent based products, and of these active agents the alkyl benzene sulphonates are by far the largest volume material in use. The main anionic agents are listed in Table IV.8, derived from reference 163.

TABLE IV.8

The Main Synthetic Detergent Active Agents

Raw material of manufacture	Surfactant material	Remarks
1. Propylene tetramer + benzene 2. Normal paraffins + benzene	Alkylbenzene sulphonate (ABS) SO_3Na (benzene ring) C_{11-14}	Main anionic active agents in domestic products
C_{12}–C_{18} linear primary alcohols	Alcohol sulphates Alcohol ether sulphates Alcohol ethoxylates	Relatively high-priced, but used in liquid products in combination with ABS for control of foaming
C_{16}–C_{18} primary alcohols	Alcohol sulphates Alcohol ethoxylates	Good detergents, but would need price parity to compete with ABS
Linear alpha olefins	C_{15}–C_{18} olefin sulphonate C_{12}–C_{14} Amine oxides etc.	Olefin supply at low enough cost limits growth. Speciality high-priced products
Normal paraffins	C_{14}–C_{18} Paraffin sulphonates C_{12}–C_{16} Paraffin sulphonates	Feedstock cheap, process costs may be high and application probably limited to liquids
Secondary alcohols	Secondary alcohol sulphates Secondary alcohol ethoxylates	Not commercially produced. C_{11}–C_{15} alcohol based products compete with 'hard' non-ionics in mainly non-domestic uses.
Linear alpha olefins	Secondary alkyl sulphates	Find only limited use.

The use of detergent intermediates is expected to increase quite rapidly,[164] but concern about pollution through active agents which do not degrade biologically in sewage works and rivers has led to the use of non-biodegradable products being restricted. Few West European countries now allow unrestricted use of materials based on dodecylbenzene manufactured from propylene tetramer, and West Germany and Spain have laws banning such products. The majority of other countries, including the United States, have voluntary agreements whereby producers make mostly or entirely biodegradable products. The expected growth in alkyl benzenes will therefore be of the linear alkyl benzenes, and the consumption of tetramer based material is expected to decline and probably disappear completely in Western Europe by about 1975 and in Eastern Europe by about 1980. Other biodegradable products are also expected to increase, notably the synthetic fatty alcohols and linear primary alcohol sulphates and the alpha olefin-based materials, particularly the alkenyl sulphonates derived from linear alpha olefins. However, none of these materials are expected to reach the high volumes forecast for the linear alkyl benzene sulphonates during the next decade.

The following sections review the production of propylene tetramer based alkyl benzene sulphonate, and the possibilities of producing biodegradable materials through the oligomerisation of propylene and other lower olefins.

4.5.1. The production of propylene tetramer (UOP Process) A simplified flow sheet of the process for producing propylene tetramer on a solid phosphoric acid catalyst (i.e. a solid support such as kieselguhr impregnated with phosphoric acid) is shown in Fig. 4.3. The propylene feedstock is derived either from a

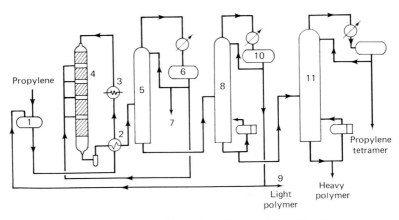

FIG. 4.3   *Production of Propylene Tetramer From Propylene*

refinery stream such as a C_3 fraction from a catalytic cracker or the C_3 fraction from a steam cracker producing ethylene. The former is typically a propane/propylene mixture containing around 65 per cent propylene, and the latter has much higher propylene contents. Before charging to the reactor hydrogen sulphide and any basic nitrogen compounds present must be removed, for example by soda washing and water scrubbing. The purified feed is mixed in the drum 1 with a light polymer recycle stream, heated by exchange with reaction product in the exchanger 2 and the temperature adjusted to that required for the reaction in the heater 3 before feeding to the reactor 4. The reactor is charged with multiple beds of solid phosphoric acid catalyst, the reaction temperature being controlled by inter-bed quench with a propane-rich stream. The reactor effluent is passed to a depropanising column 5 where unreacted propane is taken overhead and in part recycled as quench stream to the inter-bed quench points and in part discharged by the line 7 from the reflux drum 6 to remove the propane added in the feed from the system. The depropanised material from column 5 is passed to the recycle column 8 where an overhead distillate of light polymer boiling up to the initial boiling point of the tetramer product is removed. This light polymer is used principally for recycle to the combined feed drum for further polymerisation, but if required a light polymer stream can be withdrawn as a product via the line 9 from the reflux drum 10. If it is desired to make propylene trimer as a product, an additional column to fractionate out the correctly cut trimer product from this light polymer is necessary. The crude propylene tetramer product is withdrawn from the base of column 8 into the re-run column 11. Here the specification tetramer product is taken overhead as distillate, and a small amount of heavy polymer rejected as bottoms product. A slightly heavier product can be made, for example containing a certain amount of pentamer, if this is required for the subsequent processing. The fractionating system is capable of a good deal of variation, according to the detailed specifications of the trimer and tetramer products which are being manufactured. The usual main products are C_9 trimer, C_{12} tetramer polymer, C_{15} and higher polymers and the light polymer between C_4 and C_8, the last usually being blended into gasoline. The unreacted propane containing a certain amount of propylene withdrawn as a purge stream 7 is usually used as fuel or sold as liquified gas (L P G) for fuel use. The reaction conditions are in the range 14–42 atm pressure and 170–220°C. To produce a satisfactory product the feedstock must be free of isobutene. A number of references have described the process in more or less detail, and the chemistry has been discussed in the foregoing sections.[165] According to E. K. Jones,[166] the solid phosphoric acid catalyst gives at 200°C and at 500–1000 psig about 16 per cent of dimer, about 10 per cent of C_7–C_8, 53 per cent C_9, 6 per cent C_{10}–C_{11} and about 20 per cent C_{12} in the once through product. The formation of products of carbon number intermediate between the dimer, trimer and tetramer illustrates the occurrence of conjunct polymerisation reactions giving products much more complex than the simple carbonium ion mechanism of polymerisation suggests. These reactions can be reduced by reduction of the reaction temperature, but at the expense of reaction rate. A hydrocarbon type analysis of the trimer and tetramer products has been quoted[167] by the Enjay Company based on alcohols produced on hydroformylation of the oligomers. These are shown in Table IV.9: the highly branched structures of the polymers are clearly shown in these results.

TABLE IV.9

Structures of Propylene Trimer and Tetramer (Vol. Percent)

		Trimer	Tetramer
I	$RCH = CH_2$	1	2
II	$RCH = CHR^1$	14	10
III	$RR^1C = CH_2$	8	7
IV	$RR^1C = CHR^2$	35	26
V	$RR^1C = CR^2R^3$	42	55

Many plants have been operated using this process, making a range of products from narrow cut tetramers[168] to combinations of trimer, tetramer and pentamer.

A process using phosphoric acid on pumice catalyst for making trimer and tetramer in fixed bed multitubular reactors operated at 180–250°C and 400–1000 psig has been described:[169] this is said to give 70–80 per cent yields of these products.

Liquid phosphoric acid (a mixture of orthophosphoric and pyrophosphoric acid was reported to give best results) processes were also used at 180–200°C and 40–60 atm pressure in Germany,[171] giving about 90 per cent olefin conversion and, in single pass operation, about 30 per cent yields of the tetramer. Lower olefinic products could be recycled, at least in part, to increase the tetramer yield.

Phosphoric acid on quartz has also been used as catalyst: this is of considerably lower activity than the phosphoric acid on kieselguhr catalyst because of the low surface area, but is very robust and can if necessary be regenerated by burning off polymeric deposits from time to time.[166]

The great bulk of the propylene trimer and tetramer produced is now made by the phosphoric acid on kieselguhr catalyst process. Ease of control of reaction temperature by injection of the propane-rich recycle gas in reactors of simple and cheap design and relative freedom from the corrosion problems associated with the use of large volumes of hot liquid phosphoric acid are major advantages.

4.5.2. Uses of propylene tetramer   A relatively small tonnage of propylene tetramer is converted to dodecylmercaptan by addition of hydrogen sulphide. The oldest established process uses a silica-alumina catalyst of low (1–5 per cent) alumina content[172] at about 100°C and 70 atm pressure. Anhydrous HF or BF_3/phosphoric acid also catalyses the reaction,[173] and a new process using BF_3 as catalyst in a continuous packed column reactor operated at ambient temperature and close to atm pressure has been described.[174] The dodecylmercaptan is used mainly as a regulator in emulsion polymerisation processes.[175]

The tetramer, like other olefins, will react with maleic anhydride to form alkenyl succinic anhydrides: these products have been used in producing ashless dispersent additives for lubricating oils.[176]

However, by far the largest tonnage outlet for propylene tetramer is in the production of dodecylbenzene, which is sulphonated to give dodecylbenzene sulphonate, the principal large tonnage anionic active agent in domestic and industrial detergents. The first step in the production of this material is the alkylation of the tetramer to benzene, benzene being the only aromatic now used in any quantity for this production.

Although alkylation with anhydrous aluminium chloride, concentrated sulphuric acid and anhydrous hydrogen fluoride as catalysts has been described,[177] anhydrous hydrogen fluoride is much the most important catalyst for alkylation with propylene tetramer. Aluminium chloride is extensively used in producing linear alkylbenzenes, but the use of sulphuric acid catalysts has been practically discontinued.

Use of a large excess of benzene to avoid the formation of higher boiling polyalkyl benzenes, and of as low a temperature as is consistent with a commercially acceptable rate of reaction and substantially complete conversion of the olefin are required,[177–179] to ensure a high efficiency of monoalkyl benzene formation. Also the absence of isobutene from the feed to the polymerisation is necessary, since breakdown of isobutene-containing polymers during the alkylation causes reduction in yield and adversely affects the quality of the product for sulphonation.[178, 181] Some of the di- and polyalkyl benzene by-products have found uses in producing, for example, oil soluble sulphonates, and the higher dodecylbenzenes produced in the hydrogen fluoride catalysed reaction have been described in some detail in the literature.[180]

In the early 1950's excessive foaming in sewage works and rivers, especially in densely populated and highly industrialised areas of the world led to serious inconvenience and nuisance and this was ascribed to the extending use of synthetic active agents in domestic and industrial detergents. A trial held with the co-operation of the United Kingdom detergent industry at the Luton Sewage Works showed that when propylene tetramer based alkylbenzene sulphonate was largely replaced by a near-linear alkylbenzene sulphonate (with which the catchment-area for the sewage works was supplied during the period of the test), the effluent from the sewage works contained about half the residual concentration of undecomposed active agent compared with earlier operation when a very much higher proportion of tetra-propylene based material was in use in the area.[182] This finding, coupled with wide-spread investigations in other areas, and with a great deal of laboratory study, established that the highly branched propylene polymer side chain in the dodecyl-benzene sulphonate resulted in products which degraded only slowly under bio-logical attack by micro-organisms. Replacement of the biologically 'hard' products by the 'soft' linear products in the densely populated and highly industrialised parts of the world has been proceeding since that time; the amount of tetrapropylene based material used is likely to become very small by 1975 and virtually to disappear by 1980. This is illustrated by the forecast for Europe shown in Table IV.10 (see ref. 183). The situation is similar in other industrialised areas, some of which have imposed statutory limits on the use of biologically 'hard' active agents, while volun-tary schemes to eliminate their use are in operation in most others.

TABLE IV.10

Forecast of Synthetic Detergent and Active Agent Demand in Europe

	Demand, thousands of metric tons per year					
	1970		1975		1980	
	Total Europe	Comecon countries	Total Europe	Comecon countries	Total Europe	Comecon countries
Synthetic detergents	4,476	1,236	6,282	2,232	8,350	3,350
Detergent chemicals						
Linear alkylbenzenes	331	62	661	165	838	215
Dodecylbenzene	154	20	18	15	0	0
Synthetic fatty alcohols	68	13	140	45	255	95
Alpha olefins	63	3	95	10	150	30
Ethylene oxide for ethoxylates	180	50	310	100	460	150

(*Chemical Engineering News*, 16 November 1970, p. 28).

The alkylbenzene sulphonates have many advantages, low cost being an important one and technical performance another, and are likely to remain the largest volume detergent intermediate for at least the next decade, but with the highly branched tetrapropylene side chain replaced by a linear or near linear side chain. The following are the most important sources of linear materials which are likely to be so used:

(1) Thermal cracking of either highly waxy fractions from suitable crudes or from specially prepared (for example by urea adduction) predominantly linear higher paraffinic material[184, 185] to linear olefins. This cracking of de-oiled or purified waxes is carried out under mild conditions reducing unwanted secondary reactions and gives products of high linearity and with alpha-olefin contents up to 90 per cent.[184] The presence of some cyclic compounds and of small amounts of saturateds are a disadvantage in the cracked wax products. The inevitable formation in the cracking step of the whole spectrum of olefins ranging from C_5 up to about C_{20} also provides problems in finding balanced outlets for the wide range of olefins made. The C_{10}–C_{13} or the C_{11}–C_{14} fractions of these wax cracked olefins are used to produce linear alkylbenzenes.

(2) Normal paraffins up to about C_{18} are now separated in high purity from petroleum fractions by selective adsorption on 5A molecular sieves and are commercially available to provide the linear alkyl side chain in alkylbenzene sulphonates.[186, 187] The normal paraffins are converted to alkylbenzenes in three ways, applied usually to the C_{10}–C_{14} or to narrower fractions:

(a) The paraffin is chlorinated under conditions to produce the alkyl monochloride in maximum yield,[188] and this is then alkylated directly to benzene using aluminium chloride as catalyst.[189, 192]

(b) The mono-alkyl chloride formed as in (a) is dehydrochlorinated catalytically to give a mixture of all the isomeric linear olefins:[188, 191] the olefins so produced are alkylated to benzene using either aluminium chloride or anhydrous hydrogen fluoride catalysts.

(c) The normal paraffins are catalytically dehydrogenated to olefin (this again giving a mixture of all the linear isomers). The per-pass conversion in this dehydrogenation is low under conditions which avoid extensive cracking and skeletal isomerisation, and the mixture of olefin and paraffin is alkylated to benzene, the unreacted paraffin being recovered and recycled to the dehydrogenation step during recovery by fractionation of the detergent alkylate product.[193] In a later version of this process, the olefins are separated from the dehydrogenated product prior to alkylation using solid absorbents. The paraffin is then recycled to the dehydrogenation without the necessity of processing through the alkylation section of the unit.[194]

(3) Linear alpha-olefins of the appropriate carbon number produced from ethylene by the Ziegler catalyst systems described above (see also ref. 195) can be used to alkylate to benzene by the available processes. There is little advantage in using alpha-olefins in this reaction, since extensive double bond movement occurs during reaction, and the products produced with a given catalyst under constant conditions appear not to vary much with the isomer-distribution of the starting olefins.[194]

Apart from providing olefins for linear alkyl benzene production, the normal paraffins of a rather higher carbon number, for example C_{14}–C_{17}, are used to produce so called alkane sulphonates, using the sulphoxidation process with sulphur dioxide and oxygen.[196] These alkyl sulphonates are produced commercially, but do not show signs of approaching in volume the alkylbenzene sulphonates. Alkenyl sulphonates, made by direct sulphonation of olefins with sulphur trioxide,[197] are also being produced.[196, 197] For these the alpha-olefins produced by ethylene oligomerisation or by wax cracking are advantageous raw materials, but again no sign of these products becoming very large in volume of application is yet evident. The linear olefins can also be used for the production by hydration of secondary alcohols or by hydroformylation of primary alcohols. The primary alcohols in the detergent range compete with linear terminal alcohols produced by hydrogenation of natural fatty acids. Commercial production of such primary alcohols by hydroformylation of linear olefins made by wax cracking has been announced.[199] These alcohols from the wax cracked olefins are said to contain 75–80 per cent of linear alcohols, the remainder being mainly singly branched with methyl groups in the 2-position. The alcohols produced by these various methods can be sulphated or ethoxylated to give the various types of active agent shown in Table IV.8 above.

The first section of this chapter indicated a number of new oligomerisation techniques capable of giving much more linear products from propylene than the old processes using acid centre catalysts. There is a substantial patent literature indicating much industrial research activity aimed at applying these techniques to the production of biodegradable products based on propylene and/or butenes. For example, dimerising propylene on cobalt-active carbon catalyst, separating and dimerising again the resulting hexenes is said to give a predominantly C_{12} olefin which can be used to make alkylbenzenes suitable for alkylbenzene sulphonate production.[200] The dimers of straight chain C_5–C_8 olefins produced on nickel oxide-silica-alumina catalysts are also claimed to give biodegradable alkylbenzene sulphonate products,[201] and the dimers of straight chain C_4–C_8 olefins using aluminium alkyl catalysts are said to give singly branched C_8–C_{16} olefins suitable for use in preparing biodegradable detergents.[202] There is no indication that any of these methods are approaching commercialisation, and at present the use of propylene oligomers in producing detergent intermediates seems likely to fall to very low levels during the next decade. The normal paraffins, olefins produced by wax cracking and possibly the linear terminal olefins produced by ethylene oligomerisations on aluminium alkyl catalysts seem likely to provide the future raw materials for detergent manufacture.

4.6. Propylene Oligomers and Co-Oligomers as Intermediates for Plasticiser Olefin Production

The increase in the use of polyvinyl chloride (PVC) over the last twenty-five years has been enormous, and world consumption now exceeds 5 million tons. At least some of the increased consumption is dependent on the modification in properties and the increased ease of processing obtained by using plasticisers. The main plasticisers used are alkyl esters of phthalic acid and the tri-alkyl phosphates, the latter because of relatively high cost being reserved for applications where the conferred properties are particularly desirable. The use of dibasic acids other than phthalic, notably adipic and sebacic acids, for plasticiser ester production is confined on account of their high cost to speciality products. The large volume applications use almost entirely the phthalate esters. Dibutyl phthalate was one of the earliest ester plasticisers used, but has now been largely superseded by higher esters whose lower volatility gives longer durability to the plasticised polymer in use. The most important of the phthalate esters now used, with the source of their raw material, are shown in Table IV.11.

TABLE IV.11

Main Phthalate Esters in Use as PVC Plasticisers

Phthalate of	*Raw material used for alcohol production*
2-Ethylhexanol	Normal butyraldehyde (mainly from propylene hydroformylation)
Iso-octanol	Heptenes (propylene/butene co-dimer) hydroformylation
Isodecanol	Propylene trimer hydroformylation
Isononanol	Di-isobutene (from isobutene dimerisation) hydroformylation
Alphanols } Linevols }	From C_6–C_7 cracked wax olefin hydroformylation
Alfols	Hydrolysis of oxidised aluminium alkyls produced from ethylene oligomerisation with Al alkyl catalysts
Tridecanol	Propylene tetramer hydroformylation

There are quite wide variations in the proportions of the particular esters used in various parts of the world. The proportion in the United States was estimated in 1969[203] to be as shown in Table IV.12.

TABLE IV.12

USA Use of Various Plasticiser Alcohols in 1969

2-ethylhexanol	47%
Isodecanol	21%
Iso-octanol	20%
Tridecanol	4%
All others	8%

The total US demand in 1969 was 825 million pounds, estimated to grow to 1,200 million pounds by 1973. In Western Europe, apart from the United Kingdom, 2-ethylhexanol is the most popular plasticiser alcohol, with iso-octanol taking an appreciable share. In the United Kingdom, the alphanols from wax cracked olefins are largely used together with iso-octanol, 2-ethylhexanol being less prominent than in the rest of Europe and the United States.

The production of propylene trimer used for isodecanol production and of propylene tetramer used for tridecanol production have already been described above: the next section deals with the production of heptenes for iso-octanol preparation.

4.6.1. Production of heptenes for iso-octanol preparation Heptenes for hydroformylation to iso-octanol have hitherto been almost entirely produced as a byproduct of polymer gasoline production. Propylene and the butenes are co-dimerised the heptenes being separated as a C_7 fraction from the total gasoline range polymer. The feedstocks used are the propylene and butylene streams from catalytic cracking units. A simplified flow sheet of the polymer gasoline-producing process is shown in Fig. 4.4. The liquid C_3/C_4 feed is introduced into the caustic washer 1

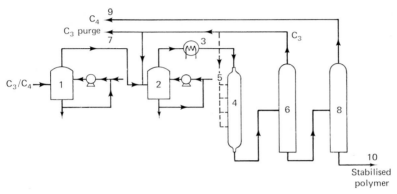

FIG. 4.4 *Polymer Gasoline Production*

where impurities such as hydrogen sulphide are removed. This is followed by a water wash in 2, which besides removing any residual caustic soda also takes out any other water soluble contaminants, particularly basic compounds. The purified liquid feed is preheated in 3 and passed to the reactor 4. As in the production of propylene trimer and tetramer, the usual phosphoric acid on kieselguhr catalysts require the water content of the feed to be controlled for most satisfactory operation.[204] With the chamber type of reactor, injection of recycle material between the catalyst beds is used to control the reaction temperature. This recycle material, which is rich in propane, is commonly fed by the line 5 from the overheads of the depropanising column 6. This recirculation of unreacted material also provides means of controlling the olefin content of the total feed passing into the reactor 4. From the reactor, the product passes to the depropaniser 6, where unconverted C_3 is taken overhead, part being recirculated to the reactor and part being removed as a purge stream from the line 7. This purge stream removes any propane fed in with the fresh feed, and also of course any unconverted propylene associated with it in this stream. The depropanised polymer product passes to the debutaniser 8, from which unreacted C_4 rich in butanes is taken overhead, leaving the stabilised polymer as residue removed by the line 10.

Hydrogen sulphide, mercaptans and basic compounds poison the catalyst and must be removed from the feedstocks. Both the feedstock and the wash water should be free of oxygen since this promotes the deposition of tarry materials on the catalyst. With the phosphoric acid on kieselguhr-catalyst, which is the most common catalyst employed for this purpose, the water content of the feed must also be controlled. Too high a water content in the feed results in softening of the catalyst and plugging of the reactor, whereas if the water content is too low, coke deposition on the catalyst which reduces the activity and increases the pressure drop becomes more marked. The optimum water content is a function of the conditions used, particularly the temperature and pressure, and is normally closely regulated.[204] Raising the reaction temperature increases the reaction rate, but also accelerates accumulation of carbonaceous deposits on the solid catalyst and shortens the catalyst life. Reactor temperature is usually held within the range 300/425°F.[205] Higher operating pressures give higher conversions, and the normal operating pressure is within the range 400–1500 psig, pressures in the region of 500 psig being normal for chamber type reactors. Higher pressure operation permits the use of lower temperatures and reduces the formation of heavy polymers and of coke on the catalyst and hence gives longer catalyst lives. Increasing contact time increases the formation of heavy polymer, too low a contact time reduces the per pass conversion to uneconomic levels. The total feed to the reactor is usually within the range 0·12–0·46 gallons per lb of catalyst per hour.[166] The relative rates of reaction of the olefins present in the feed are in the decreasing order: isobutene, butene-1, butene-2, propylene,[166] although the interactions and rearrangements in the carbonium ion reactions involved result in normal butenes reacting faster in the presence of isobutene and propylene reacting faster in the presence of butenes than these olefins do when reacted alone. In producing polymer gasoline where the maximum yield of polymer in the gasoline range is aimed at, conditions giving high product yields combined with high olefin conversions have been achieved. Operation to produce high yields of heptenes (the propylene-butene co-dimer) is with the relative reactivities in the

TABLE IV.13

Commercial Heptene Sample Composition
(Reference 206)

	Vol %
2-methylhexene-2	12
3-methylhexene-2	13
2·3-dimethylpentene-1	20
2·3-dimethylpentene-2	21
2·4-dimethylpentene-1	10
2·4-dimethylpentene-2	15
2-methyl-2-ethylbutene-1	7
4·4-dimethylpentene-2	2

order indicated clearly much more difficult while still maintaining overall conversions at an economic level. Although Ipatieff and Schaad report the formation of a product containing 40–45 per cent of heptenes (mainly olefins of the skeletal structure of 2·3-dimethylpentenes) by co-dimerisation of propylene and isobutene at 135 °C and 38 atm pressure[5] such high heptene yields appear not to be obtained in commercial operation with mixed C_4 feedstocks under economic operating conditions, and the heptene product is a good deal more complex than this in structure. An example given in the literature of the composition of the heptenes produced is shown in Table IV.13.[206] The isomers present are mainly of the 2·3- and 2·4-dimethylpentene structures, with relatively little singly branched material and virtually no straight chain olefins present.

The production of propylene trimer and propylene tetramer, which are both also used in the production of alcohols for use in plasticisers, has been described in earlier sections.

In commercial production of heptenes a second column is sometimes provided to separate out the nonenes (a form of propylene trimer).

4.6.2. Plasticiser alcohol production from propylene oligomers and co-oligomers The production of alcohols from olefins by hydroformylation is described in full detail in Chapter 9. The following briefly summarises some special aspects relating to the hydroformylation of propylene oligomers.

By the hydroformylation reaction, heptenes yield iso-octyl alcohol, propylene trimer (nonenes) yields isodecyl alcohol and propylene tetramer (dodecenes) yields tridecyl alcohol.

In the hydroformylation reaction, straight chain olefins react more rapidly than the branched chain olefins. The former, although all reacting at approximately the same rate, give a mixture of products due to double bond movement prior to reaction, and to addition of the formyl group at both sides of the double bond.

Branched chain olefins react faster when the double bond is in the terminal position. In fact a product such as 2-3 dimethyl butene-2 (I) hardly reacts at all and double bond isomerisation takes place more rapidly than addition of the formyl group. As in these substituted olefins the formyl group adds mainly at the terminal carbon atom, the main alcohol formed from such a mixture is 3–4 dimethylpentanol-1 (II).

$$
\begin{array}{c}
CH_3-C = C-CH_2 \\
\quad\ \ | \quad\ | \\
\quad\ CH_3\ CH_3
\end{array}
\qquad \text{(I)}
$$

$$
\begin{array}{c}
CH_3CH - CHCH_2CH_2OH \\
\quad\ \ | \qquad | \\
\quad\ CH_3 \quad\ CH_3
\end{array}
\qquad \text{(II)}
$$

This means in general that the alcohols produced from a mixture of olefins are less branched and contain fewer components than would have been expected from the composition of the starting material.

This may be illustrated by the composition quoted in ref[208] for the commercial iso-octanol made from heptenes and given in Table IV.14.

TABLE IV.14

Composition of Iso-octyl Alcohols from Heptenes

3·4 dimethyl hexanol-1	20%
3·5 dimethyl hexanol-1	30%
4·5 dimethyl hexanol-1	30%
3 methyl heptanol-1 ⎱	15%
5 methyl heptanol-1 ⎰	
Others	5%

4.6.3. Plasticiser alcohols from oligomers—relative performance Some of the properties of the main phthalate esters used in PVC compositions (abstracted from reference[208]) relative to 2-ethyl-hexanol and normal octyl phthalates are shown in Table IV.15.

TABLE IV.15

Properties of PVC Resins Plasticised with Various Phthalate Esters

Phthalate ester	(1) Loss %	(2) Low temp. flex Tf, °C	(3) Water extraction % loss	(4) Kerosene extra % loss	Shore hardness
Di-butyl phthalate	44·0	−41·0	0·25	8·3	62
Di-*n*-octyl phthalate	2·2	−48·2	0·01	51·5	74
Di-iso-octyl phthalate	4·3	−37	0·03	28	74
Di (2-ethylhexyl) phthalate	4·5	−39	0·01	44·3	68
Di-isodecyl phthalate	1·8	−37	0·02	74	71
Di-tridecyl phthalate	0·8	−37·4	0·07	88	83

(1) 24 hours at 87°C over active carbon.
(2) Clash-Berg.
(3) 24 hours at 50°C.
(4) 24 hours at 23°C.

These refer to compositions containing 67 parts of ester plasticiser per 100 parts resin. The table illustrates the high volatility of dibutyl phthalate from the PVC composition, which has led to the substitution of higher phthalates in all applications where this is important. They also show the superior plasticising power of the straight chain alcohol esters as evidenced by the low temperature flex figures. However the plasticising power of the straight chain alcohol products is not so superior to that of the branch esters (and can be counteracted by using more plasticiser), that the large price differential can be tolerated in most applications. Moreover the electrical properties of resins plasticised with the branched alcohol esters (2-ethyl-hexyl and iso-octyl) are somewhat superior (see also reference 209) and these have advantages in cable insulation compositions. The decreased loss in volatility for the tridecyl and isodecyl esters has recently become important for resin applications in the automobile industry, where windscreen fogging through volatilisation of plasticiser can be a problem.[208] At present several products compete for the large volume market in plasticised PVC compositions the main protagonists being:

(1) 2-ethylhexanol, the most generally accepted plasticiser in the United States and Europe, showing good plasticising power and good electrical properties.

(2) The linear primary alcohols (for example, Shell Linevols) derived by hydroformylation of olefins made by wax cracking (the 'Alfol' alcohols made by ethylene oligomerisation would be contenders in this area if their costs were lower). These linear alcohols show very good plasticising power, but are somewhat inferior in electrical properties.

(3) The branched alcohols from heptenes, nonenes and to a lesser extent the dodecenes made from propylene either alone or with butenes, show plasticising power very slightly inferior to that of the 2-ethylhexyl phthalates. The nonene derivatives have advantages in low volatility loss, and the heptene derivatives are only slightly inferior to 2-ethylhexanol.

Some doubts have been expressed (see reference 207) on the possibility of expanding or even maintaining the supply of heptenes from refinery polymer gasoline production. Should these doubts be well founded, the possibility of utilising new technology to produce olefins having greater linearity and hence greater reactivity in hydroformylation and some gain in resin properties compared with the present heptenes may arise. The patent literature provides evidence of considerable activity in this area. The nickel oxide-silica-alumina catalyst,[18, 19] the alkali metal catalysts which are capable of giving higher yields of heptenes from propylene and butenes,[210] and the complex transition metal catalysts,[211] which were discussed in the earlier sections of this chapter, have all been described as capable of use in producing plasticiser intermediates. However, there is not at present any clear indication that commercial use of these techniques is contemplated in the immediate future, nor of how the derived plasticisers will fare in competition with 2-ethyl-hexanol or with the materials derived from linear alpha-olefins using the newer hydroformylation catalysts.

REFERENCES

[1] *Hydrocarbon Processing and Pet. Refiner*, 1969, **48** (*2*), 77.

[2] Egloff, G. and Weinhart, P. S., Proc. 3rd World Petroleum Congress, The Hague, 1951, Section 14, p. 201.

[3] Whitmore, F. C., *Ind. Eng. Chem.*, 1934, **26**, 94.

[4] Terres, F., *Brennstoff Chem.*, 1953, **34**, 355; *Erdöl-u-Kohle*, 1959, **12**, 469, 547, 614.

[5] Ipatieff, V. N. and Schaad, R. C., *Ind. Eng. Chem.*, 1945, **37**, 362.

[6] Monroe, L. A. and Gilliland, E. R., *Ind. Eng. Chem.*, 1938, **30**, 58.

[7] Scott, H. et al., *J. Polym. Sci.*, 1964, A2, 3233, 5257.

[8] Balas, J. G., De La Mere, H. and Schlissler, D. O., *J. Polym. Sci.*, 1965, A3, 2243.

[9] Zachoval, J., Kalal, J., Veruovic, B. and Stefka, L., *Coll. Czech. Chem. Comm.*, 1965, **30**, 1326.

[10] Jones, J. R., et al., *J. Chem. Soc.*, 1971, 1117, 1124.

[11] Shephard, F. E., Rooney, J. J. and Kemball, C., *J. Catalysis*, 1962, **1**, 379.

[12] Holm, V. C. and Clark, A., *J. Catalysis*, 1963, **2**, 16.
[13] Finch, J. N. and Clark, A., *J. Catalysis*, 1969, **13**, 147.
[14] Borisova, M. S., Dzisko, V. A. and Cheridnik, E. N., *Kin e Kat*, 1962, **3**, 734.
[15] Hogan, J. P., Banks, R. L., Lanning, W. C. and Clark, A., *Ind. Eng. Chem.*, 1955, **47**, 752.
[16] Uchida, H. and Imai, H., *Bull. Chem. Soc. Japan*, 1962, **35**, 989, 995; 1965, **38**, 925.
[17] British Hydrocarbon Chemicals Limited, BP 978,602.
[18] Esso Research and Engineering Co., BP 1,168,940.
[19] Imperial Chemical Industries, BP's 1,155,125; 1,155,246; 1,155,532; 1,155,657.
[20] Pines, H. and Mark, V., *J. Am. Chem. Soc.*, 1956, **78**, 4316, 5946.
[21] Bush, W. V., Holzman, G. and Shaw, A. W., *J. Org. Chem.*, 1965, **30**, 3290.
[22] Benkeser, R. A., Foster, D. J. and Sauve, D. M., *Chem. Reviews*, 1957, **57**, 867.
[23] Morton, A. A., *et al.*, *J. Am. Chem. Soc.*, 1945, **67**, 2224.
[24] Pines, H. and Haag, W. O., *J. Org. Chem.*, 1958, **23**, 528.
[25] Wilkes, J. B., *J. Org. Chem.*, 1967, **32**, 3231.
[26] Wilkes, J. B., Proceedings of the 7th World Petroleum Congress, 1967 (Mexico City), **5**, 299.
[27] Hambling, J. K., *Chem. in Brit.*, 1969, **5**, 354.
[28] Pis'man, I. I. *et al.*, *Akad. Nauk SSSR*, 1968, **179**, 608.
[29] Schramm, R. M. and Langlois, G. E., *J. Am. Chem. Soc.*, 1960, **82**, 4912.
[30] Brandes, O. L., Grasse, W. A. and Lowy, A., *Ind. Eng. Chem.*, 1936, **28**, 554.
[31] Petrov, A. D., *Fette-u-Seife*, 1955, **57**, 98.
[32] Antsus, L. I. and Petrov, A. D., *Doklady Akad. Nauk SSSR*, 1950, **70**, 425.
[33] Verstappen, J. J. and Waterman, H. I., *J. Inst. Petrol.*, 1955, **41**, 343.
[34] Shaw, A. W., Bittner, C. W., Bush, W. V. and Holzman, *G. J. Org. Chem.*, 1965, **30**, 3286.
[35] Pines, H. and Schapp, L. A., *Advances in Catalysis*, 1960, **XII**, p. 117.
[36] Pines, H., Ipatieff, V. N. and Vesely, J. A., *J. Am. Chem. Soc.*, 1955, **77**, 347.
[37] Reggel, L., Friedman, S. and Wender, I., *J. Org. Chem.*, 1955, **23**, 1136.
[38] Pines, H. and Eschinazi, H., *J. Am. Chem. Soc.*, 1955, **77**, 6314; 1956, **78**, 1178, 5950.
[39] Eberhardt, G. G. and Davies, W. R., *J. Polym. Sci.*, 1956, A**3**, 3753.
[40] Langer, A. W., *Trans. New York Sci.*, 1965, **27**, 741.
[41] Asamah, T. and Sato, T., *Bull. Jap. Petrol. Inst.*, 1969, **11**, 44.
[42] Ziegler, K., *Hydrocarbon Proc. and Petrol. Refiner*, 1955, **34** (8), 111.
[43] Skinner, W. A. *et al.*, *Ind. Eng. Chem.*, 1960, **52**, 695.
[44] Ziegler, K. *et al.*, *Ann.*, 1960, **629**, 172.
[45] Lobo, P. A. *et al.*, *Chem. Eng. Prog.*, 1962, **58** (5), 85.
[46] Ziegler, K., *Ang. Chem.*, 1956, **68**, 721.
[47] Martin, H., *Ang. Chem.*, 1956, **68**, 306.
[48] Robinson, R., *Chem. Age*, 1956, **74**, 996.
[49] Stille, J. K., *Chem. Reviews*, 1958, **58**, 541.
[50] Imperial Chemical Industries Ltd., BelgP 562,638.
[51] Ziegler, K., BP 775,384.
[52] Ziegler, K., FP's 1,099,257; 1,128,369.
[53] Wilke, G. *et al.*, Proceedings 7th World Petroleum Congress, Mexico City, 1967, **5**, 115.
[54] Ewers, *Erdöl-u-Kohle*, 1968, **21** (12), 763.
[55] Wilke, G. *et al.*, *Ang. Chem. Int. Edn.*, 1955, **5**, 151: BP 1,058,680.
[56] Felibbylum, V. Sh. *et al.*, *Dokl. Akad. Nauk SSSR*, 1967, **172**, 111.
[57] Onsager, O. T., Wang, H. and Blindheim, U., *Helv. Chim. Acta*, 1968, **52**, 187, 196, 215, 224, 230.
[58] Chauvin, Y., Phung, N. H., Guichard-Loudet, N. and Lefebvre, G., *Bull. Soc. Chim. Fr.*, 1966, 3223.
[59] Jones, J. R., *J.C.S.*, 1971, 1117.
[60] Tsutsui, M. and Koyano, T., *J. Polym. Sci.*, 1967, A-1, **5**, 681.
[61] Pu, L. S., Yamamoto, A. and Ikeda, S., *J. Am. Chem. Soc.*, 1968, **90**, 7170.
[62] Cramer, R., *J. Am. Chem. Soc.*, 1965, **87**, 4717.
[63] Alderson, T., Jenner, E. L. and Lindsey, R. V., *J. Am. Chem. Soc.*, 1965, **87**, 5638.
[64] Ketley, A. D. *et al.*, *Inorg. Chem.*, 1967, **6**, 657.

[65] Cheney, H. H. et al., Ind. Eng. Chem., 1950, **42**, 2580.
[66] Schultz, R. G. et al., J. Catalysis, 1966, **6**, 685, 419; 1967, **7**, 286.
[67] British Petroleum, B P 1,024,314.
[68] Nelson, W. L., 'Petroleum Refinery Engineering', 4th Edition (McGraw Hill, New York, 1958), p. 651.
[69] Moore, R. N., Franz, R. A. and Spillane, L. J., Am. Chem. Soc., Division of Petroleum Chemistry, Minneapolis meeting, April 1969, **14** (*1*), 118.
[70] Sullivan, F. W., Jr., Ruthruff, R. F. and Kuentzel, W. E., Ind. Eng. Chem., 1935, **27**, 1072.
[71] Ziegler, K., Gilbert, H. G., Holzkamp, E. and Wilke, G., Brennst. Chem., 1954, **35**, 321.
[72] Ziegler, K. et al., Ann., 1960, **629**, 172.
[73] Barlon, M. G. et al., J. Organometal Chem., 1970, **21**, 215.
[74] Hata, G. and Mikaya, A., Chem. and Ind., 1967, 921.
[75] U O P, B P 842,136.
[76] California Research Corporation, B P 824,917; U S P 2,986,588.
[77] British Petroleum, B P's 933,253; 945,969; 1,066,113; 1,003,576; 1,005,609.
[78] Ethyl Corporation, B P's 987,590; 933,700.
[79] Hambling, J. K. and Northcott, R. P., Rubber and Plastics Age, 1968, **49**, 224.
[80] Pis'man, I. I. et al., Doklady Akad. Nauk S S S R, 1968, **179**, 608.
[81] Sparke, M. B., Turner, L. and Wenham, A. J. M., J. Catalysis, 1965, **4**, 322.
[82] British Petroleum Co., B P's 934,449; 934,450; 991,705; 1,002,894; 1,007,325.
[83] Distillers Company Ltd., B P 1,024,684.
[84] Hambling, J. K. et al., A C S Division of Petroleum Chemistry, Chicago Meeting, 1964, **9**, No. 3, 101.
[85] Watt, W. R., J. Polym. Sci., 1960, **45**, 509.
[86] Campbell, T., J. Appl. Polymer Sci., 1961, **5**, 184.
[87] I C I Booklet: T P R Methylpentene Polymers, September 1967.
[88] Friffiths, L. L., Modern Plastics, Aug. 1957, Vol. 34, p. 111.
[89] Clarke, K. J. and Palmer, B. P., S C I Monograph No. 20, p. 82.
[90] Isaacson, R. B. et al., J. Appl. Polymer Sci., 1964, **8**, 2789.
[91] Robinson, D. W., J. Sci. Instruments, 1966, **82**, p. 2.
[92] Landes, L. L., S P E Journal, Dec. 1960, p. 1329–32.
[93] Dyer, B. S., Plastics, Oct. 1968, p. 1146.
[94] Mainprize, C., Brit. Hospital and Soc. Serv. Review, Dec. 1966, p. 2355; Chem. Eng. News, 29 March 1965, p. 44.
[95] Griffith, J. H. and Ranby, B. G., J. Polym. Sci., 1960, **44**, 369.
[96] Union Carbide Corporation, B P 823,309.
[97] Celanese Corporation of America, F P 1,316,819.
[98] Socony Vacuum Oil Co., U S P 2,404,056.
[99] Horne, S. E., Ind. Eng. Chem., 1956, **48**, 784.
[100] Gaylord, H. G. and Mark, M. F., 'Linear and Stereoregular Addition Polymers: Polymerisation and Controlled Propagation' (Interscience, New York, 1959), pp. 400–6.
[101] Chemical Week, 6 May 1961, pp. 73–8.
[102] Anhorn, V. J., Pet. Refiner, 1960, **39**, 227.
[103] Mayor, R. H., Saltmann, W. M. and Pierson, R. M., Pet. Refiner, 1958, **37**, 208.
[104] Anhorn, V. J. et al., Chem. Eng. Progress, 1961, **57**, 41.
[105] Adams, H. E., Stearns, R. S., Smith, R. S. and Bender, J. L., Rubber Age, 1957, **81**, 637.
[106] Bass, C. J. et al., Brennst. Chem., 1964, **45**, 161.
[107] Hay, J. N., Hooper, P. G. and Robb, J. C., Trans. Faraday Soc., 1970, **66**, 2045, 2800.
[108] British Petroleum Co., B P 1,163,091; 962,255.
[109] British Petroleum Co., B P 990,465.
[110] Goodyear Tyre and Rubber Co., B P 841,351.
[111] British Hydrocarbon Chemicals Ltd., B P 886,716.
[112] British Petroleum Co., B P 934,450.
[113] British Petroleum Co., B P 932,748.

[114] Scientific Design, BP's 840,028; 934,783.

[115] Goodyear Tyre and Rubber Co., BP 975,454.

[116] Halcon International, USP 3,104,269.

[117] Goodyear Tyre and Rubber Co., BP 832,475.

[118] British Petroleum Co., BP 912,826.

[119] Pechiney, BP 831,249.

[120] Goodyear Tyre and Rubber Co., BP 868,566.

[121] Goodyear Tyre and Rubber Co., BP 915,447.

[122] Goodyear Tyre and Rubber Co., BP 970,469.

[123] Goodyear Tyre and Rubber Co., BP 916,133.

[124] Whitby, G. S.(ed.), '*Synthetic Rubber*' (John Wiley & Sons, Inc., New York, 1954).

[125] Gilliland, '*Science of Petroleum*', Vol. V, Part II, p. 47, ed. Brooks and Dunstan (Oxford University Press, 1953).

[126] *European Chemical News*, 2 October 1970, p. 30.

[127] Demalda, M. *et al.*, *Pet. Refiner*, 1964, **43** (7), 149.

[128] Demalda, M., *Chem. Ind.* (*Milan*), 1963, **45** (6), 665.

[129] Shachat, N. *et al.*, *J. Org. Chem.*, 1962, **27**, 1498.

[130] Petrotex Chemical, BP 988,619; USP's 3,277,207, 3,278,626.

[131] Shell International, BP 983,907.

[132] Shell, USP 2,993,841.

[133] Phillips Petroleum Co., USP 2,382,119.

[134] Swanson, R. W. *et al.*, *J. Chem. Eng. Data*, 1962, **7**, 132.

[135] Mavity, J. M. *et al.*, *Trans. Amer. Inst. Chem. Eng.*, 1944, **40**, 473.

[136] Giacomo, A. A. *et al.*, *Chem. Eng. Progr.*, 1961, **57** (5), 35.

[137] Arundale, E. *et al.*, *Chem. Rev.*, 1952, **51** (3), 505.

[138] Giraud, A., *Rev. Gen. Caoutchouc*, 1961, **38** (11), 1598.

[139] Halcon International, BP's 1,128,094, 1,128,095.

[140] *Europ. Chem. News*, 4 September 1970, p. 18.

[141] *Japan Industrial Project News*, 19 January 1971.

[142] Ockerbloom, Brownsteen and Root, *Oil and Gas International*, 1969, **9**, 79.

[143] Banks, R. L. and Regier, R. B., Symposium on New Olefin Chemistry, Division of Petroleum Chemistry, American Chemical Society, Houston Meeting 22–7 February 1970.

[144] British Petroleum Co., BP's 1,054,864; 1,056,980; 1,170,498.

[145] Bradshaw, C. P. C., Howman, E. J. and Turner, L., *J. Catalysis*, 1967, **7**, 269.

[146] Institut Francais du Pétrole, BP's 893,206; 825,034; 884,808; 884,809.

[147] Hellin, M. *et al.*, *Pet. Eng. Management*, 1959, **31**, 12.

[148] Hellin, M. *et al.*, *Bull. Soc. Chem. Fr.*, 1963, **12**, 2722, 2725.

[149] British Hydrocarbon Chemicals Ltd., BP's 826,545; 841,746; 863,136.

[150] Kurashiki Rayon, USP 3,284,534.

[151] Atlantic Refining, USP 3,154,563.

[152] Farbenfabrik Bayer, FP 1,364,615.

[153] British Petroleum Co., FP 1,556,915; Belg.P 726,745.

[154] *Chem. Eng. News*, 6 July 1970, pp. 60–2.

[155] (a) Distillers Co. Ltd., BP's 915,590; 1,099,398.
        (b) Petrotex Chemical Corp., USP's 3,270,080; 3,303,235.
        (c) Phillips Petroleum Co., USP's 3,304,342; 3,320,329.
        (d) Standard Oil of Ohio, USP's 3,251,899; 3,370,103.

[156] Dunkel, W. L. *et al.*, '*Introduction to Rubber Technology*', ed. M. Morton (Reinhold Publishing Corp., New York, 1959), Chapter 12.

[157] *Rubber Age*, 1950, **66**, p. 557.

[158] Horne, S. E. Jr., *Ind. Eng. Chem.*, 1956, **48**, 784.

[159] *Chemical Week*, 18 March 1970, pp. 56–64.

[160] *Chem. and Eng. News*, 12 January 1970.

[161] *Oil Gas J.*, 1970, **68**, 70.

[162] *Eur. Chem. News*, 7 August 1970, p. 6.

[163] *Eur. Chem. News*, 11 April 1970, pp. 4–5.

[164] *Chem. and Eng. News*, 16 November 1970, p. 28.

[165] Egloff, G. and Jones, E. K., Proc. Fourth World Petroleum Congress, Section IV, Paper 1: cf. P. Sherwood. Petroleum, May 1957, p. 183.

[166] Jones, E. K., *Advances in Catalysis* (Academic Press, New York, 1956), Vol. XIII, p. 219.

[167] Enjay Company Inc., '*Higher Oxo Alcohols*' (New York, 1956), pp. 10, 33.

[168] Kesen, F. L., *Oil Gas J.*, 1954, **52**, 203, 209.

[169] Sun Oil Co., USP 2,802,890.

[170] Jones, E. K., *Pet. Refiner*, 1954, **33** (*12*), 186.

[171] Bradner, J., Lockwood, W. H., Nagel, R. H. and Russell, K. L., FIAT Final Report no. 1141.

[172] Schultze, W. A., Lyon, J. P. and Short, G. H., *Ind. Eng. Chem.*, 1948, **40**, 2308.

[173] Phillips Petroleum Co., USP 2,454,409.

[174] Frantz, J. F. and Glass, K. I., *Chem. Eng. Prog.*, 1963, **59** (*7*), 68.

[175] Johanson, A. A. and Goldblatt, L. A., *Ind. Eng. Chem.*, 1948, **40**, 2086.

[176] See e.g. Sinclair Research, USP 3,558,743.

[177] Sharrah, M. L. and Feighner, G. C., *Ind. Eng. Chem.*, 1954, **46**, 248.

[178] California Research Corp., USP 2,477,382.

[179] Feighner, G. C., *J. Amer. Oil Chem. Soc.*, 1958, **35**, 520.

[180] Holst, E. H. *et al.*, *Ind. Eng. Chem.* (*Prod. Res. & Dev.*), 1963, **1** (*2*), 120.

[181] Birch, S. F., *J. Inst. Petrol.*, 1952, **38**, 69.

[182] Fifth Progress Report of the Standing Committee on Synthetic Detergents, 1962 (London, HMSO).

[183] *Chemical Eng. News*, 16 November 1970, p. 28.

[184] *Oil Gas J.*, 13 December 1965, pp. 102–3; *Chem. Eng. News*, 1961, **39** (*2*), 30; (*38*), 81.

[185] See e.g. Shell, USP 2,642,466.

[186] Thomas, T. L., 6th World Petroleum Congress, Frankfurt 1963, Section III, Paper 16.

[187] Avery, W. F. and Lee, M. N. Y., *Oil Gas J.*, 1962, **60** (*23*), 121.

[188] Asinger, F., *Berichte*, 1942, **75**, 664, 668, 1247.

[189] Continental Oil Co., BP 969,104.

[190] British Hydrocarbon Chemicals Ltd., BP 968,338.

[191] Friedmann, L. and Berger, J. G., *J. Am. Chem. Soc.*, 1961, **83**, 492, 500.

[192] (a) Barilli, F. *et al.*, *Ind. Eng. Chem.*, 1970, **62** (*6*), 62.
(b) *Europ. Chem. News*, 1963, **5** (*71*), 24 May 1963, p. 28.

[193] Bloch, H. S., *Oil Gas J.*, 16 January 1967, pp. 79–81.

[194] Broughton, D. B. and Berg, R. C., *Hydrocarbon Processing*, June 1969, pp. 115–17.

[195] '*Ethylene and its Industrial Derivatives*', ed. S. A. Miller (Ernest Benn, London, 1969), pp. 45, 1228–30.

[196] *Europ. Chem. News*, 1963, **3** (*74*), 31; *Chem. Age*, 1963, **89**, 869. Konecky M. S. and Fleming C. L. Paper presented at Milan Chemical Exposition, 10 June 1963. *Chem. Prod.*, 1963, **26** (*8*), 27.

[197] Marquis, D. M., *Hydrocarbon Processing*, 1968, **47** (*3*), 109.

[198] *Chem. Week.*, 1968, **103** (*6*), 35.

[199] *Europ. Chem. News*, 23 May 1969, p. 8.

[200] British Petroleum Co., BP 1,011,347.

[201] British Hydrocarbon Chemicals Ltd., BP's 978,602; 983,376; 1,132,527.

[202] (a) Phillips Petroleum Co., USP 3,492,364.
(b) Antonson, D. H., Warren, R. W. and Johnson, R. H., *Ind. Eng. Chem.* (*Prod. Res. & Dev.*), 1964, **3**, 311.

[203] *Oil, Paint and Drug Reporter*, 3 March 1969, p. 9.

[204] Nelson, W. L., '*Petroleum Refinery Engineering*', 4th Edition (McGraw-Hill, 1958), pp. 722–35, 741–2.

[205] Schmerling, L. and Ipatieff, V. N., '*Advances in Catalysis*', Vol. II (Academic Press, New York, 1950), pp. 21–78.

[206] Hurd, Y. N. and Gwynn, B. H., Proc. 5th World Petrol. Congress, Section IV, Paper 14, pp. 1–10 (1959).

[207] *Europ. Chem. News*, 11 December 1970, p. 32.
[208] Quoted in *Encyclopaedia of Chemical Technology* (Kirk-Othmer), 2nd Edition, Vol. 15, pp. 720–89 (Darby J. R. and Leans J. K.): Interscience.
[209] Fedor, W. S., *Chem. Eng. News*, **39**, 118–38.
[210] British Petroleum Co., BP's 1,170,497; 1,159,070; 1,120,515.
[211] British Petroleum Co., BP's 1,123,474; 1,171,281.

CHAPTER 5

POLYPROPYLENE

By **A. Valvassori, P. Longi and P. Parrini**

5.1. MECHANISM OF POLYMERISATION

THE vinyl double bond of propylene is known to undergo polymerisation reactions. The catalysts that promote such reactions are of cationic (Friedel-Craft) and anionic co-ordinated (Ziegler–Natta) types; in some cases they are free-radical initiators.

Prior to 1954, only cationic or free-radical catalysts were known to promote propylene polymerisation to either low (dimers, trimers, tetramers) or high (e.g. above 100,000) molecular weight polymers. It was first observed that phosphoric acid transforms propylene into oily polymeric products.[1, 2, 3, 4, 5, 6] Later on other cationic catalysts wefe found to polymerise propylene: in particular, boron trifluoride,[7, 8, 9, 10, 11] alumina and silica,[12, 13, 14] sulphuric acid,[15] hydrofluoric acid[16] and aluminum halides.[17] The polymers obtained had generally low molecular weights higher than 100,000 were obtained only by polymerisation of propylene at $-60°C$ in n-butane and in the presence of the catalyst system $AlBr_3/iC_3H_7Br$,[18, 19, 20] while polymers having molecular weights ranging from 6,000 to 880,000 were isolated by fractionation. Even polymerisation of propylene by free-radical initiators, such as dark discharges,[21] diazomethane,[22] tetraethyl lead,[23, 24, 25] ultra-violet radiations in the presence of propionic aldehyde[26] or mercury vapours,[27] generally result in liquid or oily polymers having low molecular weights; only by the use of particular 'Redox' systems and under particular experimental conditions, could low amounts of high-molecular-weight polymers be obtained.[28, 29, 30]

Propylene polymers obtained by cationic and free-radical catalysts are always amorphous owing to the existence of irregularities both with regard to the arrangement of the chains and to the chemical structure of the monomer units. In particular, the following groupings in varying amounts have been found:[31]

$$-CH-CH_2-CH-CH_2- \quad \text{1,2-polymer; head-to-tail arrangement}$$
$$| |$$
$$CH_3 CH_3$$

$$-CH_2-CH-CH-CH_2- \quad \text{1,2-polymer; head-to-head arrangement}$$
$$| |$$
$$CH_3 CH_3$$

$$-CH- \quad \text{monomer unit from 1,1-polymerisation}$$
$$|$$
$$C_2H_5$$

$$CH_3 \quad \text{monomer unit from 2,2-polymerisation}$$
$$|$$
$$-C-$$
$$|$$
$$CH_3$$

$$-CH_2-CH_2-CH_2- \quad \text{monomer unit from 1,3-polymerisation}$$

Crystalline polymers with high molecular weight, high melting point and much greater commercial interest than the polymers so far described, were firstly obtained by Natta and co-workers in 1954[32, 33, 34] with the so-called 'Ziegler catalysts'. These catalysts, prepared by reaction of transition metal compounds of the IV, V and VI group of the periodic system (in particular $TiCl_4$) with organometallic compounds of metals of the I, II and III group (in particular aluminum alkyls) had been successfully used by K. Ziegler *et al.* in the low pressure polymerisation of ethylene.[35, 36, 37] Natta *et al.* found that, in the case of the polymerisation of propylene, the catalysts were most active when prepared with a trialkyl aluminum to titanium tetrachloride molar ratio of about 2, and used soon after their preparation.

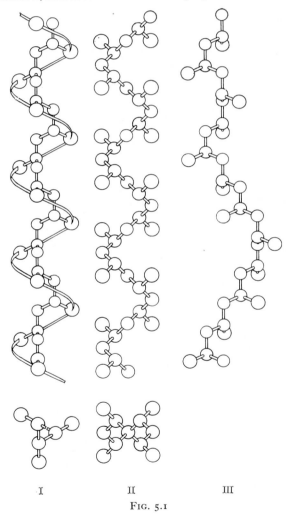

I   II   III

FIG. 5.1

At the end of 1954 Natta and co-workers discovered[38, 39] that catalysts prepared from aluminum trialkyls (or from aluminum dialkyl halides) and titanium trichloride, could give high yields of polypropylene showing a high degree of crystallinity. The catalyst systems nowadays used for commercial production of isotactic propylene are essentially the same; as a matter of fact, they are generally prepared from diethyl aluminum monochloride and titanium trichloride usually containing aluminum chloride with molar ratios $TiCl_3/AlCl_3$ ranging between 3 and 1.

Thorough examination of the structure of polypropylene obtained by the above catalysts[40, 41, 42] revealed that it consists of different amounts (depending on the catalyst used) of three types of macromolecules and, consequently, a different capacity for yielding crystal lattices.

However, unlike the polypropylene obtained by cationic catalysts, polypropylenes of the above three types show an absolute regularity of chemical structure of the single monomer units ($-CH_2-CH-$) and of their arrangement (head-to tail).

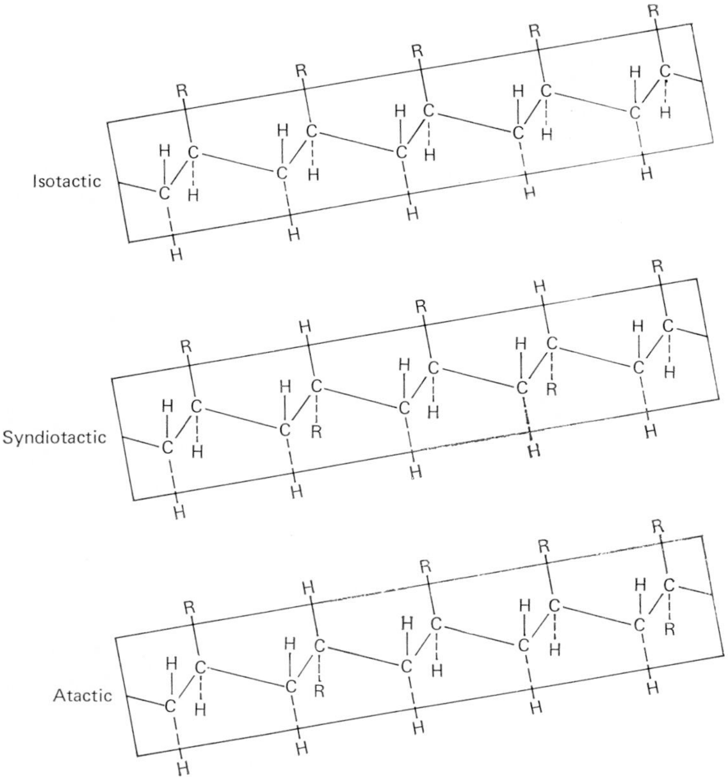

$$CH_3$$

FIG. 5.2

The first and most important of such types is the 'isotactic' polypropylene; it consists of a boiling-n-heptane-insoluble polymer, showing a high crystallinity by X-ray examination (70–80 per cent) and a melting temperature of about 175°C. Macromolecules of this fraction (which in the case of the polypropylenes obtained by catalysts prepared from titanium trichloride, is 90–95 per cent of the total polymer) were shown to have monomer units in which tertiary carbon atoms have all the same steric configuration; furthermore, such macromolecules have a helix structure with ternary symmetry (i.e. they repeat themselves every three monomer units) with an identity period of about 6·5Å (Fig. 5.1, I). The side view of the macromolecule, made linear by stretching, shows that methyl groups of the monomer units are all on the same side of the main chain (Fig. 5.2). Owing to the high steric order of its macromolecules, isotactic polypropylene not only possesses high crystallinity and a high melting temperature, but also high mechanical properties over a wide range of temperature.

When only some chain blocks of the macromolecules are isotactic and alternated with polymeric segments with no steric regularity or with a different structural regularity (e.g. tertiary carbon atoms have all the same configuration, but different from that of the neighbouring segments (see Fig. 5.3)), the macromolecule has an 'isotactic stereoblock' structure. The melting temperature and mechanical properties of the polymer consisting of such macromolecules are lower compared to the pure isotactic polymer, this being approximately in line with the lower degree of crystallinity of the polymer. This second type of polypropylene is generally soluble in boiling n-heptane.

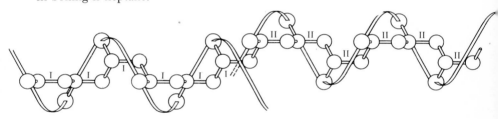

FIG. 5.3

The third type is soluble even in boiling ethyl ether and no steric order can be detected in its macromolecules (Figs. 5.1, III and 5.2). The polypropylene consisting of these macromolecules is 'atactic': owing to its steric irregularity, it cannot crystallise; its aspect and properties are those of an uncured elastomer.

More recently, Natta et al.[43, 44] found that, with the use of particular catalyst systems, such as that prepared from aluminium dialkyl chloride and vanadium tetrachloride, propylene may give macromolecules having a stereoregularity different from the isotactic one: i.e. 'syndiotactic' polypropylene, where tertiary carbon atoms are regularly arranged with opposite configurations (Figs. 5.1, II and 5.2). Polymers consisting of such macromolecules are supposed to crystallise well; actually, owing to the low steric purity of the macromolecules obtained up to now, the polymer shows a very low degree of crystallinity and, consequently, has not yet found practical applications.

Researches performed to investigate the nature of the reaction products between aluminum alkyls and titanium compounds containing halogen atoms as well as cyclopentadienyl groups have contributed a great deal to the understanding of the nature of the Ziegler–Natta catalysts. This has helped to explain the nature of the mechanism according to which these catalysts promote the polymerisation of propylene and of other olefins containing vinyl double bonds. These reaction products are actually very similar to the Ziegler–Natta catalysts but, unlike these, they are soluble, crystallisable and stable (even at room temperature). A typical example of these catalysts was prepared by Natta et al.;[45]

using titanium dicyclopentadienyldichloride and aluminum triethyl. The catalytic activity of this complex is, however, very low; hence, it can be hardly considered as a Ziegler–Natta catalyst. On the other hand, even a rather more active (and less stable) catalytic complex, arstly obtained by Breslow[46, 47] by reaction of aluminum-diethyl chloride and titanium dicyclopentadienyldichloride, was found to have the structure of a bimetallic complex

like that mentioned above.

The existence of a Ti-ethyl bond (which, as known, is rather unstable) and the decrease with time of the catalytic activity of this complex, suggest that such a Ti—C bond is responsible for the polymerisation process.

However, mainly by analogy with the mechanism postulated by K. Ziegler et al. for the ethylene addition to aluminum alkyls, early hypotheses on chain growth mechanisms for the polymerisation of propylene are outlined on p. 160.[48, 49]

The hypothesis was put forward of the formation of a complex between olefin and catalyst, followed by the insertion of the olefin molecule in the (partially polarised) bond between aluminum and macromolecular growing chain (indicated by P). The anionic nature of this mechanism was proved by:

(a) the existence, as end groups of the macromolecular chains obtained, of substituent groups initially bound to the aluminum alkyl compound used for the catalyst preparation;[50, 51]

(b) the presence of deuterium or tritium in the polymer treated, after short polymerisation times (when the polymer chain may be still linked to the catalytic complex), with deuterated or tritiated methanol;[52, 53]

(c) the inefficiency of some inhibitors of free-radical polymerisations in decreasing the activity of these catalysts.[54]

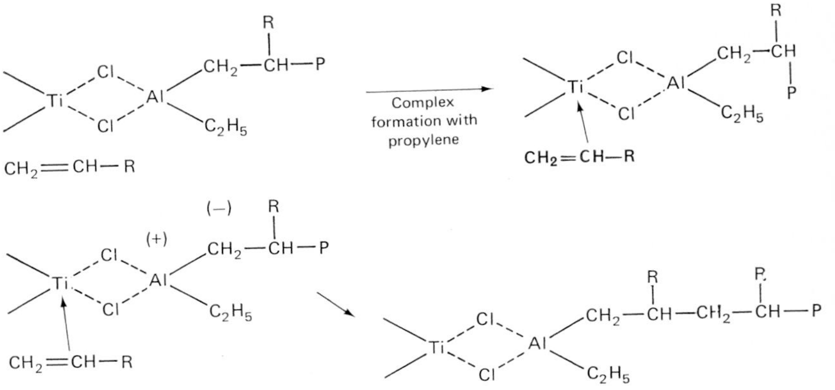

R
|
Cl   CH₂—CH—P
  \   /
   Ti   Al
  / \Cl \
         C₂H₅

CH₂=CH—R

→ Complex formation with propylene →

R
|
Cl   CH₂—CH
  \   / |
   Ti   Al   P
  / \Cl \
         C₂H₅
          ↑
CH₂=CH—R

(−)   R
|
(+)   CH₂—CH—P
Cl   /
  \ /
   Ti   Al
  / Cl \
         C₂H₅

CH₂=CH—R

↘

R   P
| |
Cl   CH₂—CH—CH₂—CH—P
  \ /
   Ti   Al
  / Cl \
         C₂H₅

This hypothesis was later modified by the same and other authors, although it was still based on an anionic polymerisation mechanism. The most significant change was made by taking into account that the chain growth on the Al—C bonds could not be proved experimentally, whereas some proofs were found supporting the growth on the Ti—C bonds: in particular (in addition to the conclusions concerning the Breslow complex) the existence of some catalysts, which exclusively consist of organometallic compounds of titanium (e.g. $Cp_2Ti(C_6H_5)_2 + TiCl_4$,[48] $CH_3TiCl_3 + TiCl_3$,[55, 56] CH_3TiCl_2,[57] tetrabenzyltitanium and titaniumtribenzyl monochloride[58]), and which can promote, even though with low catalytic activity, the stereospecific polymerisation of α-olefins. Hence, the mechanism of growth was modified as follows:[59]

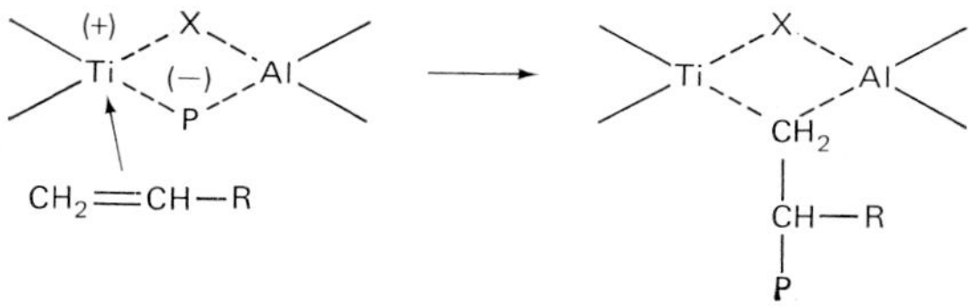

(+)   X
Ti   (−)   Al
      P
      ↑
CH₂=CH—R

→

X
Ti   Al
   CH₂
   |
   CH—R
   |
   P

where both metals participate in the chain growth process; however, the titanium atom, having the positive (or more positive) charge, was considered responsible for the anionic mechanism of the polymerisation.

In the case of the catalyst system prepared from $Al(C_2H_5)_3$ and $TiCl_3$, Boor[60] put forth a hypothesis according to which the Al atom plays the most important role:

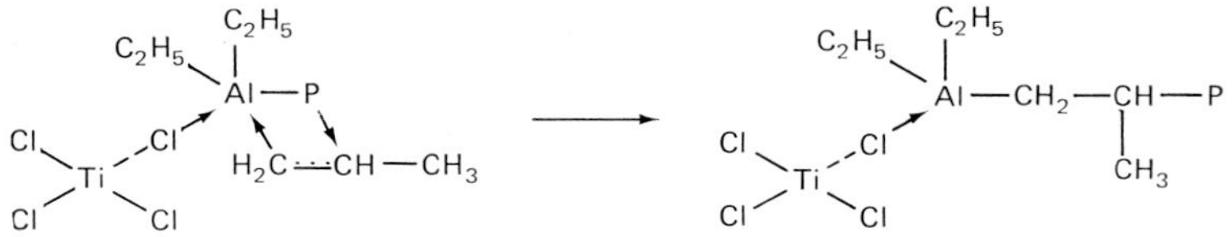

C₂H₅
|
C₂H₅   Al—P
     \
Cl   Cl H₂C⋯CH—CH₃
  \ /
   Ti
  / \
Cl   Cl

→

C₂H₅
|
C₂H₅   Al—CH₂—CH—P
     \ |
Cl   Cl   CH₃
  \ /
   Ti
  / \
Cl   Cl

On the contrary, according to Cossee and Arlman, the $Al(C_2H_5)_3$ role is only to produce on the $TiCl_3$ surface alkyl titanium ions, on which the chains grow.[61, 62, 63]

As for chain termination and chain transfer processes, the following schemes were taken into account:

$$\overset{(+)}{Cat}\!-\!CH_2\!-\!\overset{(-)}{CH}\!-\!\overset{\scriptstyle R}{|}\!\!P \longrightarrow Cat\!-\!H + CH_2 = \overset{\scriptstyle R}{\underset{|}{C}}\!-\!P \qquad (1)^{54,64}$$

$$\overset{(+)}{Cat}\!-\!CH_2\!-\!\overset{(-)}{CH}\!-\!\overset{\scriptstyle R}{|}\!P \;\; \xrightarrow[\overset{(+)}{Zn}\!-\!\overset{(-)}{C_2H_5}]{\overset{(+)}{Al}\!-\!\overset{(-)}{C_2H_5}} \;\; Cat\!-\!C_2H_5 + Al(Zn)\!-\!CH_2\!-\!\overset{\scriptstyle R}{\underset{|}{CH}}\!-\!P \qquad (2)^{54,64}$$

$$Cat\!-\!CH_2\!-\!\overset{\scriptstyle R}{\underset{|}{CH}}\!-\!P \;\; \xrightarrow{CH_2 = CH\!-\!R} \;\; Cat\!-\!CH_2\!-\!CH_2\!-\!R + CH_2 = \overset{\scriptstyle R}{\underset{|}{C}}\!-\!P \qquad (3)^{65}$$

$$Cat\!-\!CH_2\!-\!\overset{\scriptstyle R}{\underset{|}{CH}}\!-\!P \;\; \xrightarrow{H_2} \;\; Cat\!-\!H + CH_3\!-\!\overset{\scriptstyle R}{\underset{|}{CH}}\!-\!P \qquad (4)^{66}$$

$$\overset{(+)}{Cat}\!-\!CH_2\!-\!\overset{(-)}{CH}\!-\!\overset{\scriptstyle R}{|}\!P \;\; \xrightarrow{\overset{(+)}{R'}\!-\!\overset{(-)}{X}} \;\; Cat\!-\!X + R'\!-\!CH_2\!-\!\overset{\scriptstyle R}{\underset{|}{CH}}\!-\!P \qquad (5)^{67,68}$$

$$Cat\!-\!CH_2\!-\!\overset{\scriptstyle R}{\underset{|}{CH}}\!-\!P \;\; \xrightarrow[\text{high temp.}]{H_3C\!-\!CH = CH_2} \;\; Cat\!-\!CH_2\!-\!CH = CH_2 + H_3C\!-\!\overset{\scriptstyle R}{\underset{|}{CH}}\!-\!P \quad (6)^{69}$$

Process (1) refers to the mechanism of spontaneous deactivation of the growing chain, in the absence of any chain transfer or stopper. The resulting polymer contains, as end groups, vinyl double bonds (if the monomer was ethylene) or vinylidenic ones (for α-olefins).

Process (2) is brought about by the transfer of alkyl aluminium, which is always present in excess in the polymerisation medium. Zinc alkyls behave similarly and are used, sometimes, as molecular weight modifiers. The polymer obtained (after the usual purification treatment), does not contain unsaturated end groups.

Process (3), caused by transfer with the monomer, generally takes place to a less extent than processes (1) and (2) and yields a polymer containing terminal unsaturation, like process (1).

Process (4) is the main one when the polymerisation is carried out in the presence of hydrogen. The polymer obtained is free from unsaturation.

Process (5) may occur when the polymerisation medium contains polarisable compounds (e.g. alkyl halides, acid halides, compounds containing carboxylic or carbonyl groups, etc.).

Process (6) was observed for high-temperature polymerisation of propylene (above 100°C). In this case, the 'allylic' hydrogen atom of propylene is reactive enough to stop the growing chain, with formation of a polymer chain free from terminal unsaturation, and of a Cat-allyl bond able to bring on, through a subsequent process of growth, chains containing terminal vinyl-type unsaturation.

12

High-molecular-weight, essentially amorphous, propylene polymers can be also obtained by 'Phillips' and 'Standard Oil' catalysts.

The 'Phillips' catalysts are generally prepared by calcination at 500–600 °C, under an anhydrous air stream, of mixtures of silica and alumina (preferred weight ratio 90:10) containing chromic anhydride (2–3 per cent by weight).[70, 71, 72, 73] The catalytic activity seems to be connected with the presence of unstable chromium compounds (having an intermediate valency between 6 and 5) in the calcined support; addition of substances which can stabilise high-valency chromium compounds (e.g. strontium oxide) was found to improve the catalytic activity.

The 'Standard Oil' catalysts are obtained by calcination, at about 450 °C under hydrogen (or other reducing gas) atmosphere, of α-alumina containing molybdenum oxide MoO_3 (5–10 per cent) and cobalt oxide CoO (1–5 per cent).[74, 75] Also in this case, the activity seems to be connected with the presence of some molybdenum-unstable compounds.

Several investigations were carried out in order to define the relationship between polymerisation rate and propylene and catalyst concentration.

In the case of the stereospecific polymerisation of propylene by a heterogeneous catalyst system prepared from $Al(C^2H^5)^3$ and α-$TiCl_3$, Natta et al.[54, 64] proposed the following equation:

$$R_p = K_p K_a [A] [S] [M] / 1 + K_a [A] \qquad (I)$$

where R_p is the polymerisation rate, K_p the rate constant, K_a the absorption equilibrium constant of $Al(C_2H_5)_3$ on $TiCl_3$; $[A]$, $[S]$ and $[M]$ are the concentrations of, respectively, $Al(C_2H_5)_3$, active sites (on $TiCl_3$) and monomer in the liquid phase. The equation may be also derived from the Langmuir–Hinshelwood equation for catalytic reactions in the heterogeneous phase:[76]

$$R_p = K_p K_m K_A [A] [S] [M] / (1 + K_m [M] + K_A [A])^2 \qquad (II)$$

where K_m is the monomer absorption equilibrium constant on active sites. The assumption is made that R_p versus monomer concentration relationship is of the first order.

According to equation (I) and in agreement with experimental results, the polymerisation rate increases with low values of $Al(C_2H_5)_3$ concentration,[77] whereas it becomes virtually independent for higher values (> 0.05 moles/l) of such concentration.

In the presence of hydrogen, the polymerisation rate decreases according to the equation:

$$R_p{}^H = R_p - \cfrac{1}{\cfrac{A}{\sqrt{p_{H_2}}} + B} \qquad (III)$$

where $R_p{}^H$ and R_p are the polymerisation rates in the presence and absence respectively of hydrogen, p_{H_2} is the partial pressure of hydrogen and A and B are two positive numbers related to the polymerisation temperature.[66]

With regard to the molecular weight of the polypropylene obtained, Natta *et al.* proposed the following equation:

$$1/p_n = \frac{K_1 + K_2[M] + K_3[Al]^{\frac{1}{2}} + K_4[M][Ti]^{\frac{1}{2}}}{K_p[M]} \qquad (IV)$$

where p_n is the average number degree of polymerisation, K_p, K_1, K_2, K_3 and K_4 are rate constants, respectively, of propagation, normal chain termination (ion hydride transfer), transfer with the monomer, transfer with $Al(C_2H_5)_3$ and transfer with $TiCl_3$; [M], [Al] and [Ti] are concentrations in the liquid phase of, respectively, monomer, $Al(C_2H_5)_3$ and $TiCl_3$.[54, 64] In the presence of hydrogen, the term $K_5\sqrt{p_{H_2}}$ has to be introduced into equation (IV).[66]

5.2. INDUSTRIAL POLYMERISATION PROCESSES

Since the discovery of Ziegler–Natta catalyst systems a lot of published information became available, both in patents and in technical journals, about olefins (particularly, propylene) polymerisation processes.

The large number of publications clearly indicates the strong industrial interest in these processes and the large amount of research and development effort carried out in this field all over the world.

Notwithstanding the variety of processes described, it is possible to give a general description of propylene polymerisation process. The usual stages are as follows:

5.2.1. Monomer Purification

Ziegler–Natta catalysts are very sensitive to polar compounds, as for instance alcohols, water, hydrogen sulphide. They are also deactivated or inhibited by a number of other compounds, such as oxygen, carbon monoxide and dioxide, conjugated or cumulated dienes and acetylenic derivatives.

Propylene must then be carefully purified up to the right specifications, which are typical of 'polymerisation grade' propylene (purity 99·5 per cent or more).

5.2.2. Catalyst Preparation

Catalysts are now commercially available; however a number of companies prefer to prepare themselves the catalyst they need. In fact, many polymerisation characteristics, polymer qualities and economic factors depend on the catalyst performance.

Up to now, only titanium compounds are used, along with aluminum alkyls. Titanium trichlorides obtained from tetrachloride by reduction with hydrogen or with metal aluminum are preferred. The most common co-catalysts are triethylaluminum and diethylaluminum monochloride.

Another kind of catalyst is titanium chloride obtained from tetrachloride by aluminum alkyl reduction. By careful control of shape and granulometric distribution of this last type of catalyst it is possible to produce polypropylene granules of narrow particle size distribution. This kind of polymer is commercially known as 'polypropylene flakes' and may replace standard pellets in injection moulding and extrusion machines.

5.2.3. Polymerisation

Polymerisation operation consists in bringing monomer into contact with catalyst, at such a pressure, temperature and chemical environment, to obtain polypropylene with the desired tacticity and molecular weight, at the highest convenient polymerisation rate and highest yield with respect to the catalyst.

No side reactions occur in the polymerisation (other than the polymerisation to atactic polymer, which is of lower commercial interest). Polymerisation may be carried out:

—in solution;[78]
—in suspension in hydrocarbon diluent;[79]
—in suspension in liquid monomer;[80]
—in the gas-phase.[81]

Choice of the process, design of reactor, and operating conditions are largely based on engineering considerations, as heat removal, possibility of making the process continuous, possibility of simplifying subsequent operations (for instance, operating at such high yields with respect to the catalyst to avoid removal of catalyst residues), possibility of obtaining complete homogeneity within the reacting mass. Heat removal is obtained by jacket cooling, in conventional reactors, or by evaporation of part of the hydrocarbon diluent or of the monomer. Standard propylene polymerisation enthalpy is $-24 \cdot 89$ kcal/mol [$C_3H_6(g) \xrightarrow{\frac{1}{n}} (C_3H_6)_n$]; standard propylene vaporisation enthalpy is $+3 \cdot 59$ kcal/mol. For comparison, standard ethylene polymerisation enthalpy is $-25 \cdot 88$ kcal/mol [$C_2H_4(g) \xrightarrow{\frac{1}{n}} (C_2H_4)_n$].[82]

Isotactic polypropylene is insoluble in n-heptane up to its boiling point at atmospheric pressure: for this reason the polymerisation in hydrocarbon 'solvent' is, in fact, a suspension polymerisation if carried out at temperatures lower than around $100\,^{\circ}C$.

Suspension polymerisation is particularly convenient because the viscosity of the polymerisation medium is low even at high polymer concentration: ease of stirring and homogenisation is thus enhanced. Atactic polypropylene is soluble in hydrocarbons: thus for process reasons (heat exchange, ease of stirring, ease of mass transfer) it is advisable to keep as low as possible the production of atactic polymer. It is advisable to regulate polymer molecular weight at the polymerisation stage, even if methods are claimed for regulation of molecular weight by accurate thermomechanical degradation, for instance during granulation. Hydrogen is one of the most effective molecular weight regulators, while the use of certain metal alkyls (e.g. $ZnEt_2$), of hydrogen halides or of halogenated hydrocarbons is also claimed. Solution processes, which are carried out in hydrocarbon solvent at such temperatures that the isotactic polymer is dissolved (i.e. temperatures higher than about

120°C), require much less or no molecular weight modifier at all. The rate of polymerisation is strongly dependent on the type of catalyst, on monomer concentration, i.e. on its partial pressure, and on temperature.

The conclusion may be reached that each particular process, each catalytic system, requires selection of the optimum polymerisation conditions: that is for a good balance of catalyst activity, minimum atactic polymer production, reactor specific productivity (mass of polymer produced per unit volume in unit time). The temperature and pressure range from a minimum of 40–50°C to a maximum of 120–140°C (solution polymerisation) and from a few atmospheres to a maximum of about 45 atm (gas-phase polymerisation).

5.2.4. Removal of Atactic Polymer; Removal of Catalytic Residues

These may be considered typical 'purification' processes of the polymer. A great number of processes are described; almost the only common principle is that the two 'impurities' are separated from the isotactic polymer through rendering them soluble with suitable chemical agents. As has already been said, hydrocarbons dissolve atactic polymer; the simplest way of separating atactic from isotactic polymer is therefore to dissolve the atactic in hydrocarbon or halocarbon solvents and subsequent solid/liquid separation, e.g. by centrifugation. Centrifugation can be carried out quite simply on the slurry discharged from the polymerisation reactor, in the suspension process.

Catalyst components can be reacted with suitable chemicals, as for example, alcohols, and transformed in titanium and aluminum compounds soluble in hydrocarbons, or soluble in water or in alcohols.[83] The polymer is then purified from catalyst residues either by centrifuging the hydrocarbon suspension, or by extraction with water or alcohol.

5.2.5. Polymer Recovery and Drying

According to which processes is employed for the purification of the polymer from its actactic content and from catalyst residues, a solid polymer suspended in hydrocarbon diluent, or in water, or alcohol has to be handled. Conventional drying equipment and procedure may be employed to dry the polymer: steam stripping, flashing (in the case of low-boiling hydrocarbons), heating in air or nitrogen tunnels.

5.2.6. Granulation

The dry polymer powder can be used as such (or, better, after addition of stabilisers) as in the case of polymer in 'flakes' form. More generally, it is extruded and converted to pellets. During this operation thermal and anti-ultra-violet stabilisers, anti-acid agents, pigments and other additives may be added.

5.2.7. Recovery and Purification of Unreacted Monomer and Solvents

Monomer and polymerisation solvents must, of course, be carefully purified before recycling to the polymerisation section. Purification is generally carried out by conventional physical methods: distillation, absorption of polar compounds (e.g. water) on activated alumina or on molecular sieves.

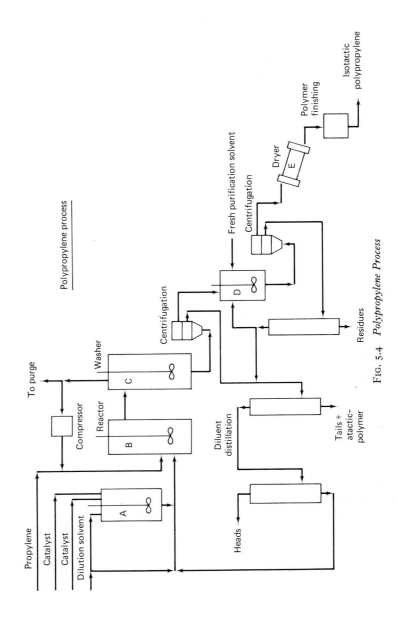

FIG. 5.4 *Polypropylene Process*

Fig. 5.4 presents a scheme of an industrial process for polypropylene production, using a suspension process in a hydrocarbon diluent. A is a catalyst preformer, wherein the two components of the catalyst, $TiCl_3$ and aluminum alkyl, are brought into contact, in a hydrocarbon medium.

Catalyst, diluent, monomer (and, if necessary, molecular weight modifier) are continuously fed to the polymerisation reactor, B. From B the polymerisation slurry is transferfed to the vessel C, a 'washer', where untreated propylene is separated by a release of pressure. The polymer slurry is treated with alcohol and then centrifuged; the solvent is separated from the atacic polymer and distilled before recycling. The solid polymer is suspended, in washer D, in 'fresh purification solvent' (for instance, an alcohol). In the second centrifuge step, the alcohol is separated from undistillable catalyst residues and sent to the recovery section; whereas the polymer is sent to the dryer E and, finally to granulation.

5.2.8. Economic Considerations

The largest single item which determines the production cost of polypropylene is the cost of the monomer. The second item is the capital cost of the plant (amortisation, interest on investment, etc.). Cost of monomer is outside the control of the polymer producer, as it depends on the cost of naphtha, how the ethylene and other co-products from the cracking operation are valued and the costs of purifying the propylene to 'polymerisation grade'. On the other hand, the polymer producer can control the costs of his plant.

Major research efforts are therefore being carried out by a number of industrial companies to simplify production technologies and processes. Another way of reducing production costs is by making use of the 'scale factor'. Thus the average size of industrial plants has regularly increased from a level of 10,000 tons polymer/year at the beginning of the sixties to 50–60,000 tons/year today.

5.3. PROCESSING THE POLYMER

Isotactic polypropylene is a typical thermoplastic polymer, which by melting may be transformed into a wide variety of articles. Soon after its discovery, because the polymers available at that time had a high molecular weight, attempts were made to transform them into fibres[84] and films[85] starting from solutions in hydrocarbon solvents such as naphtha, petroleum, etc.

This trend was soon given up once means were found of regulating the molecular weight in the polymerisation process. Specific grades for the more important applications were produced, and it became possible to prepare films and fibres by melt extrusion.

In this way, transformation costs were drastically reduced, although a thorough study was required concerning the behaviour of the polymer in the molten state, in particular its viscosity and flow characteristics under different conditions. Numerous investigations were done on the rheology of polypropylene and on the basis of the

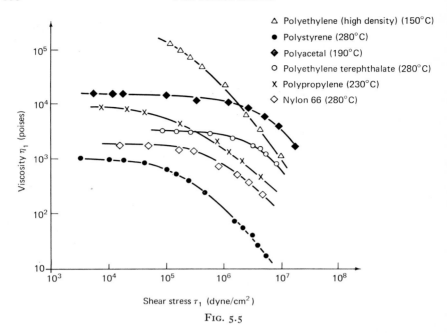

FIG. 5.5

information acquired, polypropylene products were obtained by specialised technologies, and found a very large number of applications.

For example, isotactic polypropylene may be used for the production of textile
fibres for blending with wool, very thin films for packagings and electric uses,
moulded and blow-moulded articles of large size.

The discovery of convenient stabilisation formulae preventing thermal-oxidative
degradation, allows easy polymer processing.

The degrading effect of light can be prevented: polypropylene may be conveniently employed for the productions of articles used under very severe light and
heat conditions. Finally, polypropylene may be easily coloured by thermostable
pigments to brilliant colours.

5.3.1. Rheology of the Molten Polymer

Polymers in the molten state are liquids with a more or less marked pseudoplastic
behaviour, which is particularly evident in the case of isotactic polypropylene; this
may be seen from the flow curves of Fig. 5.5. The figure shows the behaviour of
viscosity versus the velocity gradient of some of the best known polymers. Polypropylene behaves similarly to linear polyethylene and polystyrene but very dif-
ferently from polyethylene terephthalate, hexamethylenadipamide (nylon 6·6) and
polyacetal. The difference is not only due to structural reasons, but also to the
exceptionally high average molecular weight of polypropylene used for various
applications. In fact, it ranges from about 200,000 in the types used for textile
applications to about 600,000 in those for extrusion or moulding, compared with

about 20,000 in polyethyleneterephthalate used for the preparation of polyester textile fibres.[86, 87]

The pseudoplastic behaviour of polypropylene depends on molecular weight, temperature and molecular weight distribution.

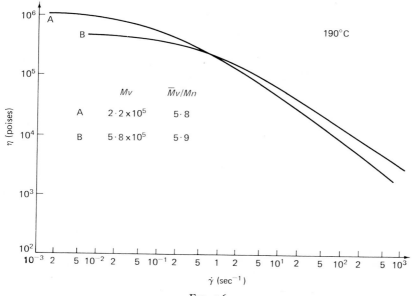

FIG. 5.6

As to the first factor, pseudoplasticity of the molten polymer increases with increasing molecular weight, as may be seen from the flow curves of Fig. 5.6 which relate to two polymers with a weight average molecular weight Mw of $2 \cdot 7 \times 10^5$ and 7×10^5 respectively and with the same molecular weight distribution ($Mv/Mn = 5 \cdot 8$ and $5 \cdot 9$).[88] Temperature exerts the same influence as in other polymers, i.e. its increase decreases the pseudo-plasticity of the molten product; however, its influence on the zero-gradient absolute viscosity, or Newtonian viscosity, is less marked than in other polymers, as may be seen from the curves of Fig. 5.7. This may explain the easy processability of polypropylene in different applications, which is one of the most important properties of the polymer. Finally, the influence of molecular weight distribution is so well defined that some authors[89] proposed a method for its determination just on the basis of rheological measurements. In this case, owing to the high regularity of steric and chemical structure, polypropylene is particularly suitable for a quantitative correlation between rheological properties in the molten state and molecular polydispersity. Such a correlation is practically impossible when these properties result from both molecular weight distribution and from the number, length and distribution of branchings. Fig. 5.8 shows some flow curves relating to polypropylene with different molecular weight distributions.[88]

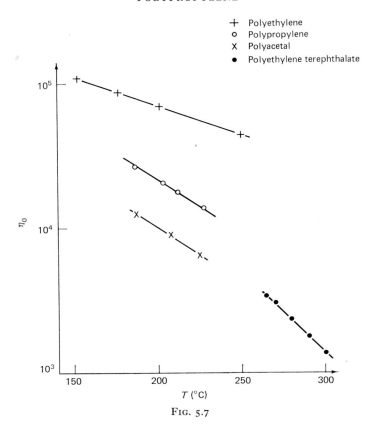

FIG. 5.7

The rheological properties, on the other hand, do not seem to be influenced by the stereoregularity of the product, at least within the limits of its normal use (isotactic content of 92–96 per cent).

Polypropylene in the molten state is not only a pseudoplastic material. Under particular conditions, the molten polymer behaves like an elastic substance so that, after compression and deformation (e.g. when passing through an extrusion nozzle or a capillary of a spinneret), it tends to recover its original shape. Recovery in polypropylene, particularly under shear stress conditions, is quite noticeable and during extrusion causes some effects at the nozzle outlet, that impair extrusion regularity and consequently the shape of the articles.

First there is a tendency to 'swell' on leaving the die: for example, it assumes the shape shown in Fig. 5.9 reported by Cappuccio *et al.*[90] and is particularly evident when spinning or in injection moulding, i.e. in those transformations requiring high material output and consequently a high output rate. In such cases the elastic force of recovery is quite high, the residence time in the compressed state in the nozzle is very short, shorter than the relaxation times of the viscous component; this

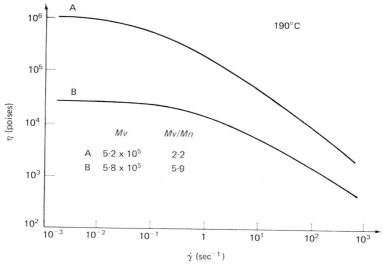

	Mv	Mv/Mn
A	5·2 x 10⁵	2·2
B	5·8 x 10⁵	5·9

FIG. 5.8

is the reason why a high elastic component still exists at the outlet and results in swelling of the material. Fairly detailed investigations on this phenomenon were carried out in the case of polypropylene, from which it was ascertained that the size increase at the nozzle outlet was inversely proportional to the residence time of the molten mass in the nozzle and is dependent on the nozzle dimensions and in particular on the ratio between hole length and diameter.[91]

This effect, sometimes called the 'Barus' effect, is reduced by increasing the ratio of length to diameter and by decreasing the average gradient of the rate in the capillary. Swelling increases with decreasing temperature, owing to the decrease of the effects of relaxation and of viscous damping.

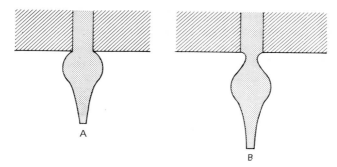

FIG. 5.9 *Die Swelling (in Spinning)*

Such drawbacks may be avoided by designing nozzles with proper dimensions, following the recommendations given in reported work. The importance of this procedure is shown from the curves in Fig. 5.10 where the influence of the length-diameter ratio on the lowest denier that can be obtained in a polypropylene fibre at various outputs is reported.[90]

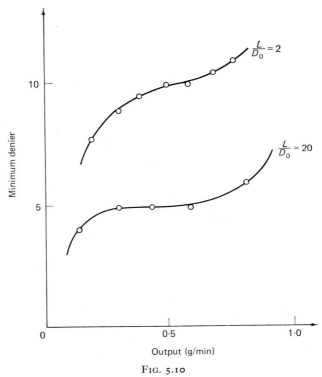

FIG. 5.10

A second and more critical effect when transforming polymer into articles is 'melt fracture', which occurs when a certain critical shear stress is exceeded. Above this value, the extruded article loses its cylindrical shape and becomes at first spiral-like and then completely irregular, like the extrusion 'B' shown in Fig. 5.11. Its surface becomes distorted and the article has no practicable value. This critical value of shear stress which corresponds to a critical pressure and critical shear rate is a function of the molecular weight, the extrusion temperature and the molecular weight distribution. It increases with increasing temperature and with decreasing molecular weight and polydispersity.[92]

The behaviour of isotactic polypropylene is here not dissimilar to that of polystyrene and branched polyethylene (critical shear stress of the order of 1.2×10^6 dyne cm² at 230°C compared with 1.4×10^6 for polystyrene from 175° to 250°C, and $1.5-2 \times 10^6$ for low-density polyethylene between 150–240°C), whereas nylon 66 or polyethylene terephthalate show critical values that are higher by about five times and practically outside the normal conditions adopted for conversion processes.[93] With polypropylene, melt fracture may take place quite easily, especially when using dies of very small diameter, or when using relatively low temperatures and high transformation rates.

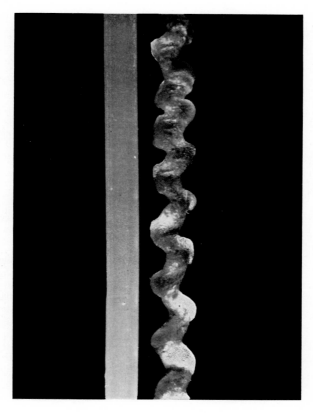

A   B

FIG. 5.11

5.3.2. Moulding

Polypropylene is moulded by techniques similar to those adopted for other polymers. In general, for this method, polymers with a medium molecular weight are preferred (melt index at 190°C and 10 kg[94] ranging between 1 and 5) over those with high molecular weight, since they yield products almost free from internal stress.

The viscosity of polypropylene varies markedly both with temperature and with injection pressure. Since maximum degree of flow is advantageous for completely filling a mould before the product cools, injection moulding should be performed at the highest possible temperature, consistent with the thermal stability of the polymer, and at high pressures. Thus, pressures in the injection cylinder should be above 1000 kg/cm².

Great care must be taken in mould design in order to produce articles with good dimensional stability, excellent surface, brightness and mechanical properties. Large sections for main and secondary channels of moulds are required since the polymer rapidly cools to the solid state, owing to its high melting point (see dilatometric curves of Fig. 5.12.)[95]

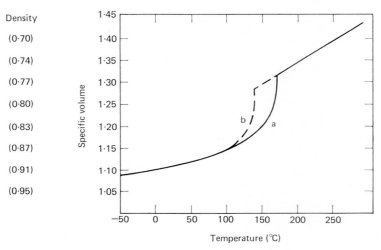

FIG. 5.12

The conversion of polypropylene can be satisfactorily carried out, bearing in mind that the average linear shrinkage of polypropylene varies between 1·5 and 2·5 per cent and its value—see Fig. 5.13[86]—depends on the thickness of the moulded sample.

Then the overall volume contraction of polypropylene from the injection temperature to that of extraction of the moulded specimen (see Fig. 5.13) is of about 20 per cent.

All types of injection moulding equipment can be used, although single-screw injection can be used only for small and thin articles. The variation of flowability of the molten polymer (expressed as a length of the filled spiral) is plotted against temperature and injection pressure and is shown in Fig. 5.14.

Normal moulding temperatures range between 200° and 300°C depending on the polymer melt index. The most commonly used pressures range between 850 kg/cm² and 1100 kg/cm² and moulding cycles between 50 and 100 seconds. High pressures are always preferred because they not only increase the area of good processability, but also decrease the moulding times, and, owing to the markedly pseudoplastic nature of the polymer, they cause a decrease of viscosity with consequent decrease of internal stresses.

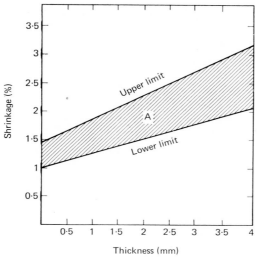

Fig. 5.13

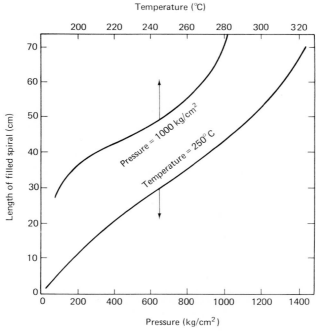

Fig. 5.14

During solidification, crystallization of the polymer starts with the formation of spherulites, the size of which depending on the cooling rate. With a high cooling rate, spherulites are much smaller; at the same time, an improvement is observed in both the transparency of the articles produced and in the impact strength, the latter depending largely on the degree of crystallinity and on the type of crystalline structures present.

A similar size reduction of the spherulites may be obtained by the use of polymers containing traces of products acting as crystallisation 'seeds'.[97, 98] In Fig. 5.15 processability areas of high-impact polystyrene, of low-density polyethylene and of polypropylene are compared.[96]

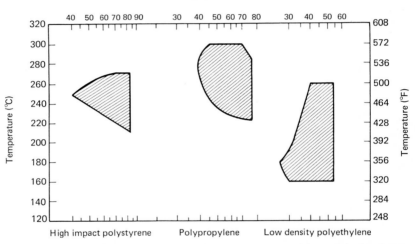

FIG. 5.15 *Moulding Cycles Determined by Same Mould and Same Injection Press*

5.3.3. Extrusion

Isotactic polypropylene, although crystalline, can be easily extruded like other thermoplastics, because its softening area above melting temperature is fairly extended. For the preparation of pipes, sheets or profile shapes the use of a low melt index polymer is preferred, stabilised and reinforced according to the expected use of the articles; on the other hand, in the insulation of electric cables, more fluid polymers must be used in order to obtain the thin coatings needed to give good adhesion to the metal. A polymer with medium molecular weight is best for the production of blow-moulded articles and of some cold-stretched articles, such as straps, slit film and monofilaments.

5.3.3.1. Normal extruded articles Either one- or two-screw extruders may be used for extrusion. For polypropylene, as for other polymers, the screw consists of three areas, although it is single-threaded and has constant pitch: i.e. feeding, compression areas. The compression area need not be very long, but must be very effective, with a minimum compression ratio of 2·8 in order to improve the fluid flow. The third area, i.e. the metering, must ensure constant flow and pressure, but it should not be too long so that the residence time of the polymer in the apparatus may be as short as possible. Extruders have usually a screw with a length-diameter ratio ranging between 20 and 26. In general filters are placed between the screw and the extrusion matrix for filtering the fluid mass.

During operation, the thermal gradient of the extruder ranges from 150°C at the first feeding area, up to 220–240°C at the screw end. The extrusion rate depends on the polymer grade and on the degree of surface regularity and brightness requested for the article.

The thermal gradient of the extruder may be varied, within a narrow range depending on the type of article to be produced. For the extrusion of sheets up to 8 mm in thickness, a temperature of 160–170°C is maintained in the first area; then it is gradually increased up to 230°C in the terminal area of the screw end of the head.

Sheets, above 8 mm in thickness, are obtained by compression moulding, laying up the required number of these extruded sheets to achieve the desired thickness and hot-pressing (above 200°C).

Cross-heads, also employed for other polymers (polyvinylchloride, polyethylene) are used for extrusion of polypropylene-coated electric cables. Owing to the nature of the polymer, the cable cannot be cooled quickly: cooling is started in air, then in warm water at 80–90°C and finally in cold water. In this way sharp shrinkages and consequent formation of internal bubbles and separation of polymer from the metal are avoided.

5.3.3.2. Stretched-extruded articles   It has long been known that stretching improves the mechanical properties of many synthetic fibres. Certain products can only be successfully prepared from polypropylene by using this technique as it gives the necessary degree of molecular orientation, crystallinity and structural regularity. By these means products (e.g. straps for packaging) could be obtained from polypropylene previously only obtainable from other materials.

Therefore, the usual extrusion processes were coupled to stretching, this imparting a uniaxial orientation to the extruded products, transforming them into articles with the desired properties. The extrusion equipment and process is the same as in the standard extrusion: only the size and the shape of the nozzles, and the cooling system are different. The latter must be particularly effective since the fibre can only be stretched in a subsequent process to the desired degree if the extruded polymer is in a smetic or at most slightly crystalline state.[99, 100]

The stretching section is designed like that for textile fibres. Stretching occurs above room temperature, usually in hot air or steam; it is followed by stabilisation and annealing, where, with shrinkage mechanically prevented, crystallisation and orientation of the polypropylene are completed.

A similar technology is also adopted for production of monofilaments (diameter of 0·5 mm and above). In this case, in order to avoid melt-fracture effects and to decrease enlargement at the outlet, the ratio of the orifice length to diameter in the spinneret must be at least 10.

Split yarn (or artificial rafia) is obtained from a plane or tubular film, whereas straps are generally obtained by multiple spinnerets of rectangular section placed on the head.

Rapid cooling is generally accomplished by cold-water immersion. In the case of film it may be carried out both by water immersion and by contact with a series of chromium-plated rolls internally cooled by water circulation. Cooling of tubular film is made at the bubble basis by air blowing.

As soon as cooled, the various extruded products are sent to the drawing oven, where at a temperature ranging between 90° and 150°C depending on the section, on the stretching ratio adopted, on the production rate and on the final properties desired, the actual stretching is carried out, by passing the products between sets of rolls revolving at different speeds.

In the case of monofilaments and of straps, the material is stabilised in another oven (to prevent any retraction) where it is kept at a temperature ranging from 40 to 140°C, depending on the material and the circumstances. For the production of split film, the film, stretched in one direction only (the direction of extrusion), is passed through equipment cutting it into a great number of longitudinal strips of the wanted width (more or less that of natural rafia 1·5–4 mm). The split film (or artificial rafia) is then annealed again.

5.3.4. Miscellaneous Mechanical Processes

Isotactic polypropylene sheets may be advantageously transformed by vacuum forming, by means of the apparatus used for other thermoplastics. However, the traditional technology must be modified, since the polymer has a high melting point and a reduced softening range resulting from its crystalline nature.

Moulding temperatures are higher than those used for other thermoplastic polymers, ranging between 165 and 175°C for times between 1 and 4 minutes, depending on the sheet thickness and on the heating system adopted.

Moulded or extruded pieces of isotactic polypropylene may be processed mechanically better than any other polyolefines, provided that particular devices are adopted, which in practice avoid overheating the specimen by mechanical friction.

Thus polypropylene may be shaped by the lathe, with tools used for light metals. No cooling of the tool is required provided that its rate does not exceed, e.g. 8 mm/min and it does not turn faster than 250–300 turns per minute.[86]

Polypropylene may be drilled or screw-threaded by manual or mechanical equipment. Cooling by a current of air is necessary. To prevent the tool becoming fouled it must work at a low rate but at a high number of turns per minute.

For printing, polypropylene requires some chemical modification of the surface, thus high-frequency electric discharges or similar treatment can give a key for the ink.

Metallisation is also possible if anchorage points are created by chemical treatment on the surface of polypropylene articles.

The polymer will stick to other materials fairly well, but not to itself; welding is necessary to join polypropylene. Welding may be performed by the same techniques used for the other thermoplastic materials: hot air welding is preferred. However, hot tool welding techniques can be used but care must be taken to insert an anti-adhesive material, such as polytetrafluoroethylene, between the hot tool and polypropylene article since, at the welding temperature (200°C), the polymer tends to stick to the hot metal.[86]

5.3.5. Spinning

Spinning is the process by which the polymer is transformed into multifilaments of different deniers. They may be used both as textiles and in industrial applications (that do not require the use of monothread).

Polypropylene is transformed into a fibre by spinning from the melt: its thermal stability, by appropriate stabilisation and when melted under an inert atmosphere, is more than sufficient for industrial-scale spinning.

By spinning followed by stretching and crimping which is similar to the process used for other fibres obtained from a melt, such as polyamides and polyesters, polypropylene yields either continuous filament or staple fibres with various characteristics.

Obviously optimum conditions exist for the various processing phases, although these are not so critical as in the case of polyamides or polyester fibres. This is due to the higher molecular weight of the polymer and the higher viscosity of the melt.

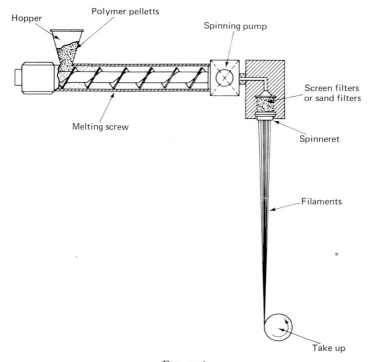

FIG. 5.16

The latter property prevents the use of the grid apparatus employed when spinning, e.g. polyesters, since in such equipment polypropylene cannot flow. The spinning apparatus for polypropylene therefore consists of a screw-extruder pushing the molten polymer under pressure toward the filters and the spinning metering pumps. A sketch of the spinning apparatus is shown in Fig. 5.16.

At the spinneret outlet, the molten polymer is transformed into filaments and becomes solid by air cooling.

Owing to the high shear stress in the pump gears and the high pressure in the spinneret, the apparent viscosity decreases and makes possible the preparation of

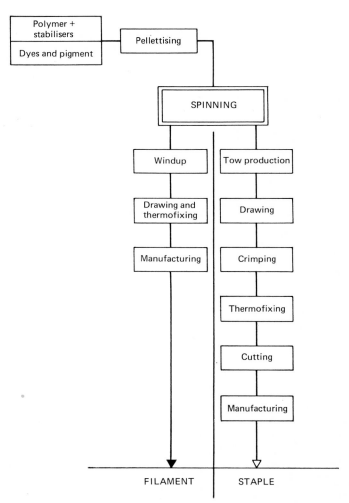

FIG. 5.17   *Scheme of Polypropylene Fibres Production*

low deniers. These may be obtained when the spinneret holes have the correct dimensions, as may be seen from the curves of Fig. 5.10.

As soon as the fibre comes out of the spinneret and during solidification, it is stretched from the take up machines at a very high rate (even higher than 1000 m/min).

Extrusion temperature at the spinneret outlet varies between 200 and 280°C, depending on the type of fibre wanted. Spinnerets for the production of staple fibres consist of some 500 or more holes.

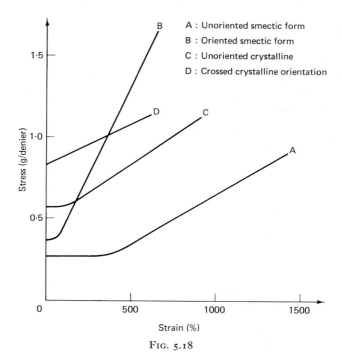

A : Unoriented smectic form
B : Oriented smectic form
C : Unoriented crystalline
D : Crossed crystalline orientation

FIG. 5.18

For the preparation of continuous filament, the fibre is collected on winding-up bobbins and then stretched by passing it through hot plates or rolls, e.g. between pairs of rolls, each pair revolving faster than the preceeding pair. The stretching ratio varies from 1:3 to 1:6 for the most common types (low, medium and high tenacity). Stretching temperature of the fibre varies from 90 to 130°C, depending on the denier and on the stretching rate adopted. After stretching, the fibre may be shrunk to a limited extent or thermofixed to a predetermined length. This operation is carried out soon after stretching, on the same machine, the fibre being heated once again to the desired temperature either on a plate or by a hot roll.

For the preparation of staple, the fibre is collected in very big ribbons, that may contain many thousands of single fibres and show a 'titre' of a few millions denier. Such ribbons are stretched in steam-heated chambers at 100 to 130°C, but at a rate that is much lower than that used for filament. After stretching, the stretched ribbons or tow may be shrunk or annealed: this treatment however, is usually done after crimping, which consists of inducing waves into the tow by heating in steam using specialised equipment.

Finally the tow passes to a machine consisting of rotating knives where fibres are cut to dimensions convenient for blending with wool or cotton.

Specialised finishing techniques are applied to the fibre at the various stages in order to improve processability.

An overall scheme for the preparation of polypropylene fibres is shown in Fig. 5.17.

By varying the extrusion conditions it is possible to obtain more or less oriented as well as more or less crystalline fibres. The influence of crystallinity in this phase of fibre preparation, may be illustrated from the curves given in Fig. 5.18 after Compostella[92] which show the stress-strain behaviour of polypropylene fibres both in the smectic (which in polypropylene somehow replaced the amorphous state) and in the crystalline state.

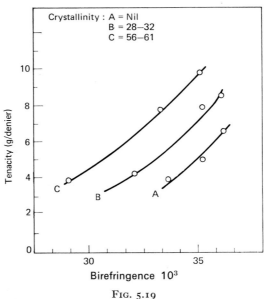

FIG. 5.19

Obviously, the characteristics of spun fibres (cf. 5.4.4.), shown in Fig. 5.18, determine the subsequent stretching treatment, during which substantial orientation of the molecules takes place with resultant increase of crystallinity, brought about by the macromolecular chains becoming more and more parallel along the fibre axis. The curves in Fig. 5.19 show[101] the influence of orientation and crystallinity on the fibre tenacity; the figures show the tenacity in g/den of fibres having different degrees of orientation, as measured by optical birefringence, at varying degrees of crystallinity.

Only special grades of polypropylene fibres may be vat-dyed. Standard grades may be dyed in bulk, by mixing thermostable dyestuffs with the polymer during pellettisation. This operation is basically fairly simple and can give excellent results; however, some snags have arisen when this operation is carried out on a bulk scale. Products to which dyeing modifiers have been added are now available.[102]

Polypropylene filaments have also been subjected to the standard texturising processes.

5.3.6. Film Formation

Both cast and oriented films—the two main types of films at present on the market—can be successfully obtained from polypropylene, whereas other polymers, such as polyethylene terephthalate, polystyrene, polyvinylchloride, give either oriented films (e.g. polyethylene terephathalate) or exclusively the cast ones.

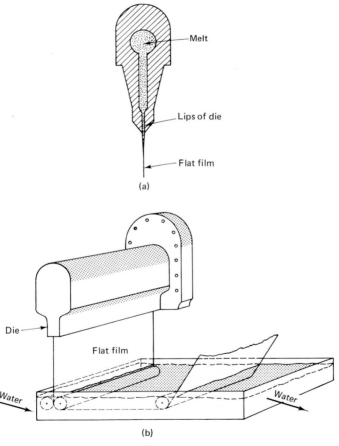

(a)

(b)

FIG. 5.20

Polypropylene used for film production generally has a medium molecular weight (melt index 5–10).

The film may be produced either by a flat-die extruder or by a tubular one. The polymer is placed into a screw extruder which is equipped either with a flat or with a circular die, and with a particularly accurate interposed filtering system.

The die width and the linear length depend on the type of film required. With a cast film, the die width will be many times greater than the film thickness, whereas the linear length will be almost that of the finished film.

If the film is to be subjected to stretching, the die width will be far larger than the film thickness (from 10 to 40 times), whereas the linear length will be much shorter than the width of the finished film.

Obviously in such an extrusion, some problems arise which originate from the rheology of the molten polymer; however, they are not so serious as when spinning low denier fibres for the following reasons:

Nozzles, i.e. slits, are fairly long and not too wide, but what is more important, they are fairly deep, since the system of lips regulating the thickness of the molten blade is relatively 'high' in the direction of extrusion, as may be seen from Fig. 5.20, where we show a flat-head section scheme.[103] In this case the swelling and 'melt-fracture' phenomena tend to produce a higher shear-rate and therefore facilitate extrusion.

Then, the extrusion rate is rather low, both for cast film and for oriented film, compared with that of fibres; hence the actual output of each hole is not as high as when spinning. Obviously, if the critical limits of the molten polymer are exceeded either by molecular weight increase or by a decrease of temperature, the rheological effects are even more noticeable than in the case of fibres, since they soon influence film planarity and its surface aspect.

5.3.6.1. Cast film   The processing scheme is fairly simple (see Fig. 5.21); the film is extruded, fast quenched, then wound up on winding-up bobbins and cut into the size wanted. The properties of the film depend largely on the method of quenching at the die outlet. At this stage, the polymer is molten and therefore completely amorphous. The purpose of fast cooling is to maintain the polymer as far as possible in the amorphous state, though this is only partly possible for polypropylene, owing to the very high rate of primary crystallisation. Actually, the film shows a lower degree of crystallinity than it would be found to have if cooling were slow; but, what is most important, the crystallites are very small. In this case, spherulites are so small that they do not interfere with the incident light that is not diffused during refraction, and the film is transparent and bright. Obviously, by regulating the cooling system, by doing it for instance with air or with metal surfaces rather than in cold water, different degrees of crystallinity may be obtained in the film although it is produced from the same type of polymer.

The properties of the films obtained by fast and by slow cooling differ considerably. In addition to transparency, the faster quenched films show: slightly lower (by 10–20 per cent) tensile strength, higher per cent elongation at break, tear resistance and resistance to repeated folds, but higher permeability to gases and vapours.

For the preparation of the flat film, cooling may be done by dipping the molten polymer blade into a water bath where two rolls drag it at the desired rate, thus regulating the film thickness (see Fig. 5.20) or by pouring the molten polymer on to a chromium-plated roll, internally cooled as shown in Fig. 5.22 (cf. 5.4.3.1.).

In the case of a tubular film, quenching is generally done by blowing cold air into the bubble and on to the outer walls. Obviously, owing to the low thermal conductivity of polypropylene ($3 \cdot 3 \times 10^4$ cal cm/cm^2 sec.) the most effective method seems that of cooling in water, even if this method does not allow high production rates. Extrusion temperatures at the die vary between 200 and 260 °C.

Production of polypropylene tubular film by air blowing is carried out by similar methods to those used for polyethylene and for polyvinyl chloride.

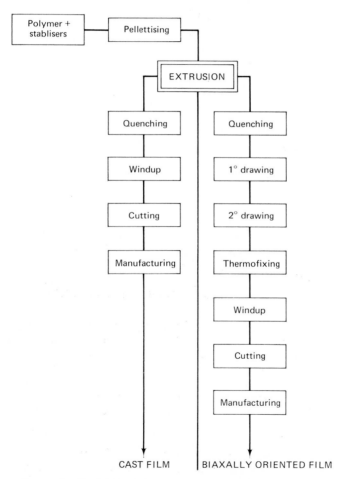

Fig. 5.21   *Production Scheme of the Two Polypropylene Films*

5.3.6.2. Bi-oriented film The technology for the preparation of bi-oriented film is much more complex, as may be also seen from the general scheme shown in Fig. 5.21. Some difficulties have had to be overcome in developing this process.

Film orientation must be planar, i.e. the film produced must be uniform in all directions (thickness being quite negligible with respect to the other dimensions). To solve this problem, simultaneous stretching in all directions is in theory required. However, stretching is generally carried out in two consecutive steps in two directions at right angles to each other.

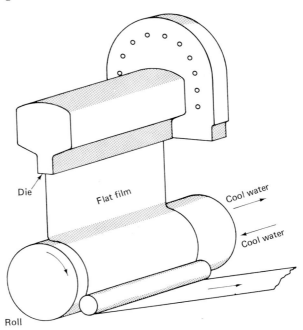

Die Flat film   Cool water
                                  Cool water

Roll

FIG. 5.22   *Extrusion of Chill-Roll Quenched Film*

Another difficulty arises from the fact that polypropylene can only be stretched with difficulty in the second step, since the first step not only causes orientation, but also induces a higher degree of crystallinity. Therefore, processing of the oriented film must be more accurate in order to obtain products with a constant thickness and with well balanced properties.

Stretching of tubular film is more difficult than that of the planar film. After forming, the film is heated and is stretched in the transverse direction by blown air. It is then cooled, squashed and then stretched longitudinally by letting it pass, while heating it again, between two pairs of rolls revolving at different rates.

Stretching of planar film can be better controlled. Some patents[104] also indicate the possibility of simultaneous stretching in two directions. However, the technology already adopted for other films, e.g. for polyethylene terephthalate, is more generally used. The planar film goes into the longitudinal stretching machine, where the first orientation takes place between two pairs of rolls rotating at different rates. The film is heated either by a hot air stream or by hot water immersion.

On coming out from the first stretching, the film goes into the transverse stretching machine. This grips the film in two chains which diverge to the desired amount, thus applying the necessary degree of stretching. The chains then continue parallel and the film is fixed at a constant width. The general layout of this machine is shown in Fig. 5.23. All these operations are performed in hot air. Crystallisation occurs during thermofixing: hence orientation becomes stable and the mechanical properties slightly improve.

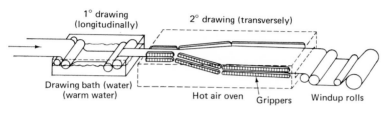

FIG. 5.23

Stretching temperatures are more or less those adopted for fibres. They depend on the film thickness and on its rate of production. Very thin films, below $10\mu m$, may be obtained from the stretched film.

Planar orientation involves a considerable change in the mechanical properties (it decreases elongation at break and markedly increases tensile strength), improves the optical properties and impermeability to gases and vapours. Tensile yield disappears, and the elastic modulus, distortion recovery and resistance to creep increase (cf. 5.4.3.2.).

The only drawback of oriented film is that it cannot be hot-welded. Cast film may be hot-welded by the usual techniques used for polyethylene film; on the other hand oriented polypropylene film, when heated to the temperature necessary for welding, loses orientation in the welding area; furthermore during its slow cooling, it undergoes a disordered crystallisation with formation of large spherulitic structures; as a result brittleness increases and breakdown in the welding area takes place.

5.3.7. Film Coating

In order to increase the field of possible applications, films may be coated with suitable substances or may be laminated to films or sheets of other materials so as to obtain composite materials of industrial interest. They may also be metallised.

Coating consists in forming on the film a very thin layer ($2–3\mu m$) of a substance which can modify the characteristics of the film. In the case of polypropylene, it is nearly always the oriented film that is coated. After such treatment the film may be hot-welded, which, as explained earlier, was otherwise not possible. If, however, a hot-weldable polymer layer is applied to the film, the properties required for welding are obtained. The only difficulty concerns the adhesion of the hot weldable substance to the polypropylene film which, owing to its saturated hydrocarbon structure, has poor affinity with other substances.

This has been overcome by first oxidising the film surface, then treating it, e.g. with a butylated urea formaldehyde resin and finally spreading a solution of a weldable resin such as a vinylidene chloride-acrylonitrile co-polymer.[105]

Coating generally is a continuous process and when the film unwinds from the bobbins it is first allowed to oxidise, then immersed successively in the baths containing the resins.

Other properties can be modified, e.g. the permeability to gases and vapours may be decreased by coating with polyvinylidene chloride solutions. Useful products can be obtained by laminating polypropylene with paper, aluminum or other films, by co-extrusion or by film coupling using techniques already well known for polyethylene.

Very thin polypropylene films are suitable for the production of small condensers having excellent dielectric properties. Condensers are often made of metallised polypropylene film. The metal layer is deposited on the film by evaporation under vacuum: aluminum is generally used, although other metals can be used (gold, silver). The difficulty lies in the low adhesion of metal to the film; however, in the case of polypropylene, it becomes quite satisfactory after suitable surface treatment.

5.4. PROPERTIES AND APPLICATIONS

The applications of the polymer are very varied both in view of the numerous methods of processing available and on its particular properties.

5.4.1. Properties

Isotactic polypropylene is a colourless, translucent to transparent solid, with a glossy surface while the non-crystalline polymer fraction looks like a viscous rubber. Table V.1 shows the main characteristics of the polymer.

5.4.1.1. Density One of the most valuable properties of polypropylene is its low density, the actual value depending on the physical state of the product. When amorphous, its density is about 0·85; when perfectly crystalline, the calculated value is about 0·94. Actually, the two extremes are not found in commercial polypropylenes. Owing to the very easy crystallisability of the isotactic polymer, an appreciable crystalline fraction is always present at room temperature; hence the density is necessarily higher than 0·85. On the other hand, since the polymer consists of alternately well crystallised and amorphous (although well ordered, owing to the high steric regularity) portions of chains, and since chain stereoregularity, although very high, is not complete, 100 per cent crystallinity is never reached. Therefore the density of manufactured articles does not exceed 0·90–0·905 and its value may indirectly indicate the physical state of the polymer. In any case, the above values indicate that polypropylene is properly rated as one of the lightest polymers known.

TABLE V.1

Polypropylene Properties

General properties		
Density	0·85 (amorphous polymer)	(162)
	0·94 (crystalline unit cell)	(163)
	0·88–0·91 (commercial polymer)	(164)
Crystallinity—Crystalline forms	α-monoclinic (stable from melt)	(163)
	β-pseudo hexagonal (in some samples	
	at high stereoregularity)	(166)
	λ-triclinic (in stereoblock polymer)	(166)
—Temperature of max. rate of crystallisation	120–130°C	(109)
Molecular weight—Relationship between $[\eta]$ and M	$[\eta] = 0·8 \times 10^{-4} \, M^{0.80}$ ($[\eta]$ in tetraline 135°C	(113)
—temperature θ	145°C (in diphenylether)	(167)

Thermal properties		
Melting temperature	170 172°C (atactic—35°C)	(168)
Glass transition temperature	−10·0°C	(114, 115)
Heat distortion temperature	90–105°C	(120)
Specific heat	0·44 cal/g°C	(117)
Enthalpy of fusion	1900 cal/unit	(118)
Thermal conductivity	$3·3 \times 10^{-4}$ cal cm/cm^2 sec °C	(119)
Brittle temperature θ	−5+10°C	(93)

Mechanical properties		
Tensile strength	350–370 kg/cm^2	(93)
Elongation at break	20–600%	(93)
Elastic modulus	$10–15·10^3$ kg/cm^2	(93)
Hardness	75 Shore D	(93)
Impact strength (notched Izod)	5–25 kg cm (25°C)	(93)

Electrical properties		
Electrical resistivity	10^{17} Ω cm (25°C)	(106) (121)
Dielectric constant	2·2 (50–10^8 cycles/sec)	(106) (121)
Loss factor, *tg* δ	$2·5 \times 10^{-4}$ (50–10^8 cycles/sec)	(106) (121)
Dielectric strength	35 kV/mm	(106) (121)

Optical properties		
Transparency	> 90% on film	(121)
Refractive index	1·50 on film	(121)

Chemical and solvent resistance		
Acid and inorganic chemical resistance	very good (except bromine)	(127) (116)
Alkali resistance	very good	(127) (116)
Organic chemicals resistance	good (poor to hydrocarbons and to chlorinated compounds)	(127) (116)

5.4.1.2. Crystallinity and structure Commercial isotactic polypropylene contains small amounts of atactic non crystalline and/or stereoblock low crystalline macromolecules. The polymer fraction essentially consisting of isotactic high crystalline macromolecules is usually expressed by the 'index of isotacticity', which is measured by the portion insoluble in boiling n-heptane; in commercial products it generally ranges between 85 and 95 per cent.

Isotactic polypropylene crystallises (from the molten state) in the monoclinic modification. A pseudo-hexagonal modification (called β) has also been found in samples with a very high regularity of structure. Finally, a third modification, triclinic and called γ, has been found in the stereoblock polymer. A syndiotactic polymer is also known.[108]

Polypropylene has a high primary and secondary crystallisation rate. After sharp cooling from the melt with liquid air, its structure is not completely amorphous but paracrystalline. By thermal treatment the paracrystalline structure becomes monoclinic again.

The highest crystallisation rate was found at about 120–130°C: 3–7 minutes.[109] During crystallisation, polypropylene forms complex structures of the spherulitic type; these play an important role in determining the properties of the article; actually, if they exceed a certain size, they may either interfere with refracted light and hence reduce the transparency of the product, or decrease their impact strength.

5.4.1.3. Solubility and molecular weight   The crystalline isotactic polymer is insoluble in all solvents at room temperature. It starts swelling and it is finally dissolved by specific solvents (e.g. tetralin, decalin, chlorobenzene, α-chloronaphthalene and diphenyl ether) only at temperatures generally higher than 100°C. On the other hand, the atactic polymer is soluble at room temperature in some solvents, such as ethyl ether, aliphatic and aromatic hydrocarbons, chlorinated hydrocarbons, both aliphatic and aromatic, esters, etc. Such different solubilities allow an easy separation of the essentially isotactic polymer from the atactic one. The small amount of steric irregularities present in the essentially isotactic polymer may be exactly determined by IR[110, 111] and NMR[112] analyses.

The molecular weights of commercial polypropylene are fairly high; actually, depending on the type of product, they normally range between 150,000 and 600–700,000 but may be higher. Molecular weights are usually determined according to the relationship between intrinsic viscosity and molecular weight: $[\eta] = 0.80 \times 10^{-4}$–$M^{0.8}$ (in tetralin at 135°C) ($[\eta]$ in 100 ml g^{-1}).[113]

Molecular weight distribution is rather wide, usually wider than that of many other polymers; it may range from a heterogeneity value calculated from the Mw/Mn ratio of 2–3 up to 10–15. Such a wide distribution of molecular weight influences considerably the polymer rheology and the properties of the articles. In general, when it decreases, the mechanical properties improve.

5.4.1.4. Thermal properties   The melting point depends first of all on the polymer stereoregularity and secondly on the thermal and mechanical treatment to which the polymer has been subjected. The commercial polymers that have not been subjected to special treatment show melting temperatures between 170 and 175°C.

The influence of both structure and crystallinity on the glass transition temperature is more complex. Glass transition is obviously connected with the amorphous

fraction of the polymer; therefore the glass transition temperature of isotactic polypropylene has been found to be in the range of -13[114] to $0°C$,[115] depending on the
measurement technique adopted. The value shown for syndiotactic polypropylene
is $0°C$.[107] Isotactic polypropylene has a specific heat of about 0.44 cal/g$°C$,[117]
enthalpy of fusion of 1900 cal/mole of monomeric units,[118] while its thermal conductivity is of the order of 3.3×10^{-4} cal cm/cm^2 sec$°C$.[119]

From the point of view of applications, the softening temperature, i.e. the
temperature above which the article distorts, loses its shape and cannot be used any
longer, is particularly important. This depends on the method of measurement: for
example, Vicat softening temperature is determined measuring the penetration of a
needle 1 mm^2 in area, under 1 kg load; the heat distortion temperature is measured
by flexing small bars under a given load. With polypropylene, the two values respectively afe $105–110°C$ and $90–105°C$.[106, 120] These figures in practice constitute
the upper limit of use of articles subjected to any mechanical stress.

Limits of use exist not only for high, but also for low temperatures. Polypropylene
does not show a high resistance to freezing as may be seen from the value of glass
transition. However, considerable differences exist depending on the physical state
of the polymer. In the case of oriented films and fibres, brittle temperatures in the
cold reach $-50°C$.[121]

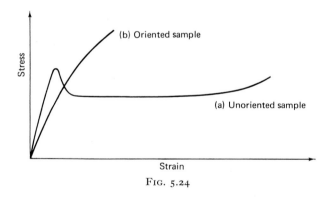

FIG. 5.24

5.4.1.5. Mechanical properties   The behaviour of isotactic polypropylene to tensile
tests varies according to the crystallinity of the polymer. Orientation obviously
influences it to a considerable extent. The stress-strain curves of a non oriented
polymer and of a polymer stretched in the machine direction are quite different.
Fig. 5.24 shows the stress-strain curves of two such samples. From the point of view
of application, the most important property is tensile yield strength, i.e. the value of
the highest tensile strength that the product may re-absorb with time, although
slowly, without undergoing permanent set. The tensile strength of articles showing
a stress-strain curve similar to that of Fig. 5.24 has no practical interest, since the
material cannot be used any longer above the value of elongation at yield.

Resistance to creep is deeply influenced by crystallinity and hence by the structural regularity as well as by the molecular weight. Creep decreases with increasing

crystallinity and molecular weight, the former exerting the greater influence. Fig. 5.25 shows creep curves of polypropylene in comparison with those of other thermoplastic polymers.[122]

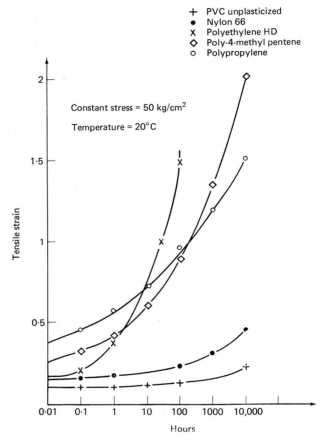

FIG. 5.25   *Tensile Creep of Some Polymers*

Flex life of isotactic polypropylene is particularly high: it depends on the crystallinity (and molecular weight) of the sample and consequently on the structural regularity.

For indexes of isotacticity from 77 to 97 per cent, stiffness varies from 7000 to 16,000 kg/cm².

The impact strength of polypropylene is not very high, especially for polymers with high isotacticity index. This property is largely influenced by crystallinity and more precisely by the size of the spherulitic structures present. If these are very large, as happens when the crystallisation temperature is high, impact strength sharply falls even in high-impact articles such as thin films.

5.4.1.6. Electrical and optical properties Being a non polar polymer, polypropylene shows excellent electrical characteristics. Further its heat resistance considerably widens possible electrical applications of this polymer.

Its volume resistivity of about $10^{17}\Omega$ cm at room temperature, falls to 10^{13} at 130°C.

Polypropylene also shows a high surface resistivity and good dielectric strength at an average value of 35 kV/mm, which falls to about 20–25 kV/mm at 130°C.[121] High frequency properties are fairly good, of the order of those of polyethylene, even at frequencies of several megacycles/second, and they remain constant at least until 130°C (see Table V.1). Therefore, this polymer may be advantageously used in electronic applications.[123]

With regard to the optical properties, the polymer is colourless and transparent when in thin films and where a microcrystalline structure may be easily obtained.

With increasing thickness, transparency of sheets decreases until they become translucent. This effect is caused by the high crystallisation rate of the polymer and by its low thermal conductivity.

Both these intrinsic properties make it difficult, if not impossible, to form a microcrystalline structure inside very thick articles.

5.4.1.7. Resistance to external agents The only effect of external agents on polypropylene, such as colour and light, is to decrease or to increase oxidation, which is the real phenomenon leading to polymer degradation.

When heated under a nitrogen atmosphere, polypropylene starts losing volatile decomposition compounds at about 310°C. At 320°C decomposition becomes fairly rapid. Under the same conditions, low-density polyethylene shows decomposition temperatures above 60–70°C.[124] In practice, oxygen is never absent; hence a partial degradation occurs even at lower temperatures owing to the peroxidic groups that are always present in the polymer. It was found that a complete degradation of a polymer occurred when absorption of less than 10 cm³ of oxygen per gram of polymer had taken place. Here, degradation is connected with the possible diffusion of gas in the polymer, which is influenced by both physical structure and thickness of the sample.

If the polymer is crystalline, oxygen penetration is slowed down and hindered in such a way that the effects of its attack are less marked. For example, the induction period at 110°C is 4·5 hours for the atactic (amorphous) polymer but 7·5 hours for the isotactic (partially crystalline) one. Oxygen attack mainly occurs in the polymer amorphous fraction and is greatly influenced by chain branching. Owing to its tertiary carbon atom, polypropylene is particularly sensitive to oxidation.

The induction period of commercial polypropylene is 20 hours, that of low-density polyethylene is 45 and that of high density polyethylene is 70.[125] However, proper stabilisation can greatly increase this period.

Many substances hinder polymer oxidation at least for a certain time and at certain temperatures. They are blended with the polymer soon after its production, to allow its easy processing at high temperatures.

The effect of light is only that of a catalyst of oxidation. Actually the process is

14

complex and consists of two phases. The first, *of initiation*, takes place during the early stages of exposing the polymer to light and, through photochemical reactions, leads to the formation of hydroperoxides along the chain. The second, *of degradation*, through photolysis of the hydroperoxides and subsequent β-scission of the tertiary alkoxy radicals, leads to polymeric chain rupture. This phase is self-maintaining. Suitable stabilisers greatly hinder photodegradation both by absorption of ultraviolet, which is responsible for this effect, and by heat dissipation and deactivation of electronic excitation energy of the photosensitised molecule. Antioxidants help photodegradation inhibition, by transforming the free peroxidic radicals that propagate the chain and the hydroperoxides, into compounds of low activity.[126] That is why the most satisfactory stabilisation formulae always involve the use of ultraviolet stabilisers with various antioxidants in order to make use of the synergistic effect.

A high degree of polymer purity makes stabilisation easier against oxidative degradation induced both by light and by heat.

5.4.1.8. Resistance to solvents and to chemical agents As already mentioned, some organic solvents at high temperature can strongly attack commercial polypropylene and dissolve it. Among organic products, polypropylene is dissolved by butyric acid at 100°C, whereas in the cold it is insoluble in organic acids in general, amines, nitriles, sulphur derivatives, nitro derivatives, aldehydes, amides and phenols. It is strongly swollen by cold halogenated compounds, among which the most effective are 1,1,2,2 tetrachloroethane and tetrachloroethylene, and by hydrocarbons; on the other hand, a milder influence is exerted by ethers and esters. It is worth mentioning the solvent power of hot anisole, dioxane, diphenyl ether and dihydropyran and of cold tetrahydrofuran and acetic acid. Among inorganic substances, polypropylene is attacked by oxidising substances, such as concentrated nitric acid, oleum, aqua regia and by-products such as bromine and gaseous chlorine. On the other hand, it is not attacked by alkalis.[116, 127]

5.4.2. Applications

The use of moulded or extruded polypropylene articles is not limited to a specific field, but concerns a wide range of different applications: reinforced articles, pipes, houseware, sanitary and industrial articles, films, fibres, etc.

5.4.2.1. Reinforced articles In the case of moulded or extruded products, polypropylene may be reinforced, quite considerably, with inorganic fillers, such as asbestos powder, or with reinforcing fillers, such as glass fibres of various lengths, or with particular dye pigments such as carbon black. With such blends, it is possible to obtain articles with an improved mechanical and flexural resistance, with a higher resistance to creep and to external agents, such as sunlight. Products reinforced with carbon black may be employed for open air applications involving long standing in hot and sunny climates, for example in irrigation pipelines and anti-hail nets.

Some blends have also been prepared in order to obtain a higher resistance to low temperatures: they are usually employed to produce pipes installed in the open exposed to winter temperatures, or for conveying liquids at a temperature far below room temperature. The Vicat softening temperature of these blends is about 90 °C, against 105–110 °C of the pure product.[106]

5.4.2.2. Anti-corrosive coatings   Owing to its satisfactory chemical properties, especially with regard to the resistance to acids and alkalis, polypropylene is used for the production of slabs, pipes, tanks and containers whenever a high resistance to corrosion combined with a good resistance to temperature are required. Prior to the discovery of polypropylene, PVC was more often used for such applications; however, its low softening point limited its use to temperatures not exceeding 50–60 °C. Polypropylene may be used for coatings of any size as well as for pipes, e.g. for carrying corrosive by-products or fumes, even at temperatures of 100 °C.

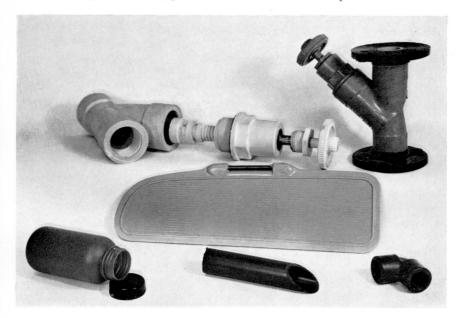

FIG. 5.26

5.4.2.3. Pipes   The production of pipes is a valuable outlet particularly because of their lightness. Even more important are polypropylene pipe fittings and valves, which have high mechanical resistance, do not jam easily and have excellent toughness. These fittings and valves with the most varied shapes (see, e.g. photos in Fig. 5.26) are often used to join pipes produced from polymers other than polypropylene, for example from polyethylene, which is sometimes preferred owing to its higher resistance to low temperatures.

Connection of pipes of the same material or of other polymers is quite easy and not expensive.

Another application of polypropylene pipes is to convey water for domestic uses; they may be employed to convey both cold and hot water.

5.4.2.4. Toys Lightness, stiffness and glossy surface make polypropylene very useful for the production of toys of the most varied shapes and sizes; in many cases it will replace polystyrene and ABS resins. Very small articles as well as very big ones (e.g. pedal cars and sledges) are produced with polypropylene. These big and glossy-coloured articles are almost always produced by moulding in one single piece, with considerable manufacturing cost savings.

5.4.2.5. Houseware, sanitary articles, packagings Soon after its discovery, polypropylene was extensively used for the production of sanitary articles, especially because it was absolutely inert to water and to aqueous solutions, and furthermore could be sterilised by boiling water. Production was started of hypodermic syringes, graduated cylinders, small basins, etc., which may be washed very quickly and dried and which are of course very light.

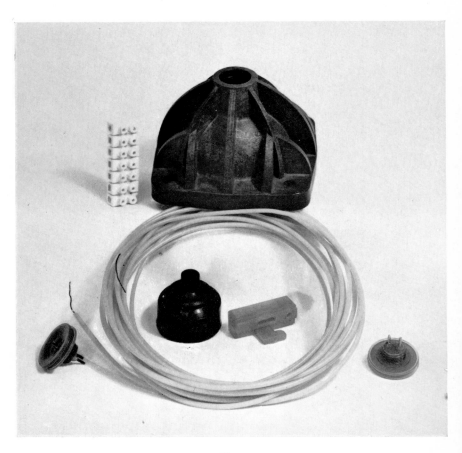

FIG. 5.27

Numerous applications exist for housewares: some require high-temperature resistance, e.g. colanders, others mechanical and thermal resistances, e.g. mixer cases and similar articles, basins of washing machines, cases for hair-driers, etc. An interesting application is in the production of air-conditioning apparatus, where many parts are made of polypropylene.

With regard to packaging, polypropylene is used for the production of cases for fish, fruit, boxes for packing up coffee, etc.

5.4.2.6. *Industrial products and electric applications*

Polypropylene is widely used also for the production of industrial articles. In the textile industry, where spinning and weaving require high humidity, metal parts easily deteriorate, especially during dyeing in acid and corrosive baths. Since polypropylene is perfectly resistant to such environment conditions, many parts, such as bobbins, shuttles, separators, containers, etc. have been made with it. Fig. 5.27 shows some applications in this field.

In the car industry, polypropylene has been used to produce sun screens and radiator fans.

Owing to its excellent dielectric properties, polypropylene is employed for a number of electric and electronic appliances, e.g. coil ignition caps and coils, electric wire insulation, frameworks for batteries. However, as will be described, it is the polypropylene film that has found the most important applications.

5.4.2.7. *Slit film and straps* (cf. 5.3.3.2.)

A particular application of polypropylene has lately developed to a very considerable extent; it is intermediate between textile and packaging uses, i.e. polypropylene slit film. It is obtained by slitting stripes of monostretched film, 1·5–4 mm wide and 20–50 μm thick. This product is at present employed to manufacture bags, previously made with jute; over which polypropylene offers great advantages: resistance to rotting and lightness.

Table V.2 shows the main properties of the different types of polypropylene slit film, compared with vegetable jute.[128, 129]

TABLE V.2

Properties of Polypropylene Split Film Compared with Vegetable Fibres[102, 130]

	Polypropylene	Jute	Vegetable raffia
Width, mm	1·5–4	—	1·5–5
Thickness, μ	20–100	—	30–50
Titre, dtex	0·4–25 × 10³	20–50 × 10³	5–10 × 10³
Tenacity, g/dtex	3·5–5	3–4	4·0
Elongation at break (%)	10–30	1·7	5·0
Elastic modulus, g/dtex	40–70	175	85
Knot strength (% tenacity)	60–80	15–20	35–40
Splitting resistance, g/μ	8–18	—	6–18
Fibrilating capacity, N°/mm	1–4	—	1·2–2
Density, g/cm³	0·90	1·5	1·5
Water absorption (%)	0·01	13·7	8·5

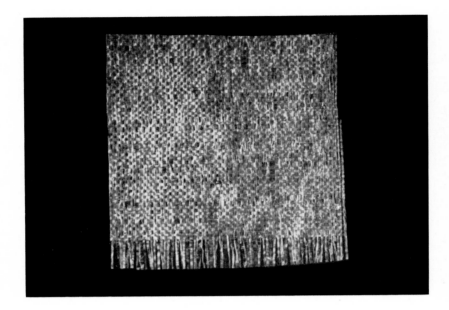

FIG. 5.28

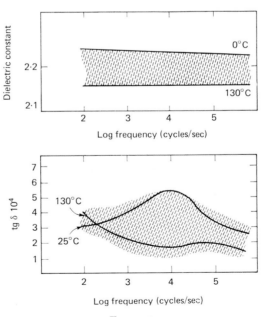

FIG. 5.29

Polypropylene slit film fabrics are also used for carpet backing, especially for those made with polypropylene fibres. Slit film easily withstands vulcanization of the rubber support required for production of tufted carpets. A polypropylene slit film fabric is shown in Fig. 5.28.

5.4.3. Films

Polypropylene may be conveniently transformed into thin films (even less than 10 μm thick) showing quite valuable properties. Many applications are possible with such a product: packaging and in electric and electronic appliances.

Two types of films are produced: cast and bioriented films. Their properties are rather different: hence also their applications are different. The main properties of the two films, compared with those of other films are shown in Table V.3. As may be seen, the mechanical properties of cast film are practically identical with those of the polymer, bearing in mind that it has a not too high microcrystalline structure. Hence elongation at break is about 400–600 per cent, tear resistance and flex life are very high.

Obviously, the bioriented film exhibits higher tenacity (about five times that of cast film) and is deformed less easily. It shows the same tear resistance and flex life, but a far higher impact strength. Thermal properties are similar to those of the polymer, except for the far lower brittle temperature of the film in the cold: about $-50°C$. Electric properties of the two films and in practice of the polymer are identical: transformation does not influence these characteristics. The behaviour of tg and dielectric constant against frequency and temperature are shown in Fig. 5.29 and thus demonstrates that polypropylene film is particularly suitable for electric appliances.

The optical properties described in Table V.3 are interesting because the very thick polymer is translucent. The two films show some difference in the haze and gloss values, which are higher in the case of the oriented film.

5.4.3.1. Cast film (cf. 5.3.6.1.) Various types of cast film are produced with a more or less high coefficient of friction which makes rapid processability possible. Cast film is mainly used for packaging, owing to its excellent transparency and gloss, and to its tenacity, which is for example higher than that of polyethylene film. Unlike cellulose films, it may be sterilised and is insensitive to water. It is used for packaging of: high-quality textiles, such as shirts, vests, etc.; numerous toilet articles; hardware; bread, cakes, etc.; sanitary articles to be subjected to sterilisation; precooked foodstuffs.

Polypropylene cast film more than 200 μm thick may be transformed into vessels of different shapes and size by vacuum forming. It may be also used for coating paper and paperboard and for the production of laminated products, such as aluminum-polypropylene; polyester-aluminum-polypropylene.

For the packaging of certain foodstuffs, the cast film coated with chlorovinylidenic compounds must be used.

5.4.3.2. Bioriented films (cf. 5.3.6.2) The high mechanical properties, transparency and good permeability properties make this film quite advantageous in the field of

TABLE V.3

Properties of Polypropylene Film Compared With Other Films

		Polypropylene cast	Polypropylene bioriented	Cellophane lacquered	PVC	Polyethylene terephthalate
Density	g/cm³	0·88-0·90	0·90	1·4-1·5	1·2-1·50	1·38
Yield	m²/kg	44-45	44·5	26·6-28·5	26·6-33·1	29
Water absorption	% (24 h)	<0·01	<0·01	50	0·5-5	0·3-0·5
Tensile strength	kg/mm²	90-150	150-350	150-300	40-280	320-420
Elongation at break	%	400-600	15-130	8-45	5-250	60-80
Elastic modulus	kg/mm²	90-150	150-350	150-300	40-280	320-420
Impact strength (falling ball)	kg cm	4	30			30-40
Tear strength (Elmendorf)	g/25μ	40-200	8-12	5-15	10-250	
Flex endurance resistence	N°	>50,000	≥70,000	500-20,000		100,000
Melting point	°C	168-170	168-170			260
Brittle temperature	°C	-32	-50			
Refractive index	—	1·50	1·50	1·52	—	—
Haze	%	≤3	≤2		—	
Gloss	%	120-130	140-150			
Volume resistivity	Ω cm	10^{17}	10^{17}	$10^{10}-10^{13}$	$10^{14}-10^{16}$	10^{19}
Dielectric strength	kV/25μ	4·4	4·2	3-3·5	3·8-3·9	4·4
Dielectric constant 1 kc	—	2·2	2·2	3·2-3·6	2·8-4·5	3·0
Loss factor 1 kc	—	$3-6 \times 10^4$	$3-6 \times 10^{-4}$	$1-4 \times 10^2$	$1-16 \times 10^{-2}$	$5·5 \times 10^{-3}$
Permeability to: H_2O	$\dfrac{g/25\mu}{m^2 \times 24\,h}$	7-8	3-5	3-8	50	16-18
N_2	$\dfrac{cm^2/25}{m^2 \times atm \times 24\,h}$	470-780	300-500	—	—	—
O_2	$\dfrac{cm^2/25}{m^2 \times atm \times 24\,h}$	3700-5300	1600-2300	80	0·08-0·4	0·0025
CO_2	$\dfrac{cm^2/25}{m^2 \times atm \times 24\,h}$	9700-15,000	4200-6500	120	0·05-0·5	0·0125

packaging, where it is widely used. Bioriented film is particularly convenient for the preparation of laminated products: polypropylene-aluminum; polypropylene-cellophane. It is also used for typewriter-ribbons and adhesive tapes.

The use of very thin films ($12 \cdot 5 \, \mu m$) for electric condensers and as an insulator for cables, transformers and motors is of particular importance. In this field, polypropylene film is even better than polyester film.

5.4.4. Fibres

Applications in the textile field depend on three factors: the properties of the fibre, their cost and their processability. The cost of the end product depends mainly on how easily the fibre is transformed into it. Polypropylene, especially for certain applications, shows to advantage in all three ways.

It is produced in six fundamental types:
(1) High tenacity continuous thread;
(2) Medium tenacity continuous thread;
(3) Texturised continuous thread for carpets;
(4) Medium tenacity staple;
(5) Low modulus staple;
(6) Staple dyeable with anionic dyestuffs for wool.

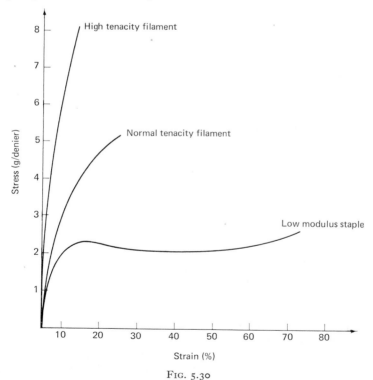

FIG. 5.30

The first five are obtained from the pure polymer, reinforced with fillers, dyeing pigments, suitable for mass-dyeing, and including stabilisers to heat and light; the sixth type is obtained from modified polymers.

One of the major drawbacks of polypropylene fibres in the textile field is their poor affinity for dyestuffs: hence until a short time ago, it was exclusively mass dyed, by incorporation of a thermostable dyestuff in the polymer during crumbling. In the sixth type, therefore, the polymer is not substantially modified as are acrylic fibres, but substances—which vary according to the specific procedures are added which show a high affinity toward the anionic dyestuffs used for wool. Obviously, the additive must not markedly alter the properties of the fibre. The modified polymer is obtained by the addition of basic substances, which are particularly stable to heat. Some producers have developed a particular form of dyeing using a chelating dyestuff, which during extrusion is bound to the saline substances incorporated in the fibre.[102]

Polypropylene fibres may exhibit quite a range of mechanical properties, as may be seen from the stress-strain curves of Fig.5.30. High-tenacity threads reach 9 g/denier, the low-tenacity ones at high elongation do not even reach 3 g/denier.

Table V.4 gives the main properties of polypropylene fibres compared with the polyamides, polyesters and polyacrylics. Just as the mechanical properties may be varied depending on the different stretching and annealing treatments (see, e.g. Fig. 5.31, which shows tenacity versus orientation and annealing,[92]) so also can the electrical properties be varied considerably.

TABLE V.4

Properties of Polypropylene Fibres Compared with the Main Synthetic Fibres

	Polypropylene	*Nylon 6·6*	*Polyester*	*Polyacrylic*
Density, g/cm$_3$	0·90	1·14	1·38	1·14–1·17
Water absorption, %	0·01	4–4·5	0·4	1–1·5
Tenacity, g/denier	4·0–6·3	4·2–7·9	4·2–6·8	2·0–5·0
Polyethylene tetraphthalate, %	15–60	16–45	11–35	16–34
Knot resistance (% tenacity)	70–90	75–80	65–85	70–75
Wet tenacity (% dry tenacity)	100	85–90	100	95–100
Elastic modulus, g/denier	34–45	30–60	50–125	45–50
Melting point, °C	170	250	260	—

High modulus fibres recover 90 per cent of a 10 per cent deformation almost immediately, without any permanent set. Medium tenacity fibres recover 90–98 per cent of a 5 per cent deformation, whereas the low-modulus ones are even more elastic: in fact, although they have a tenacity of 3 g/denier, they show a 95 per cent immediate recovery from a 50 per cent deformation and a complete recovery after 5 minutes.

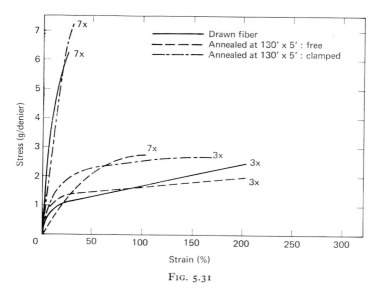

FIG. 5.31

One of the drawbacks of polypropylene fibres is their low resistance to creep, which is lower than that of polyamide and polyester fibres. Polypropylene fibres, owing to their low melting point, cannot be ironed, whereas they stand up perfectly to boiling water and therefore to vat dyeing, when possible.

Polypropylene shows a fairly high coefficient of friction particularly when it flows on metal surfaces or on china-ware. Static coefficients of friction between filaments lie between 0·32 and 0·42, and the corresponding dynamic figure is between 0·29 and 0·40.

With regard to textile properties, polypropylene, especially when obtained from a polymer with high index of isotacticity, has a pleasant touch. The covering power of polypropylene fibre is far higher than that of any other fibre owing to its low density (it is the only fibre that floats on water); it is therefore particularly suitable for the production of furnishing fabrics and carpets. Its thermal insulating power is among the highest, even though it does largely depend on the air included between the fibres. In any case, polypropylene fibre shows the lowest thermal conductivity of all fibres.

Crease resistance is good and may be varied considerably by thermal treatment of the fibre. In some cases, it is of the same order as that of wool, although unlike wool, it is not influenced by humidity.[130]

Polypropylene fibre shows good abrasion resistance when dry and even better under wet conditions. Therefore, it is particularly suitable as blends with less resistant fibres since it supplies improved abrasion properties.

Its flex life is particularly high: when blended with wool, it may also be subjected to ironing, but not above 140°C.

Polypropylene fibre may be easily washed with water or with aqueous solutions either acid or containing bleaching products up to 100°C. Dry cleaning in trichloro-ethylene is possible up to temperatures not above 50°C; above such a temperature dimensional variations may take place.

Dimensional stability of polypropylene fabric is excellent: it is not influenced by environmental conditions, by washing (except dry cleaning above a certain tempera-ture). Finally, polypropylene fibres are perfectly resistant to mildew, insects, and do not felt.

After appropriate stabilisation, its resistance to light is more than enough even under drastic conditions, such as use as ropes and fishing nets.

The main applications of polypropylene fibres are the following:

5.4.4.1. Carpets and moquettes Polypropylene has rapidly found wide applications in the production of tufted carpets and moquettes. It is soft to the touch; its cover-ing power is higher than that of any other fibre and its resistance to dirt is high. In particular dirt and spots of any type may be easily removed from polypropylene carpets by simple washing with detergents or with hot water, without any alteration of the fibre or of its colours.

In general, fibre for carpets are dyed by the mass procedure and the dyestuffs used are particularly brilliant and permanent.

Resistance to wear and to abrasion is very high; hence polypropylene carpets may be advantageoulsy used in places where large numbers of people congregate. Also static electricity does not reach dangerous values. Polypropylene carpets may be damaged by contact with lighted cigarette ends; however, they do not go on burning provided they have been correctly formulated.

Polypropylene carpets are manufactured either by the standard techniques start-ing from staple or by up-to-date procedures from high-bulk continuous thread. The fibres used in the former technique have low tenacity (about 3 g/denier), but high elongation at break (70–80 per cent) and above all excellent elastic recovery, which makes them particularly convenient for carpets and moquettes. 'Resilience' is in this case similar to that of wool.

Light stabilisation of these fibres is more than enough for indefinite indoor use.

5.4.4.2. Ropes Fibres for the production of ropes must fulfil many conditions:

(a) high tenacity under all conditions;
(b) high abrasion resistance, flex life and high stress resistance;
(c) minimum water absorption, so that its weight under wet conditions does not vary;
(d) lightness and resistance to ageing;
(e) resistance to micro-organisms, to solvents, etc.

In this application, polypropylene bears comparison with other fibres, since it may fulfil almost all these requirements. Polypropylene ropes float on water, have high abrasion resistance and are absolutely inert to water and to the attack of mildew, micro-organisms, etc. Hence they are widely used for nautical uses and for fishing.

5.4.4.3. Fishing nets The properties required for this application are similar to those required for the production of ropes. Polypropylene filament is used in competition with natural fibres (cotton, manilla, etc.) and with some synthetic ones. It has advantages over these owing to its lightness, impermeability, and a good weight-length ratio.

Ropes are made with monofilaments and nets with high-tenacity multifilaments (in general 6·5–7 g/denier and 20 per cent elongation). The products thus obtained may be handled more easily and have a high resistance to abrasion.

5.4.4.4. Hosiery Polypropylene fibres blended with wool are also employed in the production of personal hosiery. It is not only pleasant to the touch, but it also strengthens the natural fibre; the fabric thus obtained may be dried more easily, does not felt, shrink and it is light. Both staple and texturised threads are used for such applications.

5.4.4.5. Fibres from films Several producers have lately developed a procedure for the preparation of fibres obtained by fibrillation of monostretched films, similar to that for split film. The reported reduction in costs attained by this procedure is up to 25 per cent. The fibres obtained may be used for nets, ropes, packagings, etc.

5.5. ETHYLENE COPOLYMERS

By means of Ziegler–Natta catalysts, propylene can be copolymerised with ethylene to yield, under suitable experimental conditions, elastomers that have found very wide applications.[131] The synthesis procedure and the physical-mechanical characteristics of cured and uncured ethylene/propylene elastomers have been described, for instance in the book *'Ethylene and its Industrial Derivatives'*, ed. S. A. Miller (E. Benn Ltd., London, 1969).

In this section we will just consider the crystalline copolymers of propylene with a low ethylene content, obtained in the presence of the same catalysts used for the industrial production of crystalline polypropylene.

The crystalline homopolymer of propylene exhibits a large number of satisfactory physical and rheological properties; however, it cannot be used for some applications. For instance, although polypropylene shows a high degree of crystallinity, excellent stiffness and good processability, its elongation at break is low at low temperatures; consequently polypropylene articles are brittle at temperatures below 0 °C. In order to obviate such a drawback, some blendings with polyethylene, or with ethylene-propylene copolymers, or else with rubbers have been tried, although with limited success. A better result was obtained by polymerising propylene in the presence of small amounts of ethylene or of butene-1. For example, when using amounts of 1·5–3·0 per cent of ethylene, a random copolymer can be obtained, which shows an improved impact resistance at low temperatures, better optical properties and processability. However, polymer stiffness and crystallinity are very much lower than with straight polypropylene. Synthesis of crystalline block copolymers was then tried in order to overcome such drawbacks.

In particular heteroblock copolymers of α-olefins were studied; they were obtained by a multi-phase process including the polymerisation of the first monomer, elimination of its unreacted fraction, introduction of a second monomer and its polymerisation; complete elimination of the unreacted second monomer. These steps may be repeated at will using either the same or different monomers.

The average length of each block of monomer units depends on several factors among which are, chiefly, the average life-time of the growing chain, and therefore also the length of each phase, and the rate of polymerisation of each monomer.[132–136]

However, these polymers have no practical interest and do not find applications in industry owing to the difficulties involved in the process for industrial-scale production; moreover, while their impact resistance at low temperatures improves, their stiffness is not completely satisfactory.

After 1962, several companies started the commercial production of modified polymers of propylene[137–142] which all exhibit a common feature: i.e. they consist of 65–85 per cent of unmodified crystalline polypropylene, of 15 – 30 per cent of an ethylene/propylene copolymer (which may be partially crystalline), and of less than 5 per cent of stereoblock or amorphous polypropylene, with a very small amount of a product consisting of a long crystalline polypropylene chain, at the end of which a chain block containing ethylene and propylene units is bound. Such a mixture obtained directly in the polymerisation process is thoroughly homogeneous, and does not give rise to any separation or to any of the typical defects of the blends that contain exclusively polypropylene and high-density polyethylene.

The percentage of ethylene, present in the product, usually ranges from 5 to 15 per cent by weight: a low amount of ethylene is used for the production of a copolymer fraction with very low ethylene crystallinity (in this case the polymer will show a fairly good impact strength at low temperature, a good stiffness and low transparency); on the other hand, a large amount of ethylene is used in order to obtain a copolymer with a medium degree of ethylene crystallinity (in this case, the polymer shows a good impact strength, a fairly good stiffness and good transparency).

The techniques adopted by the various companies for the preparation of these products are different, but may be summarised as follows: propylene is polymerised by a continuous process as a first step; the polymerisation product is then submitted to a second step where an ethylene-propylene mixture is copolymerised with the same catalyst, so as to obtain a copolymer fraction of well defined characteristics.

As already mentioned, by varying the amount of ethylene, the degree of crystallinity and the copolymer composition, a final product with the desired properties can be obtained.[143]

These copolymers may be processed by extrusion, injection moulding and blow-moulding equipment as commonly employed for polypropylene. Their rheological properties are good.

5.6. OTHER COPOLYMERS

By using Ziegler–Natta catalysts, propylene can be copolymerised with higher α-olefins and with styrene.[144] However, little information is available, either on the characteristics or the possible applications of these copolymers. In general, for propylene contents between 30 and 70 mole per cent, these copolymers are amorphous and look like non-vulcanised elastomers.

Only the crystalline copolymers of butene-1 with a propylene content of 10–20 per cent have been sufficiently studied and appear interesting for commercial applications. In fact, they show very similar mechanical characteristics to poly-butene-1, but compared to this last, they offer the advantage that they do not crystallise after moulding into a metastable form (form II, tetragonal) which can only slowly be converted into the stable form (form I, rhombohedral).[145]

As a matter of fact, conversion rates from the second form to the first one of butene-1/propylene copolymers are of the same order as the polymer processing time.[146]

Ziegler–Natta[147] or Alfin[148] catalysts also promote the copolymerisation of propylene with conjugated diolefins, such as butadiene-1,3 and isoprene. The amorphous copolymers of this type may be vulcanised with sulphur and yield rubbers exhibiting good mechanical and dynamic properties.

Propylene has also been copolymerised, although with some difficulty, with other vinyl monomers, e.g. acrylic and methacrylic esters, acrylic and methacrylic nitriles, acrylic and methacrylic amides, vinyl acetate, vinylketone, etc.

Ziegler–Natta[149] and Friedel-Craft[150] catalysts or special kinds of radical initia-tors[151] are used for such copolymerisations. Although the products obtained are generally described as alternating copolymers, their regularity of structure is not such that crystallisation takes place. Since molecular weight and copolymer yield with respect to the catalyst used are generally very low, the practical interest in these products is not very great.

Copolymers of propylene with vinyl chloride have been prepared in the presence of radical initiators, by operating under conditions very similar to those adopted for the homopolymerisation of vinyl chloride.

The physical-mechanical properties of products with a low propylene content (2–6 mole per cent) are very similar to those of a P V C with the same intrinsic viscosity; however, the former products are characterised by a higher fluidity in the molten state and by higher thermal stability.[152] Hence manufactured articles are obtained at lower moulding temperature and with low-toxicity stabilisers, such as calcium and zinc stearates or epoxy soybean oil (owing to their low effectiveness, these stabilisers have only limited use in processing conventional P V C). Although the impact strength of these manufactured articles is not higher than that of P V C, they are expected to be used for the production of bottles and foodstuffs containers.[153]

The properties of these propylene-vinyl chloride copolymers can be explained by the presence of propylene units in the macromolecules (one to every 10–20 units of vinyl chloride) which improves the rheology of the materials and hinders the dehydrochlorination process at high temperature (typical of P V C).

The introduction of propylene units, which decreases branching phenomena in P V C macromolecules and favours the formation of stable terminal groups, such as $—CH_2—CH_2—CH$ and $—CH_2—CH = CH_2$,[152] also increases thermal stability.

Among the propylene polymers, mention should also be made of some polymers obtained by grafting other vinyl monomers on polypropylene. They are generally prepared by bringing active radicals in contact with the polypropylene chain on which chains of the monomer to be grafted grow.

The reaction is generally carried out either by contacting monomer and polymer and by heating the mixture after addition of a convenient radical initiator,[154] or else by subjecting it to the action of ionising radiations at room temperature. Another method is based on the introduction on the polypropylene chain[155] of peroxidic groups which, by heating in the presence of the monomers to be grafted, decompose to yield both grafted copolymers and homopolymers.

By using highly crystalline, peroxidised polymers of propylene either as powder or as manufactured articles, e.g. filaments, films, etc., grafted copolymers with methyl acrylate and methacrylate, styrene, acrylic acid,[156-158] 2-vinylpyridine[159] were obtained.

The mechanical properties of grafted products are almost the same of those of peroxidised polypropylene; however, depending on the monomer used, considerable variations of the surface properties of the starting article are detected.

By block polymerisation of styrene in the presence of peroxidised atactic polypropylene, polymeric compositions are obtained with high impact strength, compared with conventional polystyrene, and without considerable variations of surface hardness.[122]

5.7. STATISTICS

The first plant for commercial production of isotactic polypropylene went in operation late in 1956, by Montecatini, in Ferrara, Italy. Its capacity was about 5000 tons per annum; in the first year only about one hundred tons were produced.

In addition to Italy, the countries where polypropylene production and consumption developed at the fastest rate were the United States, Japan and United Kingdom.

TABLE V.5

Polypropylene Production in Some Countries
(tons)

Country	1964	1965	1966	1967	1968	1969
France	4,500	5,000	6,500	7,500	17,000	20,500
Germany (West)	11,000	14,000	15,500	22,500	31,000	49,000
Italy	20,000	23,000	31,000	47,000	59,000	61,000
Netherlands	2,500	6,000	12,000	14,000	15,000	28,000
UK	12,500	19,500	27,000	50,000	63,000	70,000
USA	122,000	170,000	250,000	293,000	398,000	492,000
Japan	39,000	58,000	100,000	170,000	291,000	420,000
World		310,000	450,000	610,000	1,000,000	1,300,000

It was estimated that, in 1970, the world consumption of polypropylene had reached 1·3 million tons; i.e. a little more than 16 per cent of polyolefine consumption and about 4 per cent of total plastics consumption.[160] The same source forecasts, for 1980, a world polypropylene production of 6·5 million tons (the same as that for high density polyethylene); i.e. more than 21 per cent of polyolefines consumption and about 7 per cent of total plastics consumption.

After having actually been one of the fastest growing polymers in the last decade, polypropylene is thus considered as a polymer whose consumption will grow, in the near future, faster than the average. The above mentioned statistics include the fibre applications of polypropylene.

In Table V.5 some statistical data are reported for the years 1964 to 1969, relating to world and major producing countries' production.

Table V.6 reports a breakdown of Italian polypropylene consumption in 1968 by individual uses.[161]

TABLE V.6

Breakdown of Polypropylene Use (1968, Italy)

End-use	Share %
Household and sanitary articles	28
Fibres	24
Film	13
Industrial parts	11
Toys and sporting goods	5
Tubing, panels, pipe, fittings	4
Split film	3
Electric household appliances	3
Radio and TV sets	2
Monofilament	2
Industrial containers	1
Other end-uses	4
	100

In Table V.7 a breakdown of UK uses in 1969 and 1970, by single major transformation techniques is reported: it may be observed that the most widely used transformation technology for polypropylene is injection-moulding, followed by fibre and then film production.

TABLE V. 7

Breakdown of Polypropylene Use by Major Transformation Technology (1969 and 1970, UK)

Technology	1969 Share %	1970 Share %
Injection and blow moulding	50·5	49·5
Fibres (including rope and film yarns)	30·5	30·0
Sheet and pipe (extrusion)	3·5	3·5
Film	13·5	15·0
Miscellaneous	2·0	2·0
	100·0	100·0

Film production may substantially increase, in the future, if the synthetic paper from plastic film proves successful. Extrusion may substantially increase its percentage, too, if semiexpanded polymer (synthetic wood) is applied industrially.

15

REFERENCES

1 Maass, O. and Wright, C. H., *J. Am. Chem. Soc.*, 1924, **46**, 2664.
2 Ipatieff, V. N., *Ind. Eng. Chem.*, 1935, **27**, 1067.
3 Ipatieff, V. N. and Pines, H., *Ind. Eng. Chem.*, 1936, **28**, 684.
4 Otsuka, H., *J. Soc. Chem. Ind. Japan*, 1937, **40**, Suppl. 21.
5 Monroe, L. A. and Gilliland, E. R., *Ind. Eng. Chem.*, 1938, **30**, 58.
6 Farkas, A. and Farkas, L., *Ind. Eng. Chem.*, 1942, **34**, 716.
7 Otto, M., *Brennstoff Chem.*, 1927, **8**, 321.
8 Whitmoore, F. C. and Laucius, J. F., *J. Am. Chem. Soc.*, 1939, **61**, 973.
9 Topchiev, A. V. and Paushkin, Ya. M., *Neftyanoe Kho*, 1947, **25**, 36.
10 Topchiev, A. V. and Paushkin, Ya. M., *J. Gen. Chem. USSR*, 1948, **18**, 1537.
11 Topchiev, A. V., *Doklady Akad. Nauk*, SSSR, 1948, **62**, 641.
12 Gaiyer, F., *Ind. Eng. Chem.*, 1933, **25**, 1122.
13 Thomas, C. L., *Ind. Eng. Chem.*, 1945, **37**, 343.
14 Tamelo, M., *Faraday Society Discussions*, 1950, **8**, 270.
15 Otsuka, H., *Mem. Fac. Eng. Hokkaido Jap.*, 1950, **8**, 123.
16 USP's 2,400,520; 2,421,950; 2,470,175.
17 Mayo, F. R. and Walling, Ch., *J. Am. Chem. Soc.*, 1949, **71**, 8845.
18 Fontana, C. M. and Kidder, G. A., *J. Am. Chem. Soc.*, 1948, **70**, 3745.
19 Fontana, C. M., Kidder, G. A. and Herold, R. J., *Ind. Eng. Chem.*, 1952, **44**, 1688.
20 Fontana, C. M., Herold, R. J., Kimmey, E. J. and Miller, R. C., *Ind. Eng. Chem.*, 1952, **44**, 2955.
21 Andreev, D. N., *Bull. Acad. Sci. USSR*, 1938, 1039.
22 Sickmann, D. V. and Rice, O. K., *J. Chem. Phys.*, 1936, **4**, 608.
23 Romm, F., *J. Gen. Chem. USSR*, 1940, **10**, 1784.
24 Kooijmann, P. L. and Ghijsen, W. L., *Rec. Trav. Chim.*, 1947, **66**, 673.
25 Kooymann, E. C. and Farenshorst, E., *Rec. Trav. Chim.*, 1951, **70**, 867.
26 Danby, C. and Hinshelwood, C., *Proc. Roy. Soc.*, 1941, 179.
27 Vanhaeren, L. and Jungers, J. C., *Bull. Soc. Chim. Belg.*, 1945, **54**, 236.
28 BP 804,580.
29 USP 2,666,758.
30 USP 2,478,066.
31 Natta, G., Valvassori, A., Ciampelli, F. and Mazzanti, G., *J. Polymer Sci.*, 1965, A3, 1.
32 Natta, G., *Atti Accad. Naz. Lincei (Italy)*, 1955, **4**, 61.
33 Natta, G., Pino, P., Corradini, P., Danusso, F., Mazzanti, G., Mantica, E. and Moraglio, G., *J. Am. Chem. Soc.*, 1955, **77**, 1708.
34 Natta, G., Pino, P., Mazzanti, G. and Longi, P., *Gazzetta Chim. It.*, 1957, **87**, 549.
35 Ziegler, K., *Kunstoffe*, 1955, **45**, 506.
36 Ziegler, K., Holzkamp, E., Breil, H. and Martin, H., *Angew. Chem.*, 1955, **67**, 426; 1955, **67**, 541.
37 Ziegler, K. and Martin, H., *Makromol. Chem.*, 1956, **18-19**, 186.
38 Ital. P 526,101 (4 December 1954).
39 Natta, G., Pino, P., Mazzanti, G. and Longi, P., *Gazzetta Chim. It.*, 1957, **87**, 570. See also 15.
40 Natta, G., *Chim. Ind. (Milan)*, 1957, **37**, 888.
41 Natta, G., Pino, P. and Mazzanti, G., *Chim. Ind. (Milan)*, 1955, **37**, 927.
42 Natta, G., Pino, P. and Mazzanti, G., *Gazzetta Chim. It.*, 1957, **87**, 528.
43 Natta, G., Pasquon, I. and Zambelli, A., *J. Am. Chem. Soc.*, 1962, **84**, 1488.
44 Zambelli, A., Natta, G. and Pasquon, I., *J. Polymer Sci.*, 1963, 4C, 411.
45 Natta, G., Pino, P., Mazzanti, G. and Giannini, U., *J. Am. Chem. Soc.*, 1957, **79**, 2975.
46 Breslow, D. S. and Newburg, N. R., *J. Am. Chem. Soc.*, 1957, **79**, 5073; 1959, **81**, 81.
47 Long, V. P. and Breslow, D. S., *J. Am. Chem. Soc.*, 1960, **82**, 1953.
48 Natta, G., Pino, P., Mazzanti, G. and Lanzo, R., *Chim. Ind. (Milan)*, 1957, **39**, 1032.
49 Patat, F. and Sinn, Hi., *Angew. Chem.*, 1958, **70**, 496.
50 Natta, G., Pino, P., Mantica, E., Danusso, F., Mazzanti, G. and Peraldo, M., *Chim. Ind. (Milan)*, 1956, **38**, 124.

51 Natta, G., Pino, P., Mazzanti, G., Giannini, U., Mantica, M. and Peraldo, M., *J. Polymer Sci.*, 1957, **26**, 120.
52 Kohn, E., Schuurmans, H. J. L., Cavender, J. V. and Mendelson, R. A., *J. Polymer Sci.*, 1962, **58**, 681.
53 Chien, J. C. W., *J. Am. Chem. Soc.*, 1959, **81**, 86.
54 Natta, G., *J. Polymer Sci.*, 1960, **48**, 219.
55 Karapinka, G. L., Smith, J. J. and Carrick, W. L., *J. Polymer Sci.*, 1961, **50**, 143.
56 Beerman, C. and Bestian, H., *Angew. Chem.*, 1959, **71**, 618.
57 Kühlein, K. and Clauss, K., *Makromol. Colloquim*, Freiburg, März 1969.
58 Giannini, U. and Zucchini, U., *Chem. Comm.*, 1968, 940.
59 Natta, G. and Mazzanti, G., *Tetrahedron*, 1960, **8**, 86.
60 Boor, J. Jr., *J. Polymer Sci.*, 9163, **C1**, 237.
61 Cossee, P., *Tetrahedron Letters*, 1960, no. 17, 12, 17; *J. Catalysis*, 1964, **3**, 80.
62 Arlman, E. J., *J. Catalysis*, 1964, **3**, 89.
63 Arlman, E. J. and Cossee, P., *J. Catalysis*, 1964, **3**, 99.
64 Natta, G. and Pasquon, I., *Advances in Catalysis*, Vol. II (Academic Press, New York, 1959), see also 25.
65 Natta, G., Pasquon, I. and Giachetti, E., *Makromol. Chem.*, 1957, **24**, 258.
66 Natta, G., Mazzanti, G., Longi, P. and Bernardini, F., *Chim. Ind. (Milan)*, 1959, **41**, 519.
67 Natta, G., *Chem. and Ind.*, 1957, 296.
68 Weber, H. and Kiepert, K., *Makromol. Chem.*, 1964, **70**, 54.
69 Longi, P., Mazzanti, G., Roggero, A. and Lachi, A. M., *Makromol. Chem.*, 1963, **61**, 63.
70 Clark, A., Hogan, J. P., Banks, R. L. and Lanning, W. C., *Ind. Eng. Chem.*, 1956, **48**, 1152.
71 Topchiev, A. V., Krentsel, B. A. and Perel'man, A. I., *Uspechi Khim.*, 1957, **26**, 1355.
72 Mihail, R., Corlateanu, P. and Ionesco, Al. Gh., *J. Chim. Phys.*, 1959, **56**, 568.
73 Cossee, P. and Van Reijen, L. L., *Actes Deux. Congres Intern. de Catalyse* (Paris, 1960), p. 1679.
74 Peters, E. F., Zletz, A. and Evering, B. L., *Ind. Eng. Chem.*, 1957, **49**, 1879.
75 Carroll, W. H. Mc., Katz, L. and Ward, R., *J. Am. Chem. Soc.*, 1957, **79**, 5410.
76 Eirich, F. and Mark, H., *J. Colloid Sci.*, 1956, **11**, 748.
77 Mark, H., *'Plastics Progress'*, ed. P. Morgan (New York, 1957), p. 5.
78 Eastman Kodak, USP 3,304,295 (14 Feb., 1967).
79 Montecatini, USP's 3,112,300 and 3,112,301 (26 Nov., 1963).
80 Shell Oil, USP 3,040,015 (19 June, 1962).
81 BASF, DP 6,811,052.
82 Parks, G. S. and Mosher, H. P., *J. Polymer Sci.*, 1963, PA, **1**, 1959.
83 Montecatini, BP 1,018,328 (26 Jan., 1966).
84 DiGiulio, E. and Ciampa, G. (Soc. Montecatini), Ital.P 561,304 (30 March, 1956).
85 DiGiulio, E. and Parrini, P. (Soc. Montecatini), Ital.P 562,705 (30 March, 1956).
86 Information Bulletin 'Informazioni Tecniche sul MOPLEN' of Soc. Montecatini Edison, 1970, Milano.
87 Parrini, P., Sebastiano, F. and Messina, G., *Makrom. Chemie*, 1960, **38**, 27.
88 Coen, A. and Petraglia, G., *Materie Plastiche*, 1967, **33**, 3, 269.
89 Coen, A. and Petraglia, G., *Materie Plastiche*, 1965, **31**, 10, 1057.
90 Cappuccio, V., Coen, A., Bertinotti, F. and Conti, W., *Chim. Ind. (Milan)*, 1962, **44** (5), 463.
91 Coen, A. and Petraglia, G., *Materie Plastiche*, 1965, **31**, 1270.
92 Compostella, M., *'The Stereochemistry of Macromolecules'*, ed. A. D. Ketley (M. Dekker, 1967), Vol. I, p. 327.
93 Parrini, P. and Crespi, G., *'Encyclopedia of Polymer Science and Technology'* (Interscience Publishers), 1970, 13, 86–122.
94 ASTM D1238.
95 *'Enciclopedia delle Materie Plastiche'*, ed. L'Industria Milano, p. 212.
96 Ranalli, F. and Crespi, G., *Materie Plastiche*, 1959, **24**, 273.
97 Parrini, P. and Corrieri, G., *Chim. e Ind.*, 1967, **49**, 1172.

[98] Parrini, P. and Corrieri, G., *Chim. e Ind. (Milan)*, 1970, **52**, 970.

[99] Natta, G., Peraldo, M. and Corradini, P., *Proc. Rend. Accad. Naz. Lincei*, 1959, VIII, 26, 14.

[100] Natta, G., *Makromol. Chemie*, 1960, **35**, 94.

[101] Coen, A., Bertinotti, F. and Petraglia, G., *Materie Plastiche Milano*, Atti XV Congr. Internaz., 1963, 145.

[102] Cook, J. Gordon, 'Handbook of Textile Fibres', II (Merrox Publish. Co., 1968), p. 618.

[103] Sittig, M., 'Plastic Films from Petroleum Raw Materials' (Noyes Development Corp., 1967).

[104] Kobayashi, Y. *et al.* (Asahi Chemical Ind. Co.), USP 324,875 (3 May, 1966).

[105] Imperial Chemical Industries, USP 3,250,639 (10 May, 1966).

[106] 'Enciclopedia delle Materie Plastiche', ed. L'Industria Milano, p. 212 and following.

[107] Boor, J. Jr. and Youngman, E. A., *J. Polymer Sci.*, 1966, A-**1**, **4**, 1861.

[108] Natta, G., Baccaredda, M. and Butta, E., *Chim. Ind. (Milan)*, 1959, **41**, 398.

[109] Parrini, P. and Corrieri, G., *Makromol. Chemie*, 1963, **62**, 83.

[110] McDonald, M. P. and Ward, I. W., *Polymer*, 1961, 2, 3, 344.

[111] Peraldo, M. and Cambini, M., *Spectrochimica Acta*, 1965, **21**, 1509.

[112] Heatley, F. and Zambelli, A., *Macromolecules*, 1969, **2**, 618.

[113] Parrini, P., Sebastiano, F. and Messina, G., *Makromol. Chemie*, 1960, **38**, 27.

[114] Natta, G., Mazzanti, G., Crespi, G. and Valvassori, A., *Quim. Ind., Madrid*, XXXVII Congr. Intern., 1967, 5–9 (II).

[115] Frank, F. C., Keller, A. and O'Connor, A., *Phil. Mag. A.*, 1959, 200.

[116] Parrini, P. and Pinto, L., *Rubber and Plastics*, Res. Assoc. of G. Britain, Trans., 1967, 2, 1417.

[117] Dainton, F. S., Evans, D. M., Hoare, F. E. and Melia, T. P., *Polymer*, 1962, **3**, 268.

[118] Danusso, F. and Gianotti, G., *Europ. Pol. Journal*, 1968, **4**, 165.

[119] Crespi, G. and Ranalli, F., *Materie Plastiche*, 1959,181, 253.

[120] Kresser, T. O. J., 'Polypropylene' (Reinhold Plastics App. Series, 1960), p. 10 et seq.

[121] 'Moplefan' Bollettino Informazioni su Film Polipropilenico Soc. Montecatini (1961).

[122] Ogorkiewicz, R. M., 'Engineering Properties of Thermoplastics' (Wiley, Interscience, 1970).

[123] Protospataro, F., *Materie Plastiche*, 1963, **29**, 1412.

[124] Schooten, I. V. and Wijga, P. W. O., *Soc. Chem. Ind. Monograph*, 1961, **13**, 432.

[125] Hansen, R. H., Martin, W. M. and DeBenedictis, T., *Trans. Inst. Rubber Ind.*, 1963, **39**, 301.

[126] Cicchetti, O., *Advances in Polymer Science*, 1970, **7**, 70.

[127] Parrini, P. and Pinto, L., *Rubber and Plastics*, Res. Assoc. of G. Britain, Trans., 1966, **7**, 1344.

[128] Protospataro, F., Centro Ricerche 'Polymer', Terni, unpublished data.

[129] Parrini, P., Soc. Montecatini Edison, Ferrara, unpublished data.

[130] Natta, G. and Compostella, M., *Chim. Ind. (Milan)*, 1968, **50**, 784.

[131] Natta, G., *Chim. Ind. (Milan)*, 1957, **39**, 653; Natta, G., Mazzanti, G., Valvassori, A. and Pajaro, G., *Chim. Ind. (Milan)*, 1957, **39**, 733; Mazzanti, G., Valvassori, A. and Pajaro, G., *Chim. Ind. (Milan)*, 1957, **39**, 743; idem 1957, **39**, 825; Natta, G., *J. Polymer Sci.*, 1959, **34**, 531; Carrick, W. L., Karol, F. J., Karapinka, G. L. and Smith, J. J., *J. Am. Chem. Soc.*, 1960, **82**, 1502; Bier, G., *Angew. Chem.*, 1961, **73** (b), 186; Phillips, G. W. and Carrick, W. L., *J. Am. Chem. Soc.*, 1962, **84**, 920; Wei, P. E. and Rehner, J. Jr., *Rubber Chemistry and Technology*, 1962, **35**, 133; Gladding, E. K., Fisher, B. S. and Collette, J. W., *Ind. Eng. Chem. Product Res. Develop.*, 1962, **1**, 70.

[132] Green, J. H. S., *Research*, 1958, **12**, 232–7.

[133] Natta, G., *Gazz. Chim. It.*, 1959, **89**, 52.

[134] Bier, G. *et al.*, *Plast. Inst. Trans.*, 1960, **28**, 98.

[135] Bier, G. *et al.*, *Makromol. Chemie*, 1961, **44–6**, 347–57.

[136] Gier, G. *et al.*, *Angew. Chemie*, 1961, **73**, 186–97.

[137] Anonymous, *Chem. Week*, 1962, **90**, 111.

[138] Hagemayer, H. J., *Modern Plastics*, 1962, **39**, 157, 161 and 206.

[139] Hagemayer, H. J. et al., *J. Polymer Sci.*, 1964, part **C**, 731–42.

[140] Vernullon, L., *Plastics Design-Processing*, February 1967, 16–20.

[141] Tusch, R. L., *Polymer Eng. Sci.*, 1966, **6**, 255.

[142] Ewan, *Soc. Plastics Eng. Tecn. Papers*, Philadelphia Section, 1 December 1964.

[143] Bier, G. et al., *High Polymers*, 1964, Vol. XVIII, pp. 164–91.

[144] Montecatini and K. Ziegler, BP 538,782 (6 June, 1955); Mazzanti, G., Valvassori, A., Sartori, G. and Pajaro, G., *Chim. Ind.* (*Milan*), 1960, **42**, 468; Zambelli, A., Lety, A., Tosi, C. and Pasquon, I., *Makromol. Chemie*, 1968, **115**, 73; Griffiths, I. H. and Randy, B. G., *J. Polymer Sci.*, 1960, **44**, 396; Turner-Jones, A., *Polymer*, 1965, **6**, 249.

[145] Danusso, F. and Gianotti, G., *Makromol. Chemie*, 1965, **88**, 149.

[146] Turner-Jones, A., *J. Polymer Sci.*, 1967, part **C**, 393; *Polymer*, 1966, **7**, 23.

[147] Suminoe, T. and Kagatu, K., 1963, **20** (216), 262; Asahi K.K., FP 1,505,882 (20 December, 1966); Kyowa Ferm. Ind. Co., BP 1,026,615 (20 April, 1966).

[148] Richardson Co., BP 924,654 (1 May, 1963).

[149] Kawai, W., *Koyo Kagaky Zasshi*, 1963, 66; Toyo Koatsu Kogyo Co., JP 54991/1962 (12 December, 1962).

[150] Sumitomo Chem. Co., BP 1,089,279 (29 January, 1965); Sumitomo Chem. Co., BP 1,056,236 (9 June, 1964); Sumitomo Chem. Co., FP 1,489,950 (29 July, 1966); Sumitomo Chem. Co., FP 1,487,211 (21 July, 1966).

[151] Andreev, L. N. et al., *Izv. Akad. Nauk SSSR*, 1959, 1507; Du Pont Co., USP 2,703,793 (10 July, 1951); Sumitomo Chem. Co., FP 1,528,220 (6 June, 1967); Farbwerke Hoechst AG, BP 981,241 (21 August, 1961).

[152] Heiberger, C. A. and Phillips, R., *Rubber and Plastic Age*, 1967, 636; Weintranb, L., Zufall, J. and Heiberger, C. A., *Polymer Eng. and Sci.*, 1968, 64; Air Reduction Co., US Appl. 459,446 (27 May, 1965).

[153] *Chemical Week*, 12 March, 1966, p. 77.

[154] Natta, G., Severini, F., Pegoraro, M. and Tavazzani, C., *Makromol. Chemie*, 1968, **119**, 201.

[155] Munari, S., Tealdo, G., Vigo, F. and Rossi, C., *Chim. Ind.* (*Milan*), 1965, **47**, 439.

[156] Natta, G., Berti, E. and Severini, F., *J. Polymer*, 1959, **34**, 685.

[157] Montecatini SpA, Ital.P 564,711 (14 March, 1956).

[158] Natta, G., Severini, F., Beati, E., Pegoraro, M. and Pizzotti, F., *Chim. Ind.* (*Milan*), 1965, **47**, 14.

[159] Beati, E., Toffano, S. and Severini, F., *Chim. Ind.* (*Milan*), 1963, **45**, 690.

[160] Roenitz, K. H., *British Plastics*, February 1971, **44**, 56.

[161] *Chemische Ind.*, May 1970, p. 332.

[162] Natta, G., Pasquon, I. and Zambelli, A., *J. Am. Chem. Soc.*, 1962, **84**, 1488.

[163] Natta, G., Corradini, P. and Cesari, M., *Rend. Acc. Naz. Lincei*, 1956, **21**, 365.

[164] Parrini, P., unpublished data.

[165] Turner-Jones, A., Aizlewood, J. M. and Beckett, D. R., *Makromol. Chem.*, 1964, **75**, 134.

[166] Danusso, F., *Polymer*, 1967, **8**, 281.

[167] Kinsinger, I. B. and Hughes, R. E., *J. Phys. Chem.*, 1963, **67**, 1922.

[168] Saunders, F., *J. Polymer Sci.*, 1964, **B2**, 755.

CHAPTER 6

ISOPROPYL ALCOHOL AND ACETONE

By J. C. Fielding

6.1. INTRODUCTION

ALTHOUGH isopropyl alcohol has applications of its own in industry nearly half is converted to acetone and as the latter is produced by other routes, also from propylene, acetone must be considered as a major propylene derivative in its own right.

In this chapter we will therefore first discuss the production and uses of isopropyl alcohol and then its derivatives under a major subheading (62).

To help make things clearer the more important reactions to be discussed in this chapter are shown in Fig. 6.1.

6.2. ISOPROPYL ALCOHOL

Formula $\qquad$ $CH_3–CHOH–CH_3$

Isopropanol, otherwise known as iso-propyl alcohol (IPA) and propan-2-ol is a colourless liquid with a molecular weight of 60 and a boiling point of 82·5°C. It is soluble in ether, ethanol and water and distils azeotropically from the latter at 80°C to give a constant boiling mixture (CBM) containing about 87·5 per cent wt IPA.

6.2.1. Production

Two main routes to IPA will be discussed in detail, both starting from propylene. These are the sulphuric acid process involving the reaction of propylene with sulphuric acid and subsequent sulphate hydrolysis; and the direct hydration of propylene at elevated pressures using solid catalysts.

6.2.1.1. The sulphuric acid process This process was first developed in the USA and a pilot plant was operated in 1919 by the Melco Chemical Co. The manufacturing rights were subsequently acquired by Standard Oil of New Jersey and the first commercial plant was commissioned at their Bayway (NJ) Refinery at the end of 1920. This episode is normally regarded as the starting point of the petrochemical industry. The bulk of current IPA production still uses this process.

The simple chemistry of the process can be represented by the following series of equations; they represent essentially the indirect hydration of propylene which is illustrated by:

$$H_2O + CH_2 = CH–CH_3 \xrightarrow{\ H_2SO_4\ } CH_3–CH(OH)–CH_3$$

Reaction $\begin{cases} CH_3–CH=CH_2 + H_2SO_4 \longrightarrow (CH_3)_2CH–O–SO_3H & (1) \\ \qquad\qquad\qquad\qquad\text{Propyl hydrogen sulphate (PHS)} \\ CH_3–CH=CH_2 + (CH_3)_2CH–O–SO_3H \longrightarrow (CH_3)_2CH–O–SO_2–O–CH(CH_3)_2 & (2) \\ \qquad\qquad\qquad\qquad\text{Dipropyl sulphate (DPS)} \end{cases}$

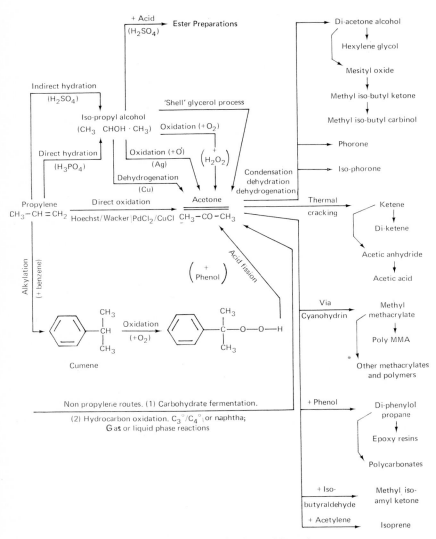

FIG. 6.1   *Acetone Production and Reactions*

Hydrolysis $\begin{cases}(CH_3)_2CH\text{-}O\text{-}SO_3H+H_2O\longrightarrow CH_3\text{-}CH(OH)\text{-}CH_3+H_2SO_4 \qquad (3)\\ \qquad\qquad\qquad\qquad\qquad IPA \\ (CH_3)_2CH\text{-}O\text{-}SO_2\text{-}O\text{-}CH\text{-}(CH_3)_2+2H_2O\longrightarrow 2CH_3\text{-}CH(OH)\text{-}CH_3+H_2SO_4 \quad (4)\end{cases}$

By-products $\begin{cases}(CH_3)_2CH\text{-}O\text{-}SO_2\text{-}O\text{-}CH(CH_3)_2+CH_3\text{-}CH(OH)\text{-}CH_3 \\ \qquad \longrightarrow (CH_3)_2CH\text{-}O\text{-}CH(CH_3)_2+(CH_3)_2CH\text{-}O\text{-}SO_3H \qquad (5) \\ \qquad\qquad\qquad (Di\text{-})isopropyl\ ether\ (IPE) \\ *n(CH_3\text{-}CH\text{=}CH_2)\longrightarrow CH_3(C_3H_6)_{n-1}CH\text{=}CH_2 \qquad\qquad\qquad (6)\end{cases}$

Reversion—$(CH_3)_2CH\text{-}O\text{-}SO_2\text{-}O\text{-}CH(CH_3)_2\longrightarrow CH_3\text{-}CH\text{=}CH_2+(CH_3)_2CH\text{-}O\text{-}SO_3H$ (7)

Recovery —$(CH_3)_2CH\text{-}O\text{-}CH(CH_3)_2+H_2O\xrightarrow[\substack{\text{Dilute}\\ H_2SO_4}]{}2CH_3\text{-}CH(OH)\text{-}CH_3$ (8)

* $n = 2,\ 3,\ 4,$ or 5, giving propylene dimer/trimer/tetramer/or pentamer. The product contains many isomers whose structures are predominantly branched.

The two feed materials are a propylene rich C_3 stream and sulphuric acid of 70–85 per cent wt strength. The propylene feed is generally 50–90 per cent propylene. Higher propylene concentrations can result in high local temperatures and pressures and, unless the equipment is specifically designed, inerts (propane) may have to be re-circulated to reduce the effective propylene concentration. Propylene contents below 50 per cent result in slow absorption, poor propylene utilisation, large equipment, and long residence times. The propane/propylene (PP) feed should contain minimum impurities; other olefins (ethylene and butylenes) result in by-product formation (ethanol and butanols), and abnormal pressure/temperature effects in the reaction section. Other constituents—propadiene and methyl acetylene—result in high polymer formation which reduces yields and complicates the acid recovery.

The reaction is normally carried out at a sufficient pressure (300–400 psig) to maintain the reactants completely liquid throughout the system at the operating temperature of 45–55 °C. The reaction stages take the form of intensely agitated vessel/cooler systems followed by phase separators. Single, two, or three stage units can be used, with counterflow of acid and hydrocarbon phases, the number of stages depending on feedstock quality, availability and cost/alternative use value. The reactors can be baffled time tanks or stirred reactors, with internal or external cooling. A simple flow diagram of the reaction, hydrolysis, and crude alcohol recovery sections is shown in Fig. 6.2. The reaction products are withdrawn from the phase separators. Designating the first stage as that where the feed PP is introduced, 'fat acid' is withdrawn as the lower phase from the first phase separator and passes to the hydrolysis section. The depleted PP is withdrawn from the last stage phase separator as the upper phase. It is distilled to recover PHS, DPS, and polymers as column bottoms which are passed to the hydrolysis section, while the tops are neutralised with caustic, water washed, and go forward to other uses. In the hydrolysis section, the combined 'fat acid' (containing H_2SO_4, DPS, PHS, IPE, polymers, and some PP), and the heavy ends from the spent PP distillation are diluted with cold water to 60–65 per cent wt based on total acid content, cooled to 35–40 °C, and depressured to about 10 psig. The hydrolysis section, which consists of one or more agitated and cooled vessels, serves to promote the hydrolysis of PHS and DPS to sulphuric acid and IPA preventing their further reaction to IPE, to minimise further polymer formation, and to de-gas the contents under controlled conditions. The recovered gas, liberated either by the pressure change or by reversion reaction, is rich in propylene and is recycled to the reaction. The hydrolysed product is then passed to the crude alcohol recovery section.

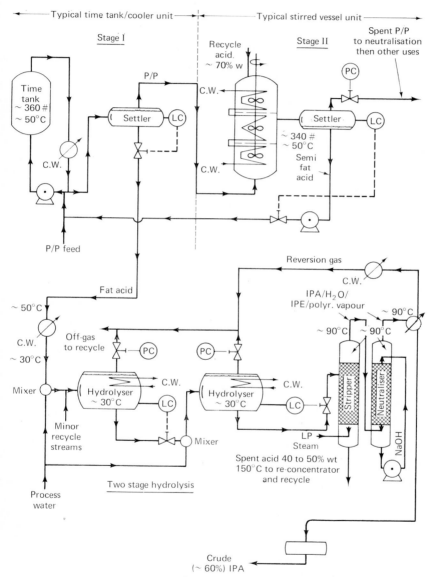

FIG. 6.2 *IPA Production by Indirect Hydration Using Sulphuric Acid*

The most critical reaction conditions are listed below with brief comments on their effects. The optimum conditions on any specific plant will be selected on the basis of prevailing feed quality, availability, and economics.

(i) *Operating temperature*   High temperatures accelerate all the reactions but have their greatest effect on by-product formation because this is less dependent on mass transfer effects. High temperatures also exacerbate corrosion problems.

(ii) *Acid concentration*   High concentrations improve olefin absorption rate some-what but again this is mainly a function of mass transfer. By-product formation is more directly affected, thus 75 per cent wt and 85 per cent wt acid at 125°C and 75°C respectively will give almost complete conversion to ether and polymer. High concentrations also increase reconcentration costs. Concentrations below about 65 per cent at reaction temperatures increase corrosion rates.

(iii) *Acid to olefin ratio*   This is always greater than the stoichiometric 0·5 required for DPS production and is generally between 0·6 and 1·0. It can be used as a balance factor with acid concentration, i.e. high ratios compensate for low concentration. High ratios also increase reconcentration costs, and equipment size and cost. In practice the ratio is adjusted to give the desired propylene absorption which is calculated from the spent PP analysis.

(iv) *Pressure*   This must be kept high enough to keep the reactants liquid to avoid two-phase flow, poor heat transfer, pump cavitation, etc. Tail gas analysis should include inerts and C_2 concentrations as these can be critical.

Alcohol is recovered by steam stripping it from the hydrolysed fat acid. Several variations are possible but the essential features are dilution of the total acid content to between 40 and 50 per cent wt, then feeding it to the top of a stripping column (packed or trayed); and the introduction of stripping steam to the base of the column either as live steam or using a reboiler built of adequately corrosion-resistant materials. The top product contains IPA, water, IPE, polymer and reversion gas, while the bottoms are sulphuric acid (40–60 per cent wt) containing traces of IPA and high boiling polymers. The overhead vapours are acidic and are scrubbed with alkali and then condensed to give crude IPA (containing about 60 per cent wt IPA and 40 per cent wt water) which is sent to the distillation section. The non-condensable reversion gas is recovered, combined with hydrolysis gas, compressed, and recycled to the reaction. The spent acid is rundown at between 130°C and 180°C to the acid reconcentration unit.

When conditions lead to excessive ether (IPE) formation and no ready outlet is available, IPA can be regenerated by heating the recovered ether with dilute sulphuric acid (2–10 per cent wt) at 90–120°C and adequate pressure to keep the system liquid phase. The reaction products are conveniently fed to the hydrolysis section for alcohol recovery.

The crude IPA is purified by distillation. There are many variations practised, but essentially the distillation is three stage; firstly the removal of light ends and the bulk of the water to produce IPA/water azeotrope or constant boiling mixture

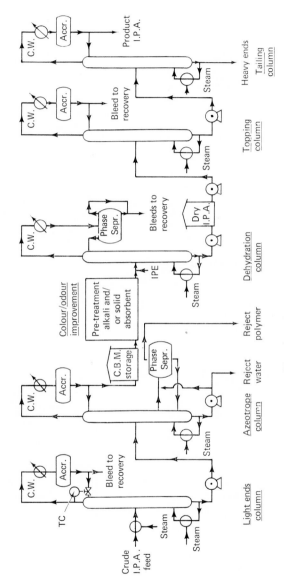

FIG. 6.3 *Crude IPA Distillation*

(CBM) (88 per cent wt IPA). The azeotrope is then extractively distilled (using IPE or benzene) to give anhydrous IPA, which is finally topped and tailed giving product IPA as the final column overheads; a simple flow scheme is shown in Fig. 6.3. The degree of purification applied depends on the market requirements which can vary over the whole range from the crude azeotrope to 'essence' grade IPA. The various methods of improving product quality include caustic treatment, fixed bed absorption treatment using activated carbon, molecular sieves, etc., and intense aqueous extractive distillation. Essence grade is often finally distilled in non-ferrous equipment, e.g. copper and bronze. Normal quality of anhydrous IPA produced is 99+ per cent while special grades are better than 99·8 per cent. The normal impurities are traces of water and acetone. Di-isopropyl ether (IPE) which is co-produced in small quantities, can be recovered from the light ends of IPA distillation, by water washing to remove IPA and topping and tailing. Recovered IPE can be used as a solvent or an extractive distillation solvent. An alternative use for light polymer and IPE is for gasoline blending on account of its high octane rating (97 for pure IPE). Heavy polymer is normally disposed of as refinery fuel.

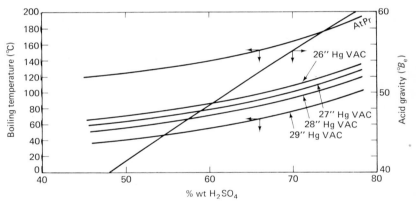

FIG. 6.4   *Sulphuric Acid Gravity and Boiling Points*

Sulphuric acid reconcentration is another major part of the sulphuric acid IPA process. Some properties of sulphuric acid at various pressures are shown in Fig. 6.4. The effluent acid from the process can be between 40 and 60 per cent wt and must be restored to 70–85 per cent wt for re-use. Taking an average figure of 1·2 tons 100 per cent acid recycled per ton IPA produced, and reconcentration from 50–75 per cent wt, requires the removal of 0·8 tons of water. As this must be achieved at high temperatures and under corrosive and fouling conditions it is clearly a major problem and cost element in IPA production. The sulphuric acid is normally recovered by one of three methods, single or multiple effect evaporation in specially designed vacuum evaporators, e.g. Mantius concentrators; using a proprietary concentration unit, e.g. the Chemico concentrator which uses hot furnace gas to strip the water; or by submerged combustion evaporation (Fig. 6.5). Another major complication is contamination of the recycled acid. Impurities in the lean acid include IPA itself, heavy polymeric material in a finely dispersed condition, and higher olefin sulphates dissolved in the acid. Removal prior to reconcentration is almost impossible and during the extreme conditions of reconcentration a major portion of the organic material becomes carbonised. Filtration will remove the larger carbon particles but leaves fine and colloidal material in the acid. Various chemical treatments for oxidising or flocculating the solids, e.g. the use of H_2S or oxidising agents, are only partially successful and can be expensive. Another method is to take an acid slipstream and concentrate it at temperatures high enough to oxidise the carbon producing CO_2 and SO_2:

$$2H_2SO_4+C \quad CO_2+2SO_2+2H_2O$$

Commercial plants normally use one or more of these carbon removal methods but it is still necessary to add fresh acid and withdraw a slipstream of contaminated acid to maintain the acid quality.

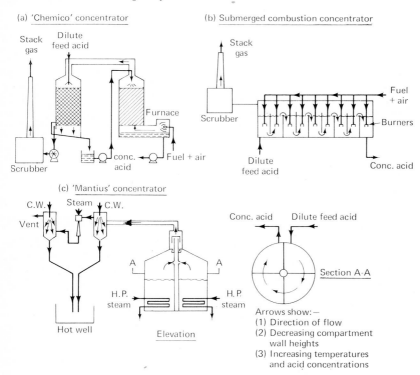

FIG. 6.5 *Types of Sulphuric Acid Reconcentrators*

Materials of construction present a major problem. Concentrated acid ($>$ 65 per cent wt) at moderate temperatures ($<$ 60°C) can be successfully handled in mild steel equipment provided precautions are taken against excessive velocities and erosion by suspended solids. However at lower concentrations and higher temperatures an increasingly severe problem must be faced. A multiplicity of materials is therefore used including lead lining (thermally protected by acid brickwork where necessary), silicon iron where thermal shock and pressures are low, tantalum, Hastalloy, teflon lining, hard rubber lining, graphite, phenolic resin filled mouldings, and many others. This represents another major contribution to capital and operating costs of this process.

Despite the disadvantages this process remains the major producer of IPA. Efficiencies are acceptable, propylene absorption normally exceeding 95 per cent and the conversion of absorbed propylene to IPA exceeding 90 per cent. A major advantage is that relatively cheap process materials can be used, i.e. sulphuric acid and relatively impure propylene. Plants producing more than 100,000 tons per year are currently in operation though such capacities may require multiple production lines particularly in the crude IPA stripping section.

6.2.1.2. Direct hydration   This reaction is essentially the direct combination of water and propylene to give IPA by the simple equation previously stated. In this case an alternative reaction can take place to give n-propanol:

$$CH_3\ CH=CH_2+H_2O \xrightarrow{\text{catalyst}} CH_3-CH_2-CH_2-OH \quad \text{(Normal propyl alcohol—NPA)}$$

Side reactions again occur and their products are similar, though in general somewhat less by-product is formed than in the sulphuric acid process; thus in the modern process a moderate amount of IPE is produced, but only trace qualities of polymeric material.

The direct process is becoming increasingly important as a source of IPA. Plants between 20,000 tons per annum and 130,000 tons per annum have been designed or built and this process shows considerably improved economics with increasing plant size, thus for plants of greater than 50,000 tons per annum capacity, this process is usually the obvious choice provided clean propylene is available.

The process can be seen as a parallel development to that of direct hydration ethanol production. Indeed, the catalyst currently favoured—supported phosphoric acid—is similar to that developed by Shell Development Co. and used in the first direct hydration ethanol plant, operated by Shell at Houston in Texas since 1947. The reaction is essentially an equilibrium one, which, under moderate conditions, is very slow, though yields are acceptable. For industrial purposes a catalyst is required which will allow a rapid approach to equilibrium, at temperatures and pressures which will give acceptable conversions and economics. The equilibrium concentration of alcohol is depressed by rising temperature, but enhanced by rising pressure as would be expected, thus the best catalyst will be one which gives rapid reaction at moderate temperatures hence maximising conversion and minimising by-products without introducing constructional problems at the high pressures used. Three development stages can be seen, firstly the early work to develop a suitable catalyst which was certainly proceeding in Germany and the UK during World War II. Subsequently the first industrial scale production appeared in the early

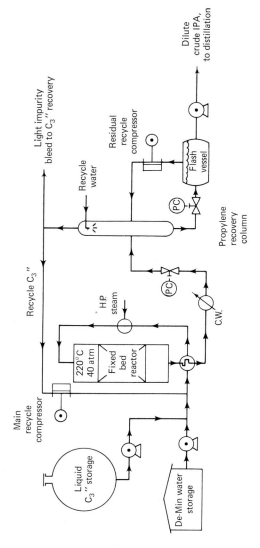

FIG. 6.6   *IPA Production by Direct Hydration: Phosphoric Acid Catalyst*

1950's. These involved fairly high temperatures (220–300°C) and high pressures (150–250 atm) and were initially characterised by considerable by-product formation. Currently, the second generation industrial plants are appearing using more moderate temperatures (170–270°C) and lower pressures (20–60 atm) with attendantly reduced by-product formation. Current designs indicate better than 97 per cent utilisation of feed propylene and better than 97 per cent conversion to IPA of consumed feed, at optimum conditions.

The processes developed in the late 1940's are typified by the ICI process which was put on stream commercially in 1951 at a capacity of 40 to 50,000 tons per annum. The catalyst used was based mainly on tungsten oxide with minor quantities of zinc and other oxides on a silica base, which was used in pelleted form in fixed beds. A typical flow scheme for the direct hydration of propylene is shown in Fig. 6.6.

The ICI process initially used reaction temperatures of 230–290°C and operated at pressures of 200–250 atm. The feeds are pure water and > 95 per cent pure propylene. The propylene/water ratio is controlled at 0·3–0·5 : 1 molar and under these conditions the reactants are partially liquified resulting in a mixed phase reaction. The liquid feeds, water and propylene, are fed to the reaction by high pressure pumps and heated to reaction temperature, together with recycle water and propylene, in two stages. The first heater is a feed/product exchanger, while the second is a process heater using high pressure steam. The reactants flow downwards over the catalyst bed at about 2 volumes per volume catalyst per hour. Conversion is low so the temperature rise through the reactor is small. The reactor product consists essentially of water, unreacted propylene, 3–5 per cent wt of IPA, and traces of IPE and polymers. After feed/product heat exchange the products are further cooled and partially depressured, into a stripping column where the bulk of the propylene is recovered, compressed, and recycled to the reaction feed. The aqueous phase is then further depressured through a water stripping column at about atmospheric pressure where some propylene is again recovered and recycled. The resulting aqueous phase contains about 5 per cent IPA and it is recovered in a similar distillation unit to that already described for acid process IPA, i.e. removal of light ends and water to give the azeotrope, extractive distillation to give anhydrous IPA, and finally topping and tailing to give > 99 per cent IPA product. A variation frequently incorporated involves removal of a side stream from the azeotrope column which contains the NPA present. This stream can be discarded, or the NPA recovered as a by-product, though the market is small. The water recovered during distillation is normally recycled except for a bleed stream taken to restrict impurity recycle and consequent build-up. Impurities also accumulate in the recycle propylene stream and a bleed is also required in this case. The propylene can be recovered and recycled if necessary.

The most critical conditions are summarised below with their major effects.

1. Water/propylene molar ratio. A low ratio increases polymer formation. However high ratio directly affects plant size, cost, and operating cost.
2. Temperature. High temperatures will increase reaction rates but depresses the equilibrium IPA concentration. It will also increase polymer and ether formation.

3. Pressure. High pressures will also increase by-product formation but the effect on equilibrium IPA concentration is favourable.
4. Residence time. Clearly this has an optimum. Increasing residence time will have a decreasing effect on IPA concentration because of the equilibrium involved, but will promote by-product formation. Short residence times will be expensive because of the recycling involved and the increasing amounts of water to be handled. Unless all these factors are carefully balanced and controlled the plant could well be inflexible.

Little information is available about the current operation of these first generation industrial plants but it is probable that catalyst improvements have been introduced which will have improved throughput, flexibility, and economics. A significant improvement could be expected from lower operating temperatures and pressures, as has been recently achieved by I C I and other companies in the case of methanol plants. It is known that problems of relatively high by-product formation initially experienced had been largely overcome in the first few years of plant operation.

The second generation plants are similar in concept to the earlier ones and the major advance appears to have been in the improved catalysts developed. Probably the best known of the current processes is that operated and licensed by Veba-Chemie A G in West Germany. Veba-Chemie incorporates the former Hibera-Chemie, and has operated the process successfully at Wanne-Eickel for about six years. Power Gas and Scientific Design have both obtained licences to design and construct plants.

Fig. 6.5 illustrates again the design concept. Published data indicate the operating pressure to be in the range 20–60 atm, and obviously the actual selection will be to some extent dictated by local conditions. Normal optimum pressure is probably in the middle of this range, say 35–45 atm. Similarly, temperatures quoted are 170–270°C but actual operation could well be in the 200–240°C bracket, where no extreme circumstances exist. The catalyst used is pelleted and the active component is phosphoric acid. Feedstocks are again water and propylene; however demineralised water is specified and propylene is preferably (though not necessarily) better than 99 per cent pure. Under the less severe operating conditions, particularly the lower pressure, the reaction is vapour phase. It is believed that the increased activity and specificity of the catalyst allows much lower water/propylene ratios within the reactor but precise data is not currently available. Similar critical conditions will again exist and their effects, though reduced, will still be present. It is probable that the more moderate conditions and improved catalyst will lead to higher conversions per pass; this claim has been made by the licensees. Published data also indicates that the current process converts less than 4 per cent of the reacted propylene to by-products and that the major side reactions are to NPA and IPE. The water recovered during distillation is not recycled in these plants (demineralised feed). A bleed of propylene is taken to maintain a low level of inerts and light by-products. The bleed stream can be purified and recycled if the economics are favourable.

If a similar mass-flow rate per unit volume of catalyst to that used in earlier processes is assumed, then IPA production is between $0 \cdot 1$ and $0 \cdot 13$ m ton/m³ catalyst/hr. This figure gives a much higher space velocity than before because of the vapour phase flow and reduced pressure. Clearly this means large reactors and it is doubtful whether, at the pressures involved, a single reactor would accommodate more than 5 or 6 m ton/hr (reactor volume about 50 m³) or 40 to 50,000 m ton per annum. For larger plants reactor duplication may be necessary.

Product purification is similar to the previously described method and flexibility will be available to draw off either IPA azeotrope, or crude or finished anhydrous IPA. Market and own use requirements for NPA will dictate whether NPA is recovered.

6.2.2. The Properties and Uses of IPA

Three grades of IPA are available commercially, though limited applications exist for intermediate grades. These grades are the normal azeotrope containing about 12·5 per cent wt. of water and about 2 per cent wt. of other impurities, and two grades of anhydrous IPA; the normal bulk product containing > 99 per cent wt of IPA, and the more refined 'essence' grade (pharmaceutical grade) containing about 99·8 per cent wt. IPA. Typical pure properties are given below because detailed variations can occur in properties depending on the production reaction and on the treating and recovery methods used.

IPA Azeotrope. (Constant boiling mixture—CBM.)

Pure azeotrope. BP 80·3°C. 12·6 per cent wt H_2O. 87·4 per cent wt. IPA

Commercial product:

    Boiling range 78–81°C

    IPA content > 85 per cent wt

    Water content < 13 per cent wt

    Impurities. IPE, polymer, acetone

    Specific gravity. 0·815 20/20°C

This grade is normally used for conversion to acetone or in other chemical intermediate roles, or is upgraded by distillation.

Anhydrous IPA (pure product). The commercial product contains traces (normally < 0·5 per cent wt.) of impurities, chiefly acetone and water.

    Boiling point. 82·4°C. Vapour pressure 21 mm Hg abs at 20°C

    Specific gravity. 0·786 20/20°C

    Expansion coefficient. $0·85 \times 10^{-3}/°C$

    Colour. Water white

    Molecular weight 60·1

    Refractive index 1·3776

    Specific heat 0·60 at 20°C

    Viscosity 2·4 cp at 20°C

    Flash point (TOC) 59°F (15°C)

    Flammability limits 2·02 and 7·99 per cent vol. in air

    Critical temperature 235°C

    Critical pressure 53 atm. abs.

    Latent heat of evaporation 160 cals/g at boiling point

The major use of IPA is the production of acetone. This has represented a decreasing percentage of IPA production over the past few years mainly as a result of the incursions of cumene/phenol process acetone. The percentage has probably fallen from about 55 per cent to about 45 per cent over the past ten years. The indications currently are that cumene/phenol capacity is adequate and since phenol demand is increasing more slowly than acetone demand (7 per cent per year against 9 per cent per year) this percentage use will stabilise and may rise slowly. Other incursions have been of a minor nature. Acetone production and uses will be dealt with in subsequent sections.

Another major use of IPA is based on its low freezing point ($-88 \cdot 5\,°C$) and the freezing point depression it causes in aqueous solutions. The following figures in Table VI.1 indicate the degree of depression of IPA and other alcohols.

TABLE VI.1

% wt. alcohol	Freezing point °C				
	10	20	30	40	100
Methyl alcohol	−6·8	−16·2	−28	−42	−97
Ethyl alcohol	−4·3	−10·0	−18·1	−28	−114
Iso-propyl alcohol	−3·4	−7·7	−13·2	−20·7	−88·5
Glycol	−3·3	−7·8	−13·5	−22·1	−15·6

The uses based on this property are of two types. Firstly to prevent ice formation or to eliminate it when formed, and secondly in systems required to operate at low temperatures.

The use of IPA as a cold weather additive to petroleum fuels to prevent carburetter icing is widespread and is practical both in automobile and aircraft fuel. An additional advantage is that IPA has good solubility in gasoline and a high octane rating. It can be used alone or with other alcoholic additives, or with surface active components of the amine or hydroxylamine type which prevent ice crystal adhesion. Its function is twofold; to prevent traces of water precipitating from the fuel at low temperatures and freezing to form crystals which block jets and passages, and to prevent water vapour present in the induced air, which is cooled by fuel evaporation, from freezing and depositing on the carefully contoured carburetter surfaces. The concentration of IPA normally used varies from 0·5 per cent wt. to 2·0 per cent wt. depending on the duty, the likely ambient temperature and humidity, and the anticipated maximum water content of the fuel.

IPA is also used as a de-icing fluid. In airfields it is used (prior to take-off) to de-ice aircraft which have been parked in freezing or snowing conditions. It has also been used to de-ice railway points. It is now also widely employed for automobile de-icing in two main forms. Firstly as a component of aerosol and manual sprays which can be used to remove ice films from the windscreen and windows of cars; and secondly as a component in proprietary packs of concentrated windscreen washer fluid where it acts to prevent the reservoir from freezing and also to prevent the wash fluid from solidifying when it contacts the cold screen.

Solutions of IPA can also be used in circulating systems which operate at sub-zero temperatures. Essentially it is used as an antifreeze in the same way as ethylene glycol in automobile cooling systems. However, its volatility limits its use to closed systems operated under pressure. These include the cooling systems of the larger engines in road vehicles, of heavy duty earth moving machinery and similar, and in large stationary engines in exposed locations. The use of IPA has the following advantages:

1. Cheapness—about half the cost of glycol antifreeze and competitive with ethanol.
2. A reduced tendency to high viscosity at low temperatures which can cause local overheating.
3. Greater stability than glycol antifreezes, hence minimising corrosion, and stabiliser concentrations.

A similar use is as the heat transfer fluid in refrigeration and chilling systems.

The other major use of IPA is in the field of chemical solvents. This is an extremely complex field and a comprehensive review is outside the scope of this work. Specialised literature in the form of manufacturers bulletins and textbooks is freely available. However, some basic elements of the topic will be discussed.

IPA is used directly as an alcoholic solvent and has special advantages resulting from its complete solubility in both water and many aliphatic and aromatic hydrocarbon solvents. Additionally, it is compatible with most other chemical solvents. It is also used as an intermediate in acetone production as mentioned above, and the latter is further processed to produce the so called 'C$_6$ solvents', di-acetone alcohol (DAA), mesityl oxide (MO), methyl-isobutyl ketone (MIBK), methyl isobutyl carbinol (MIKC), and hexylene glycol (HEG). These compounds are again valuable chemical solvents and will be subsequently described in more detail. Finally, IPA is used in the preparation of ester solvents. The most important of these is isopropyl acetate but limited quantities of other esters are also produced. Isophorone (also produced from acetone) and IPE also have limited use as chemical solvents. Thus IPA, acetone, and their derivatives, can be seen as major components in a complex system of chemical solvents. A brief but by no means exhaustive listing is given in Table VI.2.

These solvents are used either individually or in combination with other chemical or hydrocarbon solvents. Their use falls broadly into three categories: as components of finished products which are most conveniently dispensed as liquids, or liquid suspensions, e.g. paints, laquers, pharmaceuticals, and cosmetics; secondly, as process solvents in the chemical industry to assist in processes requiring their solvency or dilution, e.g. in plastics and resins plants, in chemical extraction procedures, and in hydrogenation and other catalytic reactions; finally, in the extraction of essential oils, waxes, resins and gums from naturally occurring products, e.g. seed oil and resin extraction. In general IPA is widely used in the first two categories but less so for natural product extraction for which hydrocarbon solvents are preferred.

Included in Table VI.2 is oxitol acetate which is representative of another class of chemical solvents; the glycols and their derivatives. Whilst not directly derived from IPA or acetone they are important solvents which extend the temperature range of the alcohols and esters and therefore require brief mention. The starting materials are ethylene glycol and di-ethylene glycol (derived from ethylene oxide) and propylene glycol (derived from propylene oxide). From these are derived the glycol ethers by condensation of the glycols with alcohols; thus ethylene glycol with ethanol gives ethylene glycol mono-ethyl ether ($CH_2OH-CH_2-O-CH_2-CH_3$) otherwise known as

TABLE VI.2

Boiling Points and Densities of Some Chemical Solvents

Alcohol solvents

Name	b.p. °C	Specific gravity 20/20°C
Methyl alcohol	65	·79
Ethyl alcohol	78	·79
*Iso-propyl alcohol	82	·79
*N-propyl alcohol	97	·79
Sec-butyl alcohol	100	·81
Iso-butyl alcohol	108	·81
N-butyl alcohol	118	·81
Amyl alcohol	130	·81
*Methyl iso-butyl carbinol	132	·81
2 Ethyl hexyl alcohol	184	·83
*Hexylene glycol	198	·92

Ketone solvents

Name	b.p. °C	Specific gravity 20/20°C
*Acetone	56	·79
Methyl ethyl ketone	80	·81
*Methyl iso-butyl ketone	116	·80
*Mesityl oxide	128	·86
*Di-acetone alcohol	167	·94
Ethyl amyl ketone	168	·81
Cyclohexanone	157	·95
*Isophorone	215	·92

Ester solvents

Name	b.p. °C	Specific gravity 20/20°C
Methyl acetate (1)	58	·94
Ethyl acetate	77	·90
*Iso-propyl acetate	89	·87
Sec-butyl acetate	112	·87
Iso-butyl acetate	118	·87
N-butyl acetate	126	·88
Amyl acetate	142	·88
*Methyl amyl acetate	146	·86
Oxitol acetate (2)	156	·975
2 Ethyl hexyl acetate	198	·87
Nonyl acetate	212	·97

*These solvents are derived directly from propylene through IPA and acetone. See text.
(1) The acetate radicals may derive from acetic anhydride manufactured from IPA or acetone. See text.
(2) Simplified commercial name. See text.

ethoxy ethanol, 'ethyl cellosolve', or 'oxitol'. These ether alcohols are also used to prepare esters by reaction with acids; hence 'oxitol' with acetic acid gives 'oxitol' acetate. Similarly di-ethylene glycol ($CH_2OH–CH_2–O–CH_2–CH_2OH$) yields the di-oxitols.

The widest use of the chemical solvents in general and IPA in particular is in paint and laquer preparation. A single solvent is seldom used, as mixed solvents generally give better performance and economy. Broadly, paints can be classed as those surface coatings which dry by oxidation, polymerisation, and usually partially by solvent evaporation. Laquers dry exclusively by solvent evaporation.

Paints in general contain a series of components usually including a resin or resins, fillers, pigments, plasticisers, drying oils, and the solvent. The resins, either natural or synthetic form the basic coating and may or may not be self drying. The fillers give body to the paint and allow greater 'build' (thickness). Plasticisers are used to prevent the paint becoming brittle with age. Drying oils are included if the resin is not self drying to speed the setting by oxidation. The solvent is used primarily to dissolve the resin, plasticiser and drying oil, to promote easy application, and to initiate the setting by evaporation. The main characteristics should include:

1. Good solvency for the specific components,
2. Optimum volatility so that consistency is maintained during use but evaporation is rapid enough to prevent runs during initial drying,
3. Stability and compatibility so that degradation and reaction with other constituents does not limit shelf life,
4. Odour, which, if unpleasant can deter users. This particularly applies to use in confined spaces.

A limited number of paints dry entirely by polymerisation and these include the two-component paints, and oxygen, or moisture-catalysed, paints. They thus contain no solvent except that necessary to reduce viscosity. The bulk of paints, however, dry in a series of stages. Initially the bulk of the solvent evaporates and the paint sets. Secondly, the resins and/or drying oils 'skin' either by polymerisation or oxidation to give a soft but 'touch dry' coating. Finally, over a period extending from several hours to several days the polymerisation/oxidation proceeds and the paint hardens to its final form. By this time evaporation has reduced its solvent content to a very low level.

Originally, turpentine, obtained from the resin (pine oleoresin) in coniferous trees by steam distillation was the exclusive paint solvent and the resins used were naturally occurring. Turpentine is about 85 per cent α- and β-pinenes, which are two of the many naturally occurring terpenes ($C_{10}H_{16}$).

α-pinene   β-pinene

This material is an exceptional solvent for natural resins. However, vastly increased requirements and the use of synthetic resins for paints required a supplement and substitute for natural turpentine. Originally this was found in the hydrocarbon solvents (white spirits or turpentine substitutes) which compromise between good solvency with a high aromatic content and hence strong odour, and the reverse. Subsequently, chemical solvents have been used to boost the solvency of 'white spirit' without the odour penalty, though with some cost penalty. The chemical solvents used include alcohols, ketones, glycol ethers, and esters.

Lacquers dry by solvent evaporation only and are generally quick drying, which means they achieve their final hardness in a few hours. They do not contain drying oils and the resins are usually nitrocellulose or vinyl. They are generally used on wood and metals. The best solvents for nitrocellulose lacquers are undoubtedly the ketones and glycol ethers. However, these are expensive and the method normally used is to make a high resin concentration solution using these solvents and then dilute to the required consistency with alcohols (usually IPA) and hydrocarbons.

Vinyl lacquers are more complex. They are based on the polymers, mixed polymers, or co-polymers of vinyl acetate, vinyl butyral, and vinyl chloride. Alcohols are less generally useful here and predominate only for vinyl butyral resins, while ketones are generally preferred for the other resins. The vinyl lacquers are normally more plastic than the cellulose lacquers and their use extends to cloth, paper and paperboard finishes as well as metals.

Table VI.3 gives some indication of the solvent properties of some of the solvents included in Table VI.2.

There are several similar or associated uses for IPA and other chemical solvents. Directly associated with the paint industry are the paint removers which are used to soften and denature cured paint films prior to removal. Aggressive solvents are required, e.g. methylene dichloride (CH_2Cl_2) and acetone or other ketones. In general however, such solvents are low boiling (e.g. CH_2Cl_2 b.p. $40°C$) and evaporate too quickly to act effectively. They are usually, therefore, either diluted with a higher boiling component (e.g. carbon tetrachloride, CCl_4, b.p. $77°C$; or IPA for the ketones), and/or thickened with solids.

Chemical solvents are also used in miscellaneous household and industrial formulations, e.g. spirit dyes, adhesives, printing inks and dyes, and liquid agricultural and garden preparations.

The ester solvents are prepared by refluxing together the alcohols and acids (frequently acetic acid) with up to 10 per cent of concentrated sulphuric acid as the condensation catalyst. The general reaction is shown by:

$$\underset{\text{acid}}{R_1\ COOH} + \underset{\text{alcohol}}{R_2\ OH} \underset{\text{Reflux}}{\overset{H_2SO_4}{\rightleftharpoons}} \underset{\text{ester}}{R_1\ COO\ R_2} + H_2O$$

The reaction is an equilibrium and several methods are used to increase the ester yield. The use of the anhydride instead of the acid halves the water production thus reducing the tendency to hydrolysis of the ester, and the use of concentrated sulphuric acid itself reduces the effective water content. In difficult cases an entrainer

TABLE VI.3

Solvent Powers

Solvent \ Polymer resin	Natural			Cellulose acetate	Epoxy resin Shell Epikote 1001	Formaldehyde resins	Polystyrene	Thermoplastics			Synthetic rubber (SBR)	Vinyl	
	Shellac	Rubber	Acrylate (PMMA)					Polyamide (nylon)	Polyacrylonitrile (Acrilan)	Polyethylene (HD)		Polyvinyl chloride	Polyvinyl acetate
Acetone	2	4	1	1	1	1	3	4	4	4	4	3	1
Methyl ethyl ketone	2	2	1	3	1	1	1	4	4	4	3	4	1
Methyl iso-butyl ketone	4	1	4	4	1	1	1	4	4	4	1	4	1
Mesityl oxide	1	1	1	3	1	1	1	4	4	4	1	3	1
Di-acetone alcohol	1	4	2	3	2	1	4	4	4	4	4	4	1
Isophorone	2	3	4	4	4	1	1	4	4	4	1	1	1
Iso-propyl alcohol	1	4	4	4	4	1	4	4	4	4	2	4	1
Hexylene glycol	4	4	4	4	4	2	4	4	4	4	4	4	4
Iso-propyl acetate	4	2	1	4	1	1	1	4	4	4	3	4	1
Amyl acetate	4	1	4	4	1	1	1	4	4	4	1	4	1
2-Ethyl hexyl acetate	4	1	4	4	1	1	1	4	4	4	1	4	1
Toluene*	4	1	1	4	3	2	1	4	4	2†	1	4	1

* Included as a typical aromatic hydrocarbon solvent.
† Soluble in hot solvent only.
Solvent power rating. (1) Good.
                 (2) Moderate.
                 (3) Poor.
                 (4) None.

can be added to remove water from the reflux stream. An indirect method can be used for the higher alcohol esters, involving esterification with methanol as the first step, followed by transesterification with excess of the higher alcohol and regeneration of the methanol, which is removed from the reflux stream to favour equilibrium to the higher alcohol ester. Generally a small excess ($<$ 15 per cent molar) of alcohol is used for both direct and indirect reactions. Commercially, esters are available either still containing the excess alcohol or as the pure ester.

The use of IPA as a source of acetic acid and acetic anhydride has been briefly mentioned (Table VI.2). The route to the anhydride and to the acid is via ketene which can be obtained by thermally cracking acetone, acetic acid, or IPA. The ketene is then absorbed in acetic acid to give the anhydride. IPA gives relatively poor yields of ketene so its derivative acetone is normally used. Acetic acid is used when the anhydride only is required and the process is then essentially an indirect conversion of acetic acid to acetic anhydride. The acetone process is detailed in a subsequent section on acetone uses.

Two major uses of IPA as a chemical intermediate are in the preparation of synthetic glycerol and in the chemical production of hydrogen peroxide. Further details of these processes are also included in a subsequent section on acetone production from IPA. Most other intermediate uses are small. A typical example is the production of synthetic thymol by the sulphuric acid condensation of IPA with m-cresol. Thymol, naturally found in the essential oil of thyme, is used as an antiseptic in medicine and as a perfume component.

$$CH_3- \bigcirc \quad + \quad CH_3-CH(OH)-CH_3 \quad \xrightarrow{\quad H_2SO_4 \quad} \quad CH_3- \bigcirc -CH\begin{smallmatrix} CH_3 \\ \\ CH_3 \end{smallmatrix}$$

| m-cresol | IPA | Thymol (m.p. 52 °C) |

Primary and secondary isopropylamines are produced commercially on a considerable scale. Two process routes are normally used both of which are vapour phase catalytic reactions, one being the condensation of ammonia with IPA, and the second the condensation of acetone, ammonia, and hydrogen. Both reactions yield mono, di, and tri-alkyl amines, the product ratios being to some extent controllable by varying feed ratios, pressures and temperatures. The reactions are represented by:

$$CH_3-CHOH-CH_3+NH_3 \rightarrow (CH_3)_2CH-NH_2+H_2O$$
$$\text{Isopropylamine}$$

$$CH_3-CO-CH_3+NH_2+H_2 \rightarrow (CH_3)_2CH-NH_2+H_2O$$

Di-isopropylamine has the formula $(CH_3)_2CH-NH-CH(CH_3)_2$

The IPA/ammonia reaction is carried out at about 350°C and 100–200 atm. pressure over an aluminium phosphate catalyst. The acetone/ammonia/hydrogen reaction is carried out at temperatures of 130–180°C and 5–100 atms. pressure over a copper or nickel hydrogenation catalyst. An excess of ammonia is always used to minimise secondary and tertiary amine production, so that between three and five times the stoichiometric ammonia is used. Considerable by-product is formed so yields are relatively low; the conversion of IPA or acetone is about 80 per cent per pass and the yield of desired amines represents only about 50 per cent of the converted material. The reaction products are condensed and fractionated. Recovered ammonia and IPA or acetone are recycled, and the primary and secondary amines are the normally recovered products.

Isopropylamine	b.p. 32·4 °C	Specific gravity 0·69
Di-isopropylamine	b.p. 84·0 °C	Specific gravity 0·72

These compounds find general use in the fine chemicals industry as intermediates and catalysts. Their derivatives find their major uses in the rubber and agricultural chemicals fields, a typical example being N.N. di-isopropyl-2-benzothiazole-sulphenamide which is used as a high temperature rubber curing accelerator having a useful induction period. An unusual product is di-isopropylamine nitrite or di-isopropylammonium nitrite which is used as a vaporizing corrosion inhibitor. It is particularly useful for ensuring a good shelf life for packed metal components and minimum deterioration of 'mothballed' equipment, being a solid which sublimes slowly onto the metal surfaces when enclosed with them. The two amines are also very soluble in water, and therefore find some use as solubilising agents where their 'fishy' odour does not preclude it.

IPA and other chemical solvents are also widely used in the pharmaceutical, cosmetic, and essence industries. The uses are again divided between processing uses, and incorporation in the finished products. There are so many of these minor applications that a detailed discussion is outside the scope of this review but a few typical examples will be given. A major use in pharmaceuticals is in the preparation of rubbing compounds (liniments) and sterilising fluids where the combination of solvency and antiseptic properties is essential. In the cosmetics field both IPA and its derivatives are used in product formulations as solvents and solubilising agents. IPA itself is used to solubilise synthetic detergents in high active matter formulations for shampoos and similar products. IPA esters of the fatty acids, e.g. the myristates, stearates, and palmitates, are used in lotions, oils, skin creams and lipstick where they provide non-drying neutral bases. In the essence industries, IPA and its light esters are used both as process solvents, and to a minor extent in final formulations. Other examples of diverse uses of IPA include the extraction of oil from edible fish protein, as a liquid damping fluid for sealed instruments, e.g. compasses, as cleaning and de-greasing fluids in optical instrument assembly, and the use of fatty acid esters (xanthates and oleates) in froth flotation separators and as oil additives respectively. IPA is also used to prepare aluminium tri-isopropoxide as a specialised chemical reducing agent.

6.3. ACETONE

Acetone is the simplest (symmetrical) ketone possible, otherwise known as dimethyl ketone (DMK) and propan-2-one. It is a light, sweet smelling, colourless liquid with a molecular weight of 58 and a boiling point of 56 °C. It is an excellent solvent for many organic species ranging from acetylene gas to high molecular weight resins and lacquers. It is also soluble in all proportions in water, ethanol and ether. The classical laboratory preparation is by the thermal decomposition of calcium acetate as follows:

$$Ca(CO_2-CH_3)_2 \xrightarrow{\text{Heat}} CH_3\ CO\ CH_3 + CaCO_3$$

This was also the basis of early industrial production. Calcium acetate was produced by treating pyroligneous acid (the aqueous phase condensed during the destructive distillation of wood) with lime.

Increasing demands have led to the development of more sophisticated production methods, almost invariably based on hydrocarbon feedstocks, the predominant one being propylene derived from petroleum. Acetone is now produced in plants rated in 10,000's of tons per annum and is a major chemical solvent and intermediate.

6.3.1. Production—Older Methods

Before passing to the current high tonnage production routes some other processes which have been used commercially should be mentioned.

6.3.1.1. By carbohydrate fermentation This method was developed to augment acetone production by Weizmann in 1911 and made use of the bacteria clostridium acetobutylicum which co-produces acetone and n-butanol in 1:2 ratio. It became significant during World War I when acetone was required as a gelatinising agent to produce cordite explosive (introduced in the UK in 1890). In post-war years the method, based on molasses, continued to be used mainly for n-butanol production. Small amounts are still produced by this route.

6.3.1.2. From acetic acid This method is basically an adaptation of the calcium acetate route where a fixed catalyst bed is used. The catalysts used include calcium oxide, cerium oxide, or manganous oxide on pumice, or their combinations. Reaction temperatures are between 300°C and 400°C. The reaction proceeds by in situ formation of acetates and their subsequent decomposition, the overall reaction being:

$$2CH_3\ COOH \longrightarrow CH_3-CO-CH_3 + CO_2 + H_2O$$

This method became important where acetic acid was readily available, usually based on acetylene chemistry or fermentation. The acetylene is first hydrated to acetaldehyde, which is subsequently oxidised to peracetic acid and reacts further with acetaldehyde to give acetic acid. This process was used for tonnage production in USA and Germany in the 1930's and 1940's.

Another route based on acetylene is by the direct reaction with water (steam) over a catalyst. The catalyst used is refined zinc oxide and temperatures are in the 350–450°C range. The overall reaction is represented by:

$$2C_2H_2 + 3H_2O \longrightarrow CH_3-CO-CH_3 + CO_2 + 2H_2$$

6.3.1.3. From ethanol This is again a high temperature catalytic method. The catalysts used include zinc chromite, and iron oxide activated with calcium oxide or barium oxide. The reaction temperature ranges from 425–500°C. Steam and ethanol vapour react to produce acetone, CO_2, and hydrogen according to the equation:

$$2C_2H_5OH + H_2O \xrightarrow{500°C} CH_3 \ CO \ CH_3 + CO_2 + 4H_2$$

This route was also used commercially in the 1930's and 1940's but has probably not been used in the last fifteen to twenty years.

6.3.1.4. By oxidation of natural gas This route was developed in the USA in the 1930's and has only been used on a limited scale. The C_3/C_4 fraction of natural gas is the normal feedstock ('lean' natural gas) which is oxidised with air or oxygen under carefully controlled conditions of temperature and pressure with or without a catalyst. The reaction products are complex but their ratios can be partially controlled by varying the O_2/hydrocarbon ratio, the hydrocarbon composition, and the temperature. They are condensed, the product gas stream is scrubbed with water and recycled, and the liquid products fractionated. Products comprise a wide range of oxygenated compounds including formaldehyde, acetone, and the butanols. The method is still used to a limited extent and produces about 15,000 tons per annum acetone in the USA.

6.3.2. Acetone from Isopropanol (IPA)

Acetone production by IPA dehydrogenation over metal or metal oxide catalysts was introduced about 1920. This rapidly became and remains the major source of acetone, accounting for between 40 and 70 per cent of the production in developed countries. This process followed directly from the 'new' availability of propylene-derived IPA which occurred in 1920. Three main routes to acetone will be discussed here. The direct dehydrogenation and oxidation routes from IPA will be treated first, while a further section will discuss routes where the conversion of IPA to acetone is complementary to a main process route.

6.3.2.1. Direct dehydrogenation These routes are similar in chemistry and in technology and differ mainly in their by-products which are hydrogen and water respectively. The simple reaction equations are:

$$CH_3-CHOH-CH_3 \xrightarrow{Cu} CH_3-CO-CH_3 + H_2 - 16 \text{ k cals at } 330°C$$

$$2CH_3-CHOH-CH_3 + O_2 \xrightarrow{Cu} 2CH_3-CO-CH_3 + 2H_2O + 42 \text{ k cals at } 300°C$$

These reactions are carried out in furnace-heated catalyst-packed tubes in the vapour phase at high temperatures and moderate pressures. In general high temperatures favour the equilibrium value particularly in the endothermic dehydrogenation reaction and also accelerate the equilibrium approach. However, high temperatures also promote thermal decomposition resulting directly in carbon deposition and catalyst deterioration. Low pressures favour the equilibrium value but compromise is necessary to minimise capital costs and achieve stable operation. By-product formation is low even at the highest temperatures used and is chiefly IPE and propylene, thus yields > 95 per cent molar are normal.

A typical flow scheme for the production and purification of acetone is shown in Fig. 6.7. The dehydrogenation route is most common and is described first. The feed is generally IPA/CBM containing 88 per cent IPA and trace quantities (about 2 per cent wt.) of polymers. The water component acts as a diluent in the furnace tubes and reduces the tendency for hot spots to form causing local coking. To achieve maximum throughput and economy for a given furnace size, the feed is preheated in three exchangers, two of which are feed/product exchangers. In the first exchanger the condensing product heats the feed to close to its boiling point. In the second exchanger the feed is vaporized using HP steam (15–25 atm.), while the third exchanger superheats the feed by desuperheating the product. The reaction products are further cooled to about 40°C and pass to an accumulator, while the superheated feed enters the furnace tubes at 300–350°C.

The furnace layout can take several forms; the one illustrated is a 'Petrochem' furnace of circular section which is particularly well suited to this process. The feed vapours are divided at a manifold into as many streams as there are furnace tubes, and flow downwards through them. The flow is counter current to minimise fuel costs and furnace temperatures and three separate zones are apparent. The upper preheat zone is a convection section and finned or studded tubes are used for maximum efficiency. The middle reaction zone is a radiant section with a central refractory ceramic cone, and the lowest reaction section relies on direct combustion gas heating. The feed to each tube is flow controlled to ensure even feed distribution, either using sonic flow orifices, or flow control, or temperature/flow control instruments. The reaction products leave the furnace at 400–500°C and are collected in a manifold.

The product accumulator serves as a phase separator, the crude acetone passing to intermediate storage, while the hydrogen passes through a water scrubber to the consumer units. The reaction pressure control acts on the scrubbed hydrogen and is set to maintain a pressure of 5–7 atm. gauge in the reaction tubes. The product hydrogen is of good quality containing only traces of CO and CH_4, and is generally better than 95 per cent vol. hydrogen.

The classical laboratory preparation using copper catalyst is carried out at 300–350°C at which temperatures equilibrium represents 97 per cent conversion. Industrially to achieve economic reaction rates and space velocities higher temperatures are required and 380–400°C is commonly used. However, as the catalyst activity declines, higher temperatures are required to achieve the desired conversion and the temperature is gradually raised to 500°C before the reaction is stopped for catalyst regeneration. This reactor cycle time can vary from five to thirty days and major efforts are made to maximise it. The formation of carbon deposits (coking) is the main reason for catalyst deactivation and is itself a function of temperature, being slow at 400°C but increasingly rapid up to 500°C. The presence of water vapour inhibits coking and further improvement results from hydrogen recycle.

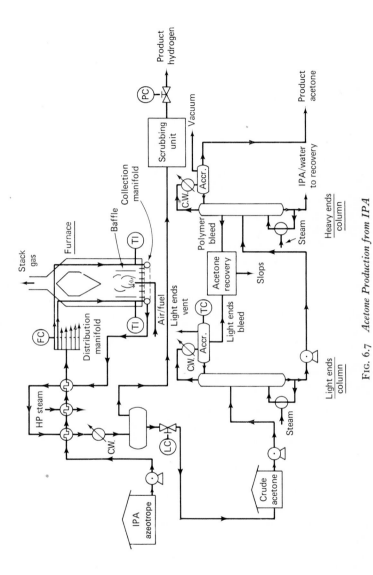

FIG. 6.7   *Acetone Production from IPA*

Periodic dosing with sulphur compounds also prolongs the cycle (cf. the use of CS_2 in ketene furnaces); this mechanism is not clear but recent publications indicate that sulphur and sulphides may themselves have a catalytic action. Even heat distribution and hence tube temperatures are also important and are further discussed below.

Various catalysts are used including metals, metal oxides and metal salts, with or without catalyst supports. The metals used are zinc and/or copper. The predominant catalysts are zinc/copper alloys (e.g. friable quenched spelter) and zinc and copper oxides on silica or pumice supports. The former has the decided advantage of good heat transmission. The catalysts require regeneration or replacement at fairly frequent intervals as discussed above. Regeneration is normally in situ using the furnace to keep tube temperatures at about 500 °C while steam and air (2 per cent vol O_2) are blown through the tubes to burn off the coke, followed by a short hydrogenation cycle. For metal catalysts periodic removal from the reactor and acid pickling are advantageous. Catalyst life can vary from six to fifteen months.

The main process variables are pressure, temperature, space velocity, and catalyst particle size. The selected operating condition is a compromise to achieve adequate conversion with an acceptable cycle time and economically sized equipment. The reaction section is normally sized to give 70–85 per cent molar conversion at about 400 °C and 5 atm. gauge which requires about 2 seconds residence time in the reaction zone. Some effects of these variables are discussed below.

1. *Temperature and space velocity* The reaction is endothermic so that temperature within a reaction tube will decrease from the wall to the centre (tube diameters between 3 and 6 in. are used) and high skin temperatures will be required to maintain an acceptable bulk temperature. Thus the reactor has only limited flexibility. Low throughputs and space velocities are easily accommodated by reducing the furnace firing to maintain the desired conversion. However, as throughput and space velocity increase, so must the firing severity with adverse effects on cross-tube temperature gradients and coking speed. This effect is aggravated by the reduced residence time, and the overall effect of high space velocities is to reduce cycle time. These reactors are therefore frequently designed to operate at one fixed feed rate. The main variable is then temperature and normal operating technique is to adjust firing/temperature to give a desired conversion. Individual tube outlet temperatures are monitored and the reactor is shut down when the first tube outlet temperature reaches 500 °C.

Clearly heat distribution to the tubes (ten to fifty in number) must be good and individual tube performances are important. Great care must be exercised in packing the tubes to ensure equal catalyst charges and consistent quality. If a tube or tubes performs badly relative to the others the whole furnace performance can be affected in a progressively adverse way. Tubes giving limited conversion reduce the effective conversion of the whole unit and firing severity is increased; also these tubes tend to run hot and coking is consequently more rapid than in the rest of the tubes and performance, already poor, rapidly deteriorates. This phenomenon results in premature run termination and, in severe cases, the furnace must be shut down and the poor performers repacked to prevent catalyst fusion. These effects are exacerbated by the sonic orifice method of flow distribution which, though inexpensive, is also inflexible. Manual adjustment of the flows to individual tubes is sometimes practised but becomes complex in large units with many tubes. Cascaded exit temperature/flow control is satisfactory but expensive. Provision can also be made to bleed steam or hydrogen into defective tubes to reduce their operating and outlet temperatures. Nevertheless, temperature distribution remains the major factor in determining reactor cycle time.

2. *Pressure*   Operating pressure in the reaction section is generally 5–7 atm. g. (75–105 p.s.i.g.) and is seldom varied on operational units. Small pressure increases can be used to compensate for increased space velocities but this tends to have an adverse effect on equilibrium conversions. In many commercial plants pressure changes are difficult especially where sonic feed flow control is used, and the by-product hydrogen is required elsewhere for hydrogenation.

3. *Catalyst particle size*   Ideally, very small particles should be used to give maximum surface area, turbulence, and even heat distribution. However, tube velocities are high (about 5 ft/sec superficial) and pressure drop increases rapidly as particle size is reduced. In practice particles between $\frac{1}{2}$ in. and 1 in. are used.

6.3.2.2. Oxidation   The oxidation route to acetone is of less significance as it is used in only a minority of cases. There are two main reasons for this, firstly the yield is slightly but significantly lower than in the dehydrogenation reaction. Secondly, the hydrogen from dehydrogenation is usually a valuable and desired by-product because of its purity. Although oxidation reaction temperatures are higher (generally 500–600°C) the fuel requirements are similar because the oxidation reaction is exothermic. Reactor control can be equally or more critical because of the high heat release. However, where hydrogen has no value the oxidation reaction is used and the hazard and handling of hydrogen is eliminated.

The reaction is performed in similar equipment to that already discussed for dehydrogenation. The catalyst however is different and is either metallic copper or silver. It is likely that the primary reaction is again the dehydrogenation of IPA but the hydrogen produced is immediately removed by reaction to water thus enhancing the conversion to ketone. Less than the stoichiometric quantity of air, or preferably oxygen, is used to ensure its complete consumption within the reactor and non explosive conditions. The furnace layout can be reversed so that the flow in the tubes is co-current and the reaction zone is in the coolest part of the furnace.

The reaction variables are the same as those considered for dehydrogenation except that the temperature gradient in the tubes is reversed. Even heat distribution is equally important but in this case overheating cannot take place provided thorough gas mixing and equal flow distribution are ensured, because the oxygen present is all consumed. The by-products are again mainly IPE and polymer with traces of methane, CO, and hydrogen.

In both systems the recovery section is similar. The reactor product is collected in a manifold, cooled against the feed and then against cooling water and passes to an accumulator where gas and liquid separate. The liquid is passed to storage while the vent gas is scrubbed with water to remove IPA and acetone which are subsequently recovered. The vent gas is then either passed to other consuming units or used as fuel.

The crude liquid product which contains water, IPA, acetone, polymer, and trace by-products (e.g. aldehydes and light hydrocarbons) is fractionated in two columns. The first column removes light ends (IPE etc.) as overhead product. To minimise light ends in the final product either a high top bleed is taken from this column, or the condenser is of the dephlegmator type and controls the vent temperature at 40–45 °C to encourage aldehyde venting. The bottom product from the first column feeds the second column which is operated at reduced pressure (about 5 p.s.i.g.) to assist the separation. Finished acetone is recovered as column overheads and pumped to storage. IPA, heavy polymer, and water are recovered as bottoms and are passed to crude IPA storage for recovery. Light polymer accumulates in the middle of the column and is bled away as a sidestream for recovery together with the tops from the first column.

6.3.2.3. IPA conversion to acetone by secondary reaction   The processes where this occurs were largely developed by Shell Development Co. in the USA during research into synthetic glycerol manufacture. Further development work has since been carried out by Shell Research at Amsterdam. These processes have been operated in the USA at Norco refinery since 1948 and in Europe at Shell's Pernis site since 1952. Full discussion of all the stages involved in the complex chemistry of this topic is beyond the scope of this section and only those reactions where acetone is a by-product will be considered in any detail. However, for clarification of how these particular reactions are involved, a diagramatic presentation of synthetic glycerol chemistry is shown in Fig. 6.8. The details of these processes are further discussed in Chapters 7 and 10.

The first reaction which produces acetone from IPA is the direct oxidation of IPA. The reaction almost certainly proceeds via the hydroxy-hydroperoxide and may be likened to the cumene/phenol, acetone process. The reactions involved are:

$$CH_3—CHOH—CH_3 + O_2 \longrightarrow CH_3\ HOOCOH\ CH_3$$
$$\text{2.2. hydroxy hydroperoxy propane}$$

$$\overset{\text{acid}}{CH_3\ HOOCOH\ CH_3 \longrightarrow CH_3—CO—CH_3 + H_2O_2}$$

The reaction is carried out in a single tower reactor into which feed and recycle IPA are fed at the top, while oxygen is fed in near the base. The reaction product is continuously withdrawn from the tower base, and the reaction is carried out at about 40 p.s.i.g. and between 95 and 140°C. The reaction is exothermic and is cooled by refluxing the acetone and IPA present. The actual operating temperature depends partly on the conversion allowed in the reactor and hence on the composition and boiling point of the reaction mixture. The conversion can be varied between fairly broad limits depending on the critical product. If hydrogen peroxide is the primary product the conversion is kept low and acetone production minimised. At higher conversions the peroxide tends to be consumed in a secondary reaction giving more acetone, so high conversions are only used if acetone is the primary product. The

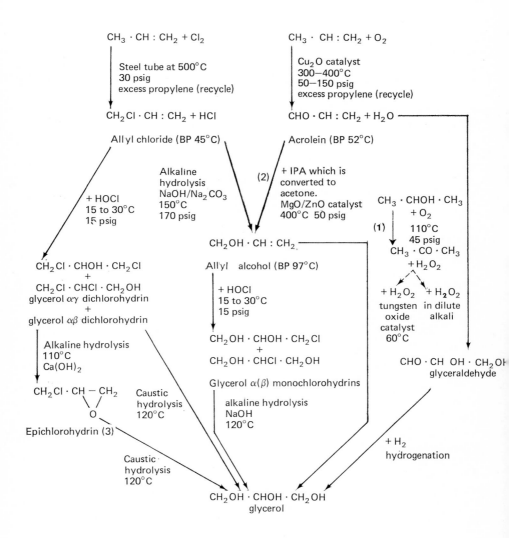

Reactions (1), (2) and (3) discussed in accompanying test

FIG. 6.8 *Synthetic Glycerol Reactions*

normal conversion for hydrogen peroxide production is 20–30 per cent. This secondary reaction is represented by:

$$CH_3—CHOH—CH_3 + H_2O_2 \longrightarrow CH_3—CO—CH_3 + 2H_2O$$

The reaction product consists of IPA, acetone, hydrogen peroxide and traces of hydroxyhydroperoxide. As no specific precautions are taken to prevent trace acid formation in the reactor or to neutralise it, the reaction conditions are slightly acidic so that the hydroperoxide reaches an equilibrium concentration, the acidic conditions favouring the cleavage of the hydroperoxide. At moderate conversions the yield of acetone is better than 95 per cent molar while the yield of hydrogen peroxide varies between 85 and 90 per cent molar.

The reaction products are diluted with water to prevent further reaction and to create dilute acid conditions to promote cleavage of residual hydroperoxide. The reaction products are separated by distillation to give a crude acetone stream which is sent forward for upgrading to saleable acetone, an IPA/CBM stream which is recirculated to the reactor, and a bottoms stream consisting of hydrogen peroxide and water which is used in a subsequent reaction stage.

The second reaction producing acetone from IPA is the trans-hydrogenation reaction which converts acrolein (acraldehyde) to allyl alcohol. The reaction was developed to allow hydrogenation of the aldehyde group to alcohol without significant hydrogenation of the adjacent double bond. The reaction is as follows:

$$\underset{\text{acrolein}}{CHO—CH=CH_2} + \underset{\text{IPA}}{CH_3—CHOH—CH_3} \longrightarrow \underset{\text{allyl alcohol}}{CHOH—CH=CH_2} + \underset{\text{acetone}}{CH_3—CO—CH_3}$$

The reaction can be regarded as a variation of the Meerwein-Ponndorf-Verly reaction without aluminium isopropoxide which is carried out at high temperature (about 400°C) and is only mildly endothermic. The catalyst bed used is a mixture of magnesium oxide and some zinc oxide on a silica support. The reaction pressure is moderate (30–60 p.s.i.g.) and the reaction is carried out completely in the vapour phase. A large excess of IPA (two or three times stoichiometric) is used to maximise the conversion to allyl alcohol.

The reaction products are a complex mixture of the four reactants and the by-products and are not generally completely separated. The product stream is distilled to give four fractions. The lightest contains unreacted acrolein and is recycled. crude acetone is combined with that from the previous reaction for upgrading. third stream consisting mainly of the excess IPA is recycled to the catalytic reactor, while the heavy stream containing the allyl alcohol is transferred to a stirred reactor where it is reacted with the previously separated hydrogen peroxide to yi ld glycerol (probably via the epoxide and a hydrolysis reaction). The yield of allyl alcohol on acrolein is between 75 and 80 per cent molar while the acetone yield on IPA is better than 95 per cent molar.

The third marked item on Fig. 6.8 is the production of epichlorohydrin. While this is not directly associated with acetone production it is mentioned here because af its use in making epoxide resins. These will be mentioned later under the uses of acetone. The basic reactants in the production of epoxide resins are epichlorohydrin and di-phenylol propane. As illustrated above the production of epichlorohydrin is by an indirect route from propylene. Di-phenylol propane is also an indirect product of propylene, being produced by the condensation reaction of acetone and phenol.

6.3.2.4. The cumene process   This route is a relatively recent development in acetone production and has been developed mainly as a route to phenol with acetone as a valuable by-product. The basic chemistry was developed during World War II in Germany but no industrial process was developed. In postwar years several companies (notably Distillers Co. in the UK and Hercules Powder Co. in the USA) developed commercial processes. The incentives were the increasing demand for phenol for use in nylon and thermosetting resin production, and the difficulty of conventional phenol routes via benzene sulphonate or chlorobenzene which involve the handling of very corrosive, acidic media under severe temperature and/or pressure conditions. The simple chemistry of the route is summarised in the following equations:

$$
\text{Benzene} + CH_3\!-\!CH\!=\!CH_2 \;\xrightarrow[\text{or PO}_4^-]{\text{AlCl}_3}\; \text{Cumene (Iso propyl Benzene)}
$$

$$
\underset{\text{CH}_3}{\overset{\text{CH}_3}{\text{CH}}} + O_2 \;\xrightarrow{\text{alkaline}}\; \underset{\text{CH}_3}{\overset{\text{CH}_3}{\text{C}\!-\!OOH}} \;\xrightarrow{\text{acid}}\; \text{—OH} + CH_3\!-\!CO\!-\!CH_3
$$

Cumene hydroperoxide   Phenol + Acetone

The first reaction represents a simple alkylation reaction to combine benzene with propylene. Both the reaction and the catalyst used represent an application of a typical Friedel-Crafts' synthesis reaction on an industrial scale. This reaction was known and used in the petroleum industry several years before its adoption by the petrochemical industry. It was used on a large scale during World War II to produce cumene as a high octane blending component for aviation fuel. The research method (F1) octane rating of cumene is 113 while the motor method (F2) is 100. Several commercial processes have been developed and three are currently used for cumene production. Two are liquid phase processes operating at moderate pressures and temperatures, and using either hydrogen fluoride (HF) or aluminium chloride (AlCl$_3$) as catalyst. The third is a higher pressure and temperature vapour-phase reaction carried out using a supported catalyst whose active component is phosphoric acid or metal phosphates.

These processes are operated either exclusively to provide cumene for phenol preparation, or to provide this as a main product while providing a cruder cumene for gasoline blending, and α-methyl styrene for co-polymerisation purposes. Thus the precise conditions of operation vary from plant to plant. Furthermore, this synthesis route is particularly sensitive to the catalytic agent used, the concentration of catalyst used, the ratio of benzene to propylene, purity, the reaction temperature, and the reaction pressure. Within fairly wide limits these several variables can be adjusted to best suit the conditions at a particular location while still giving an acceptable yield of cumene. In general the liquid phase reactions using HF or supported $AlCl_3$ are carried out at low temperatures (0–80°C) and moderate pressures (10–100 p.s.i.g.). The vapour phase reaction using supported phosphoric catalyst is carried out at 50–150 p.s.i.g. and temperatures in the range 250–350°C.

This process suffers from the normal disadvantage of Friedel Crafts' reactions in that specific reactions are impossible because the substituted product (in this case cumene) undergoes secondary substitution more readily than the feed benzene undergoes primary substitution. Of necessity, therefore, a fairly large excess of benzene is used to minimise di-substitution products, which results in high recycle flows. Typical figures would indicate a 3 to 1 benzene to propylene molar ratio in the reaction feed. Propylene conversion however is correspondingly good and recycle is only employed where the propylene content of the feed is low. Yields are not exceptional because of these secondary reactions, and figures of 88 to 91 per cent molar based on both propylene and benzene are typical provided the propylene content of the C_3 feed is high. By-product distribution is very much dependent on catalyst type and on operating conditions, thus, using identical equipment, changes of catalyst or operating temperature will affect the by-products. In general these by-products are mainly di-isopropyl benzene, tri-isopropyl benzene, n-propyl benzene, and traces of polymeric compounds. p-di-isopropyl benzene can be a valuable by-product as starting material for the production of terephthalic acid by oxidation:

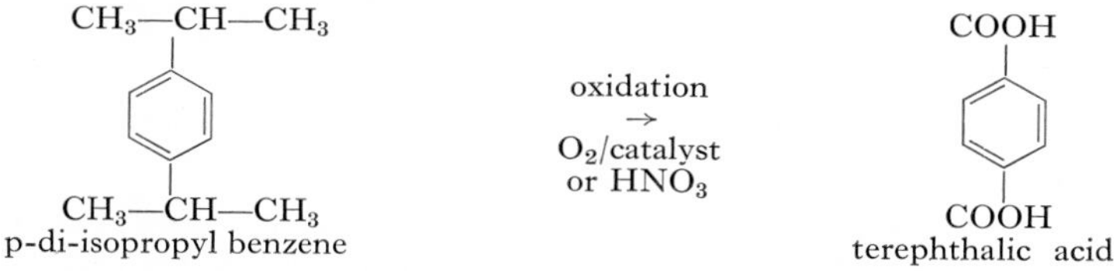

$$CH_3-CH-CH_3$$

oxidation
$\rightarrow$
O_2/catalyst
or HNO_3

$$COOH$$

$$CH_3-CH-CH_3$$
p-di-isopropyl benzene

$$COOH$$
terephthalic acid

However, the para compound is seldom exclusively produced. To maximise yields of para compound the isomerising action of these catalysts can be exploited and the non para compounds recycled.

A simplified flowscheme is shown in Fig. 6.9. This represents the outline process where supported catalysts are used, when the reactor takes the form of a series of packed beds. If a liquid catalyst (HF) is used the reactor takes the form of a stirred vessel and the catalyst is recovered by combined phase separation and distillation techniques (b.p. of HF at normal pressure is 19·4°C). Other variations also depend on the type of catalyst used; thus in the case of the $AlCl_3$ and HF catalysts precautions must be taken to minimise water contents which can result in catalyst hydrolysis and/or corrosion problems. The reaction as shown consists essentially of four stages:

(a)   Alkylation reaction
(b) C_3 and light ends removal (with or without recycle)
(c)   Benzene recovery and recycle
(d)   Cumene/heavy ends separation.

The cumene used for phenol manufacture should be > 99 per cent pure and some properties of pure cumene are listed on page 246:

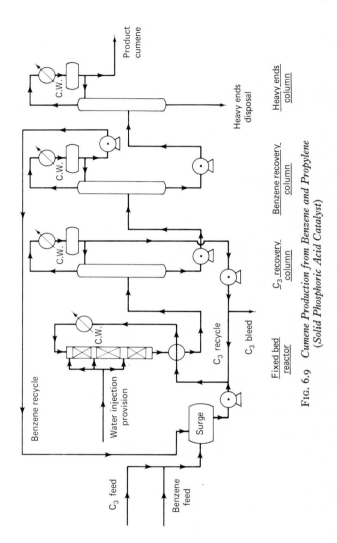

FIG. 6.9   *Cumene Production from Benzene and Propylene*
(Solid Phosphoric Acid Catalyst)

Purity	> 99·5 per cent
Boiling point	152·4°C
Specific gravity	0·862 at 20°C
Specific heat	0·37 at 25°C
Flash point (TCC)	35°C
Explosive limits	1·1 per cent vol. and 8·0 per cent vol. in air
Heat of formation	−9·9 kcal/mole at 25°C
Molecular weight	120·2
Viscosity	0·8 cp at 20°C.

The production of phenol and acetone from cumene is represented diagramatically in Fig. 6.10.

The basic steps are as follows:

(a) Oxidation (air or oxygen)
(b) Cumene hydroperoxide concentration
(c) Cumene hydroperoxide cleavage
(d) Phenol/acetone separation and acetone recovery
(e) Phenol recovery
(f) Tar and heavy ends treatment and phenol recovery

The main variations between processes commercially available are in the use of air or oxygen in the reaction section, the method of phenol recovery, and the handling of by-products and effluents. These aspects can have a considerable influence on the process selected for a given location, and on the degree of by-product and effluent treatment incorporated. Phenol and its associated compounds always represent a severe effluent problem, and where this is critical, considerable effort must be directed to minimising and treating effluent. The choice of air or oxygen can also be governed by effluent criteria. The use of oxygen is superficially more expensive, but utilisation is good and the amount of vent gas is minimal. Using air, the basic material is cheap but the utilisation is poor (50–60 per cent), the compression costs high, the vent gas flows (mainly nitrogen) are high, and the facilities to recover cumene vapour from the vent gases are extensive and costly.

The oxidation section is invariably multi-stage to allow maximum and optimum control, to maximise conversion to hydroperoxide, and to simplify the considerable heat removal problem. The heat of reaction is 20 kcals/mole cumene. The number of stages can vary between three and five, and they are arranged in series with the feed to the base of the first stage; subsequent interstage transfers being by overflow from one vessel to the base of the next. The feed consists of fresh and recycle cumene with a small amount of emulsifier and some dilute alkali solution (caustic or carbonate) added as a peroxide stabiliser. Provision is also made for interstage alkali solution addition to maintain the desired alkalinity throughout the reaction section. The alkali stabilises the hydroperoxide by preventing traces of acids formed causing acidic conditions which promote premature hydroperoxide cleavage. The reactors themselves can be either sparged tower reactors or stirred reactors, the choice generally depending on whether air or oxygen is to be used respectively. When air is used the passage of the inert nitrogen provides adequate mixing and towers are favoured, but

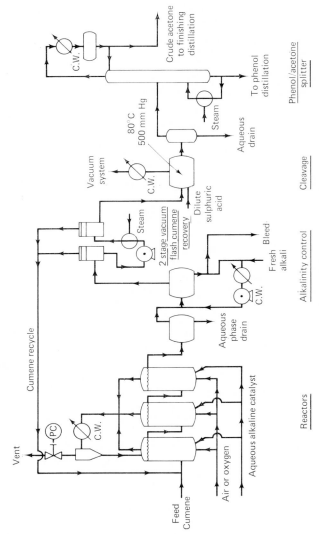

Fig. 6.10   *Cumene Oxidation to Phenol and Acetone*

when oxygen is used the induced mixing is small and is augmented by stirrers. Heat removal is mainly by the use of coils and/or jackets on the vessels, though in the air case the stripping and cold recycle of cumene contribute significantly. The oxidation rates are balanced in all stages. In the oxygen case this is achieved by accurately metering the gas feed and vent, while in the air case it is more accurately assessed by metering the feed gas and analysing the vent gas for oxygen content. In the air case the vent gas must be chilled and scrubbed to remove cumene vapours. The tendency to secondary reactions increases with hydroperoxide concentration, and can be limited to some extent by having a reducing temperature profile through the stages. Operating temperatures are generally between 100 and 130°C and often between 110 and 120°C. Operating pressure is dependent on whether air or oxygen is used; in the former case pressures between 60 and 100 p.s.i.g. are normal while in the latter case pressures a few pounds above atmospheric are quite adequate. The total cumene conversion per pass varies between 20 and 25 per cent molar and is generally better than 92 per cent to hydroperoxide. The main by-products are:

$$OH$$
$$CH_3{-}\underset{|}{\overset{|}{C}}{-}CH_3 \qquad\qquad O = C{-}CH_3$$

Phenyl di-methyl Carbinol (b.p. 215–220°C)   Acetophenone (b.p. 202°C at 750 mm Hg)

Acetophenone can be recovered as an economic by-product, while the carbinol can be hydrogenated directly to cumene, or dehydrated to α-methyl styrene and then hydrogenated to cumene. The other by-products are generally high molecular weight polymeric materials which appear as a tar. Acetophenone finds application as a high boiling solvent, as a constituent and intermediate in perfumes, and as an intermediate in pharmaceuticals, rubber chemicals, dyestuffs, and resins.

The next major step is to concentrate the hydroperoxide by removing the bulk of the cumene. The reaction product is cooled and phase separated to remove the acid salts formed in the reaction, more alkali being added if required. The separated organic layer is then contacted with fresh alkali/carbonate solution to achieve precise alkalinity control, and fed through a two or three stage heater/flasher system to reduce the cumene content. To avoid high temperatures the flashers are run at low pressures (50–150 mm Hg abs) and the recovered cumene is recycled. The stream is concentrated to 65–80 per cent wt. hydroperoxide (i.e. about 85–90 per cent of the unreacted cumene is recovered here). The hydroperoxide has a very low vapour pressure but the stripping is preferably carried out at < 120°C to avoid its decomposition.

The concentrated cumene hydroperoxide is first cooled and then passes to the cleavage vessel where it contacts a dilute sulphuric acid aqueous phase (5–20 per cent wt. acid). The cleavage reaction is very rapid being completed in a few seconds accompanied by a large exotherm. The cleavage reaction is cooled either by circulation through an external cooler or by allowing the acetone produced to reflux back into the vessel. The first system is superficially the more complex, but to achieve an acceptable temperature the acetone reflux system must be operated at reduced

pressure (400–500 mm Hg abs. to maintain the temperature at < 80 °C) and very efficient vapour cooling is necessary. The cleaved product is cooled further and phase separated to remove the acid aqueous phase and then passes to the distillation recovery section.

The reaction products are very complex and have a wide range of boiling points from acetone, b.p. 56 °C, to high molecular weight tars. Only acetone recovery will be described in detail here as the phenol purification is complex and not directly relevant. The reaction product is first split by distillation into two cuts. The light cut contains most of the acetone, some residual cumene, and some α-methyl styrene (AMS) as its main components. The bottoms are then distilled at low pressure and temperature to separate phenol with some AMS from the tars, carbinol, and acetophenone. The phenol rich stream is then purified by steam distillation, dehydration, and topping and tailing. The heavy ends are subjected to hydrogenation, cracking, and distillation to recover acetophenone, phenol and cumene, and are then discarded.

The light stream from the first splitter is crude DMK. This is distilled to remove light ends using a high reflux ratio and a temperature controlled condenser system to allow free venting of the light cut material. A small overhead bleed stream is withdrawn, water extracted, and discarded. The bottom product contains AMS and is kept alkaline (by caustic injection) to prevent polymerisation. The bottoms are then fractionated in a second column operated at reduced pressure and temperature to minimise by-product formation. Acetone is withdrawn as the overhead stream and run down to product tankage, while the bottoms, containing cumene, AMS, and water are cooled and phase separated. The aqueous phase is discarded and the organic phase, combined with other hydrocarbon streams is recovered by hydrogenation and distillation.

6.3.2.5. Acetone direct from propylene   The direct route from propylene to acetone is the most recent manufacturing route to be developed having been first used commerically in about 1963. It is a development of the parallel process for converting ethylene to acetaldehyde which was developed in Germany by Consortium für Electrochemische Industrie and first used industrially in 1961. It is now offered as a commercial process by Hoechst, Wacker, and Uhde. The process is, however, based on a reaction which has been known since the last decade of the nineteenth century. The process, while chemically simple, is difficult to commercialise because of its highly corrosive nature involving the handling of hydrochloric acid and chlorides in large quantities at fairly high temperatures and under oxidising conditions. This was probably the major impediment to its earlier development.

The process is not widely employed and probably produces less than 5 per cent of the total non 'Soviet-bloc' production. The yields are good (generally better than 92 per cent molar when rich propylene is available) but the capital investment costs and operating costs are relatively high compared to the IPA route. The major restraining influence on its expansion has been the cumene/phenol/acetone process which has developed from a minor supplier of acetone in 1960 (about 10 per cent of total supplies) to one taking nearly 50 per cent of the market in 1970. These plants have been built to meet phenol demands and the by-product acetone co-produced

has virtually met the needs of the expanding market, so little exclusively acetone-producing plant has been built since the direct route became commercial.

The process can be operated either in one stage or in two stages depending upon the desired product and the feed quality. Generally, the acetaldehyde from ethylene process is operated single stage, while the acetone from propylene and methyl ethyl ketone from n-butylenes processes are operated in two stages which have the advantage of allowing optimum conditions in each. Other major reasons for using the two stage process are that less pure feedstocks can be compensated for by modifying process conditions in the first stage, and air instead of oxygen can be used in stage II. The process involves firstly the oxidation of propylene with palladium chloride in the presence of water which produces metallic palladium, hydrogen chloride and acetone. To prevent depletion of the palladium chloride the palladium is reconverted to chloride by oxidation with air followed by reaction with the hydrogen chloride present. This last oxidation is normally slow but is promoted by the presence of chlorides of the dual valency metals, of which copper has been shown to be the most effective. The process is represented by the following equations:

$$CH_3-CH=CH_2+PdCl_2+H_2O \longrightarrow CH_3-CO-CH_3+Pd+2HCl$$

$$Pd+2CuCl_2 \xrightarrow{\ H_2O\ } PdCl_2+2CuCl$$

$$4CuCl+4HCl+O_2 \longrightarrow 4CuCl_2+2H_2O$$

There is some uncertainty about the initial reaction. The accepted mechanism is that the propylene forms a complex with the weakly ionic palladium chloride which is unstable in the presence of water and decomposes rapidly as shown. The metallic palladium so produced is very finely divided and in the optimum state for oxidation or chlorination.

An outline flow diagram of the reaction system is shown in Fig. 6.11. The first stage is a simple tower (reactor) with propylene and regenerated catalyst fed to the base. The tower is liquid full of a water solution of the soluble catalysts also containing suspended palladium and cuprous chloride. The propylene, if pure (98 per cent +) is rapidly absorbed and reacted and the only vent flow is the propylene impurities and traces of one and two carbon atom compounds produced as by-products. As the catalyst becomes depleted towards the top of the tower the soluble catalysts are largely replaced by the insolubles, palladium and cuprous chloride. The process is operated at moderate pressures (between 130 and 200 p.s.i.g.) to ensure solution of the propylene prior to reaction and evaporative cooling is not, therefore, employed as in the single stage ethylene process. The considerable heat of reaction (about 60 kcal/mole at 30°C) is, therefore, removed by jacket and/or coil cooling to maintain the temperature at 100–120°C. The reaction product is withdrawn from the top of the reactor together with the vent gases, and the pressure reduced to atmospheric.

The reaction rate is controlled by the temperature and the catalyst concentration. The propylene concentration has some effect but at the required concentrations of > 90 per cent, this is marginal and can be compensated by pressure. The optimum temperature has been selected to minimise by-product formation and corrosion. Catalyst concentrations are generally in the ranges 300–400 mg/l for palladium and

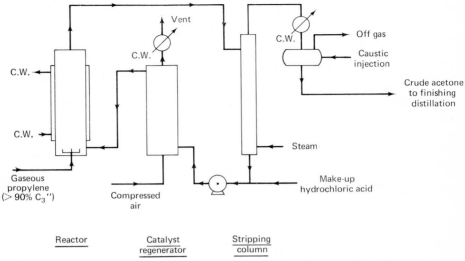

FIG. 6.11   *Acetone by Direct Propylene Oxidation*

50–100 g/l for cupric chloride. The palladium concentration governs the reaction rate while the cupric chloride concentration governs the ultimate conversion. As the cupric chloride concentration becomes depleted the rate of formation and the concentration of palladium chloride will fall and reaction rates will follow them. The reaction rate for the initial concentration of palladium chloride is about 130 g/l/hr but the overall reaction rate under steady state conditions is about half this, governed by the catalyst regeneration and recirculation rate. This corresponds to about 45 m³ of reaction volume for a production of 25,000 tons per annum[1,2] and is probably the economic limit for a single reaction train.

Reaction product is fed to a column where most of the organic products are removed by steam stripping, the overhead product being condensed, neutralised, separated from the non-condensables which are vented, and passed to the distillation train. The stripped bottom product consists of water, palladium chloride, cupric chloride, hydrochloric acid, some palladium and trace quantities of the heavier organic products. They are fed under pressure to the base of the catalyst regeneration vessel by a pump, together with a small quantity of make-up hydrochloric acid which replaces that removed by hydrochlorination and chlorination reactions. The regeneration vessel is a tower reactor operated at a pressure sufficiently above the reaction pressure to allow regenerated catalyst to flow from it to the reactor. Air is sparged into the base of the regeneration tower where it oxidises the cuprous chloride to cupric chloride, absorbing most of the hydrochloric acid present. The regenerated catalyst is taken off from the top of the regenerator. The flow of air is controlled to give minimum oxygen in the vent gas which leaves the top of the reactor, is cooled, neutralised, and vented.

The acetone distillation is conventional. The neutralised crude acetone is topped

to remove light ends and traces of gases. A light ends bleed is removed, water washed to recover acetone which is recycled, and discarded as fuel or by-product. The bottoms containing water and heavy organic impurities are distilled in a second column which produces acetone as the overhead product and rejects water and heavy ends from the base. The heavy ends are mostly chlorinated products including mono- and di-chloro acetone. The low boiling by-products contain propionaldehyde (b.p. 45 °C) and traces of isopropyl chloride (b.p. 36·5 °C).

In the reaction sections clearly corrosion is a major problem with operating temperatures around the 110 °C level and dilute hydrochloric acid present. The problem is aggravated by the presence of oxygen in the catalyst regeneration section, and generally by the presence of varying amounts of solids which add erosion to the already severe corrosion problem. This necessitates the use of exotic materials such as tantalum, titanium, teflon linings, and high nickel alloys; and makes capital and maintenance costs high. It also influences the design of the reaction sections so that pumps, stirrers, and coolers etc. are avoided where possible.

6.3.2.6. Other routes to acetone   The routes already discussed represent the major sources of acetone in commercial use. The two methods mentioned in the introductory remarks, and still in use, i.e. acetone production by fermentation of molasses, and by direct oxidation of lean natural gas, represent a small and diminishing fraction of overall production. The second is operated only in the USA and their production represents less than 2 per cent each of the total.

The one remaining source of acetone of any industrial significance is as a co-product of the direct oxidation of naphtha. This process has been developed in a few European countries, and notably by BP Chemicals in the UK. Few process details are available about this route, which produces acetic acid as the main product, and co-produces formic and propionic acids, and acetone. It is believed to be a similar process to that used for lean natural gas oxidation. BP's existing plant at Salt End near Hull has an acetic acid capacity of 90,000 tons per annum and a second plant is planned for the same location. The product distribution can be varied significantly by adjusting operating conditions and changing feedstock quality and ratios. If it is assumed that the process produces acetone equivalent to 10 per cent wt of the acetic acid production the capacity of the 90,000 tons per annum acetic acid plant represents about 5 per cent of the UK acetone market. Acetone production will undoubtedly be minimised in favour of the acids because of the higher netbacks for the latter.

A flow diagram for the reaction and recovery sections of a typical hydrocarbon direct oxidation process is shown in Fig. 6.12. The hydrocarbon feed consists of a mixture of fresh and recycle material, the saturated fresh feed varying from a C_3/C_4 fraction to a C_6/C_8 fraction. The oxidation gas can be air, or oxygen, or a mixture of oxygen and recycled inert gases. The oxygen flow is kept low enough to keep the feed composition well above the explosive limit (generally about 7 per cent volume hydrocarbon vapour in air), and to ensure complete consumption in the reactor as indicated by tail gas analysis. The hydrocarbon/gas mixture is preheated to an adequate temperature to initiate the reaction which is exothermic, so that the reactor exit temperature is significantly higher than the inlet temperature. Feed

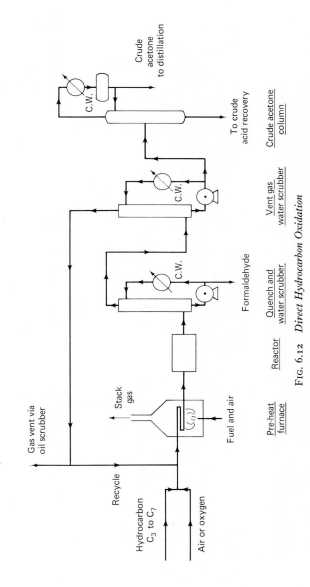

FIG. 6.12 *Direct Hydrocarbon Oxidation*

product-heat exchange is used to minimise heating costs. The operating temperatures and pressures are selected to best suit the process and feed to be used. The reaction can be catalytic or non-catalytic and this has a major effect on the temperature used. For catalytic reactions the temperature range is 130–240°C while for non-catalytic reactions it is 280–380°C. The pressure range is also wide; in general pressure has little effect on the product composition and equilibrium reactions are only involved to a limited extent. The operating pressures can be between 75 p.s.i.g. and 1800 p.s.i.g. The flow regime within the reactor can be vapour, liquid, or mixed phase flow. In general, liquid and mixed phase flow is selected for catalytic processes while vapour phase flow can be either catalytic or not.

The oxygen attack generally occurs either at the second C atom of a linear molecule or at a tertiary C atom in a branched molecule. Thus a predominant component of the reaction is a formic acid/formaldehyde mixture. Dimethyl C atoms generally yield acetone. Aldehydes can be recycled and converted into acids and the higher ketones behave similarly.

The catalysts used vary over a wide range and generally have some specificity. Metals include platinum, silver, and copper. Oxides and organic acid salts include nickel, chromium, iron, manganese, and copper. The catalysts supports can be alumina, silica, or pumice. Catalyst modifiers can also be added including aluminium phosphate and alkaline earth oxides or organic salts.

The reaction has a variable induction period depending on the temperature, pressure, and catalyst, if any. The induction period is a function of the time required to establish free radicals and peroxy radicals. Thus the reactor is a combination of dwell volume to allow the induction period to elapse and a reaction volume providing a given residence time at reacting conditions.

The reaction products form a complex mixture containing CO and CO_2, a wide range of oxygenated products, a major fraction of unreacted hydrocarbon, water, and nitrogen or inert gas if these are in the feed. The products, after heat exchange with the feed, are cooled and condensed and then phase separated into aqueous, organic, and gaseous phases. The gaseous phase is water-scrubbed to remove water-soluble components and oil-scrubbed to remove hydrocarbons before being vented or used as inert gas. The organic phase, together with the hydrocarbon recovered from the scrubbing oil, are recycled. The aqueous phase is separated into the various products by a complex distillation procedure from which non-marketable streams can be discarded or recycled to the reaction stage.

The product recovery section is outside the scope of this section but can involve distillation, azeotropic distillation, extraction, and absorption. The products normally recovered for sale or internal use include formaldehyde, formic acid, acetaldehyde and acetic acid, propionaldehyde, acetone, and propionic acid, and many higher boiling components.

A further route to acetone which is attractive and has been the subject of several patents involves the direct oxidation of isobutyraldehyde or isobutanol to IPA and/or acetone. This type of process does not appear to be commercialised as yet but is attractive since it uses as feeds two materials which are normally overproduced and rejected from C_3 oxo processes. A typical example is for liquid phase, catalysed, air oxidation of isobutyraldehyde at 110–150°C and pressures < 45 p.s.i.g.[3]

6.3.3. The Properties of Acetone

Acetone is a simple (symmetrical) ketone. Formula CH_3–CO–CH_3.

Molecular weight	58·1
Boiling point	56·2°C. Vapour pressure at 20°C. 185 mm Hg abs.
Freezing point	−94·4°C
Critical temperature	235°C. Critical pressure 47 atm.
Flash point (TCC)	−18°C
Explosive limits	2·56–13·0 per cent vol. in air
Specific gravity	0·791 20°/20°
Expansion coefficient	$1·139 \times 10^{-3}/°C$
Colour	Water white
Refractive index $n_D^{20°}$	1·359
Specific heat	0·514 at 20°C
Viscosity	0·32 cp at 20°C
Latent heat of evaporation	125 cals/g at 56°C

Acetone is completely soluble in water in all proportions.

It does not form an azeotrope with water.

It forms an azeotrope with IPE. B.p. 54·2°C. Acetone/IPE = 61·9/38·1 per cent wt.

It forms a temary azeotrope with IPE/Acetone/Water. B.p. 53·9°C. IPE/Acetone/ Water = 42·8/55·3/1·9 per cent wt.

Neither of these azeotropes phase separates on liquefaction.

Acetone is completely soluble in many hydrocarbon and oxygenated solvents.

It is very resistant to oxidation.

It is fairly easily hydrogenated to IPA.

Acetone is condensed in alkaline and acid conditions to diacetone alcohol and mesityl oxide/phorone respectively (see next Section).

6.3.4. Acetone Conversion to C_6 and C_9 Solvents

Acetone's biggest single outlet is as higher molecular weight solvents; the C_6 solvents being derived from the condensation of two molecules of acetone, while (iso) phorone (C_9) is formed by condensation of three acetone molecules. The reactions may be represented by:

$$2CH_3-CO-CH_3 \underset{\longrightarrow}{\overset{\text{Alkali}}{\longleftarrow}} (CH_3)_2COH-CH_2-CO-CH_3 \quad (1)$$
$$\text{Di-acetone alcohol (DDA)}$$

$$(CH_3)_2COH-CH_2-CO-CH_3 \xrightarrow{\text{Acid}} (CH_3)_2C=CH-CO-CH_3+H_2O \quad (2)$$
$$\text{Mesityl oxide (MO)}$$

$$(CH_3)_2C=CH-CO-CH_3+H_2 \xrightarrow{\text{Catalyst}} (CH_3)_2CH-CH_2-CO-CH_3 \quad (3)$$
$$\text{Methyl isobutyl ketone (MIBK)}$$

$$(CH_3)_2CH-CH_2-CO-CH_3+H_2 \xrightarrow{\text{Catalyst}} (CH_3)_2CH-CH_2-CHOH-CH_3 \quad (4)$$
$$\text{Methyl isobutyl carbinol (MIBC)}$$

$$(CH_3)_2COH—CH_2—CO—CH_3+H_2 \xrightarrow{\text{Catalyst}} (CH_3)_2COH—CH_2—CHOH—CH_3 \quad (5)$$
$$\text{Hexylene glycol (HG)}$$

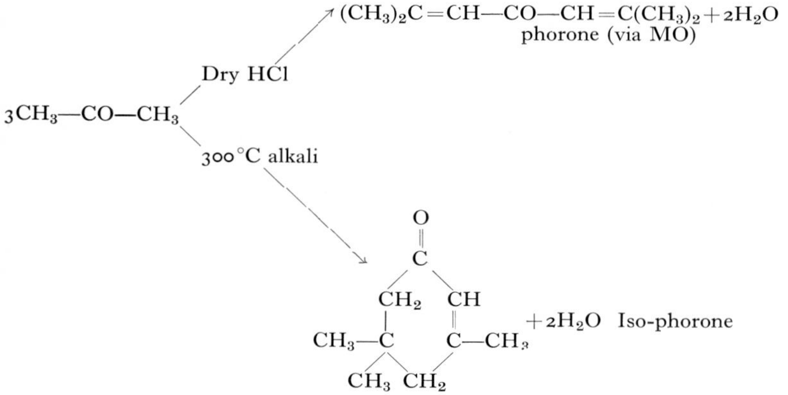

$$(CH_3)_2C=CH—CO—CH=C(CH_3)_2+2H_2O \quad (6)$$
$$\text{phorone (via MO)}$$

Dry HCl

$$3CH_3—CO—CH_3$$

300 °C alkali

$$+2H_2O \text{ Iso-phorone}$$

The first stage is the production of DAA by self-condensation of pure acetone under alkaline conditions. Unfortunately, the equilibrium is unfavourably influenced by rising temperatures and the reaction is exothermic. Furthermore, both acetone and DAA tend to dissolve the alkali catalysts. The laboratory preparation avoids these problems by using insoluble barium hydroxide as catalyst in a refluxing Soxhlet extraction unit. The boiling liquid thus becomes progressively richer in DAA because this material does not distil so the reverse reaction is prevented. Industrially, the technique used is to operate at much lower temperatures where the equilibrium is more favourable, though the reaction rate is correspondingly slower. The catalyst is usually a supported caustic material e.g. soda-lime. Temperatures between − 10°C and 20°C are used with either single or multi-stage catalyst beds. The single catalyst bed requires a lower inlet temperature to achieve a given conversion while the multistage unit can include intercoolers. In either case large beds and long residence times are necessary and the conversion is not usually above about 50 per cent. By-product formation is small.

The reactor product is a mixture of acetone, DAA, and traces of catalyst. To avoid DAA reversion this must be filtered and neutralised before the product is heated. A weak organic acid is used and the pH is carefully controlled at 4 to 5. The product is then stable but the conditions not sufficiently acidic to produce mesityl oxide (see below). The crude product is distilled in a simple vacuum distillation which recovers acetone overhead which is recycled, while the bottom product (crude DAA) is either processed further or topped and tailed by further vacuum distillation to give pure DAA which has the following properties:

Molecular weight	116·2
Boiling point	168°C. Vapour pressure at 20°C 0·8 mm Hg abs.
Viscosity	3·5 cp at 20°C
Specific gravity	0·94 20°/20°
Water solubility	Complete

18

DAA finds limited application as a solvent in the surface coating field. It is also used as a component in hydraulic oils by virtue of its low vapour pressure and low viscosity. Its mutual water/hydrocarbon solubility is used in specialised solubilising applications. It finds some application as a chemical intermediate.

DAA can be hydrogenated under fairly severe conditions to give hexylene glycol (2 methyl pentane 2·4 diol); see equation 5 above. Hydrogenation is normally carried out trickle phase over a fixed supported nickel catalyst bed at about 200°C. The product is purified by vacuum distillation and has the following properties:

Molecular weight	118·2
Boiling point	198°C
Viscosity	38 cp at 20°C
Water solubility	Complete

Hexylene glycol is also used in hydraulic fluid formulations and its solubilising power and low vapour pressure make it a useful cosmetics component.

DAA is however, normally processed via mesityl oxide (MO) to MIBK. The conversion to MO is easily achieved by subjecting the DAA to acid conditions at moderate temperatures. Crude DAA is normally used as reaction feed and the acid conditions are normally achieved by adding to the feed DAA 1–2 per cent wt. of a 1 per cent wt. sulphuric acid solution. The acid is soluble in the DAA so no agitation is required. The reaction is carried out at 100–120°C and is complete in a few minutes. The product MO is again purified by distillation. The product is tailed to remove tarry heavy ends and then topped to remove traces of acetone. The bottom product from the second column consists of MO and water. The aqueous phase is separated and steam stripped to recover MO, while the upper phase consisting mainly of MO is sent for further processing. MO has the following properties:

Molecular weight	98
Boiling point	130°C
Viscosity	0·6 cp at 20°C
Specific gravity	0·86 20°/20°
Water solubility	3·0 per cent wt at 20°C

Mesityl oxide is used mainly as a medium boiling chemical solvent. It also finds minor outlets as a pharmaceutical intermediate, and as a specialised comonomer by virtue of its vinyl group.

Partial hydrogenation of MO gives MIBK while complete hydrogenation gives MIBC; see reactions 3 and 4 above. In practice it is difficult to find conditions which are specific to MIBK and some MIBC is nearly always co-produced. The usual practice to maximise MIBK production is to use only a moderately active catalyst, to use only a small excess of hydrogen, and to control the temperature of the exothermic reaction carefully. The catalyst is normally a supported nickel catalyst and the reactor consists of either a multiplicity of packed tubes in a cooling shell for once through operation, or a single fixed bed when MIBC is recirculated to act as a heat sink. The reaction is carried out in the vapour phase and typical operating conditions are temperatures between 150 and 190°C and pressures between 50 and 100 p.s.i.g.

Higher pressures would favour the hydrogenation but would require higher temperatures which would adversely affect the product split between MIBK and MIBC. Feed/product heat exchange is used to minimise heat loads and the reaction system normally only requires heat input at startup. The crude product is further cooled and condensed and rundown to a product accumulator where excess hydrogen is vented.

In the event that more than the minimum MIBC is required two techniques are adopted. The MO hydrogenation can be carried out under more severe conditions resulting in a higher MIBC/MIBK ratio; the normal method is to increase the hydrogen: MO ratio and increase the reaction temperatures. Alternatively, the crude MIBK/MIBC produced from the first reaction can be subjected to a more severe second hydrogenation using a much higher hydrogen ratio, which allows somewhat higher temperatures and much higher pressures to be used. In this way high conversions to MIBC can be achieved without complicating the recovery system.

The reactor products consist of MO, MIBK, MIBC and water. When the primary product is MIBK the recovery is by distillation in two columns. In the first column light ends are taken overhead while a side stream of MIBK and water is withdrawn, phase separated, and the MIBK recycled as feed. The bottoms (MIBK, MIBC and MO) pass to the second (vacuum) column where MIBK is recovered as top product while MIBC, MO and some MIBK are recovered from the base. When MIBC is the main product two columns are again required, both operating at reduced pressure. The tops of the first column consists of MIBK, MO and water and the bottoms pass to the second column when MIBC is recovered overhead and the heavy ends are recovered as bottoms. MIBK and MIBC have the following properties:

	MIBK	MIBC
Molecular weight	100·2	102·2
Boiling point °C	116	131·5
Viscosity cp at 20°C	0·6	6·1
Specific gravity 20°/20°	0·800	0·806
Solubility in water 20°C per cent wt	2·04	1·9

Generally the market requires more MIBK than MIBC. Their uses are mainly as solvents where MIBK is particularly useful because of its good solvency, especially for resins and lacquers, and its relatively low volatility. MIBK is thus widely used in lacquer surface coatings, particularly those based on nitrocellulose and the vinyl resins. It is also widely used in the chemical and pharmaceutical industries as a process solvent. Another major outlet is as a de-waxing solvent in the de-waxing of lubricating oil base stock, though methyl ethyl ketone (MEK) is often preferred.

Recently (October 1970) Shell announced that a new process route from IPA to MIBK would be used in forthcoming plant expansions.[4] No details of the process have been released as yet but it is stated that two of the normal process steps have been eliminated. This could well lead to reduced manufacturing cost and wider utilisation.

The remaining products are C_9 solvents and are included as an extension of acetone condensation products (see reactions 6 above). Acetone with dry hydrogen chloride gives phorone and MO. Phorone can be recovered by vacuum distillation and MO can be recycled for further condensation. It finds some use as a chemical intermediate and high boiling solvent (b.p. 198 °C). By a similar mild hydrogenation to that of MO however it yields di-isobutyl ketone (b.p. 166 °C) which finds wider use as a high boiling solvent. The condensation of three molecules of acetone can give a very different product, i.e. isophorone. This is produced by catalytic reaction at high temperatures (200–350 °C) and moderate pressures over an alkaline catalyst. The reaction can be liquid or vapour phase and the catalyst is usually a mixture of $Ca(OH)_2$, CaO, and CaC_2. The product isophorone (b.p. 215 °C) is purified by vacuum distillation, and finds uses as a high boiling solvent particularly for nitrocellulose, and as a specialised plasticiser.

6.3.5. Acetone Conversion to Methacrylic Acid Derivatives

6.3.5.1. Monomer production   Acetone is the starting material for the commercial production of methyl methacrylate (MMA) which is polymerised to polymethylmethacrylate (PMMA), more commonly known as 'Perspex', 'Plexiglass', 'Lucite', etc. MMA readily undergoes vinyl polymerisation so that it is frequently used as a copolymer with other vinyl monomers to give a range of properties, which can be further extended by the use of other esters of methacrylic acid. Esters other than methyl are normally obtained by trans-esterification of the methyl ester. The simple chemistry of the reaction to MMA is shown below:

$$CH_3-CO-CH_3 + HCN \xrightarrow[\text{Acetone cyanohydrin}]{\text{Alkali catalyst}} (CH_3)_2COH-C\equiv N$$

$$(CH_3)_2COH\ C\equiv N + H_2SO_4 \xrightarrow{} CH_2=C(CH_3)-CONH_3HSO_4$$
$$\text{Methacrylamide sulphate (MAS)}$$

$$CH_2=C(CH_3)-CONH_3\ HSO_4 + CH_3\ OH \xrightarrow{H_2SO_4} CH_2=C(CH_3)-COO.CH_3 + NH_4HSO_4$$
$$\text{Methyl methacrylate}$$

The ability of acrylic compounds to polymerise has been known since the 1870's. Experimental work to commercialise the process was in hand about 1900 and a major contribution was made by Dr Rohm. Development was slow because of inadequate monomer processes until the early 1930's when three companies (Rohm and Haas with Du Pont in Germany and the USA and ICI in the UK) started producing MMA from acetone. The process currently used was developed by ICI and is illustrated in Fig. 6.13. All stages of the reaction are normally continuous and polymerisation inhibitors are present in all except the first stage. In the first stage hydrogen cyanide (b.p. 26 °C) is reacted with acetone in the presence of an alkali catalyst at fairly low temperatures (15–30 °C). The reaction is an equilibrium and to ensure an approach to completion and maximum fixing of the poisonous HCN an excess of acetone is used. The product acetone cyanohydrin boils at 83 °C at 25 mm Hg. It is reacted immediately with concentrated sulphuric acid in the second stage to give methacrylamide sulphate, the reaction temperature again being below 40 °C. The stable sulphate is flashed to remove water and excess acetone and is then esterified with excess methanol in the third reaction stage to give the methyl ester (MMA) b.p. 100 °C. Water generated during esterification is absorbed in the ammonium salt produced.

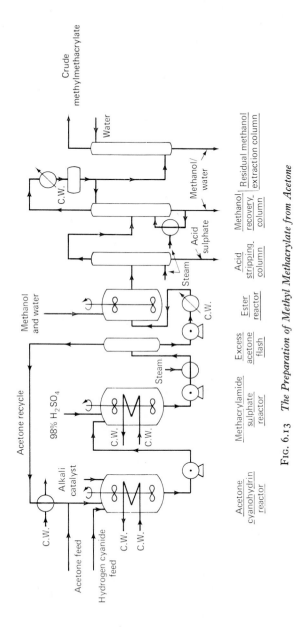

Crude methylmethacrylate

Water

Methanol/ water

Acetone recycle

Methanol and water

98% H₂SO₄

Alkali catalyst

Acetone feed

Hydrogen cyanide feed

C.W.

Steam

Acid sulphate

Acetone cyanohydrin reactor | Methacrylamide sulphate reactor | Excess acetone flash | Ester reactor | Acid stripping column | Methanol recovery column | Residual methanol extraction column

FIG. 6.13   *The Preparation of Methyl Methacrylate from Acetone*

The reaction product is stripped with steam to remove MMA and excess methanol as overheads. The ammonium salts are discarded from the base of the stripper. The overhead vapours are caustic scrubbed to remove traces of acid, condensed and water extracted. The aqueous phase containing methanol is sent to methanol recovery while the MMA contained in the organic stream is distilled to give pure MMA.

Stabilisers are used to prevent polymerisation both in the reaction stages and in storage. Polymerisation is initiated by traces of oxygen or peroxides. Typical stabilisers are hydroquinone and derivatives, and other phenolic compounds, e.g. pyrogallol and 3·5-di-tertiary butyl 4-methyl phenol. During polymerisations the stabiliser content must be accurately known so that allowance can be made when adding the initiator. MMA and its derivatives are used almost exclusively as monomers in polymerisation reactions.

6.3.5.2. Homo polymerisation of methyl methacrylate   Four methods of polymerisation are normally used, bulk, suspension, solution and emulsion. Bulk polymerisation gives the most widely known 'Perspex' type products. The constituents used are the monomer, usually MMA, the initiator, usually a peroxy component, and pigments if required. The premixed constituents are used completely to fill a mould formed by two sheets of plate glass separated by a peripheral flexible gasket. The two sheets are compressed together by springs so that the reactants are always under slight pressure and the shrinkage (about 20 per cent volume) which occurs during polymerisation does not lead to voidage. The reaction is very exothermic (13·5 kcal/mole) and great care is taken to avoid runaway reactions, which can also cause voidage by vapour bubbles. The filled mould is subjected to a temperature cycle starting at about 40 °C and rising slowly to about 75 °C. In this way the polymerisation is satisfactorily initiated, and the temperature is low enough to allow adequate cooling at the maximum exotherm. The heat release to cooling surface area ratio is a direct function of the polymer sheet thickness and must be considered when setting the cycle time. Thus sheets 1/16 in. thick require 8–12 hr to cure, but this can be extended to days for sheets 1 in. thick and over. Bulk polymerisation is also used to produce tubes (by centrifugal casting) and rods.

Suspension polymerisations are carried out by dispersing the monomer in an aqueous medium, and yield easily handled small globules of polymer which are ideal for injection moulding or dissolving in a solvent. They can also be used to prepare ion exchange resins without further physical treatment. Critical factors to give satisfactory suspension (bead) polymerisation include aqueous phase density which is adjusted by adding inorganic salts, to about 1·05 to be mid-way between the monomer and polymer density; correct agitation to give the desired bead size but prevent initial floating or final settling; suspension (thickening) agents, e.g. starches, water-soluble polymers, clays, which prevent globule agglomeration; an initiator which is monomer soluble but is not leached out by water; and additives to increase the interfacial tension and thus stabilise the globules. The reactor is prepared by purging with nitrogen, the aqueous phase is prepared in the reactor and cooled to about 15 °C, and finally the cold monomer containing the initiator and pigments is added. When the dispersion is correct the temperature is raised at a controlled rate to about 65 °C and the polymerisation is complete within 90 min. The bead product is recovered, washed and dried.

Solution polymerisation is also carried out in a stirred reactor with reflux facilities. The desired solvent is charged to a nitrogen-purged reactor and cooled to 15–20 °C. The monomer, containing the correct amount of peroxide initiator is then added and the temperature raised to initiate the reaction. Maximum temperature is easily controlled by setting the pressure so that the solvent refluxes at the correct temperature. The reaction is usually complete in less than 30 min. The solvent can be that required in the final resin solution, or can be chosen such that, on cooling, the resin precipitates and can be recovered, washed and dried.

Emulsion polymerisation also requires a stirred reactor, and is particularly useful where an emulsion is eventually required. Again, the vessel is purged and the aqueous phase prepared and cooled to about 15 °C. The aqueous phase normally contains ferrous sulphate and an emulsifier. The cold monomer containing initiator is added and when the emulsion has the correct consistency the temperature is raised in a controlled cycle to about 70 °C. The polymerisation is complete in a few minutes and after screening to remove agglomerates the emulsion is ready for further processing.

The polymers produced by these methods can have widely varying properties which can be selected to best suit the end use. Bulk polymerisation is considerably less flexible than the others. The two most important variants are molecular weight and tacticity. Molecular weight can be fairly closely controlled by adjusting the temperature and the initiator concentration. Using high initiator concentrations and high temperatures leads to low molecular weight (LMW) products and conversely. LMW products have lower hardness, lower melting and softening points and lower second order glass transition temperatures.

Tacticity (polymer regularity) is mainly a function of the initiator used, and the temperature which controls the reaction rate. Isotactic polymers have their side chains orientated to one side of the main polymer chain or spine. The molecular is not however, linear, the spine taking up a spiral configuration with the side chains radiating from it regularly in the manner of the treads of a spiral staircase. Syndiotactic polymers have adjacent side chains on alternate sides of the polymer spine leading to a less compact and more flexible molecule. Heterotactic or atactic molecules have a random arrangement. Generally, syndiotactic polymers are harder and more brittle, have higher softening and melting points and higher glass transition temperatures. In practice, polymer molecules are neither purely isotactic nor purely syndiotactic and polymer properties are correspondingly intermediate. Peroxy initiators favour syndiotacticity. By special preparation methods ranges of tacticity have been studied for PMMA which indicate second order glass transition temperatures of about 50 °C for the isotactic polymers and 145 °C for syndiotactics.

The already wide range of product properties available in the polymethacrylates is broadened by the use of different ester groupings. The original MMA can be transesterified with excess of another alcohol to give a wide range of esters up to C_{14} (Stearyl MA), and including hydroxy, amino, epoxy, and di-esters. These esters can be likewise polymerised to give the poly-ester methacrylates. The trans-esterification is slightly complicated by the reactivity of the double-bond and is usually carried out at low temperatures and pressures in the presence of an inhibitor, and often with the aid of an inert water-entrainer. Esterification catalysts include sulphuric acid, aluminium tri-iso-propoxide, sodium ethoxide and methoxide and toluene sulphonic acids.

6.3.5.3. Co-polymerisation of methyl methacrylate   Polymers so far discussed have been homopolymers, i.e. using a pure monomer starting material. Co-polymers, using mixed monomer materials are also widely made. The esters discussed above can form the components of the mixed monomer, or these can be esters of acrylic acid. The polymerisation of methacrylic esters is vinyl, so co-polymerisation can obviously be induced with other vinyl monomers. Typical examples include the acrylates, styrene, vinyl acetate, vinyl chloride, ethylene and propylene. Another variation can be introduced by co-polymerising methacrylates with formed or partially formed polymers from other monomers. Thus a second monomer can be added to polymerised MMA and the polymerisation continued, or MMA can be polymerised in the presence of a polymer, these techniques leading to block and graft co-polymers with dual properties. Finally, suitable monomer selection can give a polymer which can be subsequently vulcanised (crosslinked) to produce rubbers and elastomers.

Extensive work has been done to establish co-polymerisation characteristics. The approach is to prepare co-polymers under carefully controlled conditions and estimate the relative 'reactivity rates' of the monomers. Thus if P_{12} is the tendency of monomer 1 and monomer 2 to react, and P_{11} is the tendency of monomer 1 to react with itself, a reactivity rate for monomer 1 can be stated:

$$r_1 = P_{11}/P_{12} \qquad (r_1 \text{ is reactivity within the system of monomer 1})$$

and a similar rate r_2 can be established for monomer 2. These reactivity rates give a good guide to the co-polymerisation limits. More specific properties can be derived from reactivity rates in the form of the Price-Alfrey Q and e values which give the best guide to co-polymerisation properties.

It will be clear from the foregoing that a tremendous range of properties can be achieved by correct selection of the degree of polymerisation, the tacticity, the monomers and their relative concentrations. The properties can be selected to give optimum end use qualities, typical criteria being tensile strength, hardness, low temperature compatability, solubility, water absorption, impact resistance, stress corrosivity, softening point, electrical properties and so on. Detailed discussion of these aspects is beyond the scope of this work but is widely available in the literature. To give some idea of the range of methacrylates available a few of those commercially supplied are listed in Table VI.4.

The uses of poly methacrylates and co-polymers cover a wide range as would be expected from their properties, which can be tailored to suit most requirements. The major outlet is for cast MMA sheet in various forms where its glass-like transparency and weathering resistance, allied to its relatively light weight (specific gravity = 1·19) make it an ideal glass substitute in light fittings, illuminated displays, glazing panels and similar uses. Two grades of cast sheet are available—an aircraft grade which is of very uniform thickness and accurately flat, and a cheaper grade for normal commercial applications. The cast sheet is also ideal for moulding, by heating, and flowing or blowing into moulds. Transparent, coloured, and translucent mouldings of more complex contours can be produced by injection moulding, a technique which is ideal for small instrument cases, light fittings and decorative motifs. Co-polymers offer advantages here where toughness and impact resistance can be improved and material cost is of less significance.

TABLE VI.4

	Boiling point °C*	Heat of polymerisation kcal/mole
Methacrylic acid†	160	15·7
Methyl methacrylate‡	101	13·5
Ethyl methacrylate	118	13·9
Butyl methacrylate	164–169	13·7
Iso-butyl methacrylate	156	14·3
Hydroxyethyl methacrylate	84 (at 5 mm Hg)	12·0
Hydroxypropyl methacrylate	92 (at 8 mm Hg)	12·1
Lauryl methacrylate	270–340§	—
Stearyl methacrylate	310–370§	—
Glycidyl methacrylate	80 (at 15 mm Hg)	—
Ethylene di-methacrylate	96 (at 4 mm Hg)	—

* Physical properties are difficult to establish precisely because of reactivity.
† Other properties, m.p. 15°C   Specific gravity   1·017 20/20
                           Specific heat   0·49 at 20°C
‡ Other properties, m.p. −48°C   Specific gravity   0·941 20/20
                           Specific heat   0·44 at 20°C
   Solubility in water 1·60 at 20°C; 1·43 at 50°C pts/100 pts water.
   Water solubility in MMA 1·15 at 20°C; 1·80 at 50°C pts/100 pts MMA.
   § Depends on purity of alcohol used.

Another major use is in the surface coatings and impregnation fields. The main application is in acrylic paints where methacrylic resins are used both in solvent and emulsion paints. Exceptional colour fastness and weather resistance make them particularly valuable for both interior and exterior use, and they are compatible with most materials, i.e. wood, metal and stonework. Some stoving enamels also incorporate methacrylates. They are also used for coatings on cloth, paper, and other fabrics. Such coatings can be impregnants, true surface coatings, or backings enhancing crease resistance, colour fastness, and antistatic quality. In the expanding field of non-woven fabrics they are used as the fibre binding adhesives. The wearing qualities of natural leather are also improved by impregnation and surface coating to prevent scuffing and improve flexibility. Methacrylates are also increasingly used as adhesives, particularly heat-cured adhesives.

Other minor uses include di-vinyl benzene crosslinked methacrylates which are available as ion exchange resins, and high molecular weight hydroxy resins used as flocculation and flotation agents in water treatment while olefin/methacrylate co-polymers are used as synthetic rubbers. LMW polymers are used as thickening agents in paints and similar products, and polymers of long chain alcohol esters are used as viscosity index improvers in lubricaintg oils when they also enhance detergency and improve pour points.

6.3.6. Other Uses of Acetone

6.3.6.1. As a solvent   The C_6 and C_9 solvents and the methacrylates account for more than half the acetone produced. Another major use, mentioned in 6.3.2.2., is the use of acetone as a chemical solvent in the surface coating industry. Acetone is particularly useful in the lacquer field though its high volatility means it is usually incorporated in a solvent blend of less volatile components. Acetone is also used generally as a chemical solvent in both finished products and in processing; once again individual applications are too numerous to elaborate here but some specialised uses will be mentioned. Acetone is the solvent used in the cellulose acetate spinning process and this was once its major use. It was also a major source of the acetic acid used in making the acetate. However, acetate production has been declining for the last twenty years under pressure from the newer synthetics and this now represents a minor outlet.* Another specialised use of acetone as a solvent is for handling acetylene in cylinders. Acetylene itself, when liquefied is spontaneously explosive, and a concentrated solution in acetone is also dangerous to handle when it can move freely and impact in its container. This problem has been solved by using a cylinder shell packed with a suitable absorbent material, e.g. asbestos wool, which is normally used. The cylinder is then partially filled with acetone and pressured with acetylene which dissolves in the acetone. The solubility is about 25 vol. acetylene/vol. acetone/atm, or about 250 vol. acetylene/vol. acetone at 10 atm. abs. Thus a cylinder at 10 atm. abs (150 p.s.i.a.) contains acetylene equivalent to a pressure of about 1,900 p.s.i.g. if the cylinder is 50 per cent vol. acetone filled. Clearly this requires that the cylinder is operated in the vertical position and that acetone is lost as the acetylene is withdrawn. At 10 atm. abs. the acetylene released contains about 5·2 per cent wt. acetone while at 4 atm. abs. it contains about 12·2 per cent wt. acetone which affects the heating value of the gas and results in significant losses of acetone. About 15 per cent of the stored acetylene is left in the cylinder when it is 'empty'. The weight of the acetone and filler is compensated by the lighter cylinder which can be used. A substitute for acetylene has recently been introduced, as 'Mapp' gas by Dow Chemicals and as 'Apachi' by Air Products Ltd. These gases are similar, consisting mainly of methyl acetylene and propadiene with some propylene and propane. The gas is stable (i.e. does not detonate), can be stored as a liquid and therefore in bulk, and containers can be drained completely before re-charging. Provided no other use is found for these gases (e.g. hydration of methyl acetylene to acetone using dilute sulphuric acid and mercuric sulphate) considerable inroads could be made on this acetylene and acetone market.

6.3.6.2. Use in production of diphenylol propane   The other major uses of acetone are as a chemical intermediate, for example in the production of di-phenylol propane

---

* Methylene dichloride has partly taken its place as a solvent.

(Biophenol A) which is the condensation product of acetone and phenol according to the equation:

di-phenylol propane (DPP).

The condensation is normally carried out in the presence of strong acid (70–75 per cent wt sulphuric, or anhydrous hydrogen chloride) using an organic sulphur compound as catalyst. The major outlets for DPP are as a base chemical in the manu-

facture of epoxy resins by reaction with epichlorhydrin $CH_2Cl.CH.CH_2$ (with epoxide O), and

of polycarbonates by reaction with phosgene (carbonyl chloride, $Cl_2=C=O$).

6.3.6.3. Conversion to ketene

Acetone is also widely converted to ketene which itself is further converted. The reaction is represented by:

$$CH_3—CO—CH_3 \xrightarrow{\text{Thermal cracking}} CH_2=C=O+CH_4$$
$$\text{Ketene (b.p. } -41\,°C)$$

The cracking is achieved by passing acetone vapour through alloy steel tubes at about 700°C with a residence time < 0·5 sec, at a small positive pressure. The ketene yield is 80–90 per cent molar based on 20 per cent converted acetone. Carbon disulphide is used to minimise carbon formation and the products are quenched and absorbed in glacial acetic acid to yield acetic anhydride, thus:

$$CH_3—COOH+CH_2=C=O \longrightarrow CH_3—CO—O—OC—CH_3$$
$$\text{Acetic anhydride}$$

Diketene is sometimes produced. The reaction occurs spontaneously in either acetone or diketene as solvent, and is highly exothermic. As both ketene and diketene are unstable at high temperatures the ketene reaction products are cooled and absorbed in a circulating acetone and/or diketene stream at −10 to 10°C. The main use for diketene is to prepare aceto-acetic acid, and its derivatives via its esters. The reaction is direct and rapid between diketene and an alcohol, e.g.:

$$2CH_2=C=O \longrightarrow \begin{array}{c} CH_2=C \ — \ O \\ | \qquad\quad | \\ CH_2—C=O \end{array}$$
$$\text{Diketene (b.p. } 127\,°C)$$

$$\begin{array}{c} CH_2=C \ — \ O \\ | \qquad\quad | \\ CH_2—C=O \end{array} + CH_3—CH_2—OH \longrightarrow CH_3—CO—CH_2—CO—O—CH_2—CH_3$$
$$\text{Ethyl-aceto-acetate (b.p. } 181\,°C)$$

Ethyl-aceto-acetate or methyl-aceto-acetate are usually prepared directly from the respective alcohols. Other derivatives are prepared by transformation to yield such products as aceto-acetanilide, toluidide, xylidide, chloroanilide, or anisidide. These products find wide applications as intermediates in the pharmaceutical and dyestuffs industries.

6.3.6.4. Conversion to isoprene   The isoprene process reacts acetone and acetylene to give 2-methyl-3-butyn-2-ol, which is hydrogenated to 2-methyl-3-buten-2-ol, this being subsequently dehydrated to isoprene thus:

$$CH_3—CO—CH_3 + CH \equiv CH \rightarrow CH_3—C(CH_3)(OH)—C \equiv CH \qquad (1)$$
$$\text{2-methyl-3-butyn-2-ol}$$

$$CH_3—C(CH_3)(OH) C \equiv CH + H_2 \rightarrow CH_3—C(CH_3)(OH)—CH = CH_2 \qquad (2)$$
$$\text{2-methyl-3-buten-2-ol}$$

$$CH_3—C(CH_3)(OH)—CH = CH_2 \rightarrow CH_2 = C(CH_3) CH = CH_2 + H_2O \qquad (3)$$
$$\text{Isoprene (2-methyl butadiene)}$$

This process has been known for a considerable time having been developed chemically by Chaim Weizmann in the UK during World War II, and subsequently process developed by SNAM in Italy. The first commercial process is now (1971) under construction by SNAM for ANIC at Ravenna with a capacity of 30,000 tons per annum. No details of this plant are available but previously published information indicates the following process conditions. Reaction (1) is carried out at 250–300 p.s.i.g. and 10–50°C, using excess acetylene, and liquid ammonia as solvent. The catalyst is alkaline, preferably sodium or potassium hydroxide, and yields are > 95 per cent molar. The reaction is terminated by neutralising the alkali, the pressure is reduced, and the acetylene, acetone and ammonia separated and recycled, the reaction product being distilled to remove heavy ends. The product is hydrogenated at about 50°C and 90–110 p.s.i.g. using colloidal palladium which is subsequently filtered out and recycled. The final dehydration is vapour phase over alumina at about 290°C and the finished product is water washed and topped and tailed if necessary. Yields are claimed as > 90 per cent molar on acetylene and > 85 per cent molar on acetone.[5, 6]

6.3.6.5. Conversion to methyl-iso-amyl ketone   The process to methyl-iso-amyl ketone uses acetone and isobutyraldehyde (freely available as a propylene/Oxo by-product). No process details are available but it is likely that the route is similar to that used for the C_6 solvent methyl-iso-butyl ketone, i.e. the condensation to the keto-alcohol, dehydration to the unsaturated ketone, and finally hydrogenation to the saturated ketone. The reactions could be:

$$\overset{\text{Alkali}}{(CH_3)_2CH—CHO + CH_3—CO—CH_3 \longrightarrow (CH_3)_2CH—CH(OH)—CH_2—CO—CH_3}$$
Isobutyraldehyde   keto alcohol

weak acid $(-H_2O)$

$$\overset{H_2}{(CH_3)_2CH—CH = CH—CO—CH_3 \longrightarrow (CH_3)_2CH—CH_2—CH_2—CO—CH_3}$$
Unsaturated ketone   Methyl-iso-amyl ketone
                                                 (b.p. 142°C)

The process has been developed by Mitsubishi Chemicals and a 6,000 tons per annum plant is planned by them on Monshu. If the above route is used significant co-production of C_6 products from acetone, and C_8 products from iso-butyraldehyde must be assumed. The process is made attractive by its use of iso-butyraldehyde which is currently produced in excess and therefore fuelled.

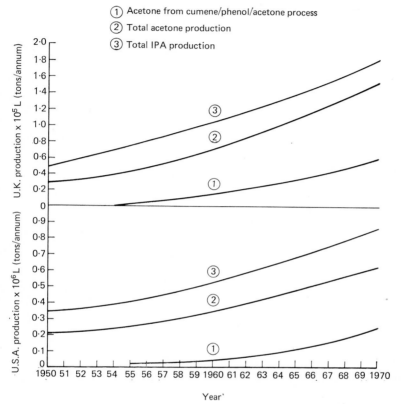

① Acetone from cumene/phenol/acetone process

② Total acetone production

③ Total IPA production

FIG. 6.14   *IPA and Acetone Production 1950–1970, UK and USA*
(*1*) *Acetone from Cumene/Phenol/Acetone Process*
(*2*) *Total Acetone Production*
(*3*) *Total IPA Production*

6.3.7. Production Statistics[7, 8, 9]

The production of IPA and acetone have risen progressively since World War II, as have plant capacities. Since 40–60 per cent of IPA produced is usually converted to acetone, these two products are closely interrelated, and their mutual production capacities are drastically affected by any other processes which produce one product only. This is currently the situation of the cumene/phenol/acetone process, which has developed since 1955 from a minor supplier of acetone to, in some countries, the major source in 1970. The main reason for this rapid build-up of by-product acetone production is the phenol supply situation. Prior to about 1950 the major source of phenol was the coal carbonisation industry. Rapidly increasing demands for phenol arose from the phenolic resin and the Nylon 6.6 industries. Simultaneously, the coal industry was in general declining, and petrochemical routes to phenol were developed, culminating in the cumene process introduced in 1954. The extent to which cumene/acetone production has affected IPA/acetone production is therefore clearly a function of how much chemical phenol was already in production when the cumene route became available; itself clearly a function of the oil/coal energy balance which varied from country to country. Thus, in the USA, the cumene route now produces some 30–40 per cent of the total phenol production, while in European countries this is usually 55–75 per cent, and is even higher in the USSR.

This situation makes it difficult to present generalised production statistics, while productions for individual countries would require extended tabulations, and would also be incomplete through lack of published production data in many cases. The general trends are therefore illustrated by Fig. 6.14 showing production in the USA and the UK of IPA, acetone, and cumene derived acetone. The curves are drawn to show the production trends and ignore low or high (annual) figures caused by temporary market fluctuations and supply inbalances.

The USA figures clearly indicate the declining share of the acetone market met by IPA conversion (direct and indirect). Thus while the incremental tonnage increases in annual production are similar, the percentages are about 4·7 per cent per year for IPA and about 5·8 per cent per year for acetone, and the increased acetone production is almost entirely represented by cumene derived acetone. At the end of World War II about 87 per cent of acetone was produced from IPA, and this increased to about 96 per cent by the early 1950's as a result of declining fermentation production. Since then the figure has been steadily eroded, to about 80 per cent in 1959–61, and about 50 per cent in 1970, while the cumene derived acetone has risen to about 40 per cent between 1954 and 1970.

TABLE VI.5

(a) Chemical process uses	%	IPA uses (b) Physical uses	%
Acetone conversion	35–60	Surface coatings	5–15
Glycerol process	0– 5	Antiseptics, etc.	3– 8
IPA acetate	2– 4	Anti-icing	0– 6
Hydrogen peroxide	0– 6	Cosmetic/Pharmaceutical	1– 4
Other esters and solvents	4–10	Other solvent uses	5–20
Total	50–70	Total	30–50
(a) Chemical process uses	%	Acetate uses (b) Physical uses	%
To C$_6$ and C$_9$ solvents	0–40	Surface coatings	10–30
To methacrylic derivatives	0–20	Cellulose acetate	5–25
To acetic acid, etc.	0–15	Acetylene	4– 8
To DPP and others	0–20	Other solvent uses	10–20

The U K figures show similar trends. Two interesting differences are that acetone tonnage/IPA tonnage has varied from about 0·5 to about 0·85 compared to 0·60 to 0·72 in the USA, and that acetone and IPA production increases are about 7·2 per cent and 6·3 per cent per year respectively, the latter indicating a still expanding market. Between 1950 and 1970 the percentage of acetone produced from IPA has fallen progressively from about 80 per cent to about 40 per cent, while cumene derived acetone has risen to about 38 per cent between 1954 and 1970.

The use patterns of IPA and acetone also vary widely from country to country and again detailed statistics for individual countries will not be quoted. The listings opposite (Table VI.5), however, give the broad patterns of use. More detailed statistics for individual countries can often be obtained from trade literature.

TABLE VI.6

Selling Prices—U S Cents/lb*

	†	UK		Germany		USA	
		Early 1969	End 1970	Early 1969	End 1970	Early 1969	End 1970
Isopropyl alcohol	‡	7	7	5·5	6	7·5	7·5
Acetone	‡	7	7	5·5	5	6·5	6
Glacial acetic acid		8	8	9	10	13	13
Acetic anhydride		11	11	12	12·5	14	14
Cumene		9	9·5	11·5	13	—	—
Di-acetone alcohol		14	14	13	13	13·5	13·5
Synthetic glycerol		23·5	23·5	23	20	24	22
Hexylene glycol		16·5	16·5	17	18·5	15·5	16·5
35% hydrogen peroxide		11·5	12·5	11	12	17	16
Isophorone		18·5	18·5	19·5	20·5	18	18
Isopropyl acetate		10	10	16	15	11·5	11·5
Isopropylamine		42·5	42·5	49	54	31	31
Methyl iso-amyl ketone		24·5	24·5	25·5	28	17·5	17·5
Methyl iso-butyl ketone		13	13	11·5	11	14	11
Phenol		9	9	8	9	9·5	9·5
Crystal polymethyl methacrylate		40	40	36·5	40·5	49·5	49·5
Sulphuric acid 66° Be		1·5	1·5	1·5	1·5	1·6	1·4

* Cents/lb quoted for easy reference to other currencies (to nearest 0,5). Spot delivered prices are quoted for 1 ton quantities or 10–20 tons bulk loads.

† First quarter 1969 and last quarter 1970 averages indicate price stability even during a period of severe inflation.

‡ Acetone and IPA prices have been stable at about these levels for about twenty years.

Table VI.6 shows how prices for isopropyl alcohol, acetone, their derivatives and co-products compare in U K, Germany and the U S A.

6.3.8. Plant Capacities and Costs

IPA plant capacities can vary from 10,000 tons per annum to 120,000 tons per annum depending on the local demand. In the more highly developed countries plants are seldom smaller than 50,000 tons per annum. Plant costs for direct hydration based on average European conditions are estimated at $3, 4·3, and 5·1 × 10⁶ for capacities of 50, 90, and 120,000 tons per annum, but these estimates are based on only very limited available information. Indirect hydration (sulphuric acid process) plants of similar capacities and including acid recovery facilities are estimated at $3·6, 5·4, and 6·6 × 10⁶.

IPA/acetone conversion facilities are normally smaller scale units of about 20,000 tons per annum capacity, and the estimated plant cost is $600,000. For the direct conversion of propylene to acetone by the Hoechst-Wacker Pd/Cu catalysed process, plant costs are estimated at $4·1 and 6·2 × 10⁶ for plants with capacities of 20 and 50,000 tons per annum. These estimates are again based on limited information and installed capacities to date are believed to be < 50,000 tons per annum. Finally, the costs of cumene based phenol/acetone plants, not including the cumene plant, are estimated at $5·8, 8·5, and 10·2 × 10⁶ for plants of 50,100 and 150,000 tons per annum phenol and co-producing 30, 60 and 90,000 tons per annum of acetone.

REFERENCES

1 Thomas, Charles, L., *Catalytic Processes and Proven Catalysts* (Academic Press, London, New York).

2 *Hydroc. Proc.*, 1967, **46**, 137 and 203.

3 Melle Bezons, DP 69–17134.

4 *Chem. and Ind.*, 19 October 1970, p. 1326.

5 *Chem. Eng.*, 9 October 1967, pp. 206–8.

6 *Chem. Eng.*, 14 December 1970, p. 91.

7 *Eur. Chem. News* (*Any issue*).

8 *The Oil, Paint, and Drug Reporter* (*Any issue*).

9 *UK Chemical Industry Statistics*, Handbook 1970, compiled by the Chemical Industries Association.

Refs. 7, 8 and 9 are Production Statistics.

CHAPTER 7

PROPYLENE OXIDE
By A. J. Gait

7.1. INTRODUCTION

1,2-EPOXYPROPANE, commonly called propylene oxide, $\begin{matrix} CH_3CH{-}CH_2 \\ \diagdown\diagup \\ O \end{matrix}$ is the immediate homologue of ethylene oxide; its preparation was first reported by B. Oser, a worker in Wurz's laboratory, in 1860[1] but it was not until the commercial development of ethylene oxide and its derivatives by the Union Carbide company in the USA about 1925 that the possibilities of propylene oxide as an industrial chemical received serious attention. It first began to be regarded as a major industrial chemical soon after the end of World War II and, since that time, production has been established in most industrial countries throughout the world. Production capacity and output have grown rapidly to keep pace with the increasing demand and total world consumption in 1970, excluding the Communist countries, is estimated to be nearly one million tons.

7.2. PROPERTIES

Propylene oxide can exist in the form of two optical isomers and it has been found[2] that, when optically active catalysts are used, stereospecific polymerisation of one or other of the isomers can take place. These reactions are not, so far, of commercial importance and the product used in industry is a racemic mixture. Values for the more common physical properties are given in Table VII.I.[3]

TABLE VII.I

Property	Value
Boiling point at 760 mm °C	34·23
Vapour pressure mm Hg at 25°C	569
Specific gravity at 25/25°C	0·826
Density g/ml at 25°C	0·823
Refractive index n_D^{25}	1·36322[38]
Viscosity at 25°C cp	0·28
Autoignition temperature in air at 760 mm °C	465
Explosive limits in air—vol %	3·1–27·5

The propylene oxide of commerce is a colourless, low boiling, flammable liquid; it is less sensitive and more easily transported and stored than ethylene oxide but it still needs to be handled with some care. All major manufacturers issue instructions and advice for the safe storage and handling of the material and these should be followed. The specification for a typical commercial product is given in Table VII.2.[4]

TABLE VII.2

Property	Specified limit
Propylene oxide % wt	99·0 min.
Ethylene oxide % wt	0·005 max.
Colour (Hazen units)	5 max.
Relative density at 20/20°C	0·829–0,831
Water % wt	0·03 max.
Distillation range at 760 mm Hg	
IBP °C	33 min.
DP °C	35 max.
Acidity (as acetic acid) % wt	0·001 max.
Residue on evaporation % wt	0·001 max.
Aldehyde (as propionaldehyde) % wt	0·004 max.

Propylene oxide is only partially miscible with water and forms a two-layer system in which the propylene oxide content of the upper and lower layers respectively is 89·6 and 40·8 wt per cent at 10°C and 86·6 and 39·9 wt per cent at 24·5°C.[5] The existence of a hydrate, $C_3H_6O.16H_2O$ melting at -3°C has been reported.[6] The oxide forms azeotropes with methylene chloride, ether and a number of the C_5 and C_6 hydrocarbons.[7]

7.3. COMMERCIAL MANUFACTURING PROCESSES

7.3.1. Chlorohydrin Processes

The first synthesis of propylene oxide[1] was by hydrolysis of propylene chlorohydrin with alkali and, until recently, this reaction has remained the basis of all commercial processes throughout the world. The successful development of large scale ethylene oxide production from ethylene chlorohydrin naturally suggested extension of the process to propylene and it was found that virtually identical plant could be used for the conversion of both olefins. It has always been claimed as an advantage for the chlorohydrin process that olefins diluted with substantial quantities of saturated hydrocarbons can be used as feedstocks. This does, however, reduce the efficiency of utilisation of the olefin and it has usually been found in practice that, when the olefin is expensive or in short supply, the extra cost of separating a relatively pure olefin is justified. Mixed olefin feeds have been used, so shifting the problems of separation from the olefins to the oxides.

As direct oxidation of the olefin replaced the chlorohydrin route to ethylene oxide it was quickly found that this process could not be so easily extended to propylene. The superior economics of the oxidation route to ethylene oxide, however, meant that much chlorohydrin production plant became redundant. This provided substantial capacity for manufacture of propylene oxide at virtually no additional capital investment, so keeping production costs low, and undoubtedly stimulated market growth for the oxide and its derivatives. At the present time the chlorohydrin route to ethylene oxide has been almost completely abandoned, but the growth in demand for the propylene homologue has been so rapid that all the redundant capacity has been taken up and it has been necessary to build additional plant specifically for propylene oxide.

One of the earliest processes for manufacture of chlorohydrins depended on the addition of olefin and acid, usually carbon dioxide, to an aqueous solution of sodium or calcium hypochlorite and was probably used in Germany during World War I as a step in the production of mustard gas.[8]

$$NaOCl + CO_2 + H_2O + CH_3CH = CH_2 \rightarrow CH_3CHOHCH_2Cl + NaHCO_3$$

1-chloropropanol-2, as shown in the equation, is the major product but the isomer, 2-chloropropanol-1 $CH_3CHClCH_2OH$, is also formed.

All modern processes are now based on the addition of olefin to a mixture of chlorine and water under carefully controlled conditions, followed by decomposition of the chlorohydrin so formed with alkali. In practice the process may be regarded as taking place in three stages—formation of the chlorohydrin, a decomposition step, usually called saponification, and recovery of pure propylene oxide. A review of the process as carried out in the United Kingdom has been published by Fyvie[6] which suggests the following reaction mechanisms:

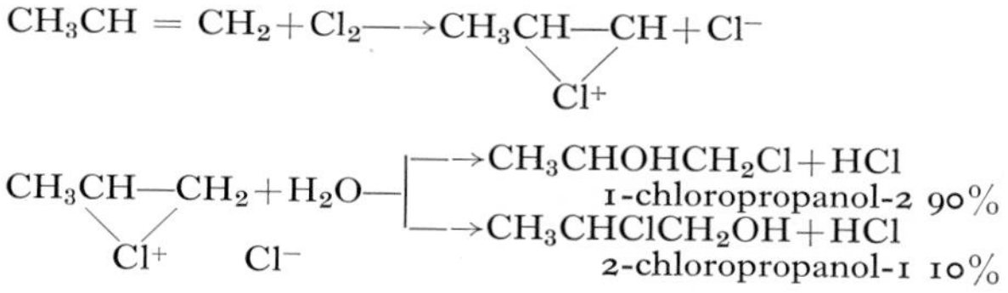

$$CH_3CH = CH_2 + Cl_2 \longrightarrow CH_3CH\!-\!\underset{\underset{Cl^+}{|}}{CH} + Cl^-$$

$$CH_3CH\!-\!\underset{\underset{Cl^+}{|}}{CH_2} + H_2O\underset{Cl^-}{} \left[\begin{array}{l}\longrightarrow CH_3CHOHCH_2Cl + HCl \\ \quad \text{1-chloropropanol-2 } 90\% \\ \longrightarrow CH_3CHClCH_2OH + HCl \\ \quad \text{2-chloropropanol-1 } 10\%\end{array}\right.$$

In addition to the desired chlorohydrin product, a number of chlorinated byproducts are produced. Propylene dichloride, the normal product of direct reaction between chlorine and propylene at low temperatures appears in quantities up to 9–10 per cent of propylene oxide production. Its formation may be represented:

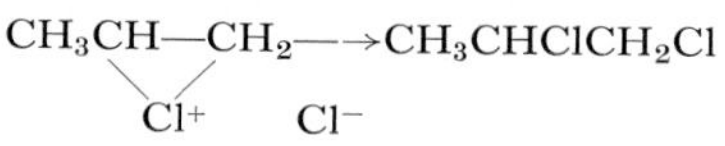

$$CH_3CH\!-\!\underset{\underset{Cl^+}{|}}{CH_2}\underset{Cl^-}{} \longrightarrow CH_3CHClCH_2Cl$$

A mixture of isomers of dichloro-dipropyl ether is also produced up to about 2 per cent of the propylene oxide and these probably arise from propylene chlorohydrin as follows:

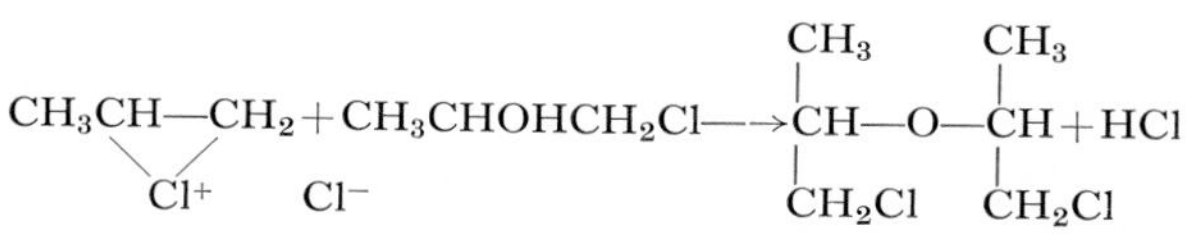

$$CH_3CH\!-\!\underset{\underset{Cl^+}{|}}{CH_2}\underset{Cl^-}{} + CH_3CHOHCH_2Cl \longrightarrow \overset{CH_3}{\underset{CH_2Cl}{\overset{|}{\underset{|}{CH}}}}\!-\!O\!-\!\overset{CH_3}{\underset{CH_2Cl}{\overset{|}{\underset{|}{CH}}}} + HCl$$

Chloroacetone is also produced in small amount through dehydration of propylene chlorohydrin.

These by-products are a waste of expensive chlorine and, in commercial operations, every effort is made to minimise the amounts produced. Formation of the dichloride is limited by ensuring that, as far as possible, chlorine and propylene do not come into direct contact and formation of the dichloroethers, since it is obviously dependent on chlorohydrin concentration, may be minimised by working in dilute solution. Propylene dichloride is only slightly soluble in the aqueous reaction mixture and it is necessary to avoid the formation of a separate dichloride phase in which propylene, chlorine and propylene chlorohydrin will readily dissolve and react. This is an additional reason for working with dilute solutions and commercial processes usually operate with reaction mixtures containing 4–4·5 per cent of propylene chlorohydrin.

7.3.1.1. Chlorohydrination Chlorohydrins may be prepared in batch reactors but continuous operation is the rule in modern large scale plants. Commercial chlorohydrin reactors usually take the form of a tower provided with a chlorine distributor plate at the bottom, an olefin distributor plate about half way up, a recirculation pipe which allows chlorohydrin solution to be recycled from the top to the bottom of the tower, a water feed into the recirculation pipe, an overflow pipe for chlorohydrin solution and an effluent gas offtake pipe. In a version of the process, originally developed by the Societe Carbochimique in Belgium, the tower has a separate leg into which the chlorine is injected as shown in Fig. 7.1.

The propylene and chlorine feeds are set by a ratio flow controller. It is important to ensure that no free gaseous chlorine remains at the point where the propylene feed enters the tower and the recirculation of liquor from the top of the tower helps to dissolve the chlorine; the gas lift effect of the gaseous feeds provides the motive force for the recirculation. Chlorohydrination takes place as the bubbles of hydrocarbon and chlorine solution pass up the main tower and it is also important that no appreciable amount of free chlorine should remain in the effluent gas as this could cause explosive reactions. The effluent gas is scrubbed with caustic soda solution, a sufficient amount is bled off to prevent accumulation of inert gases and the remainder is mixed with fresh propylene and fed back to the reactor. Fresh water is fed into the recirculation leg of the tower at a rate sufficient to maintain the chlorohydrin concentration in the circulating liquor at 4–4·5 per cent and an equivalent amount of chlorohydrin solution overflows to the saponification section. The reactions are exothermic and the temperature will normally settle down at 30–40°C without external heating or cooling.

An important controlling factor in the system, apart from the obvious ones of olefin and chlorine feed rates, is the inert gas concentration in the gas fed to the tower and this, in turn, depends on the purity of the fresh propylene feed and on the amount of effluent gas bled from the system. The inert gas helps to maintain liquor circulation, is an important factor in temperature control and prevents the formation of a separate non-aqueous phase by carrying off substantial amounts of propylene dichloride vapour.

Since the formation of the chlorohydrin is accompanied by the production of an equimolar quantity of hydrogen chloride the tower liquor is strongly acid and corrosive. The first chlorohydrin towers were built of stoneware or of mild steel lined with rubber and ceramic tiles. More recently corrosion resistant reinforced plastics have been found to give good results but their use limits operating pressures to approximately atmospheric.

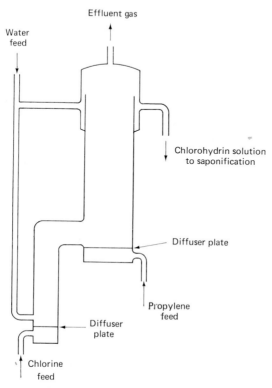

FIG. 7.1 *Reaction Tower for Production of Chlorohydrins*

7.3.1.2. Saponification The chlorohydrin is hydrolysed with the cheapest alkali available, which is normally milk of lime. The solution overflowing from the tower is immediately thoroughly mixed with a 10–15 per cent excess of the alkali. The hydrolysis proceeds according to the equation:

$$\left.\begin{array}{l}CH_3CHOHCH_2Cl\\CH_3CHClCH_2OH\end{array}\right\} + Ca(OH)_2 \longrightarrow CH_3CH\!\!-\!\!CH_2 + CaCl_2 + H_2O$$
$$\underset{O}{\diagdown\diagup}$$

Calcium chloride is also produced, of course, by neutralisation of the hydrogen chloride present in the tower liquor.

Important controlling factors in this part of the process are (1) the rate of hydrolysis, since propylene chlorohydrin forms an azeotrope with water and may be volatilised and lost in the propylene oxide distillate and (2) the rate of removal of propylene oxide, which is rapidly converted to glycol under the alkaline conditions existing in the saponifier, and may thus be lost in the effluent. The earlier plants usually employed fairly elaborate saponifiers, divided into compartments by baffles, through which the chlorohydrin liquor was forced to follow a serpentine path, with steam supplied to each compartment. Modern plants use either a simple stripping tower, with steam and chlorohydrin liquor/milk of lime mixture fed in at the bottom, or a tube reactor, in which the hydrolysis takes place, followed by a stripper. Mild steel may be used for equipment after the point of milk of lime addition.

The rate of hydrolysis is much faster for propylene chlorohydrin than for the ethylene compound and this has been made the basis of a process for using mixed ethylene propylene feedstocks and separating the oxides at the saponification stage.[9] Only sufficient alkali is added to the mixed chlorohydrins to saponify the propylene chlorohydrin in the first instance and the propylene oxide is recovered. Additional alkali is then added for recovery of the ethylene oxide. The separation is only about 90 per cent complete and each oxide fraction contains about 10 per cent of the other.

7.3.1.3. Product recovery The distillate from the saponifier consists of propylene oxide and water with some chlorinated by-products. It is passed to a distillation column from which propylene oxide is taken overhead. The residue is passed through a phase separator for recovery of chlorinated products and thence to a lime recovery system before being sent to the drain. The effluent, which contains about 5 per cent of calcium chloride, some free lime in solution and small quantities of propylene glycol and organic chlorine compounds, may present a major disposal problem. There is about forty times as much effluent as propylene oxide produced.

Chlorinated by-products from the propylene oxide tower, the tail gas washing system and, by carbon adsorption, from the bleed gas are combined and separated by distillation.

7.3.1.4. Economics The propylene chlorohydrin process has developed under specially favourable conditions of capital investment, since it has been carried out on existing ethylene chlorohydrin plant, largely written off, made available by the switch to the direct oxidation route to ethylene oxide. Virtually all of this cheap capacity has been taken up and it is now necessary to meet further increases in demand by building plant specifically for propylene oxide. Although plant for the production of chlorohydrins is relatively simple, by the time provision has been made for storage and handling of chlorine and lime, for lime recovery and for the disposal of effluent it is claimed that, for ethylene oxide, the direct oxidation process can be installed at lower capital cost per unit ton of capacity. There is thus a strong incentive for the development of a direct oxidation process for propylene oxide.

In operating costs the chlorohydrin route to propylene oxide suffers from the same disadvantages which forced the switch to direct oxidation for ethylene oxide. Approximately 1·5 tons of chlorine and 1·1 tons of lime (as CaO) are required per ton of propylene oxide made. None of the chlorine appears in the final product; most of it and all of the lime are discharged from the saponifiers, giving an effluent which is unsuitable for disposal direct into natural drainage and which will not be accepted into municipal sewage systems without expensive pre-treatment. The rest of the chlorine appears in chlorinated by-products, amounting to about 0·1 ton per ton of propylene oxide, for which the market is limited and which may themselves be a disposal problem.

Costs and prices for raw materials and finished products vary so widely according to local conditions and are also liable, at the present time, to change so quickly that actual figures may well be misleading. It may be said, however, that the price for bulk deliveries of propylene oxide in the USA, which has more than half the total world capacity outside the Communist countries, is around 9c/lb, possibly even less, which is equivalent to £84 per long ton. The cost of the chlorine and lime required for one ton of oxide is £45/46 at the present prices; it can be seen, therefore, what an important factor these raw materials are in the overall economics of the chloro-hydrin process.

7.3.2. Oxidation Processes

7.3.2.1. General The operating cost factors discussed in the previous section are so important that, in spite of the favourable capital cost position, it is probable that a switch to an oxidation process would already have taken place if a suitable process had been available. Unfortunately, when the simple catalytic oxidation with air or oxygen, which works so well for ethylene oxide, is applied to propylene the methyl group is preferentially attacked and the major product is acrolein instead of propylene oxide.

$$CH_3CH = CH_2 + O_2 \qquad\qquad CH_2 = CHCHO + H_2O$$

This has become an established process in the USA for the production of acrolein for further conversion to allyl alcohol and glycerol.

In some early work with propylene/oxygen mixtures Lenher[10] found that the first product of oxidation appeared to be a peroxide compound which then decomposed. He found propylene oxide, monopropylene glycol, propionaldehyde, acetic acid, acetaldehyde, formic acid and formaldehyde among the products of the oxidation. Newitt and Mene,[11] using an excess of propylene, obtained similar results but with higher yields of propylene oxide and glycol; they also found allyl alcohol and glycerol among the oxidation products. Many attempts have been made to develop a commercially viable vapour phase oxidation process but the maximum yields of propylene oxide obtainable are around 25 per cent and the yield of acetaldehyde generally exceeds that of propylene oxide; substantial amounts of other aldehydic and acidic products are also formed. Even if the low molal yields of oxide could be accepted, in view of the disadvantages of the chlorohydrin process, the disposal of these by-products would present an almost insuperable obstacle to the commercial establishment of the process. This is in sharp contrast to the direct oxidation process for ethylene oxide where the only by-products are carbon dioxide and water, which can be disposed of to the atmosphere, and a little ethylene glycol, for production of which much of the ethylene oxide is destined anyway.

There is, in fact, one vapour phase oxidation plant for hydrocarbons operating in which propylene oxide is one of the products. This is the propane/butane oxidation process, operated by the Celanese Corporation of America, in which mixtures of air with an excess of hydrocarbon are passed through a reaction chamber at about 400°C and 20 atm pressure. A wide spectrum of products is obtained, for which an ingenious and expensive separation plant has been designed. The main purpose of the operation is the production of the lower oxygenated aliphatic compounds such as methanol, formaldehyde, acetaldehyde and acetic acid. Propylene oxide is one of the minor products, total output being about 10,000 tons annually, and the process could never become a major source of supply for the oxide.

7.3.2.2. Liquid phase processes Vapour phase oxidation processes generally operate at fairly high temperatures and, in an attempt to improve yields of the oxide and avoid excessive by-product formation, many liquid phase processes have been devised which can operate at lower temperatures; a summary of these processes has been published,[12] which shows that they may be conveniently divided into four groups.

(a) *Processes Based on Air or Oxygen*

These are basically adaptations of the uncatalysed homogeneous gas phase oxidations to liquid phase operation. They operate at temperatures below 200°C and at pressures up to 50 atm, with or without catalysts. They give better yields than the simple gas phase oxidations but there is still substantial formation of by-products. Typical examples are the processes described in two patents granted to Escambia Chemical Corporation.[13, 14] In the first, air and propylene are passed into benzene containing some nitrobenzene to inhibit polymer formation. A yield of 26 per cent of propylene oxide and 9 per cent of monopropylene glycol is claimed. The second process is similar except that the nitrobenzene is omitted and manganese propionate is used as a catalyst. In this case the yield is said to be 40 per cent oxide and 20 per cent glycol; in both processes methyl formate, acids and carbon dioxide appear as by-products. There have been many other patents for this type of process but none of them has been developed commercially or appears likely to achieve large scale operation in the near future.

One liquid phase catalytic oxidation process deserving special mention was described by Haruo Shingu in 1957.[15] The oxidation is carried out in inert solvents such as diallyl phthalate, trialkyl or triaryl phosphates, containing a catalyst such as finely divided silver in suspension, at temperatures of 160–180°C. The gas stream, containing propylene and oxygen is intimately mixed with the liquid catalyst system; operating at 160°C Shingu claims a yield of 87 per cent propylene oxide and 10 per cent acrolein. So far as is known, no effort has been made to develop this process on a large scale although, in view of its apparent simplicity and excellent yields, it would appear to be ideally suited to commercial development.

Another process of this type is a two step operation[16] in which a propylene/

cuprous chloride complex is oxidised to form propylene oxide and copper oxy-chloride. The oxychloride is reconverted to cuprous chloride by treatment with hydrochloric acid.

(b) *Electrochemical Processes*

The principle on which these processes work is that propylene is fed to the anode of a cell in which aqueous sodium chloride is undergoing electrolysis, thus forming propylene chlorohydrin. The sodium hydroxide produced at the cathode both saponifies the chlorohydrin and neutralises the co-produced hydrochloric acid so reforming sodium chloride for electrolysis. Hydrogen is produced as a by-product and some propylene dichloride is formed thus causing a loss of chlorine to the system. Some salt make-up is therefore required and steps must be taken to remove the excess caustic soda which would otherwise accumulate in the cell. Both the Kellogg Corporation and Bayer have done work in this field but there is little detailed information available on these processes and their commercial possibilities.

(c) *Oxidation with Hydrogen Peroxide or Per-acids*

It has been known for many years that olefins can be successfully epoxidised with per-acids[17] and, in recent years, several claims have been made that a commercially viable process has been developed. Many of these processes centre round the use of peracetic acid as the oxidising agent. In the two step process, as in the British Celanese patent,[18] the peracetic acid is first formed by vapour phase catalytic oxidation of acetaldehyde with air and is then reacted with excess propylene in an inert solvent such as acetic acid or mixtures of acetic acid with acetone or methylal.

In the one step process the peroxide is formed and used *in situ* by working with mixtures of acetaldehyde and propylene. An I C I patent[19] describes the oxidation of a mixture of acetaldehyde and propylene, in solution in ethyl acetate containing cobalt naphthenate as catalyst, with a gas stream containing 8 per cent oxygen and 92 per cent nitrogen. A yield of 71·5 per cent on reacted acetaldehyde is claimed. Union Carbide are also active in this field and one of their patents describes the formation of a peroxide by oxidation of acetaldehyde at temperatures below 15 °C and its subsequent use in epoxidation of olefins.[20]

A common feature of all these processes is that the peroxide carrier is converted to a final product in approximately equal molal quantities to the propylene oxide produced; in the examples given the co-product is, of course, acetic acid. So far as the author knows, no process of this type has yet achieved commercial operation and limitations of the market for the co-product will almost certainly prevent their providing more than a part of the propylene oxide capacity required. It is known, however, that the technology of using peracetic acid now allows propylene oxide yields of 90 per cent or better on the peracid to be obtained. For companies like British Celanese (now a subsidiary of Courtaulds), I C I or Union Carbide, with large outlets for acetic acid for captive use or sale, a process of this type could be attractive, so that commercial development in the future cannot be ruled out.

It is known that epoxides can be produced in good yield by direct reaction between olefins and hydrogen peroxide. It seems unlikely, however, that such a process could compete economically with either the processes based on peracetic acid described above or with the processes to be described in the next section.

(d) *Oxidation with Organic Hydroperoxides*

Possibly the first large scale use of an organic hydroperoxide in industrial synthesis is the well known air oxidation of cumene to the hydroperoxide as an intermediate for production of phenol and acetone, developed from about 1950 onwards. In the past two decades a vast amount of research into oxidation processes has been carried out and the Scientific Design Company in the USA has been prominent in this field, especially in oxidation of benzene and cyclohexane. This company was also one of the pioneers in production of ethylene oxide by direct oxidation before 1950 and one of the first indications that they had been successful with propylene oxide was a press announcement in 1965 that this oxide could be made by oxidation of propylene with ethyl benzene hydroperoxide.[21] In 1966 it was announced[22] that a new company, the Oxirane Corporation, had been formed jointly by Halcon International, a subsidiary of Scientific Design, and Atlantic Richfield Corporation to exploit the two companies know-how in epoxidation. It was also stated that the new company would build a large scale plant for propylene oxide on the Gulf Coast. The first patents appear to have been granted to Halcon International covering the use of ethyl benzene hydroperoxide and tertiary butyl hydroperoxide[23, 24] and to Atlantic Refining Co., now part of Atlantic Richfield, covering tertiary butyl hydroperoxide,[25] in Belgium and France. These events are the beginning of what may be a fundamental breakthrough in propylene oxide production and patents have now been filed by both Halcon and Atlantic Richfield in many countries to protect their position. More recently, however, patents have been filed by Shell in Belgium[26] and elsewhere which indicate that alternatives to the Halcon/Atlantic Richfield methods may be practicable. It is clear, in any case that Halcon/Atlantic Richfield have a considerable lead in industrial development of their processes.

The present situation is that the Oxirane Corporation plant announced in 1966 came on stream at the beginning of 1968, and is said to be working extremely well, and to have exceeded its design capacity of 71,000 tons per annum. It is planned to erect a second plant on the same site at Bayport, Texas. In 1968 Oxirane set up a Dutch affiliate, Oxirane NV, to build a plant near Rotterdam and this is reported to have come on stream in February 1972. Capacity and cost of the plant were originally given as 72,000 tons per annum and 12·5 million pounds respectively,[27] but a later statement[28] gave the capacity as 155,000 tons per annum and the cost as 16 million pounds (at mid-1972 exchange rates) and also announced plans for a second plant in the EEC area to come on in 1976. It is understood that most of the output is sold to Shell and ICI with a small proportion for the general market. An agreement between Halcon, Atlantic Richfield and Alcudia to build a plant in Spain was announced in 1966. A joint company called Montoro SA was set up but no information is available on capacity or completion date. A joint company called Nihon Oxirane has been set up in Japan in which Halcon and Atlantic Richfield each hold 25%, Sumitomo Chemical 30% and Showa Denko 20%. Production is scheduled to start in 1975 but capacity and cost have not yet been stated.

This technology is so new that there is virtually no information, other than that contained in the patents, on how the processes are actually operated. The Bayport, Texas, plant and the one shortly to come on stream in Holland have been designed to work on tertiary butyl hydroperoxide, starting from isobutane. The chemistry of this route may be shown as follows:

$$2CH_3-\underset{\underset{\text{Isobutane}}{\overset{|}{CH_3}}}{\overset{\overset{CH_3}{|}}{C}}-H + 1\tfrac{1}{2}O_2 \rightarrow CH_3-\underset{\underset{\substack{\text{tert-butyl}\\\text{hydroperoxide}}}{\overset{|}{CH_3}}}{\overset{\overset{CH_3}{|}}{C}}-O-O-H + CH_3-\underset{\underset{\text{tert-butanol}}{\overset{|}{CH_3}}}{\overset{\overset{CH_3}{|}}{C}}-OH$$

$$CH_3-\underset{\overset{|}{CH_3}}{\overset{\overset{CH_3}{|}}{C}}-O-O-H + CH_3CH = CH_2 \rightarrow CH_3CH\underset{\diagdown O \diagup}{-}CH_2 + CH_3-\underset{\overset{|}{CH_3}}{\overset{\overset{CH_3}{|}}{C}}-OH$$

It is believed that the formation of the hydroperoxide takes place without catalyst but that a catalyst, such as a molybdenum salt, is used for the epoxidation stage.

If ethyl benzene is the starting point the hydroperoxide is first formed as above; the second stage of the reaction may be written:

$$\underset{\underset{\text{Ethyl benzene hydroperoxide}}{\overset{|}{CH_3}}}{C_6H_5\overset{\overset{O-O-H}{\diagup}}{CH}} + CH_3CH = CH_2 \rightarrow CH_3CH\underset{\diagdown O \diagup}{-}CH_2 + \underset{\substack{\text{Phenyl methyl}\\\text{carbinol}}}{C_6H_5CHOHCH_3}$$

Overall yields are believed to be about 80–85 per cent on both starting materials. It will be noted that, as in the per-acid based processes, the oxygen carrier, in this case the hydroperoxide, is converted to a final product in equimolar quantities to the propylene oxide produced; markets must therefore be found for these co-products or for derivatives of them. In the examples given tert-butanol may be sold as such for addition to motor gasoline or it may be dehydrated to isobutylene, for which there are large outlets in production of butyl rubber and for alkylation in the oil refining industry; phenyl methyl carbinol gives good yields of styrene by dehydration over a titanium dioxide catalyst at 180–280°C.

The economics of these processes will obviously be closely related to the price obtainable for the co-product. It will be seen that, on a stoichiometric basis, there will be 2·5 tons of tert.-butanol or 1·8 tons of styrene per ton of propylene oxide and the quantities of these materials thrown on to the market by any large scale switch of propylene oxide manufacture to these routes could not fail, in the short term, to bring about price disturbances which would react unfavourably on the process economics. So far the Texas plant, which starts from isobutane, appears to have disposed of its co-products without difficulty and the Rotterdam plant which will use the same route, should also have no difficulty. The Spanish plant and the Japanese plants, if they are built, are understood to use the ethyl benzene route. The markets for both isobutylene and styrene are very large and are growing so that, in the long term, substantial amounts of these materials coming from a new source, such as propylene oxidation, could probably be absorbed without difficulty.

Further, other hydroperoxides are potentially available so that there may be some flexibility in choosing the co-product to be made. For example, Scientific Design have a patent[29] covering the conversion of a secondary alcohol, such as isopropanol or cyclohexanol, to a hydroperoxide which is then reacted with propylene in phthalic anhydride solution to form propylene oxide, a ketone and phthalic acid. The phthalic acid may be readily reconverted to the anhydride and the ketone may be rehydrogenated to the secondary alcohol or sold. In the example given the ketones are acetone or cyclohexanone, for both of which the markets are large and growing. Hydrogenation of the ketones, of course, reproduces the starting material and could equally be applied to isobutylene and styrene, but it is always a relatively expensive process and could well destroy the economic advantages of the new technology.

7.4. REACTIONS OF PROPYLENE OXIDE

7.4.1 General.

The oxirane ring is under strain and reacts readily with many reagents, especially those containing active hydrogen atoms. Reactions with ammonia and with inorganic acids take place at ordinary temperatures without catalyst. Other reactions require elevated temperatures or a catalyst, which may be an alkali metal hydroxide or an acid, such as oxalic acid. The epoxide ring may break on either side giving either primary or secondary alcohol derivatives. Secondary alcohols are favoured by alkaline conditions whereas, under acid conditions, the reaction can go either way.

$$CH_3CH-CH_2 + ROH \begin{cases} \longrightarrow CH_3CHOHCH_2OR \\ \longrightarrow CH_3CHORCH_2OH \end{cases}$$

The products from the reaction of propylene oxide with water, alcohols, and ammonia, the propylene glycols, ethers and polyethers and isopropanolamines respectively, are the most important industrial derivatives of the oxide and are dealt more fully under these headings.

7.4.1.1. Reactions with organic acids The primary reaction is the formation of monoesters of propylene glycol in which either of the hydroxyl groups may be esterified.

$$CH_3CH-CH_2 + RCOOH \begin{cases} \longrightarrow CH_3CHOHCH_2COOR \\ \longrightarrow CH_3CHCH_2OH \\ | \\ OOCR \end{cases}$$

The two isomers are produced in approximately equal quantities in the first instance. In practice the reaction is more complicated since monoesters of dipropylene glycol are also formed and ester exchange reactions take place to give mixtures of monopropylene glycol and its mono and diesters.

7.4.1.2. Reactions with hydrogen sulphide and thio-compounds These reactions are analogous to those with water and alcohols and produce mercaptopropanols and thioethers.

$$CH_3CH—CH_2—\begin{cases} —+H_2S —\to CH_3CHOHCH_2SH + CH_3CHSHCH_2OH \\ —+RSH—\to CH_3CHOHCH_2SR + CH_3CHSRCH_2OH \end{cases}$$

7.4.1.3. Isomerisation Propylene oxide is isomeric with acetcne, propionaldehyde and allyl alcohol and may be converted to these compounds under suitable conditions. Ipatiev and Leontowitsch[30] found that isomerisation with acids or by passing the vapour of the oxide over an aluminium oxide catalyst yielded mainly propionaldehyde with some acetone. With a lithium phosphate catalyst[31] however, almost quantitative conversion to allyl alcohol takes place. In the USA this reaction forms the basis of a commercial route for synthetic glycerol[32] and, if propylene oxide were cheap enough, this route might replace the more usual ones via acrolein or allyl chloride.

7.4.1.4. Reaction with carbon dioxide Carbon dioxide will add on to propylene oxide, in the presence of a quaternary ammonium halide as catalyst, to form propylene carbonate.[32]

$$CH_3CH—CH_2 + CO_2 \longrightarrow \begin{array}{c} CH_3CH—O \\ | \qquad \backslash \\ \qquad \qquad C=O \\ | \qquad / \\ CH_2—O \end{array}$$

The carbonate is a colourless, fairly stable liquid with useful solvent properties especially for vinyl resins, and may be used instead of the oxide in reactions with hydroxy compounds. Its physical properties are shown in Table VII.3.

TABLE VII.3

Physical Properties of Propylene Carbonate

Property	Value
Melting point °C	−49·2
Boiling point °C at 760 mm	241·7
Specific gravity at 20/4°C	1·2057
n_D^{20}	1·4209

7.4.1.5. Miscellaneous reactions Catalytic hydrogenation of propylene oxide converts it to n-propanol with some acetone produced by simultaneous isomerisation.

In the presence of Friedel-Crafts catalysts, aromatic hydrocarbons will add on to propylene oxide to form 1-aryl,2-propanols of the type $CH_3CHOHCH_2Ar$. With aldehydes and ketones acetals and ketals are formed and, with Grignard reagents, compounds of the type $RCH_2CHOHCH_3$.

With halogen acids monosubstituted propanols are formed and, with phosphorus trichloride, one, two or all three of the chlorine atoms can react:

$$PCl_3 + CH_3CH-CH_2 \qquad \begin{array}{l} Cl_2POCH_2CHClCH_3 \\ ClP(OCH_2CHClCH_3)_2 \\ P(OCH_2CHClCH_3)_3 \end{array}$$

Reactions with phosphorus oxychloride and with the chlorides of arsenic, antimony, titanium and silicon are similar. The complex formed with two molecules of propylene oxide and ferric chloride is similar to that with phosphorus trichloride and is a catalyst for the formation of high molecular weight stereo regular polymers of the type described in reference [2].

None of the miscellaneous reactions described has any commercial application so far as is known.

7.4.2. Industrial Uses of Propylene Oxide

There are no known important direct uses of the oxide in industrial applications and it is in the form of its derivatives, polyethers, propylene glycol, polypropylene glycols and propanolamines, that the major uses are found. The polyethers, also referred to as polyols, constitute the largest single outlet and account for rather more than one third of total production. Propylene glycol probably accounts for rather less than one third of total output and the remainder goes into a wide variety of uses, including the propanolamines. These uses are dealt with in greater detail under the appropriate headings below.

7.4.3. World Capacity and Output

Changes are taking place so rapidly in the propylene oxide field that assessment of capacity and output is a matter of some uncertainty. Figures published in 1967–8[33] suggest that capacity in the USA in 1967 was approximately 400,000 tons per annum and that total world capacity in 1966, excluding the Communist countries, was about 620,000 tons per annum.

Since that time there have been many announcements of the intention to build new plants, or to enlarge capacity and there has been a steady switch of existing ethylene chlorohydrin capacity to propylene oxide production, which is not always publicly announced. In addition there have been claims that a direct oxidation process would be exploited commercially, all of which, with the exception of the Oxirane plants in Texas and Holland, have failed to materialise.

Landau and Deprez, of Halcon International/Scientific Design,[34] have estimated the total 1969 production capacity of the three main producing areas—USA, Europe and Japan—as 942,000 long tons for the chlorohydrin process and 71,000 for direct oxidation. Their estimates for 1971 are 1,210,000 tons per annum and 406,000 tons per annum respectively. Attainment of the 1971 estimates clearly depends on the completion of a number of projects already announced. At the present time (mid 1971) there is a considerable slowing down of investment in petrochemicals and several major projects have been abandoned or deferred. There is no doubt, however, that propylene oxide output will continue to grow at a high average rate for some years yet and the estimated outputs seem certain to be reached, although it may be somewhat later than 1971 as suggested.

Production of ethylene chlorohydrin and then of propylene chlorohydrin has tended to be concentrated in companies producing chlorine, where it can be integrated into the chlorine/caustic soda balancing problems but must depend on bought-in hydrocarbons, or in petrochemicals producers associated with the oil industry where it improves the overall economics of hydrocarbon utilisation but must depend on bought-in chlorine and alkali. It is interesting to note that Landau and Deprez still anticipate growth in chlorohydrin capacity.

The largest US producer is Dow Chemical Co., closely followed by Union Carbide and both of these companies have large interests in Europe and elsewhere. In the UK ICI and Shell Chemicals UK are the major producers; BP Chemicals, although a small UK producer, has large production in West Germany through its interest in Erdölchemie. BASF, one of the major West German producers also has a substantial stake in the US market through its interest in Wyandotte Chemical Corporation. It is expected that the US share of world capacity, currently well over 50 per cent, will gradually decrease as capacity in Europe and Japan increases.

It is, perhaps, too early yet to say that a large scale change over to oxidation processes has begun. The tertiary butyl hydroperoxide process has certainly been commercially proven and, if other hydroperoxide processes giving a choice of co-products are equally successful, then the way is open for a large scale switch. The primarily hydrocarbon based producers will certainly wish to be rid of the chlorine based process as soon as possible. The primarily chlorine based producers may find that the decrease in chlorine demand as the chlorohydrin process is phased out may improve or worsen their chlorine/caustic soda balance position so that, for some companies, the chlorohydrin route may continue to be worthwhile for some time to come.

7.4.4. Industrial Derivatives of Propylene Oxide

There are three major groups of products with industrial applications, propylene glycols, polypropylene glycols and allied products and isopropanolamines.

7.4.5. Propylene Glycols

Propylene oxide will react readily with water to form monopropylene glycol:

$$CH_3CH\text{---}CH_2 + H_2O \longrightarrow CH_3CHOHCH_2OH$$
$$\underset{O}{\diagdown\diagup}$$

The reaction is accelerated by traces of alkali or organic acids as catalysts but, in commercial practice, it is carried out without catalyst, under pressure and at temperatures up to 200°C: this avoids contaminating the reaction mixture with catalyst residues and makes the recovery of pure products easier. The reaction does not stop short at the production of the monoglycol; some propylene oxide reacts with the hydroxyl groups of the monoglycol to form di and tripropylene glycols and quite small quantities of higher glycols.

$$CH_3CHOHCH_2OH + CH_3CH\text{---}CH_2 \longrightarrow CH_3CHOHCH_2\text{---}O\text{---}CHCH_2OH$$

with the epoxide $\overset{\diagdown\diagup}{OH_3}$ and side group CH_3

The proportions of mono and higher glycols are controlled by the molar ratio of oxide and water in the initial reaction mixture, the greater the dilution the higher the proportion of monoglycol and the higher the cost of recovering pure products from the solution. Markets for the di- and triglycols are limited and commercial plants usually find the optimum is around 15 moles of water per mole of oxide, when the di and triglycols form about 13 per cent and 1·5 per cent respectively of the total recovered products.

The reaction is exothermic and, for large scale production, is usually carried out in a jacketed pipe system so that the temperature may be controlled. A steady state is attained, with propylene oxide and water being added and dilute mixed glycols withdrawn at appropriate rates to maintain a constant composition in the reactor. The reactor product is first dehydrated in a vacuum evaporator and the water condensate, which contains some glycol, is returned with fresh water to the glycol reactor. The anhydrous mixed glycols are then separated in a multicolumn distillation system to give mono, di and triglycol products and a small amount of residual higher glycols which is discarded.

7.4.5.1. Properties of propylene glycols Monopropylene glycol is a colourless, slightly viscous liquid which has a slightly sweet taste and is miscible with water in all proportions. Its boiling point, 187·3 °C at 760 mm, is lower than that of ethylene glycol (197·6 °C at 760 mm) and it sets to a glass-like solid below -60 °C. Density at 20/20 °C is 1·038 and refractive index n_D^{20} 1·4329.

The higher glycols are produced as mixtures of possible isomers; 3 diglycols are possible, 4 tetra, 10 penta and so on. The three diglycols have been separated and the proportions in which they are produced determined.[35]

$$HOCH_2CH\text{---}O\text{---}CHCH_2OH \qquad CH_3CHCH\text{---}O\text{---}CHCH_3 \qquad CH_3CHCH\text{---}O\text{---}CHCH_2OH$$

with substituents $CH_3 \quad CH_3$ (4%), $OH \quad OH$ (43%), and $OH \quad CH_3$ (53%)

The individual isomers are not important and commercial dipropylene glycol is a mixture of them. It is a colourless liquid, rather more viscous than the monoglycol, boiling at 232 °C at 760 mm and completely miscible with water. The commercial tripropylene glycol is also a mixture of isomers; it is not an important material and the small quantities produced are disposed of to outlets where they are acceptable rather than where the triglycol is a specific requirement. The most notable difference between the propylene glycols and the corresponding ethylene compounds is their relative lack of toxicity. The monoglycol is almost completely non-toxic and, in consequence, has many uses in the food and pharmaceutical industries; the higher glycols are slightly more toxic than the mono compound, but still considerably less so than their ethylene analogues.

7.4.5.2. Reactions of propylene glycols The mono and higher glycols react normally as diols. Esterification with acids produces mono or diesters according to the molar ratio of glycol and acid present in the reaction mixture. With short chain diacids cyclic esters may be produced but, when there are several carbon atoms in the chain between the carboxyl groups linear polyesters are formed. These reactions are often carried out by ester interchange. For example, with diethyl oxalate the monoglycol forms 4-methyl-1·4-dioxan-2·3-dione:

$$CH_3CHOHCH_2OH \quad
\begin{array}{l} C_2H_5OC=O \\ | \\ C_2H_5OC=O \end{array}
\quad
\begin{array}{c} O \\ / \backslash \\ H_2C \quad C=O \\ H \quad | \quad | \\ \backslash \quad C \quad C=O \\ H_3C \; / \; \backslash \; / \\ O \end{array}
\quad +2C_2H_5OH$$

However, when the diacid is of the type $HOOCRCOOH$, where R may be an aromatic nucleus or a chain of CH_2 groups, a long chain linear polyester is formed:

$$nCH_3CHOHCH_2OH + nHOOCRCOOH$$
$$\rightarrow HOHCCH_2(OOCRCOOHCCH_2)_{n-1}OOCRCOOH$$
$$\qquad | \qquad\qquad\qquad\qquad\qquad | $$
$$\qquad CH_3 \qquad\qquad\qquad\qquad\qquad CH_3$$

By the use of an excess of glycol or acid ester chains with both terminal groups either hydroxyl or carboxyl may be produced as required. The polyesters are the most important industrial outlet for propylene glycol and will be referred to again in the section on applications.

Methyl dioxolanes are formed when the glycol reacts with aliphatic aldehydes:

$$CH_3CHOHCH_2OH + CH_3CHO \longrightarrow
\begin{array}{c} H_3C \\ \backslash \\ C-O \\ / \quad | \quad \backslash \\ H \quad | \quad CHCH_3 \\ H_2C-O \end{array}$$

Monoalkyl ethers of monopropylene glycol are formed when the glycol is treated with an alkyl sulphate in the presence of an alkaline catalyst but are more easily prepared by direct reaction of the oxide with the appropriate alcohol, as described later.

7.4.5.3. Industrial applications The largest outlet for monopropylene glycol is in manufacture of unsaturated polyesters. If an unsaturated diacid, such as maleic acid, is used as part or all of the acid component in the manufacture of the polyester, double bonds are introduced into the polymer chains, which may then be crosslinked by the use of a suitable crosslinking agent such as styrene. These unsaturated polyesters are used in large quantities in the manufacture of surface coatings and reinforced plastics, especially those reinforced with glass fibres. Their formulation is a complex business which cannot be treated at length in this chapter. It may be said, however, that the use of a substantial proportion of monopropylene glycol in polyester formulations has been found to confer desirable properties on the end products.

Other important outlets for the glycols are in the formulation of performance products—that is to say products which depend for their effectiveness on physical properties rather than on specific chemical structure. The monoglycol is an excellent solvent and, due to its freedom from toxicity, it is widely used in the food industry as a solvent for food flavourings and colouring matters, as a solvent for drugs and as an ingredient in the formulation of many pharmaceutical and cosmetic preparations. The general low toxicity of propylene glycols makes them suitable for use as lubricants for food processing machinery and as humectants for tobacco and other materials. They are also used as ingredients in hydraulic fluids, mould release agents, alkyd resins and as plasticisers for some resin systems. When used as a substitute for ethylene glycol as a plasticiser for regenerated cellulose film and in anti-freeze formulations, the lower boiling point is a disadvantage and is the cause of slightly increased evaporation losses. In anti-freeze formulations the additional CH_3 group, compared with ethylene glycol, means that a correspondingly increased weight is needed to produce the same lowering of freezing point. Individually the applications listed in this paragraph are small but, there are so many of them that they add up to many tens of thousands of tons annually.

7.4.5.4. World capacity and output Propylene glycol plant is fairly simple and variation in the propylene oxide water ratio in the reaction gives some flexibility in total capacity so that estimates may be subject to errors. It has been estimated that about 30 per cent of the USA output of propylene oxide goes into glycols[36] and this could be expected to produce around 170,000 tons per annum of mixed glycols. The proportion of propylene oxide going into glycols may be somewhat lower in other areas than in the USA and total world output, excluding the Communist countries, is probably in the 300–350,000 tons per annum bracket. This certainly qualifies the propylene glycols to be regarded as major industrial chemicals.

7.4.6. Polypropylene Glycols, Glycol Ethers and Polyethers

This heading covers a large group of products which are used in some form in almost every industry. They all arise from the reaction of propylene oxide with hydroxyl compounds in much the same way as it reacts with water to produce the monoglycol.

7.4.7. Polypropylene Glycols

The formation of di and tripropylene glycols during manufacture of the monoglycol has already been described. The reaction need not stop at this point and, under suitable conditions, propylene oxide will react with the terminal hydroxy groups; as each molecule of the oxide adds on to the chain it always leaves a free hydroxy group which will react with a further molecule of oxide and the process can continue until very high molecular weights are reached. The existence of isomerism in di and tripropylene glycols has already been noted and, as molecular weight increases the number of isomers possible increases rapidly. The polyglycols are thus complex mixtures of isomers but, in practice, variations in isomer composition do not appear to be important.

Manufacture of the polypropylene glycols is carried out batchwise and is started with a small quantity of water or monoglycol mixed with the catalyst, usually an alkali metal hydroxide. Some heating may be necessary to start the reaction as propylene oxide is fed in; once started the exothermic reaction is controlled partly by the rate of propylene oxide addition and partly by external cooling. Checks of average molecular weight are made as the reaction proceeds until the required value is reached. Factors which tend to limit the molecular weight which may be reached are the increasing viscosity of the product as molecular weight rises, which makes proper mixing of the incoming oxide and temperature control difficult, dilution of the catalyst as the volume of the reaction mixture increases and the size of the reactor itself.

Mixed polyethylene/polypropylene glycols may be made by the method described above and these form an important group of products. They may be made from a mixture of ethylene and propylene oxides, when a random co-polymer is formed, or separate block additions of one or other of the oxides may be made so that the polyglycol molecules consist of alternating lengths of ethylene and propylene oxide groups with either ethylene or propylene terminal groups as required.

7.4.7.1. Properties of polyglycols Industrial polypropylene glycols are produced in four standard grades of average molecular weights: 400, 750, 1,500, and 2,000; they all have specific gravities very slightly over 1·000. In contrast with the polyethylene glycols, which are waxy solids from molecular weights of about 600 upwards, they are all liquids which become increasingly viscous with increasing molecular weight. Solid polymers may be made by stereospecific polymerisation with optically active catalysts, as already mentioned, but these are not of commercial importance.

Chemically the polyglycols are diols with the hydroxyl groups at the ends of the molecular chains. The use of an alkaline catalyst ensures that these are almost exclusively secondary hydroxyl groups. Mixed copolymers will have some molecules terminated by the more reactive primary hydroxyl groups from ethylene oxide and block copolymers terminated with ethylene oxide will have exclusively primary terminal hydroxyl groups. The terminal hydroxyl groups may be esterified or etherified by normal chemical methods and many modified polyglycols are produced in this way for specialised applications. The polyglycols are used mainly as performance products and the purpose of the modification may be to bring about a change in physical properties or merely to block off the hydroxyl groups to prevent them reacting with some other ingredient in the final product.

7.4.7.2. Application of polyglycols Straight polyglycols have a limited use in polyurethanes, though polyethers are now preferred (Section 7.4.9.).

A more specific application is to make use of their water-insoluble characteristics to enlarge the hydrophobic part of the molecule in surfactants and the molecule may then be terminated with ethylene oxide to provide the hydrophilic properties required. Special compounds of this type are the Pluronics, produced by the

Wyandotte Chemical Corporation which are block copolymers of ethylene and propylene oxides. The polymer is finished with ethylene oxide to provide hydrophilic properties and the hydrophobic/hydrophilic balance is controlled by the relative amounts of the two oxides used.

7.4.8. Glycol Ethers

When propylene oxide and an alcohol react the primary product is a monoether of propylene glycol.

$$CH_3CH—CH_2 + RCH_2OH \longrightarrow CH_3CHOHCH_2—O—CH_2R$$
$$\underset{O}{\diagdown\diagup}$$

Secondary alcohols and phenols will react in the same way but less readily. The reaction may be catalysed by an alkali, such as caustic soda, or by an acid catalyst such as boron trifluoride. With an alkaline catalyst the isomer with a residual secondary alcohol group is formed almost exclusively but, when boron trifluoride is the catalyst, roughly half of the product may be of the type $CH_3CH(OCH_2R)CH_2OH$.

As with the reaction between propylene oxide and water, the reaction does not stop at the monoglycol ether but some propylene oxide reacts with the residual hydroxyl group to form monoethers of di and tripropylene glycols. Formation of the monoether of the diglycol may be written:

$$CH_3CHOHCH_2—O—CH_2R + CH_3CH—CH_2 \quad CH_3CHOHCH_2—O—\underset{\underset{CH_3}{|}}{CHCH_2}—O—CH_2R$$

The proportion of the monoglycol ether is increased by using an excess of the alcohol and, in commercial production, an economic balance must be struck based on the cost of recovering the excess alcohol and the markets available for the di and triglycol ethers.

7.4.8.1. Properties of glycol ethers   The ethers react as hydroxy compounds and the residual hydroxyl group may be etherified or esterified by normal methods. The ethers are used almost entirely in the formulation of mixed solvents and other performance products and the purpose of such modifications is to change physical properties.

The published values for physical properties vary somewhat, possibly due to the unsuspected presence of isomers. The methyl and ethyl primary ethers (i.e. those with a residual secondary hydroxyl group) of monopropylene glycol boil at 121°C and 133°C respectively at 760 mm. The corresponding secondary ethers have somewhat higher boiling points, viz. 130° and 140°C at 760 mm respectively. The difference between the boiling points of the primary and secondary ethers decreases as the molecular weight of the alcohol increases.

7.4.8.2. Applications of glycol ethers   The lower monoalkyl ethers of mono and dipropylene glycol are excellent solvents though they are not used to the same extent as the corresponding ethylene glycol derivatives.

High molecular weight glycol ethers have been marketed by Dow and Union

Carbide as synthetic lubricants and it is as lubricants, hydraulic, and automotive brake fluids, mould release agents, coupling agents and similar performance products that these products have innumerable uses.

Straight polyglycols as well as ether esters are used for those applications which together absorb some 20 per cent of the world propylene oxide output.

7.4.9. Polyethers

In the production of monoethers described above propylene oxide can continue to add on to the residual hydroxyl group to form an ether of polypropylene glycol. This is one form of polyether in which one end of the chain has a terminal hydroxyl group. If the reaction is started with a di or polyhydric alcohol polyoxypropylene chains will form on each hydroxyl group, and the final polyether will have the same number of hydroxyl groups as the original starting material. Thus polypropylene glycol is a polyether based on a diol, but it is the polyethers based on tri and polyhydric alcohols which have achieved outstanding commercial importance in the past two decades, and which have contributed largely to the phenomenal growth in propylene oxide output. Because they have three or more hydroxyl groups they are often referred to as polyols and their main use is in the manufacture of polyurethanes, a group of high molecular weight materials which, though virtually unknown twenty-five years ago, have now become familiar to everybody through their extensive applications in manufacture of industrial and domestic products. This is primarily a chemical use, depending on the presence of the hydroxyl groups, but the high molecular weight of the materials is also important and the polyurethanes are ultimately performance products. A very brief treatment of their chemistry is necessary in order that the role of the polyethers may be understood.

Urethanes are derivatives of carbamic acid NH_2COOH, which is not known in a free state, and arise from reactions between isocyanates and alcohols:

$$RN = C = O + R'OH \longrightarrow RNHCOR'$$
$$\overset{\|}{O}$$

If such a reaction is carried out with a di-isocyanate and a diol it is clear that long chain compounds may be formed in the same way as a polyester is formed from propylene glycol and a diacid. It is important to note that the active hydrogen of the OH group is not eliminated but becomes attached to the nitrogen of the urethane grouping, where it is still active and may provide a reactive point by which the polyurethane chains can be crosslinked and bring about a still greater increase in molecular weight. This crosslinking is partly promoted by further reaction between isocyanate groups and the active hydrogen of the urethane group. Isocyanates will also react with water to form an amine and carbon dioxide.

$$RN = C = O + H_2O \longrightarrow RNH_2 + CO_2$$

The amine will immediately react with more isocyanate and the complete reaction system becomes quite complicated.

Some of the first polyurethanes were made by reaction between a di-isocyanate (tolylene di-isocyanate) and either polyethylene or polypropylene glycols as diols.

It was found that, if an excess of di-isocyanate and some water were added to the reaction mixture, the small bubbles of carbon dioxide produced by the isocyanate/ water reaction caused the whole mass to foam up while polymerisation continued through the various crosslinking reactions. By suitable choice of reactants and conditions it was found that these foams could be made rigid or flexible and that they had large potential outlets in industry.

It was found that the early foams based on polyether diols had some disadvantages, that propylene oxide was generally preferable to ethylene oxide as a polyether component and that polyethers made from polyhydric alcohols would give superior foams. One of the polyhydric alcohols most readily available was glycerol and it is on glycerol/propylene oxide polyethers that the polyurethane industry has mainly been built. The use of a triol or polyol clearly complicates the processes of chain lengthening and crosslinking outlined above. Control of the properties of the finished product depends on a choice of reactants and reaction conditions so that the processes of chain lengthening, crosslinking and foaming take place in the proper sequence and at the proper rates. The glycerol/propylene oxide adduct may be made by simultaneous reaction between polypropylene glycol, glycerol and di-isocyanate to form a prepolymer, which is subsequently polymerised further, or it may be prepared by direct addition of propylene oxide to glycerol to form the polyether, which is then reacted with the di-isocyanate and other reagents added to control reaction rates and foam properties.

For flexible foams an oxide/glycerol adduct of molecular weight about 3,000 is used while for rigid foams the molecular weight of the adduct is likely to be nearer 300. This provides more crosslinking points per unit weight of adduct and favours crosslinking rather than chain lengthening. Although glycerol is the most widely used polyhydroxy compound, other polyols such as trimethylol propane, pentaerythritol, sorbitol and some sugars are also used to a considerable extent. It is probable that about 35 per cent or nearly 350,000 tons per annum of the propylene oxide world output goes into this application, and the proportion is growing. This isocyanate—diol/polyol reaction is quite general and, in addition to foams, thermosetting resins, elastomers, fibres and surface coating materials may be produced.

7.4.10. Isopropanolamines

This is the third, and least important of the three groups of compounds which account almost completely for the industrial applications. Propylene oxide and ammonia will react spontaneously at room temperature to form 1-amino-2-propanols. One, two or three molecules of oxide will react per molecule of ammonia as follows:

$$NH_3 + CH_3CH-CH_2 \begin{cases} \rightarrow NH_2CH_2CHOHCH_3 & \text{monoisopropanolamine} \\ \rightarrow NH(CH_2CHOHCH_3)_2 & \text{di-isopropanolamine} \\ \rightarrow N(CH_2CHOHCH_3)_3 & \text{tri-isopropanolamine} \end{cases}$$

Unlike triethanolamine, tri-isopropanolamine will not add a further molecule of oxide to form a quaternary ammonium compound.

The reaction is strongly exothermic and is usually carried out in a tube reactor with aqueous ammonia at temperatures up to $100\,°C$ and pressures up to 300 p.s.i.

The presence of considerable quantities of water is said to aid temperature control. The yield of the mono compound is increased by using excess ammonia, with accompanying problems of ammonia recovery. Lowe *et al.*[37] have developed a process requiring only small amounts of water in the system which operates at temperatures of 80–300°C at pressures sufficient to keep the reactants liquid. Ammonia/propylene oxide ratios of 4:1 to 10:1 are used; it is claimed that ammonia ratios above 10:1 show little increase in yield of the mono compound. The products are separated by distillation.

7.4.10.1. Properties of the isopropanolamines   Monoisopropanolamine is a colourless liquid at room temperature while the di and tri compounds are white solids. They are miscible in all proportions with water, alcohol and ether, have a mild ammoniacal odour and are markedly hygroscopic. Their aqueous solutions, like those of the ethanolamines, readily absorb carbon dioxide and hydrogen sulphide.

Two products are normally commercially available—fairly pure monoisopropanolamine boiling at 158–166°C at 760 mm and with specific gravity at 20/20°C 0·959–0·969—and a mixed isopropanolamines fraction. The mixed fraction is the material remaining after recovery of most of the mono and contains 10–15 per cent mono, 40–50 per cent di and 40–50 per cent tri.

7.4.10.2. Reactions of isopropanolamines   Chemically the isopropanolamines react both as alcohols and as amines. For example, when heated with fatty acids to about 150°C amides, amine esters and amide esters are formed:

Amides $CH_3CHOHCH_2NHCOR$

Amine esters $RCOOCHCH_2NH_2$
$$\underset{\text{CH}_3}{|}$$

Amide esters $RCOOCHCH_2NHCOR$
$$\underset{\text{CH}_3}{|}$$

With di-isopropanolamines both mono and diesters may be formed. Tri-isopropanolamine cannot form amides as there is no hydrogen attached to the nitrogen atom. When heated with fatty acids mono, di or triesters may be formed. The triester is of the type $N(CH_2CHOOCR)_3.$
$$\underset{\text{CH}_3}{|}$$

7.4.10.3. Industrial uses of isopropanolamines   Like many other propylene oxide derivatives the isopropanolamines are used largely in formulation of a wide variety of performance products. The mono and diisopropanolamides with fatty acids are among the most important derivatives and have a multitude of uses in the detergent and cosmetic industries as detergents, foam stabilisers, shampoos and similar applications. They are also used to produce water in oil emulsions for pharmaceutical and agricultural preparations and in textile processing.

Alkanolamine soaps give almost neutral solutions in water and hydrocarbons.

Very stable oil in water emulsions may be made with them and the soaps are used in many cosmetic products.

The isopropanolamines have low toxicity and present no special problems in industrial handling except that copper or brass plant and fittings must be avoided and the material should not be passed through rubber hose. Storage tanks should have an inert gas blanket.

7.5. FUTURE OF PROPYLENE OXIDE AND DERIVATIVES

It will be clear from the various sections of this Chapter that manufacturing capacity and output of propylene oxide and its derivatives have grown at a phenomenal rate over the last two decades. Estimates of future growth over the next ten years vary between about 9 per cent and 15 per cent compound per annum. In the author's view the actual figure attained is likely to be nearer the lower limit.

It will also be clear that the growth in output is heavily dependent on two groups of products—unsaturated polyesters and polyurethanes. Worldwide there is still plenty of room for growth in these two fields and, especially in polyurethanes, substantial technological advance is still possible. Moreover there is no obvious competitor at present which seems likely to replace propylene oxide derivatives on a price/performance basis.

It is to be expected that the chlorohydrin process will be gradually abandoned in favour of an oxidation process—either one of those described or one developed from a new breakthrough. This should not, however, affect the growth in output or demand for the oxide and its derivatives.

REFERENCES

[1] Oser, B., *Bull. Soc. chim. (Fr.)*, 1860, 235.
[2] Tsurata, T., Inone, S., Yoshida, N. and Yurukawa, J., *Makromol. Chem.*, 1962, **55**, 230.
[3] Alkene Oxide Brochure, Dow Chemical Co., Midland, Mich., USA.
[4] Shell Chemicals UK Ltd. Specification.
[5] Widert, J. N., Tamplin, W. and Shank, R. L., *Chem. Eng. Prog. Symposium*, 1952, Series 48, 92.
[6] Fyvie, A. C., *Chem. and Ind.*, 7 March 1964, p. 384.
[7] Horsley, L. H., *Advan. Chem. Ser.* 6, 1952, **35**, 1962.
[8] Norris, J. F., *Ind. Eng. Chem.*, 1919, **11**, 817.
[9] Wyandotte Chemical Corporation, USP 2,417,865.
[10] Lenher, S., *J. Am. Chem. Soc.*, 1931, **53**, 3737, 3752.
[11] Newitt, D. M. and Mene, S., *J. Chem. Soc.*, 1946, **97**.
[12] *Eur. Chem. News*, 29 April 1966.
[13] USP 2,780,635.
[14] USP 2,784,202.
[15] USP 2,985,668.
[16] Goldstein, R. F. and Waddams, A. L., *The Petroleum Chemicals Industry*, 3rd ed., p. 161. USP 2,649,463.
[17] Prilschajev, N., *Ber.*, 1909, **42**, 4811.
[18] BP 900,836.
[19] BP 963,430.
[20] BP 735,974.

[21] *Chem. Week*, 12 June 1965, p. 105.

[22] *Eur. Chem. News*, 29 July 1966, p. 20.

[23] Belg.P 657,838.

[24] FP 1,421,285.

[25] Belg.P 680,302.

[26] Belg.P's 748,314–748,318.

[27] *Chem. Age*, 14 March 1968.

[28] *Chem. Age*, 12 May, 1972.

[29] USP 3,251,862.

[30] Ipatiev, W. and Leontowitsch, W., *Ber.*, 1903, **36**, 2017.

[31] Sergeev, P. G., Bukreeva, L. M. and Polkovnikov, A. G., *Khim. Navka i Prom.*, 1957, **2**, 133; *Chem. Abs.*, 1958, **52**, 6150.

[32] Peppel, W. J., *Ind. Eng. Chem.*, 1958, **50**, 767.

[33] Othmer, Kirk, *Encyclopaedia of Chemical Technology*, Vol. 16, p. 604.

[34] *Chem. Age*, 30 October 1970. The reference to '000 tons has been read as a misprint for million pounds.

[35] Sexton, A. R. and Britton, E. C., *J. Am. Chem. Soc.*, 1953, **75**, 4357.

[36] *Chem. Age*, 2 October 1970, p. 13.

[37] USP 2,823,236.

[38] Kobe, R. A., Ravicz, A. E. and Vohra, S. P., *J. Chem. Eng. Data.*, 1, 50, 1956.

CHAPTER 8

PERCHLORINATION OF PROPYLENE

By G. Stahl and J. C. Strini

8.1. INTRODUCTION AND HISTORICAL

WHEN chlorination of propylene is pushed to its limit the C_3 structure breaks down; the main products are carbon tetrachloride and perchloroethylene both important industrially. McBee[1] has coined the word 'chlorinolysis' to describe the type of reaction involving rupture of the molecule. The less picturesque, but better understood term 'perchlorination' will however be used. The reaction can be generalised to other compounds such as propane and 1,2-dichloropropane which show a similar behaviour; numerous references to these products will be necessarily found in the chapter in so far as they can be of use for our purposes. The various products formed in the chlorination of propylene are shown in Table VIII.1.

The complete chlorination of methane to carbon tetrachloride was discovered by J. B. Dumas in 1840.[2] Much later, in 1911 St. Tolloczko and K. Kling[3] carried out the complete chlorination of natural gas; this was repeated by Jones and Allison in 1919.[4] The first possibility of a rupture of the C_3 molecule appears in a thesis by Prins (Delft, 1912) which was devoted to the decomposition of heptachloropropane ($CCl_3CCl_2CHCl_2$) in the presence of aluminium chloride. At 80°C the following two reactions took place:

$$C_3HCl_7 \leftrightharpoons CHCl_3 + C_2Cl_4$$
$$C_3HCl_7 \leftrightharpoons C_3Cl_6 + HCl$$

J. Boeseken and his colleagues[5] developed the work further using aluminium chloride between 65 and 200°C in sealed tubes and then carried out studies by passing vapours of heptachloropropane through tubes heated to 190–410°C in presence of copper, zinc and barium chlorides. The products of the reaction were always the same: perchloroethylene, chloroform and hexachloropropane. Perchlorination of propane with rupture of the C_3 molecule into C_1 and C_2, anticipated from the foregoing work, was actually carried out by J. H. Reilly at Dow in 1933.[6] Ethane or propane containing butane together with chlorine were passed into a bath of fused salts at 250–300°C. At about the same time Grebe, Reilly and Wiley[7] described the conversion of 1.2 chloropropane into perchloroethylene by passing it over porous catalysts impregnated with cupric chloride. Shortly afterwards the first patents appeared:

I. G. Farben Industrie[8] in 1939 claimed the chlorination of propane and aliphatic hydrocarbons, whether chlorinated or not, over active carbon or catalysts with a large surface area at 400–650°C.

Donau Chemie[9] covered in 1942 the chlorination of propane, propylene and other hydrocarbons at 500–600°C preferably in the presence of actinic light.

Dow Chemical Co.[10-13] in 1945, 1948 and 1951 covered the chlorination of propane or methane with diluents at 500–600°C.

McBee et al. published work in 1941[1] on the perchlorination of chloropropane and on that of propane in 1949.[14]

However, industrial development of the process seemed to get going in the fifties judged by the number of patents filed by important companies (Dow,[15] Stauffer,[16, 17] Progil,[18] Ugine,[19] Chempatents Inc.,[20, 21] Solvay,[22, 23] Halcon[24] and I C I[25]).

Complete chlorination of methane and perchlorination of C_3 derivatives is gradually replacing the older processes in which perchloroethylene starting from acetylene,[26-28] and carbon tetrachloride by chlorination of carbon disulphide[29, 30] is produced.

With cheaper raw materials it is advantageous economically but the operation of the process raises problems of safety and technology which are only resolved with some difficulty. It is essentially a problem of keeping under control a highly exothermic reaction. Without suitable dilution of the reagents and without effective heat transfer, the reaction temperature increases very rapidly causing complete degradation of the reaction products into hydrogen chloride and carbon, the latter blocking the equipment. This situation arises from a rapid exothermic reaction which takes place in a more and more restricted area (hot point) so that a sufficient rate of removal of heat can no longer be maintained. Further, mixtures of chlorine and propylene or propane can produce explosions particularly at the moment when they are mixed or during the starting up of the reactor.

In practice the technique of perchlorination has been well mastered and very large units (over 100,000 tons per annum perchloroethylene and carbon tetrachloride) are being operated. The choice of raw materials, methane, propane, propylene or dichloropropane is mainly conditioned by the siting of the plant. It should be noted that various by-products and bleed-off streams from works producing vinyl chloride, 1·1·1-trichloroethane and trichloroethylene are able to play an important part as raw materials; perchlorination is a simple means of upgrading them. Another recent trend[31] is the search for a method of producing trichloroethylene side by side with perchloroethylene and carbon tetrachloride.

8.2. THE REACTION

Perchlorination is generally carried out in the vapour phase, sometimes in presence of a catalyst[6-8, 19] but more often by a purely thermal method at temperatures between 400 and 700°C. Under the latter conditions it has been known for a long time that a chain-radical reaction takes place.[32-34]

Recent work of Tanaka[35] and Shinoda[36-38] have made the various stages of the reaction quite clear. They are:

(1) Progressive chlorination of the C_3 molecule
(2) Pyrolysis and rupture of the chlorinated molecule.
(3) The equilibrium reactions are:

$$C_2Cl_6 \leftrightharpoons C_2Cl_4 + Cl_2$$
$$CCl_4 \leftrightharpoons \tfrac{1}{2}C_2Cl_4 + Cl_2.$$

All together this can be expressed thus:

$$C_3H_6+(2a+2b+3c+3)Cl_2 \rightarrow a\ CCl_4+b\ C_2Cl_4+c\ C_2Cl_6+6HCl$$

Enthalpy of the reaction for the production of carbon tetrachloride alone is:

$$\Delta H_{R\,25}{}^\circ C = -226 \cdot 5 \text{ kcal/mole } C_3H_6 \text{ (exothermic)}.$$

Enthalpy of the reaction for the production of perchloroethylene alone is:

$$\Delta H_{R\,25}{}^\circ C = -142 \cdot 8 \text{ kcal/mole } C_3H_6$$

(see Table VIII.2 for the thermodynamic constants).

To complete the picture the secondary reactions must also be taken into account particularly those leading to the formation of hexachlorobutadiene and chlorobenzenes.

A major part of Shinoda's work is devoted to the perchloration of propane but the results obtained by this author with propylene[38] are similar. Actually, as a result of the dechlorination of chloropropanes at high temperatures, chloropropylenes are rapidly formed so whether propane or propylene is used as the starting material is largely immaterial. This remains true for the 1.2-dichloropropane used sometimes instead of propylene.[12] This chloro-compound is prepared by the liquid phase chlorination of propylene in presence of ferric chloride.

8.2.1. Chlorination of Propylene

The preliminary stages of this reaction have been studied, particularly by Groll and Hearne,[39] Rust and Vaughan,[33] Hearne *et al.*[30] and McBee and Pierce,[41] c.f. also De la Mare and Vaughan[42] and Houben-Weyl.[43] The reaction is as follows:

$$Cl_2+M \rightleftharpoons 2Cl\cdot +M$$
$$CH_2 = CH-CH_3+Cl\cdot \rightarrow CH_2 = CH-\dot{C}H_2+HCl$$
$$CH_2 = CH-\dot{C}H_2+Cl_2 \longrightarrow CH_2 = CH-CH_2Cl+Cl\cdot$$
$$CH_2 = CH-CH_2Cl+Cl\cdot \rightarrow CH_2 = CH-\dot{C}HCl+HCl$$
$$\uparrow \downarrow$$
$$\dot{C}H_2 - CH = CHCl$$
$$CH_2 = CH-C\cdot HCl+Cl_2 \rightarrow CH_2 = CH-CHCl_2+Cl\cdot$$
$$C\cdot H_2-CH = CHCl+Cl_2 \rightarrow CH_2Cl-CH = CH_2Cl+Cl\cdot$$

The mechanism as shown above is in agreement with the experimental results according to which the high temperature chlorination of propylene leads almost exclusively to allyl chloride. Dichloropropane would not appear to be an intermediate product since cracking of the latter gives a mixture of allyl chloride (60 per cent) and *cis* and *trans* 1-chloropropylene (35 per cent).[44] Moreover the existence of tautomeric forms of the allyl radical resonance stabilised, explains why the chlorination of allyl chloride and that of 1-chloropropylene lead to the same ratio of dichloropropylene.[40]

The progressive development of the chlorination is shown in Table VIII.1.

Beyond the trichloro stage there is no certainty about the products and the proportions in which they are formed. The final product of the chlorination process is hexachloropropylene.

TABLE VIII.I

Chlorination of Propylene

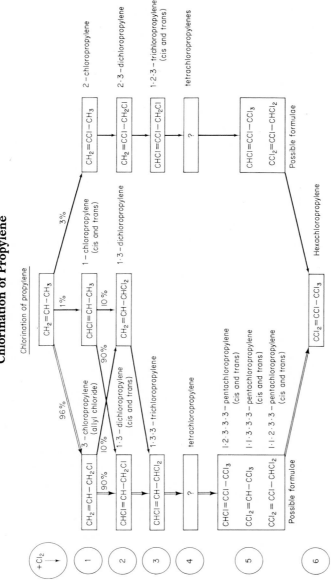

Under the conditions normally used on the large scale (time in the reactor more than 1 second, temperature above 500°C) chlorinated C_3 derivatives disappear except when the reaction is carried out with low ratios of chlorine, low residence times and reduced temperatures, this was shown in the work of Shinoda (Fig. 8.1)[37] and Tanaka.[35] The latter worker obtained at 435°C up to 50 per cent of 'heavy products' such as tetra, penta and hexachloropropylene. It appears therefore that the chlorination of propylene is a very rapid reaction culminating in derivatives which pyrolise readily.

8.2.2. Pyrolysis of Chlorinated Derivatives of Propylene

Krynitsky[45] has been able to convert hexachloropropylene to carbon tetrachloride, C_2Cl_4 and C_2Cl_6 by heating to 400°C. In presence of chlorine the temperature at which cleavage becomes noticeable falls to 300°C.

It is well known that chlorinated derivatives of propane break down more readily the higher the proportion of chlorine. Shinoda[37] has determined the kinetics of the pyrolysis of various penta-, hexa- and heptachloropropanes. Thus according to all probability the tetra- and pentachloropropylenes are pyrolised less easily than hexachloropropylene; it remains to be shown at what degree of chlorination the chloropropylenes begin to break down under the normal reaction conditions.

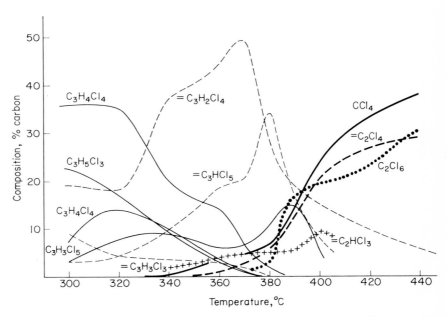

FIG. 8.1 *Shinoda.*[37] *Chlorination of Propane. Product Composition* vs. *Temperature Space Velocity = 300 ml h/mole*

The curves in Fig. 8.1 show that the pentachloropropylenes are not stable above about 400°C for a residence time of 20 sec and that there is no hexachloropropylene. It can be concluded that cleavage functions mainly at the pentachloropropylene stage. Further, the presence of certain by-products such as $CHCl = CCl_2$, $CHCl_3$[35, 37] also shows that cleavage begins before all the hydrogen atoms have been replaced. It can readily be appreciated that trichlorethylene and chloroform, easily convertible to C_2Cl_4 and CCl_4 are formed particularly where low ratios of chlorine are present, the reaction times are very short or where the reaction temperatures are low. (Fig. 8.1.) Under such conditions Tanaka has been able to obtain yields of 30 per cent of trichloroethylene which can be very valuable for industrial production of trichloroethylene starting from propylene or propane. A recent patent by Solvay[31] reinforces this idea.

On pyrolysis the C_3 molecule breaks down into one C_1 radical and one C_2; however, the radicals so formed can recombine (e.g. $2 \dot{C}Cl_3 \rightarrow C_2Cl_6$) and equilibria set up between C_2Cl_4, C_2Cl_6 and CCl_4 modify the picture in such a way that the pyrolysis only yields a 50/50 mixture of C_1 and C_2 if the residence time is sufficiently short and the temperature only moderate. Under these conditions a large excess of chlorine prevents the C_1 radicals from recombining without however appreciably influencing the equilibrium reactions.

Thus McBee[1] by perchlorination of chloropropanes at 460°C with 8 sec residence time obtained 46 per cent of carbon tetrachloride. Shinoda[37] obtained 50 per cent of carbon tetrachloride at very low residence times starting with molar ratios of Cl_2/C_3H_8 of 30. Pyrolysis of hexachloropropylene at 350–400°C also yields 50 per cent of CCl_4.[45]

8.2.2.1. Formation of by-products   The by-products most frequently mentioned in the literature are hexachlorobutadiene and hexachlorobenzene. The former, doubtless arises from reactions of the C_2 radicals but the origin of the latter is less obvious. Schmeisser[46] has found dichloroacetylene side by side with C_4Cl_6 and C_6Cl_6 from the pyrolysis of CCl_4 and C_2Cl_4 under vacuum. Further, it is known[35] that pyrolysis of trichloroethylene at temperatures approaching 660°C yields appreciable quantities of hexachlorobenzene; hence trichloroethylene and dichloroacetylene could be intermediates in the formation of hexachlorobenzene. The production of these heavy compounds is reduced when the reaction is carried out in excess of chlorine[35, 36] at moderate temperatures[11, 35, 36] or in the presence of substantial quantities of oxygen.[38] The degradation reaction to carbon is avoided by using diluants, e.g. C_2Cl_4, CCl_4 and chlorine and by limitation of the temperature of reaction to below 650°C.

8.2.3. Equilibrium Reactions

Equilibrium reactions between perchloroethylene, hexachloroethane and carbon tetrachloride have been known for a long time. Thus in 1821 Faraday[47] prepared perchloroethylene by heating C_2Cl_6. In 1839 Regnault[48] studied the decomposition of CCl_4 and obtained C_2Cl_6 besides other substances such as C_2Cl_4, C_2Cl_2, C_6Cl_6 and carbon the proportion of carbon increasing with rise in temperature. Weiser and

Wightman (1915), then Strosacker and Schwegler[50] (1933) became interested in the conversion of CCl_4 into C_2Cl_4 and C_2Cl_6. Wimmer[51] on the other hand worked on the production of CCl_4 by heating perchloroethylene with a large excess of chlorine.

Since the equilibrium reactions above are factors affecting the composition of final products in numerous perchlorinations of hydrocarbons it is important to examine their equilibrium constants and their kinetics in more detail. Some of the thermodynamic functions of the compounds concerned are shown in Table VIII.2.

8.2.3.1. The reaction $C_2Cl_6 \leftrightharpoons C_2Cl_4 + Cl_2$ While some work has been done on the reaction by McBee *et al.*[52] and Dainton and Ivin,[53] a thesis by Puyo[54] supplies the most complete information on the subject.

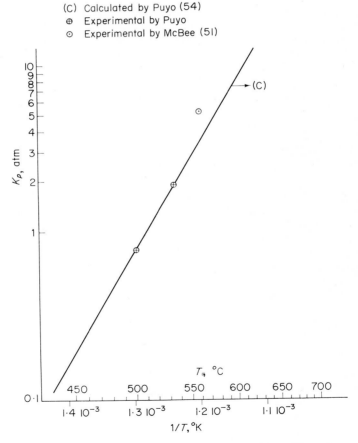

FIG. 8 2 $C_2Cl_6 \leftrightharpoons C_2Cl_4 + Cl_2$. *Equilibrium Constant vs. Temperature*

TABLE VIII.2
Thermodynamic Functions

Compound	Cl_2	H_2	C	C_3H_8	C_3H_6	$C_3H_6Cl_2$	CCl_4	C_2HCl_3	C_2Cl_4	C_2Cl_6	C_6Cl_6
State	diatomic gas	diatomic gas	solid	gas	gas	gas	gas	gas	gas	gas	gas
$Cp°$ heat capacity, cal/mole °K											
298 °K	8·111	6·892	2·066	17·57	15·27	23·47	20·02	19·17	22·69	32·59	41·9
800 °K	8·878	7·078	4·75	37·08	30·68	41·97	25	26·94	29·29	41·29	59·96
900 °K	8·922	7·139	4·98	39·61	32·7	44·18	25·29	27·61	29·73	41·76	61·31
$S°$ entropy, cal/mole °K											
298 °K	53·289	31·211	1·361	64·51	63·8	84·8	74·12	77·63	81·46	94·77	105·45
800 °K	61·747	38·108	4·74	90·95	85·89	116·92	96·75	100·66	107·48	131·96	156·45
900 °K	62·796	38·946	5·314	95·47	89·73	122	99·72	103·88	110·96	136·85	163·6
ΔH_3 f enthalpy of formation, kcal/mole											
298 °K	0	0	0	+24·82	4·88	-39·6	-24	-1·4	-3·4	-33·8	-8·1
800 °K	0	0	0	-30·11	0·77	-42·87	-22·79	-1·29	-2·23	-31·2	-5·4
900 °K	0	0	0	-30·58	0·35	-43·04	-22·54	-1·23	-2·03	-30·69	-4·93
$\Delta G_f°$ free energy of formation, kcal/mole											
298 °K	0	0	0	-5·61	14·99	-19·86	-13·92	4·75	4·9	-13·58	10·56
800 °K	0	0	0	30·45	34·82	15·83	2·4	15·11	18·17	19·02	40·39
900 °K	0	0	0	38·05	39·1	23·18	5·53	17·16	20·71	25·26	46·08
ΔH_v enthalpy of vaporisation at boiling point, kcal/mole	1·562	0·216		4·487	4·402	7·59	7·17	7·52	8·3	12·2 sublimat.	23·2 sublimat.

* From D. R. Stull, E. F. Westrum and G. C. Sinke, 'The Chemical Thermodynamics of Organic Compounds' (Wiley & Sons, 1969). $\Delta S_f°_T$ (standard entropy of formation, T °K) $= S°_T$ (standard entropy of compound) $- \Sigma S°_T$ (standard entropy of elements). $\Delta G_f°_T = \Delta Hf°_T - T \Delta S_f°_T$. Standard state for gas = ideal gas state at 1 atm.

Enthalpy of the reaction Puyo has calculated this enthalpy to be 31·9 kcal/mole (endothermic); this figure is in excellent agreement with that calculated starting from experimental equilibrium constants and the variation in the standard entropy ($\Delta H_{25} = -4·57 \log Kp - T \Delta S_{25}$).

Equilibrium constant Puyo has determined experimentally two values of the equilibrium constant shown in Fig. 8.2. Thermet[60] has established for

$$K_p = \frac{Pc_2cl_4 \cdot Pcl_2}{Pc_2cl_6}$$

a relation starting from the enthalpies of formation and the entropies of the compounds; the relation found:

$$\log K_p = \frac{7000}{T} + 9·5$$

is in agreement with the line passing through the experimental points.

Kinetics of the reactions The experimental relationship established by Puyo for the speed of the reaction $C_2Cl_6 \rightarrow C_2Cl_4 + Cl_2$ is the following:

$$v = 2·5 \times 10^7 e^{\frac{-45,000}{RT}} \quad Pc_2cl_6^{2/3}$$

$$\text{(mole/l/mm)} \qquad\qquad\qquad \text{(mm Hg)}$$

Thus by studying the decomposition of pure C_2Cl_6 at a pressure of 100 mm Puyo has confirmed that the concentration of C_2Cl_6 for a reaction time of 10 sec is 11 per cent at 503°C and only about 0·5 per cent at 590°C. This reduction shows a very rapid displacement of the equilibrium towards the formation of perchloroethylene when the temperature is increased so that under the usual conditions of perchlorination (temperatures above 500°C and residence time a few seconds) very little hexachloroethane can be detected.

A rapid quench of the gas and exclusion of light is necessary, otherwise the equilibrium reverts towards hexachloroethane.[21, 49, 55]

According to Thizy[56] the velocity of chlorination of C_2Cl_4 and C_2Cl_6 is of the first order with reference to chlorine and the energy of activation from experimental data is 15·4 kcal/mole. The difference between the figure and the energy of activation of the reverse reaction obtained by Puyo (45 kcal/mole) gives a value of 29·6 kcal/mole for the enthalpy of the reaction which is in fair agreement with the value of 31·9 kcal/mole cited above and obtained through the enthalpies of formation.

8.2.3.2. Reaction $CCl_4 \rightleftharpoons \frac{1}{2}C_2Cl_4 + Cl_2$

Enthalpy of the reaction at 25°C The value calculated from the most recent enthalpies of formation[57] of C_2Cl_4 and CCl_4 (Table VIII.2) is $+22·3$ kcal/mole; the decomposition of CCl_4 is endothermic and is thus favoured by rising temperature.

Equilibrium constant   The first equilibrium constants had been calculated by Fink and Bonilla[58] using older experimental results of Weiser[40] (1919). These figures, very different from those obtained by more recent observations are no longer of interest. Khorshak[58] studied the decomposition of CCl_4 and C_2Cl_6 but his paper does not give the complete composition of the reaction product and so does not enable the equilibrium constants to be calculated. In Fig. 8.3 we give various relations in graphical form between

$$K_p = \frac{P_{C_2Cl_4}^{\frac{1}{2}} . P_{Cl_2}}{P_{CCl_4}}$$

and the temperature:

(a)   a curve (P) of which the equation is $\log Kp = \frac{4\ 430}{T} + 4 \cdot 05$ obtained from the

values of Kp calculated by Puyo[54] starting from enthalpies of formation and standard entropies.

(b)   a curve (T) with equation $\log Kp = -\frac{5\ 125}{T} + 4 \cdot 4$ put forward by Thermet

starting from slightly different thermodynamic data.

(c)   a curve (S) corresponding to Kp values calculated by Shinoda[38] also starting from thermodynamic data.

It appears that these three relations give very different values of Kp; no published work exists enabling us to choose between them. Fig. 8.3 shows values of the ratio

$$r = \frac{P_{C_2Cl_4}^{\frac{1}{2}} . P_{Cl_2}}{P_{CCl_4}}$$

obtained from experimental data of various workers using long residence times and high temperatures. Within these conditions it is possible to estimate approximate figures for the equilibrium constant. The points corresponding to temperatures above 730 °C from the experiments by Puyo on the decomposition of hexachloroethane have been retained. Khundkar and Mullick[61] have described the thermal decomposition of carbon tetrachloride by a continuous method without stating the residence time in the reactor. However, the experimental results show that an increase in the flow does not alter the proportion of carbon tetrachloride decomposed, this proportion is only affected by the temperature. The values of r have been calculated starting from the quantity of CCl_4 decomposed by assuming that the conversion of C_2Cl_4 to C_2Cl_6 has taken place outside the reaction tube.

Other values of r from Fig. 8.3 have been obtained from Shinoda's experiments[37] in which the residence time is longer than 10 sec.

Finally, the shaded area confines the greater number of the experimental points near the equilibrium and restricts the probable limits of variation of Kp as a function of the temperature.

Influence of the pressure on the equilibrium:

$$Kp = \frac{P_{C_2Cl_4}^{\frac{1}{2}} . P_{C_2Cl_4}}{P_{CCl_4}} = P^{\frac{1}{2}} \frac{Y_{C_2Cl_4}^{\frac{1}{2}} . Y_{Cl_2}}{Y_{CCl_4}}$$

$P_{C_2Cl_4}$ etc. $=$ partial pressures of these compounds.

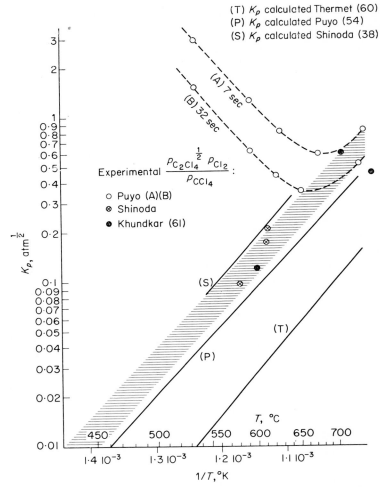

FIG. 8.3 $CCl_4 \rightleftharpoons \frac{1}{2}C_2Cl_4 + Cl_2$. *Equilibrium Constant* vs. *Temperature*

P = total pressure.

$Y_{C_2Cl_4}$ etc. = molar fraction of each component.

When the total pressure increases the formation of CCl_4 is favoured in conformity with the Le Chatelier principle.

Velocity of the reaction No complete equation for the kinetics of the reaction has been published; however by examination of the change in the value of the ratio r with temperature and residence time it becomes clear that the equilibrium reactions are much slower than those involved in the chlorination of propylene previously described. Thus, the curves A and B of Fig. 8.3 drawn from the work of Puyo make it clear that a reaction time as long as 32 sec at 530°C is not sufficient to reach equilibrium, in other words the value of r falls when the temperature rises from 530–630°C and it is only above 650°C that an increase with temperature is noticeable which enables r to be integrated with the equilibrium constant. The work done by Shinoda shows that 15–20 sec residence time at 610°C is needed to approach equilibrium.

Thermet[60] put forward the equation $v = k(C_2Cl_4)(Cl_2)^{3/2}$ for the speed of formation of CCl_4 under conditions far removed from equilibrium. He did not however, define the values of k. According to that author the following mechanism serves to explain the equation:

$$\text{(a)} \quad C_2Cl_4 + Cl_2 \rightleftharpoons C_2Cl_6$$
$$(C_2Cl_6) = K_a (C_2Cl_4)(Cl_2)$$

A state of equilibrium is rapidly reached.

$$\text{(b)} \quad Cl_2 \rightleftharpoons 2\ Cl\cdot$$
$$(Cl\cdot) = K_b^{\frac{1}{2}} (Cl_2)^{\frac{1}{2}}$$
$$\text{(c)} \quad C_2Cl_6 + Cl\cdot \rightarrow CCl_4 + \dot{C}Cl_3$$
$$\dot{C}Cl_3 + Cl_2 \rightarrow CCl_4 + Cl\cdot$$

When the system is stationary the speeds of the two reactions are identical (Vc). It follows that the speed of formation of CCl_4 is:

$$v = 2\ v_c c = 2\ k_c(C_2Cl_6)(Cl\cdot) = 2\ k_c\ K_a\text{–}K_b^{\frac{1}{2}} (C_2Cl_4)(Cl_2)^{3/2}$$

8.3. THE INDUSTRIAL PROCESSES

In working out an industrial process there are two main aims to be achieved:

(1) to bring about the maximum yield and the highest purity of the finished products.
(2) to produce carbon tetrachloride and perchloroethylene in the desired ratio according to the demands of the market.

Chlorinated C_3 by-products are easily avoided by operating at above 450°C with an excess of chlorine. Trichloroethylene and chloroform in the product indicate insufficient residence time and insufficient excess of chlorine. It is more difficult to avoid the formation of heavy products such as hexachlorobutadiene and hexachlorobenzene which are the cause of the more important losses in yield. These losses can be limited by operating at moderate temperatures and with a sufficient excess of chlorine.

The carbon tetrachloride and perchloroethylene ratios obtained are fixed by the equilibrium reaction and the final preparation is a function of the factors regulating the equilibrium and the kinetics of this reaction. This is true whatever the starting material; however, when this is a C_3 compound the 50–50 ratio is easily obtained without interference from the equilibrium reaction by operating at relatively low temperatures and short residence times. If it is desired to produce more perchloro-ethylene an increase in temperature both displaces the equilibrium in the desired direction and increases the overall rate of the reaction. The operation is carried out with only a slight excess of chlorine in the presence of carbon tetrachloride. If on the other hand carbon tetrachloride is desired as the main product of the reaction a low temperature is favourable from the standpoint of thermodynamics, but the reaction velocity is low so that a greater residence time is necessary. In practice the operation is carried out at high temperature and the equilibrium is shifted by using excess of chlorine and operating in presence of perchloroethylene. Increase in pressure also tends both to shift the equilibrium in the direction of carbon tetrachloride and to speed up the reaction by increasing the partial pressure of the reactants.

To minimise the formation of hexachloroethane through the equilibrium $C_2Cl_4 + Cl_2 \rightleftharpoons C_2Cl_6$ a rapid quench of gas must be carried out immediately after the reaction.

There are some thirty patents covering the various possibilities of carrying out the industrial process such as the choice of reaction system whether catalytic or purely thermal and the means of separating the products.

8.3.1. The Reaction System

8.3.1.1. Catalytic reactions

Catalysis in the liquid phase   The catalyst consists of a mixture of molten salts. This type of catalyst has been studied mainly by Dow Chemical 9.[7, 10] In the first patent[7] the C_3 hydrocarbon and the chlorine react by passing in the gaseous state into a bath of molten metallic salts consisting of a mixture of 60 parts by weight of aluminium chloride, 30 of sodium chloride and 10 of ferric chloride at a temperature of 380°C. In the second patent[10] the perchlorination is carried out in two stages to avoid production of highly exothermic reactions (risk of explosion). In the first stage the hydrocarbon is chlorinated in the liquid phase in the presence of actinic light until tetra or pentachloropropane has been formed. In the second stage the chloro-propanes are perchlorinated in presence of an excess of chlorine by passing through a bath of molten salts maintained at 350–400°C. The excess chlorine diluted by the hydrochloric acid is then used in the first stage.

This type of process does not give very high yields on the chlorine and the carbon. Further it seems unnecessarily complicated and requires technology which is not simple, thus the reactor containing the molten salts bath must be resistant to a very high degree of corrosion.

Catalysis in the heterogeneous phase   The catalyst is formed of a more or less porous support impregnated with metallic salts. In Dow Chemical's patent[67] a fixed bed reactor is used and the catalyst is active carbon. The temperature of the reaction is about 350°C, this is controlled by a fluid circulating around the reaction tubes and by the flow of chlorine and propane.

In the Ugine process[19] the catalyst is made of porous carbon impregnated with cupric chloride and barium chloride. The reaction takes place in a fluidised bed which makes a uniform temperature and good heat exchange possible. To limit still further the exothermic nature of the perchlorination the reaction between chlorine and propane can be associated with the decomposition of hexachloroethane into C_2Cl_4 and Cl_2 so that the overall reaction is only slightly exothermic. According to the authors the use of C_2Cl_6 has a further advantage of reducing the excess of chlorine necessary.

8.3.1.2. Purely thermal reactions This type of process, relatively simple in principle, is covered by a large number of patents while many industrial installations are based on it. Its success depends on the design and use of the reactor and the exact operating conditions.

The reaction vessel Many patent specifications claim specific designs of reactors. The main aim in these is to minimise the risk of explosion, reduce to a minimum the formation of heavy by-products and carbon. Further they prevent premature reaction outside the vessel resulting from disturbances in the flow of feedstock.

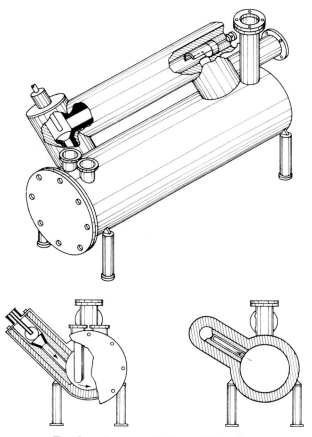

FIG. 8.4 *Reactor with External Recycling*

The raw materials can be premixed before being injected into the reactor or can
be introduced by means of a system of nozzles so arranged that mixing takes place
inside the reactor. In both cases the raw materials must be fed in at a substantial
velocity, 30–100 m/sec and at a sufficiently low temperature. The reactors claimed
in the specifications, adapted specially to the perchlorination, can be divided into
two types:

(a) *Recycling reactors*   According to those recommending this type of reactor the
formation of partially chlorinated secondary products (chloroform and trichloroethylene) and of heavy products (hexachlorobutadiene and hexachlorobenzene) are
minimised by forming an intimate mixture between the gas stream entering the
reactor and the hot gases produced in the reaction. To attain this objective the
kinetic energy of the reactants is used to bring about ready recycling of these hot
gases.

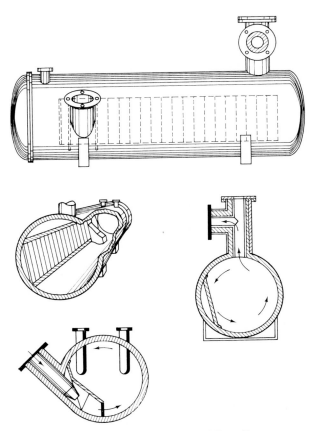

FIG. 8.5   *Reactor with Internal Recycling*

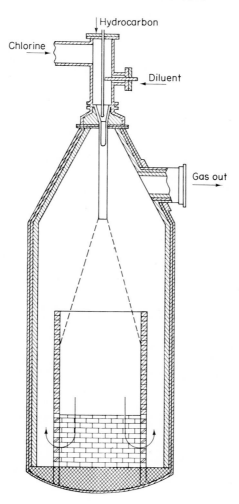

Chlorine →

Hydrocarbon

Diluent ←

Gas out →

FIG. 8.6   *Steel Recycling Reactor*

Dow Chemical[62] describe two reactors represented in Figs. 8.4 and 8.5. That in Fig. 8.4 has external recycling. It is formed of a cylinder of steel coated on the inside with a thermally insulating material. The flow of recycled material is controlled by a gate valve. Internal recycling can be obtained by arrangement such as that represented in Fig. 8.5. In both these reactors any alteration in the flow of starting materials brings about a simultaneous change in the flow of the recycled material.

The reactor shown in Fig. 8.6[63] is a steel one coated with spun glass or highly calcined alumina. A kind of cylindrical chimney perforated at the base is fitted inside to allow passage of the gas. The starting materials can be introduced in either liquid or gaseous form, the gases are injected in the form of a cone. The reactors shown in Figs. 8.7 and 8.8[64] work on the same principle. In Fig. 8.9 a rather different type of reactor is shown, it is a reactor with external recycling devices but in which one of the reagents, actually the hydrocarbon, is injected at various points along the reactor (points 2, 3, 4 and 5). At point 1 chlorine and where necessary a diluent is introduced.

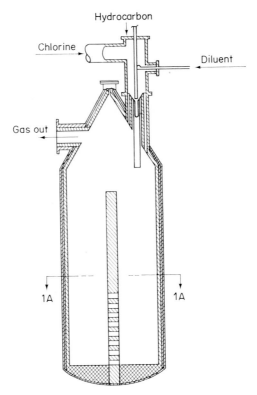

FIG. 8.7   *Recycling Reactor*

(b) *Reactors without recycling arrangements*   In this type of reactor excess of chlorine is used to minimise the formation of by-products. There are several reactors in series and the whole of the chlorine is introduced into the first reactor.

Thus according to the Progil process[18] there are two or three reactors in series. The residence time in each of the reactors is low, of the order of 1 sec, and the final excess of chlorine is about 5 per cent.

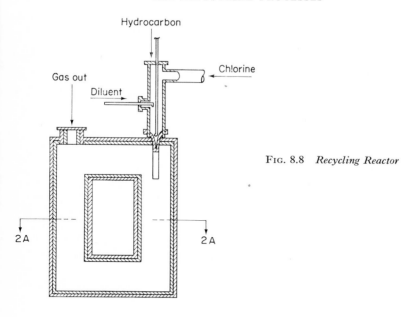

FIG. 8.8 *Recycling Reactor*

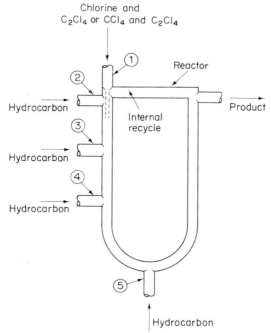

FIG. 8.9 *Different Type of External Recycling Reactor*

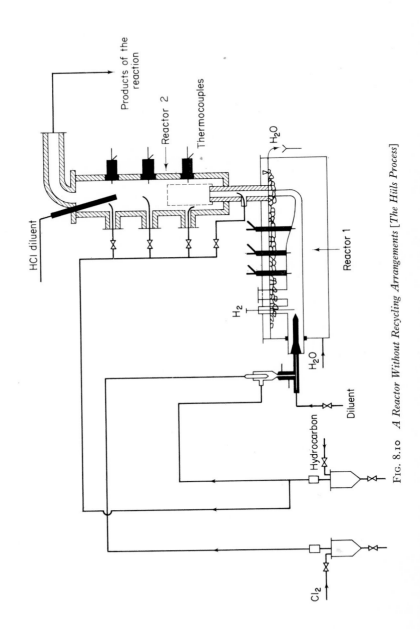

FIG. 8.10 *A Reactor Without Recycling Arrangements [The Hüls Process]*

In the Hüls process[66] the reaction is carried out in two reactors in series (Fig. 8.10). The whole of the chlorine and 70 per cent of the charge of hydrocarbons is introduced into the first reactor while the remainder is introduced at various feed points in the second reactor. Preheating of the reactor is carried out by the chlorine-hydrogen reaction.

Heat build up control The very great evolution of heat in the perchlorination reaction cannot be got rid of in the majority of cases through the reactor walls without using a complicated design. Other solutions have therefore been worked out.

Introduction of a diluent This can be either gaseous or liquid, inert (e.g. hydrogen chloride[20]) or formed from the reaction products: CCl_4 and C_2Cl_4.[12] The injection of a liquid diluent[16, 55] allows the temperature of the reaction to be controlled with the minimum quantity of diluent. Further by-products decrease as a result of a higher partial pressure of chlorine.

Multistage perchlorination Starting with propylene or its unsaturated chlorine derivatives the reaction is carried out in two stages:[22] a liquid phase addition chlorination, carried out in the presence of $FeCl_3$, followed by a fluid bed perchlorination (sand particles) at 440–500°C. Excess chlorine from the second stage is used in the first.

If propane and its chlorinated derivatives are the starting material[23] the first stage consists of a substitution reaction carried out in a fluid bed at 350°C. The second stage is a perchlorination carried out under the same conditions as in the case of propylene.

Operating Conditions

Temperature With the exception of the two stage processes[22, 23] the temperature varies between 530°C and 650°C according to the process used.

Excess of chlorine This can perhaps be defined as the ratio between the non-reacted chlorine and the total quantity introduced. It varies according to the different processes between 5 and 30 per cent.

Time of reaction This is in general rather short, 1 sec in the Progil process[18] to 10–13 sec in the Stauffer process.[16] The I C I process[25] is unusual as the residence time may reach 45–55 sec, the reaction is carried out in a fluidised bed of sand at 610°C under 4 atm with a feed ratio C_2Cl_4/C_3H_8 of 1·7. 91 per cent CCl_4 is obtained under these conditions.

Diluent With a CCl_4/C_3H_8 ratio of 1·71 Dow Chemical Co.[13] claim an output containing 97·5 per cent of C_2Cl_4. 60 per cent of CCl_4 is obtained with a 1·4 C_2Cl_4/C_3H_8 ratio.[15]

Pressure This is used in both the I C I[25] and Stauffer[17] processes. It makes it possible not only to direct production towards the production of carbon tetrachloride but also to produce a hydrogen chloride which is easily purified and used directly in other pressure processes such as oxychlorination. In the I C I process the pressure is about 4 atm and a yield of 91 per cent of carbon tetrachloride is obtained. The Stauffer process is operated at 568°C under 8·75 atm, a residence time of 6·8 sec and a Cl_2/C_3H_6 feed ratio of 10·2 and a C_2Cl_4/C_3H_8 of 2·25 and nearly 100 per cent of CCl_4.

Recycling the reaction by-products   The recycling of chloroform, trichloroethylene and hexachloroethane does not present any problem since all these by-products yield carbon tetrachloride or perchloroethylene under the normal reaction conditions. Hexachlorobenzene is never recycled since it is stable under these conditions.

The case of hexachlorobutadiene is rather different since it is liable to be converted to CCl_4 and C_2Cl_4.[67]

8.3.2. Separation of the Reaction Products

The first stage is to quench the hot gases leaving the reactor to prevent any further reaction taking place. This quench is also used in many instances to separate the hydrochloric acid and the excess chlorine from the solvents. This is then followed by recovery of the chlorine in the hydrochloric acid and distillation of the solvents.

8.3.2.1. The quench stage   Two quenching systems have been developed for practical use:

(1) a quench with water or more specifically a solution of hydrochloric acid.
(2) a solvent or 'dry' quench.

A further alternative must however, be mentioned described in a patent by Halcon[24] which enables the heat to be recovered in the form of steam. A fluidised bed of glass beads is used. The removal of the calories is achieved by means of tubular steam generator. In this way the temperature of the gases is reduced from 500–600°C to 150–200°C.

The hydrochloric acid solution quench has been developed and patented by Scientific Design[55] and Chempatents.[21] The gases leaving the reactor are rapidly cooled by contact with an aqueous solution containing 21–36 per cent wt of hydrochloric acid. The heat is removed and the absorption of the hydrochloric acid gas is effected by a solution of hydrochloric acid which is cooled by means of heat exchangers. This type of quenching system enables the separation of hydrochloric acid, chlorine and solvents in a single stage but it implies the solvents being dry and the use of a very costly and corrosion resistant material (graphite, glass, teflon, etc.).

In the 'dry' quench the gases are cooled by contact with a solvent. The solvent is not necessarily a product of the reaction thus a low boiling liquid may be used such as 1·1 dichloro 1·2·2·2 tetrafluoro ethane.[68] But cooling is more frequently carried out by the reaction products.[12, 17, 69] In the system patented by Halcon[69] the gases are cooled by contact with a CCl_4/C_2Cl_4 mixture. In Fig. 8.11 the whole scheme is shown covering the quench-recovery of hydrochloric acid and chlorine. The gases leaving the quench D101 are cooled further and washed in a tower in column D102 by a CCl_4/C_2Cl_4 mixture maintained at a temperature between 0 and −30°C so as to absorb the chlorine. The solvent leaving the base of D102 is mixed with the product liquid leaving the quench and the whole is stripped in D103 to remove the remainder of the chlorine and the hydrochloric acid. The chlorine and the hydrochloric acid leaving the top of the column are recycled to the reactor. The liquid leaving the base of D103 feeds column D102 and the section where the perchloroethylene and the carbon tetrachloride are distilled, not shown in the layout. The advantage of such a scheme is clearly the absence of water so that traditional materials of construction can be used.

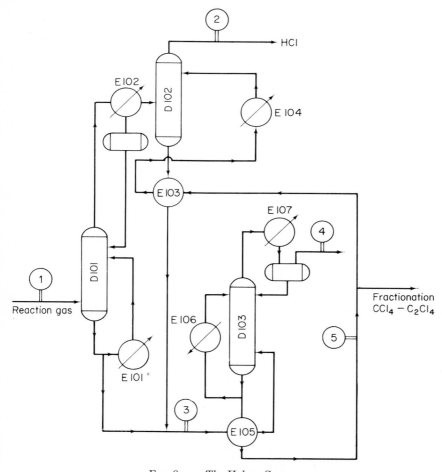

FIG. 8.11   *The Halcon System*

8.3.2.2. Separation of the chlorine and the hydrochloric acid   The recovery of the chlorine in the hydrochloric acid is not only dependant on the quench process used but on the specification which the hydrochloric acid produced must satisfy.

In the case of a water quench the wet chlorine is dried (e.g. by strongly cooling) and then recycled to the reactor either in the form of a carbon tetrachloride solution or by means of a compressor.

Where a dry quench is used the chlorine can be separated by absorbing the hydrogen chloride by a wet route in which case the problem mentioned earlier has to be faced. On the other hand, the chlorine can be absorbed in cooled solutions of perchloroethylene and carbon tetrachloride. In this case we must draw attention to the Halcon patent[70] which covers the chlorination of propylene and dichloropropane by a mixture of chlorine and hydrochloric acid and recycling the dichloropropane formed to the perchlorination reactor.

8.3.2.3. Separation of the solvents   This is carried out by the well-known method including a series of fractional distillations which usually start with a stripping column to eliminate the hydrochloric acid, chlorine and eventually the water dissolved in the solvents. This is followed by a series of columns which separate the light products (chloroform and trichloroethylene) and the heavy products (hexachloroethane, hexachlorobenzene, hexachlorobutadiene).

8.3.3. Examples of Processes

8.3.3.1. Ugine process   This represents the catalytic type of process.[19]

Process scheme (Fig. 8.12)   Reactor EK101 contains two catalytic beds, a lower bed, slightly fluidised, consisting of particles of catalyst, the diameter of which vary between 0·5 and 1·25 mm, and an upper fixed bed deposited on a grid consisting of particles of the same catalyst but varying between 3 and 6 mm in diameter. The excess calories are removed by a heat exchanger buried in the fluidised bed.

A propane-propylene mixture, chlorine and recycle products are fed into the lower part of the reactor while perchloroethylene is fed under the grid supporting the fixed bed in order to reduce the temperature of the gas.

At its exit from the reaction chamber the gas passes through a cyclone, then through heat exchangers where the organic liquid phase is condensed. No precise information is given as to how the hydrochloric acid and the excess chlorine are recovered.

The condensed solvents are stripped (column D101) to remove dissolved hydrochloric acid and chlorine, then fractionated in a series of four columns.

D102 separation of C_2Cl_4—CCl_4
D103   ,,   ,, C_2HCl_3—CCl_4
D104   ,,   ,, C_2Cl_4—heavy ends
D105 recovery of the heavy ends for recycling.

It should be noted that the production of chloroform is not mentioned in this process. If this compound were to be formed it would be necessary to separate it from the CCl_4 in a supplementary column.

Reaction conditions   Temperature of the fluid mass: 500°C.
Absolute pressure inside the reactor 920 mm Hg.
Velocity of gas at the reaction pressure and temperature—22 cm/sec.
Residence time about 10 sec.

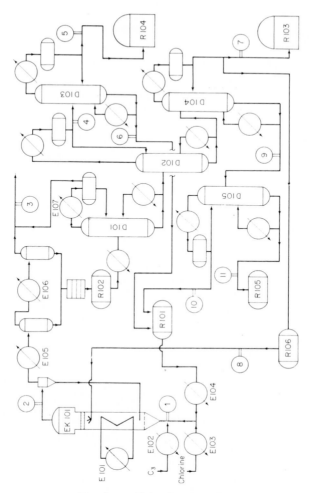

FIG. 8.12   *Ugine Process Scheme*

Temperature of gas leaving the reaction chamber (after injection of liquid C_2Cl_4): 400°C.

Ratio Cl_2/C_3 is 7·7 and C_2Cl_4/C_3 is 3·4.

Material balance (Table VIII.3)   The various figures shown in Fig. 8.12 have been calculated from those given in FP 1,147,756.

Under the above conditions a mixture is obtained containing 55·2 wt per cent of carbon tetrachloride and 44·8 wt per cent perchloroethylene. To obtain a ton of this mixture 132·5 kg of propane-propylene (50–50) mixture, and 1,685 kg of chlorine are required while 790 kg of hydrochloric acid are recovered. The yield on carbon is

22

99·3 per cent, the yield on chlorine 98·8 per cent and the ratio of by-products expressed in wt per cent of a CCl_4–C_2Cl_4 mixture is 1·5 per cent.

TABLE VIII.3

Material Balance in Ugine Process Flow in kg/h

Reference points	1	2	3	4	5	6	7	8	9	10	11
Propane-propylene	500										
Chlorine	6360	26	26								
Hydrochloric acid		2984	2984								
Perchloroethylene	6650	9680	7	437		437	1693	1330	6213	6213	
Carbon tetrachloride	38	2130	12	2118	2080	38					
Trichloroethylene	114	115	1	114		114					
Hexachloroethane	1022	1041							1041	1023	19
Hexachlorobutadiene	19	38							38	19	19
Hexachlorobenzene		19							19		19

8.3.3.2. Scientific Design process   This reaction is on a purely thermal perchlorination.[55] Actually four plants produce perchloroethylene and carbon tetrachloride according to the Scientific Design method, the largest of which produce 45,000 tons per annum of solvents. The four are the Rhône-Progil plant in France, I C I's in Australia, Kureha's in Japan and the Pemex plant in Mexico. A fifth plant, with a capacity of 40,000 tons per annum is under construction in Italy by Montedison.

Process scheme (Fig. 8.13)   No typical scheme exists in view of the wide possibilities resulting from the production ratio of C_2Cl_4/CCl_4, the re-use of excess chlorine, the quenching system and the method of separating the heavy ends.

Propylene, chlorine and recycle products, the latter consisting mainly of the carbon tetrachloride, which has been used to absorb the excess chlorine, are fed to the reactor. The recycle products can be injected either as a gas or a liquid.

The gases leaving the reactor are rapidly cooled by a 35 wt per cent solution of hydrochloric acid in a quench from which the products pass to a separating vessel, where the aqueous phase is separated from the heavier solvent phase. Part of the aqueous phase is stripped to remove chlorine and dissolved solvents (D103), then distilled (D104) to produce gaseous hydrochloric acid and the azeotropic solution (22 wt per cent). The other part is cooled and recycled to the quench. The gases leaving the quench are passed to column D102 where the hydrochloric acid is absorbed by means of a cooled azeotropic solution. The concentrated solution obtained at the base of this column is fed to the quench. The gases leaving the top of D102 consisting of chlorine, inerts and chlorinated solvents are cooled. The chlorine is recovered by absorption in carbon tetrachloride (D105) and then recycled to the reactor. The solvents are dried and stripped (D201) then fractionated through a series of five columns to separate chloroform, trichloroethylene and heavy by-products.

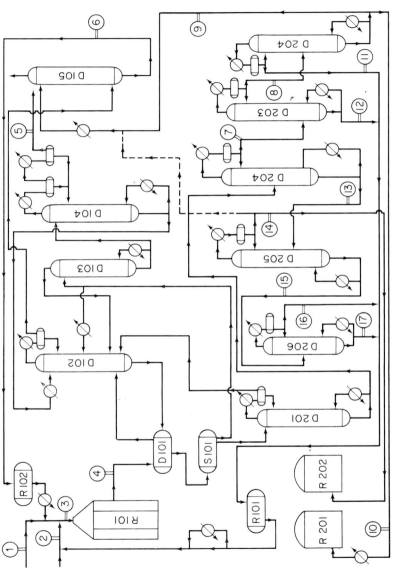

Fig. 8.13 *Scientific Design Process Scheme*

TABLE VIII.4

Material Balance in Scientific Design Process

Flow in kg/h

Reference points	1	2	3	4	5	6	7	8	9	10	11	12	13	14	15	16	17
Propylene		500	500														
Chlorine	5683		6228	545		545											
Hydrochloric acid				2607	2607												
Chloroform		15	15	15			15	15			15						
Carbon tetrachloride		275	3500	4877		3225	4877	4857	3225	1377	255	20					
Trichloroethylene		30	30	30			30					30					
Tetrachloroethylene		400	400	2545			200					200	2345	2145	200	200	
Hexachloroethane		35	35	39									39	39	39	35	4
Hexachlorobutadiene		10	10	35									35		35	10	25
Hexachlorobenzene				25									25		25	25	25

Reaction conditions Precise conditions such as the chlorine hydrocarbon ratio, the recycle products-hydrocarbon ratio cannot be given. The hydrocarbon starting material can be propane, propylene or dichloropropane. The reactor temperature lies within the range 530–610 °C.

Material balance (Tables VIII.4) This has been drawn up solely to illustrate the process and to show the importance of the various streams. It must not be used to make comparisons with other processes, particularly as regards the yields obtained.

8.4. THE PRODUCTS OF THE REACTION

8.4.1. Physical Properties

TABLE VIII.5

	Perchloroethylene	*Carbonte trachloride*
Molecular weight	165·85	153·8
Density of the liquid at 20/4 °C	1·623	1·592
Density of the vapour at boiling point and under 1 atm	4·9	5·24
Boiling point in °C	121·2	76·7
Melting point in °C	−22·35	−22·9
Vapour pressure at	13 °C 10 mm	−20·8 °C 10 mm
,, ,, ,,	30 °C 24 mm	10 °C 55·9 mm
,, ,, ,,	40 °C 39 mm	22·5 °C 100 mm
,, ,, ,,	50 °C 63 mm	30 °C 143 mm
,, ,, ,,	60 °C 95 mm	40 °C 215·8 mm
,, ,, ,,	70 °C 150 mm	176 °C under 10 atm
Specific heat cal/g/°C		
Liquid at 20 °C	0·216	0·2028
Vapour at 1 atm	0·154 at 100 °C	0·1440 at 76·7 °C
Critical temperature °C	347·1	283·7
Critical pressure in atmospheres	46 (est)	46·5
Viscosity of liquid in centipoises at 20 °C	0·902	0·965
Refractive index of liquid ($n_D 20$)	1·5058	1·4607
Surface tension in air at 20 °C in dyne/cm	32·32	27·37
Coefficient of cubic expansion	0·00103	0·00124
Dielectric constant of the liquid at 25 °C	2·30	2·25
Magnetic susceptibility	$−0·508 \times 10^{-6}$	$−0·429 \times 10^{-6}$
Molar refraction	26·51	26·51
Flammability	non flammable	non flammable
Solubility in water at 25 °C in g per 100 g	0·015	0·08
Solubility of water in the chlorinated hydrocarbon 20 °C in g per 100 g	0·0105	0·013

Binary azeotropes of CCl₄   A great number of these exist[71] of which important ones are described in Table VIII.6.

TABLE VIII.6

Compound	Boiling point of the azeotrope in °C	wt % of CCl₄
Water*	66	95·1
Methanol	55·7	79·4
Ethanol	65·1	84·1
Isopropyl alcohol	68·95	82·0
Formic acid	66·69	81·5
1-2 Dichloroethane	75·30	78·4
Acetone	56·1	11·5
Acetonitrile	65·1	83·0
Nitromethane	71·3	83·0

* heteroazeotrope

Binary azeotropes of C_2Cl_4   Over sixty exist and some of these are shown in Table VIII.7.

TABLE VIII.7

Compound	Boiling point of the azeotrope in °C	Wt % at C_2Cl_4
Water*	87·1	84·1
Methanol	63·75	36·5
Ethanol	76·75	37·0
n-Propyl alcohol	94·05	52·0
Formic acid	88·15	50·0
Acetic acid	107·35	61·5
1·1·2 trichloroethane	112·00	57·0

* heteroazeotrope.

8.4.2. Chemical Properties

8.4.2.1. Carbon tetrachloride   The thermal decomposition of carbon tetrachloride at temperatures above 700°C leads to the formation of hexachlorobenzene. Between 600 and 700°C perchloroethylene and chlorine are obtained.

It is the least resistant of the chlorinated methanes to oxidation. In presence of air and iron it is converted to phosgene. Heated to 250°C with excess water, carbon dioxide and hydrochloric acid are obtained; phosgene is obtained in presence of limited quantities of water.

It only reacts slowly with copper and lead but explosive reactions can take place with aluminium and its alloys.

Carbon tetrachloride reacts with hydrofluoric acid giving mainly dichloro-difluoromethane.

In presence of various catalytic systems it reacts with ethylene to give polymers of the general formula CCl_3—$(CH_2$—$CH_2)_n$—Cl in which n is relatively small.

8.4.2.2. Perchloroethylene Perchloroethylene in contact with water gradually decomposes to trichloroacetic acid and hydrochloric acid. In the absence of light it does not react with oxygen. Ozone decomposes it into phosgene and trichloroacetyl chloride.

When it is irradiated with ultra-violet light perchloroethylene is oxidised by air to trichloroacetyl chloride. The first stage in the reaction is the formation of peroxy compounds:

$$\begin{array}{c}
CCl_2 \\
\parallel \quad + O_2 \\
CCl_2
\end{array}
\left\{
\begin{array}{l}
\nearrow \quad
\begin{array}{c}
Cl_2C \\
\mid \quad \diagdown \\
Cl_2C \quad \diagup O\!-\!O
\end{array} \quad A \\[2ex]
\searrow \quad
\begin{array}{c}
Cl_2C\!-\!O \\
\mid \quad \mid \\
Cl_2C\!-\!O
\end{array} \quad B
\end{array}
\right.$$

Compound A rearranges itself to produce, via an epoxy compound, trichloroacetyl chloride. Compound B is converted to phosgene. The reaction leading to trichloro-acetyl chloride is much faster so that phosgene is only produced in a relatively small quantity during the oxidation of perchloroethylene. Its slow decomposition in presence of air results from this reaction. The photochlorination of tetrachloro-ethylene leads to hexachloroethane.

At 225–400°C in presence of a catalyst tetrachloroethylene reacts with a mixture of hydrofluoric acid and chlorine to give 1·2·2 trichloro 1·1·2 trifluoroethane.

Pyrolysis at temperatures above 700°C, yield hexachloroethane and hexachloro-benzene.

8.4.3. Toxicological Properties

8.4.3.1. Carbon tetrachloride This is a particularly dangerous solvent because of its rapid and serious action on the kidneys. The penetration on the product into the body takes place by inhalation, through the skin or more rarely by the mouth.

Acute poisoning starts with a narcotic effect together with various symptoms such as eye, nose and throat irritation, vomiting and dizziness. After sometimes a fairly long latent period symptoms of kidney trouble appear.

Chronic poisoning shows itself in subjects who have been exposed to a low con-centration of carbon tetrachloride over a long period of time. Nervous and kidney troubles develop insiduously and are worse in those with a predisposition to kidney complaints or in chronic alcoholics. Dermatitis has also been noted in persons sub-ject to prolonged or repeated contact of the product with the skin.

The maximum concentration of carbon tetrachloride vapour permissible in air has been fixed by US specialists in 1966 at 10 p.p.m. or 65 mg/m³.

8.4.3.2. Perchloroethylene There is little published information on the toxicity of perchloroethylene. It is certainly a narcotic. It has been said to cause conjunctivitis and a kind of dermatitis similar to eczema. Finally its use has been noted to cause gastro-intestinal complaints and hepatitis. The USA in 1966 fixed the permissible maximum concentration in the air at 100 p.p.m. or 670 mg/m^3.

8.4.4. Uses

During the period 1930–40 carbon tetrachloride was used mainly for dry cleaning. But it was rapidly replaced by perchloroethylene on account of its toxicity, its powerful corrosive action on metals and its volatility so that by 1956 it was only being used to the extent of 7 per cent. On the other hand, the growing demand for fluorinated hydrocarbons of which carbon tetrachloride is one of the starting materials has maintained the demand for the latter product. In the USA it is used to make mainly dichlorodifluoromethane (69 per cent) and trichloro fluoromethane (26 per cent).

Perchloroethylene is mainly used as a dry-cleaning solvent (80 per cent of its consumption in the USA), in the degreasing of metals in the vapour phase (10 per cent) and as a chemical intermediate, particularly in the production of fluorinated hydrocarbons (10 per cent). A new end-use is making itself felt for perchloroethylene; the treatment of textile fibres in non aqueous media. If this kind of textile treatment develops as a result of the anti-pollution drive, which is expected on all sides, the demand for perchloroethylene would increase very rapidly and in a dozen years from now the textile industry might become the largest consumer of chlorinated solvents.

8.4.5. Stabilisation

For use as discussed in the previous paragraph the perchloroethylene must be stabilised to:

inhibit the action of air, light and moisture,
protect the solvent from the action of metals and its salts,
prevent any local decomposition.

Good stabilisation can be obtained by using various compounds carefully chosen to effect one or more of the actions mentioned above. Further, each compound added must be sufficiently volatile to be present both in the liquid and the vapour phase and capable of being recovered by distillation. It must not react with the numerous and varied contaminants with which it comes in contact during the degreasing of pieces of metal and at the same time be compatible with any other stabiliser in the system.

Very often the stabiliser will not completely prevent the reaction but will at least reduce its velocity. Thus in the case of degreasing in the vapour phase, if a small quantity of hydrochloric acid is formed due to moisture, metallic salts such as ferric and aluminium chloride are produced which accelerate the decomposition of perchloroethylene. To combat this type of decomposition two kinds of compounds are added:—acid acceptors and metal deactivators.

Acid acceptors react with hydrochloric acid. Epoxides can be used, while a synergistic effect can be obtained by combining certain epoxides with certain amines.

Metal deactivators are alcohols, esters and ethers and certain sulphur compounds which chelate metallic salts.

8.4.6. Economic Data

8.4.6.1. Carbon tetrachloride   Table VIII.8 shows the annual production in various countries. The chief companies making carbon tetrachloride are also given.

TABLE VIII.8

All Figures in Thousands of Tons

Country	Years						
	1965	1966	1967	1968	1969	1970	1971
West Germany Hoechst Hüls Dow (projected)	40	42	45	53	52	55	
France Progelec Solvay Rhône Progil	35·3	40	47·1	48·2	66·6	68·5	
Italy Rumianca Solvay Montedison	36·8	45·1	52·1	50·9	52	54	
EEC	113·1	132·6	151·7	159·1	188·6	217·5*	
UK Albright and Wilson ICI	25	25	24	20	15		
Japan Asahi Kureha Nipon Soda	17·7	26·4	27·7	28·7	32·3		40·6*
USSR				30*			
USA Dow Vulcan Allied Stauffer		268	299	345	386	440	460·8*

* Estimate.

TABLE VIII.9

All Figures in Thousands of Tons

Country	Years						
	1965	1966	1967	1968	1969	1970	1971
West Germany Dynamit A G Hüls Wacker Chemie Dow (projected)	65·1	74·9	74·7	81·7	82·3	88	
France Progelec Pechiney St Gobain Solvay	49·2	56·5	56·5	58	62	57*	
Italy Montedison Solvay Rumianca	22·7	35·5	37·8	41·4	42·2	50	
Benelux Solvay	7·1	5·8	5·5	4	12	25·5*	
EEC	140	170	175	185	198	215	
UK ICI	62	65	70	70	72		
Japan Asahi Kureha Toa Gosei Kanto Denka	21·3	29·5	29·8	35·8	43·9	51	57*
USSR				30*			
USA Diamond Dupont Dow Ethyl Vulcan Hooker PPG Stauffer Detrex	192	220	240	290	285	277	

* Estimate.

The proportion of carbon tetrachloride produced by perchlorination of light hydrocarbons (C_2 and C_3) in the USA is 25–30 per cent of the total, the remainder being obtained by chlorination of methane and chlorination of carbon disulphide.

The increase in production between 1958 and 1968 has been 9·3 per cent per annum. The rate of increase estimated to 1973 is 6·5 per cent per annum; it is clear that this rate will be dependent on the expansion of the fluorinated hydrocarbons F 11 and F 12. It is now possible that F 11 can compete in certain fields with mixtures of fluorinated hydrocarbons not derived from CCl_4 as starting material (F 115 and F 22).

8.4.6.2. Perchloroethylene The production and producers for various countries is shown in Table VIII.9.

Perchloroethylene is made by various processes of which the perchlorination of C_3 only forms about 40 per cent of the total production.

Production capacities are estimated in 1970 as follows:

USA	410,000	tons per annum
West Europe	300,000	,, ,, ,,
Japan	70,000	,, ,, ,,

The rate of growth of production in the USA was 12·5 per cent per annum between 1959 and 1969, it may not exceed 7 per cent per annum in the future up till 1975. Although the consumption in 1970 increased by 9 per cent in relation to that of 1969, certain companies like Dupont de Nemours, who first produced perchloroethylene thirty-six years ago, and Detrex Chemical Industries have decided to stop production due to the low profitability of this product.

REFERENCES

[1] McBee, E. T., *et al.*, *Ind. Eng. Chem.*, 1941, **33**, 178.
[2] Dumas, J. B., *Annalen*, 1840, **33**, 187.
[3] Tolloczko, S. T. and Kling, R., *Chem. Zent.*, 1913, **11**, 98.
[4] Jones and Allison, *Ind. Eng. Chem.*, 1919, **11**, 639.
[5] Böeseken, J. *et al.*, *Rec. Trav. Chim. des Pays-Bas*, 1915, **34**, 78. [CA 1915, **9**, 1765.]
[6] Reilly, J. H. (Dow Chem. Co.), USP 1,947,491 (1934).
[7] Grebe, J. J. *et al.* (Dow Chem. Co.), USP 2,034,292 (1936).
[8] Farbenindustrie, I. G. AG, FP 836,979 (1939).
[9] Donau Chemie, Belg.P 447,044 (1942).
[10] Brown, E. T. *et al.* (Dow Chem. Co.), USP 2,377,699 (1945).
[11] Heitz, R. G. *et al.* (Dow Chem. Co.), USP 2,442,323 (1948).
[12] Heitz, R. G. *et al.* (Dow Chem. Co.), USP 2,442,324 (1948).
[13] Warren (Dow Chem. Co.), USP 2,577,388 (1951).
[14] McBee, E. T. and Devaney, L. W., *Ind. Eng. Chem.*, 1919, **41**, 803.
[15] Dow Chem. Co., USP 2,727,076 (1955).
[16] Obrecht, R. P. (Stauffer Chem. Co.), USP 2,857,438 (1958).
[17] Stauffer Chem. Co., FP 2,020,428 (1969).
[18] Progil, FP 1,179,977 (1956), BP 819,987 (1959).
[19] Thermet, R. and Parvi, L. (Soc. d'Ugine), BP 794,478 (1956), FP 1,147,756 (1957), USP 2,938,931 (1960), FP 1,166,211 (1959), USP 2,919,296 (1959).
[20] Chempatents Inc., USP 2,839,589 (1958).
[21] Brown, D. *et al.* (Chempatents Inc.), USP 2,746,998 (1956), BP 772,126 (1957).

[22] Solvay and Cie, FP 1,292,994 (1961).

[23] Solvay and Cie, FP 1,373,709 (1963).

[24] Spitz (Halcon International), USP 3,177,632 (1965).

[25] ICI Ltd., FP ,1510,705 (1967).

[26] Donau Chemie, GP 734,024 (1943).

[27] Kons. für Elektrochem, BP 25,967 (1910).

[28] BIOS 1056. Fiat 843.

[29] Müller and Dubois, GP 72,999 (1893).

[30] Molyneux, *Brit. Chem. Eng.*, 1963, **8**, 696.

[31] Solvay and Cie, FP 2,018,910 (1970).

[32] Rust and Vaughan, *J. Org. Chem.*, 1940, **5**, 449; 1940, **5**, 472.

[33] Hass and McBee, *Ind. Eng. Chem.*, 1936, **28**, 330.

[34] Mamedaliev, Yu. G. and Huseinov, M. M., Congrès de la Catalyse, Paris 1960— Section III n° 122.

[35] Tanaka, R., *Yuki Gosei Kagaku*, 1962, **20**, 482.

[36] Shinoda, K. and Watanabe, H., *Kogyo Kagaku Zasshi* (*J. Chem. Soc. Japan*; *Ind. Chem. Sect.*), 1967, **70**, 1316.

[37] Shinoda, K. and Watanabe, H., *Kogyo Kagaku Zasshi* (*J. Chem. Soc. Japan*; *Ind. Chem. Sect.*), 1967, **70**, 1320.

[38] Shinoda, K. and Watanabe, H., *Kogyo Kagaku Zasshi* (*J. Chem. Soc. Japan*; *Ind. Chem. Sect.*), 1967, **70**, 1482.

[39] Groll and Hearne, *Ind. Eng. Chem.*, 1939, **31**, 1530.

[40] Hearne, G. W. *et al.*, *J. Am. Chem. Soc.*, 1953, **75**, 1392.

[41] McBee, E. T. and Pierce, *Halogenation*, Ref. 53, **7**, 347.

[42] De La Mare and Vaughan, *J. Chem. Educ.*, 1957, **34** (*1*), 10; 1957, **34** (2), 64.

[43] Weyl, Houben, '*Methoden der Organischen Chemie*', V (3) (E. Müller, 1962).

[44] Williams, E. C., *Trans. Am. Inst. Chem. Eng.*, 1941, **37**, 157.

[45] Krynitsky, J. A. and Carhart, H. W., *J. Am. Chem. Soc.*, 1949, **71**, 816.

[46] Schmeisser, M. *et al.*, *Ber.*, 1962, **95**, 1648.

[47] *Faraday Ann.*, 1821, **18** (2), 48.

[48] Regnault, V., *Ann. Chim. Phys.*, 1839, **71** (2), 385.

[49] Weiser, J. and Wightman, G., *J. Phys. Chem.*, 1919, **23**, 415.

[50] Strosacker and Schwegler (Dow Chem. Co.), USP 1,930,350 (1933).

[51] Wimmer, H., USP 2,305,821 (1938).

[52] McBee, E. T. *et al.*, *Ind. Eng. Chem.*, 1941, **33**, 276.

[53] Dainton, F. S. and Ivin, K. J., *Trans. Faraday Soc.*, 1950, **46**, 295.

[54] Puyo, J., These Faculté des Sciences Nancy (1961)—Pyrolyse de l'hexachloréthane.

[55] Brown (Scientific Design), FP 1,356,229 (1962).

[56] Thizy cited by E. Charles, Bulletin Association Francaise des techniciens du pétrole— Section Chimie, April 1958, p. 381.

[57] Stull, D. R. *et al.*, '*The Chemical Thermodynamics of Organic Compounds*', ed. J. Wiley (1969).

[58] Fink, C. G. and Bonilla, C. F., *J. Phys. Chem.*, 1933, **37**, 1135.

[59] Khorshak, V. V., *J. Gen. Chem. USSR*, 1947, **17**, 1626.

[60] Thermet, R., *Chim. & Ind.*, 1957, **78**, 351.

[61] Khundkar, M. H. and Mullick, S. U., *Pakistan J. Sci.*, 1961, **13**, 279.

[62] Bender, H. *et al.* (Dow Chem. Co.), USP 2,441,528 (1948).

[63] Bender, H. (R. P. Obrecht), USP 2,806,768 (1957).

[64] Stauffer Chem. Co., FP 1,182,343 (1957).

[65] Halcon International, BP 1,047,258 (1963).

[66] Chemische Werke Hüls, GP 1,074,025 (1958).

[67] Chemische Werke Hüls, FP 1,333,947 (1962).

[68] Chempatents, BP 794,408 (1958).

[69] Yildirim, U. (Halcon International), FP 1,493,131 (1967).

[70] Halcon International, Belg.P 699,715 (1966).

[71] Azeotropic Data, Advances in Chemistry Sciences, *Am. Chem. Soc.* (1952).

CHAPTER 9

HYDROFORMYLATION (OXO-PROCESS)

By J. Falbe

9.1. GENERAL REMARKS AND HISTORICAL BACKGROUND

THE reaction of olefins with carbon monoxide and hydrogen in presence of catalysts was discovered by O. Roelen in the laboratories of Ruhrchemie AG, Oberhausen-Holten, Germany, in 1938[1, 2, 3] when he studied the recycle of olefins to the Fischer-Tropsch synthesis reactor. Roelen obtained propionaldehyde and diethyl-ketone from ethylene, carbon monoxide and hydrogen under pressure at a temperature of 150°C.

$$H_2C = CH_2 + CO/H_2 \longrightarrow H_3C-CH_2-CHO$$

$$2H_2C = CH_2 + CO/H_2 \longrightarrow H_3C-CH_2-\overset{\parallel}{\underset{O}{C}}-CH_2-CH_3$$

Since both an aldehyde and a ketone were formed the reaction was named the 'Oxo reaction'. However, in the course of time it became apparent that ketone formation was restricted to ethylene under special reaction conditions while all other olefins were converted almost solely to aldehydes. Thus 'hydroformylation' seems to be the better name for this reaction. However, the term 'Oxo reaction' had already become firmly established. Today both names are used. Sometimes the reaction is also called the 'Roelen reaction' after its discoverer.

The first Oxo plant was brought on stream early in 1945 in the Ruhrchemie works at Oberhausen-Holten with a capacity of 10,000 tons/annum of detergent alcohols with 12 to 14 carbon atoms. Production ceased at the end of World War II in mid-1945.

At present the Oxo process is one of the most important petrochemical processes. The estimated world capacity of Oxo plants in 1970 was more than 3 million tons/annum (see Chapter 3.7). The rapid growth was assisted by two factors:

a. the increasing demand of alcohols made from 'Oxo aldehydes' as plasticiser raw material, base materials for detergents and solvents.

b. the fact that the development of petrochemistry gave access both to low cost olefins and synthesis gas.

Many different olefins are used as feedstock in the oxo plants, the C-numbers ranging from 2 to 18. However, propylene is clearly number one in volume. Under normal reaction conditions (e.g. temperatures of 110–180°C, pressures of 200–350 at. in presence of cobalt catalysts) the reaction product consists mainly of n- and iso-butyraldehyde and of smaller amounts of some other products formed through side or consecutive reactions.

$$H_3C-CH = CH_2 + CO/H_2 \overset{cat.}{\longrightarrow} H_3C-CH_2-CH_2-CHO$$

$$H_3C\text{---}CH = CH_2 + CO/H_2 \xrightarrow{\text{cat.}} H_3C\text{---}\underset{\underset{CH_3}{\mid}}{CH}\text{---}CHO$$

The ratio of straight chain to branched aldehyde depends on the reaction conditions (see Chapter 2.4). Usually it is in the range of 3–4 parts of n-butyraldehyde to one of iso-butyraldehyde.

Extensive reviews on the Oxo reaction have been published recently.[4, 5, 6]

9.2. THERMODYNAMICS, KINETICS AND REACTION MECHANISM OF THE HYDROFORMYLATION REACTION

9.2.1. General Remarks

Although the technical application and commercialisation of the hydroformylation progressed rapidly there were no systematic investigations carried out either on the kinetics or the reaction mechanism in the early years. For quite some time only a number of emperical rules[7, 8] resulting from a great number of experimental data were the basis for the understanding of the reaction itself and of the composition of the reaction products.

Unfortunately a number of confusing analytical results gave rise to many mis-interpretations. For a long time hydroformylation was believed to be a heterogeneous catalytic reaction and this view was still supported by some authors in the sixties.[9, 10] A critical discussion of such papers was published by Macho, Mistrick and Ciha[11] in 1964. It was some time before the homogeneous nature of the catalytic reaction in hydroformylation was confirmed.[12–16]

Although even at present questions concerning the mechanism are still under discussion it appears that due to the extensive research carried out by a great number of scientists in recent years the basic steps are now sufficiently well defined.

9.2.2. Thermodynamics

The hydroformylation of propylene is exothermic:

$$H_3C\text{---}CH = CH_2 + CO + H_2 \longrightarrow H_3C\text{---}CH_2\text{---}CH_2\text{---}CHO- \ 32,476 \ kcal*$$
$$H_3C\text{---}CH = CH_2 + CO + H_2 \longrightarrow H_3C\text{---}\underset{\underset{CH_3}{\mid}}{CH}\text{---}CHO- \ 33,838 \ kcal*$$

Calculations of Natta, Pino and Mantica[17] indicated that the thermodynamics are favourable for the reaction under normal pressure and at low temperature. However, the reaction proceeds adequately only at temperatures above 90°C and pressures around 50 atm.[4, 5] Thus it appears that kinetic factors are controlling the reaction rather than thermodynamics.

9.2.3. Kinetics

Kinetic measurements by Natta et al.[18–20] led to the following rate equation which may be used with good approximation even on an industrial scale.

* Calculated by Wicke (Ruhrchemie).

$$\frac{d[\text{aldehyde}]}{dt} = k \, [\text{olefin}] \, [\text{Co}] \, [p_{H_2}] \, [p_{CO}]^{-1}$$

Recently Wysokinskii *et al.* published additional kinetic measurements[21] supporting the above equation. Provided the reaction conditions guarantee the stability of the cobalt carbonyls the hydroformylation reaction is of first order with respect to the cobalt dissolved.* However, the catalyst concentration cannot be increased at will since given carbon monoxide and hydrogen partial pressures correspond to a maximum concentration of cobalt carbonyls at a defined temperature (see Fig. 9.1).

Under the hydroformylation conditions usually applied, the reaction velocity decreases with increasing carbon monoxide partial pressure and increases with increasing hydrogen partial pressure. Due to these two counter-effects the reaction rate is nearly independent of the total pressure when using equimolar mixtures of carbon monoxide and hydrogen.

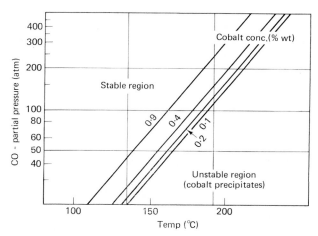

Fig. 9.1[22]   *Stability of Cobalt Carbonyl Catalyst $(Co_2(CO)_8 + HCo(CO)_4)$ as a Function of Reaction Temperature and Carbon Monoxide Partial Pressure in the Liquid Phase.*
(*Data for 0·1 and 0·2 cited from original literature, data for 0·4 and 0·9 wt per cent cobalt added by Falbe and Cornils, Ruhrchemie AG, based on Ruhrchemie experiments*)

9.2.4. Reaction Mechanism

Summarising the results of various research teams on the reaction mechanism[23–31] the reaction leading to n-butyraldehyde follows the mechanism given by equations 1–5, demonstrated with cobalt as hydroformylation catalyst.†

* Besides cobalt, rhodium and ruthenium may be used as catalysts.
† Other catalysts may be used in the Oxo reaction such as rhodium, ruthenium and modified cobalt catalysts. Since cobalt is the most common catalyst all following discussions are based on cobalt if not mentioned otherwise. With other metals the reaction mechanism is more or less similar to the one given in the above reaction sequence.

$$2\text{Co (metal, oxide, hydroxide or salt)} \xrightarrow[-\text{H}_2]{\overset{8\text{CO}\qquad+\text{H}_2}{}} \text{Co}_2(\text{CO})_8 \rightleftharpoons 2\text{HCo}(\text{CO})_4 . 2\text{HCo}(\text{CO})_3 + 2\text{CO} \quad (1)$$

$$\text{CH}_3\text{—CH} = \text{CH}_2 + \text{HCo}(\text{CO})_3 \longrightarrow \underset{\underset{\text{HCo}(\text{CO})_3}{\downarrow}}{\text{CH}_3\text{—CH} = \text{CH}_2} \quad (2)$$

$$\underset{\underset{\text{HCo}(\text{CO})_3}{\downarrow}}{\text{CH}_3\text{—CH} = \text{CH}_2} \longrightarrow \text{CH}_3\text{—CH}_2\text{—CH}_2\text{—Co}(\text{CO})_3 \underset{-\text{CO}}{\overset{+\text{CO}}{\rightleftharpoons}} \text{CH}_3\text{—CH}_2\text{—CH}_2\text{—Co}(\text{CO})_4 \quad (3)$$

$$\text{CH}_3\text{—CH}_2\text{—CH}_2\text{—Co}(\text{CO})_4 \longrightarrow \quad (4)$$
$$\text{CH}_3\text{—CH}_2\text{—CH}_2\text{—CO—Co}(\text{CO})_3 \underset{-\text{CO}}{\overset{+\text{CO}}{\rightleftharpoons}} \text{CH}_3\text{—CH}_2\text{—CH}_2\text{—CO—Co}(\text{CO})_4$$

$$\text{CH}_3\text{—CH}_2\text{—CH}_2\text{—CO—Co}(\text{CO})_{3,\,4} \xrightarrow{+\text{H}_2} \text{CH}_3\text{—CH}_2\text{—CH}_2\text{—CHO} + \text{HCo}(\text{CO})_{3,\,4} \quad (5)$$

In equation 3 cobalt hydrocarbonyl may also add across the double bond such that the cobalt atom will be bound to the carbon atom no. 2. From this addition another reaction path leads to iso-butyraldehyde via the steps (3a)–(5a).

$$\underset{\underset{\text{HCo}(\text{CO})_3}{\downarrow}}{\text{H}_3\text{C—CH} = \text{CH}_2} \longrightarrow \underset{\underset{\text{Co}(\text{CO})_3}{|}}{\text{H}_3\text{C—CH—CH}_3} \underset{-\text{CO}}{\overset{+\text{CO}}{\rightleftharpoons}} \underset{\underset{\text{Co}(\text{CO})_4}{|}}{\text{H}_3\text{C—CH—CH}_3} \quad (3a)$$

$$\underset{\underset{\text{Co}(\text{CO})_4}{|}}{\text{H}_3\text{C—CH—CH}_3} \rightleftharpoons \underset{\underset{\underset{\text{Co}(\text{CO})_3}{|}}{\text{CO}}}{\overset{|}{\text{H}_3\text{C—CH—CH}_3}} \underset{-\text{CO}}{\overset{+\text{CO}}{\rightleftharpoons}} \underset{\underset{\underset{\text{Co}(\text{CO})_4}{|}}{\text{CO}}}{\overset{|}{\text{H}_3\text{C—CH—CH}_3}} \quad (4a)$$

$$\underset{\underset{\underset{\text{Co}(\text{CO})_{3,\,4}}{|}}{\text{CO}}}{\overset{|}{\text{H}_3\text{C—CH—CH}_3}} \xrightarrow{+\text{H}_2} \underset{\underset{\text{CHO}}{|}}{\text{H}_3\text{C—CH—CH}_3} + \text{HCo}(\text{CO})_{3,\,4} \quad (5a)$$

Since the reaction mechanism looks rather complicated on the first sight some explanations are added.

Cobalt is fed to the Oxo reactor as metal, Raney cobalt, cobalt hydroxide, oxide, carbonate, sulphate, as salt of fatty acids or in form of aqueous cobalt salt solutions.[4, 5, 32-42] Under the conditions of the Oxo reaction cobalt metal or said cobalt compounds form a mixture of cobalt hydrocarbonyl and dicobaltoctacarbonyl. The ratio of both depends on the reaction temperature and the hydrogen partial pressure. According to Gankin, Krinkin and Rudkowskii[43] the ratio is determined by the following equation:

$$\frac{\text{C}^2 \text{ (hydrocarbonyl)}}{\text{C (octacarbonyl)}} = K \cdot p_{\text{H}_2}$$

The temperature dependence of the equilibrium constant is given by:

$$K = 1 \cdot 365 - \frac{1900}{T}$$

Measurements of Gankin, Krinkin and Rudkowskii[43] indicate about equal quantities of octacarbonyl and hydrocarbonyl in the Oxo reactor under the usually applied Oxo-conditions in industrial operations.

Results of Wender, Sternberg and Orchin[27] and a number of other observations strongly support the assumption that the hydrocarbonyl rather than the dicobalt-octacarbonyl is the active catalyst, since stoichiometric amounts of cobalthydro-carbonyl react with olefins at atmospheric pressure and low temperatures to yield the same reaction products as are obtained under the conditions applying in industrial production.

The formation of a π-complex is assumed to be the first stage in the Oxo reaction, [4, 5, 18, 44-47]

There are two possible ways for a π-complex to be formed from propylene and cobalthydro-carbonyl: (a) by direct attack of the olefin on the central atom of the hydrocarbonyl with formation of an intermediate complex in which the coordination number of the Co-atom is enlarged by one, followed by elimination of 1 mole of carbon monoxide.

$$HCo(CO)_4 \xrightarrow{\text{olefin}} HCo(CO)_4 \text{ olefin} \longrightarrow HCo(CO)_3 \text{ olefin} + CO$$
$$(S_{n2}\text{—or associative mechanism})$$

and (b) by elimination of 1 mole of carbon monoxide from the hydrocarbonyl with formation of an intermediate complex having a coordination number which is lowered by one compared to the hydrotetracarbonyl, followed by addition of the olefin (S_{N1}– or dissociative mechanism).

Heck and Breslow[30, 48] favour the second way and exclude the first one. Quite a number of related reactions support this assumption,[49-58] e.g. the reaction of cobalthydrocarbonyl with phosphines [50-52]

$$HCo(CO)_4 + P(C_6H_5)_3 \longrightarrow HCo(CO)_3P(C_6H_5)_3 + CO$$

and of manganesehydrocarbonyl with ^{14}C labelled CO[47-50]

$$HMn(CO)_5 + 5\,^{14}CO \longrightarrow HMn(^{14}CO)_5 + 5CO$$

The S_{N1}– mechanism also provides a better explanation for the well-established fact that the rate of reaction decreases with increasing carbon monoxide partial pressure. It has to be mentioned at this stage of the discussion that the final proof for the formation of such π-complexes during the course of the hydroformylation and also the formation of alkylcobaltcarbonyls according to equation 3 under the conditions of the catalytic Oxo reaction are still lacking.[59, 60]

The subsequent steps of the reaction sequence appear sufficiently certain. It could be shown that alkylcobalttetracarbonyls are in equilibrium with acylcobalt-tricarbonyls[31] and the latter with acylcobalttetracarbonyls.[61]* Using ^{13}CO as

---

* See equation 4 on p. 336.

23

carbonylating agent Nagy-Magos, Bor and Marko[62] showed that the acyl group formation proceeds through migration of the alkyl group from the cobalt atom to a complex bonded carbonyl ligand. Carbon monoxide from the gas phase is taken up as a new ligand to the cobalt atom to take the position which was previously held by the alkyl group and to restore the original coordination number. These results are in line with earlier findings with related manganese carbonyls.[63-66]

The hydrogenation of the acyl compounds yields aldehydes (equation 5). This hydrogenation can either be effected by molecular hydrogen[31] as shown in equation 5 on page 336 or by cobalthydrocarbonyl[25, 26, 31, 61] as shown by the following equation:

$$CH_3{-}CH_2{-}CH_2{-}CO{-}Co(CO)_{3,\,4} \begin{cases} \xrightarrow{\;\;H_2\;\;} CH_3{-}CH_2{-}CH_2{-}CHO + HCo(CO)_{3,\,4} \\[2mm] \xrightarrow[HCo(CO)_4]{\quad\quad\quad} CH_3{-}CH_2{-}CH_2{-}CHO + Co_2(CO)_8 \end{cases}$$

It is very likely that under the conditions of the technical Oxo process (high temperature, high pressure, low catalyst concentration) the hydrogenation is mainly effected by molecular hydrogen, since the concentration of hydrocarbonyl is very low during the main reaction.[67] However, at the end of the Oxo reaction when there is no unconverted olefin left to trap the hydrocarbonyl but only acylcarbonyls, the hydrocarbonyl may to a certain extent act as hydrogenation agent. This assumption is supported by Marko's investigations[67] who could not find any cobalthydrocarbonyls in the crude Oxo product as long as acylcobaltcarbonyl was present.

Whether the formation of aldehydes in equations 5 or 5a respectively proceeds via the acylcobalttri- or -tetracarbonyls is not yet clear. Heck and Breslow[30, 31] as well as Karapinka and Orchin[68] assume that it proceeds via the tricarbonyls since it was demonstrated in experiments at low temperature that the hydrogenation of acylcarbonyls is inhibited by a CO atmosphere.[30, 64, 68] However, from the data available so far a reaction via the acyltetracarbonyls cannot be completely excluded.

It was already mentioned that both n- and iso-butyraldehyde are formed in the hydroformylation of propylene. The ratio of both strongly depends on the reaction conditions: the higher the CO-partial pressure and the lower the temperature the more straight chain aldehyde is formed. In industrial operations a ratio of 3–4 parts of n-butyraldehyde for every 1 of iso-butyraldehyde are formed at CO partial pressures of 140–150 atm (total pressure 280–300 atm) and temperatures of 140–160°C.

Both steric and electronic factors account for the fact that the formation of the straight chain aldehyde is preferred.[4, 5, 69, 70, 59, 71, 72] Isomerisation reactions were postulated to be the reason for the different n/iso-ratios obtained at different reaction conditions. For more details see ref. 14.

The assumption that the catalyst concentration is responsible for the isomerisation[73] was shown to be wrong.[74] It was already mentioned that modified metal carbonyls may also be used as hydroformylation catalysts and that the reaction mechanism follows essentially the same route with these catalysts. However, there are some differences which will be mentioned briefly.

Suitable modifiers are nitriles, tertiary amines, phosphines, phosphites, and other compounds with trivalent phosphorous atoms, arsines and stibines. Especially the named phosphorous compounds[53, 75-96] show valuable properties which e.g. have lead to their industrial application (Shell Oxo process, see also Chapter 3.3).

The differences between modified and unmodified hydrocarbonyls have been extensively discussed during the last years and mechanisms have been proposed.[4, 53, 80, 97-110, 91, 52]

Replacing a CO-molecule in the hydrocarbonyl by a phosphine ligand increases the thermal stability of the catalyst:

$$HCo(CO)_4 \xrightarrow[+PR_3]{-CO} HCo(CO)_3PR_3$$

because the trivalent phosphorous ligands are better σ-donors than CO but poorer π-acceptors.[59, 103, 111] As a consequence the remaining CO-ligands are more strongly bound[53] since the cobalt atom has the tendency to transfer the increased negative charge obtained from the phosphorous to the CO-ligands through π-back donation.[19, 20, 63, 111] The increased strength in the bond of the CO-ligands is the reason for the higher thermal stability and also for the lower reactivity since the dissociation of the hydrocarbonyl—which is the first step in the Oxo reaction sequence—will be lower than in cobalt hydrocarbonyl.

Phospine modified complexes exhibit an increased stereoselectivity towards the formation of straight chain aldehyde. This may be explained by two factors: some of the negative charge transferred from the phosphorous atom to the cobalt atom will be transferred to the hydrogen atom in this complex giving it a more hydridic character than in the cobalt hydrocarbonyl[4, 5, 59, 91, 103, 112] which leads to a preferred addition of this hydrogen atom to the more electropositive non-terminal C-atom in the olefin, whereas the $Co(CO)_3PR_3$ complex adds preferably to the terminal C-atom.[4, 5, 59, 103, 112] This electronic effect is supported by a steric effect. It is well established that the modified hydrocarbonyl is a trigonal bipyramid (dsp[3] —hybrid) the three CO-ligands being in the equatorial position and hydrogen and phosphorous in the axial position.[5, 59] The whole molecule is more bulky than cobalt hydrocarbonyl resulting in hindered formation of branched alkyl-$Co(CO)_3PR_3$.[5, 59, 71, 110]

9.3. INDUSTRIAL PROCESSES

9.3.1. General Remarks

The Oxo reaction is applied world-wide in about forty plants with a present world production of over 3·0 million tons per year. The largest plants, namely those of Ruhrchemie at Oberhausen (Germany) and of BASF at Ludwigshafen (Germany) have capacities well above 600 million pounds per year each (see also Chapter 3.8, Table III.2).

The operating plants produce aldehydes in the range of C_3-C_{18} which are either hydrogenated to the corresponding alcohols or subjected to aldol condensation prior to hydrogenation, resulting in alcohols which contain double the number of the carbon atoms as the starting aldehyde. Thus, e.g. 2-ethylhexanol is obtained by aldol condensation of 2 moles of n-butyraldehyde via 2-ethylhexenal.

Often smaller quantities of Oxo aldehydes are oxidised to carboxylic acids in adjacent plants.

Although nearly all olefins are suitable as feedstock for industrial purposes the most important starting material is propylene either as monomer or as oligomer. With only a single exception all commercial Oxo processes follow more or less the technique developed by Ruhrchemie AG in Oberhausen-Holten (Germany) in co-operation with BASF.[113–119] The single exception is the Shell process which will be discussed separately.

The first industrial application dates back to 1940 when an Oxo plant was designed by Ruhrchemie at Oberhausen-Holten for the production of detergent alcohols in the range of C_{10}–C_{18}. Today all Oxo plants are operated fully continuously. In general they consist of six sections:

1. Hydroformylation reactor
2. Catalyst removal section
3. Catalyst work-up and make-up section
4. Aldehyde distillation section
5. Aldehyde hydrogenation reactor
6. Alcohol distillation section.

As already mentioned, the plants may contain in addition aldolisation and oxidation units.

9.3.2. The Ruhrchemie Process and Related Processes

The Ruhrchemie process was the first technical Oxo process and except from the Shell process the flow sheets of all other processes are very similar. This fact may be explained by the loss of the Ruhrchemie Oxo patent rights as a consequence of World War II. Moreover, after the War Ruhrchemie at Holten was investigated by an Allied Control Commission and the special know-how of the process was laid open to everybody since it was published in BIOS reports,[120, 121] making possible a rapid development of industrial Oxo plants by other companies throughout the world using free of license the processing principle pioneered by Ruhrchemie.

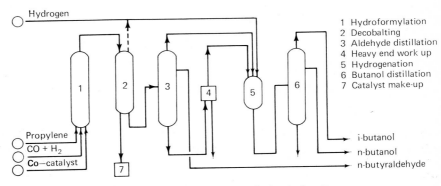

FIG. 9.2[5, 125]   *Flow Sheet of the Ruhrchemie Oxo Process*

The flow sheet of the Ruhrchemie plant at Oberhausen-Holten may serve to demonstrate the reaction principle: the flow sheets of the B A S F and of Mitsubishi processes are very similar.[4, 5, 122–124]

9.3.2.1. The catalyst

All Oxo plants use cobalt carbonyls or modified Co-carbonyls as catalyst. The use of the corresponding carbonyls of rhodium and ruthenium which do also catalyse the hydroformylation is, for economic reasons, so far restricted to the production of special high priced products or to pilot plant operations. From the excellent results obtained on small-scale trials using unmodified rhodium catalysts, their use on an industrial scale looks at least promising.[70, 40, 126–143] Ruthenium offers smaller advantages.[90, 134, 144–149]

In Oxo-operations the cobalt carbonyls or hydrocarbonyls respectively which are used as catalysts are generally formed *in situ* by feeding cobalt in the form of metal, as oxide, hydroxide or salt of an organic or inorganic acid, either in solution or suspension in the olefin itself or in high boiling distillation residues or water to the reactor. In special cases the carbonyls may also be preformed in a small carbonyl generating reactor from the same catalyst precursors. From this catalyst generating reactor the catalyst is then fed to the hydroformylation reactor.[150]

9.3.2.2. The hydroformylation reaction

Propylene and synthesis gas are fed through separate lines. Synthesis gas may either be obtained by gasification of coal, by partial combustion of heavy oil, or by steam reforming of methane or light hydrocarbons.[151, 152]

Since sulphur compounds such as H_2S and COS are catalyst poisons the synthesis gas has to be desulphurised prior to its use in Oxo processes. For the same reason propylene has also to be largely free of sulphur containing compounds. Other poisons which should only be present in very low concentrations are: oxygen, carbon dioxide and nitrogen compounds with dissociation constants larger than 10^{-8}, such as butylamine, piperidine and triethylamine; furthermore acetylene and metals like lead, mercury, bismuth and zinc must be absent.

Common reaction conditions in the Oxo reactor are 110–180°C, 200–300 atm, carbon monoxide to hydrogen ratios of 1 : 1 to 1 : 1·2 and residence times of 1–2 hours; the usual catalyst concentrations are 0·1–1·0 wt per cent of cobalt based on olefin. Solvents are generally not used in industrial processes.

The Oxo reactors are made from stainless steel. The most common reactor is the backmixed type which is also used in the production of ammonia or in liquid phase hydrogenation reactions. Some plants have plug-flow narrow tubular reactors instead.

The heat of reaction is removed either by internal cooling with cooling water under pressure or with evaporative cooling or in a few plants by circulating the reactor contents over external cooling systems when using the backmixed reactor type. The heat has to be exclusively removed through the wall in the case of the tubular reactor type.

9.3.2.3. Removal of catalyst

The reactor effluent is sent to a cobalt removal system. The main difference between the various processes derived from the Ruhrchemie process lies in the catalyst recovery technique.[4, 5] There are two different decobalting methods used in a number of variations: one is the thermal decomposition of the carbonyls after releasing the product from the reactor with simultaneous reduction of the carbon monoxide partial pressure. The decomposition of the carbonyls is effected by recycling cobalt-free hot reaction product or by introducing hot water or steam to the cobalt removal system.[153] An inert stripping gas may be applied simultaneously. The solid cobalt metal, cobalt oxide or cobalt hydroxide separating is mechanically removed.[4, 5, 154, 155] The other method is the removal of cobalt from the crude Oxo product by chemical treatment. This may be achieved by reaction with aqueous acids or salts with or without simultaneous application of an oxidising agent such as air. The resulting cobalt containing acid solution is worked-up and recycled to the reactor.[156, 157] Union Carbide, however, removes cobalt at elevated temperatures with sulphuric or acetic acid.[158] A variation of this method was developed by Kuhlmann who extracts cobalt hydrocarbonyl from the crude Oxo product with a dilute sodium bicarbonate solution. Subsequent acidification of the water layer reforms the hydrocarbonyl which then is recycled to the Oxo reactor as such.[159, 160] The decobaltation may be accelerated by oxidising agents such as oxygen or air.[161] Thus, B A S F feeds a 120°C hot Oxo raw-product to a small reactor in which it is mixed with dilute acetic acid and air under a pressure of 10 atm.[162]

9.3.2.4. Separation and working up of product   The product leaving the cobalt removal section which mainly consists of aldehydes besides smaller amounts of alcohols, formates and aldehyde condensation products[4, 5] is then fed to the aldehyde distillation section from which n- and iso-butyraldehyde are obtained in high purity by distillation. For corrosion reasons the aldehyde distillation columns are made of stainless steel. For specifications of the butyraldehydes, see Chapter 4.

The aldehydes may be hydrogenated in the hydrogenation section either in the gas or in the liquid phase using fixed bed nickel or copper catalysts. The resulting alcohols are then purified in the alcohol distillation section. As to the properties of n- and isobutanol see Chapter 5.1 and 5.2.

Yield and isomer distribution of the aldehydes produced by the propylene hydroformylation are shown in Table IX.1 in Section 9.3.4.

9.3.3. The Shell Process

As mentioned above the only Oxo process which differs considerably from the Ruhrchemie technique is the process developed by Shell.[82, 99] Shell uses a special catalyst system (see also Chapter 2.4 on the reaction mechanism). Cobalt salts of organic acids (e.g. 2-ethylhexanoic acid) trialkyl phosphines (e.g. tributylphosphine) and alkali, such as potassium hydroxide, are fed to the Oxo reactor as catalyst precursors. Under the conditions of the Shell Oxo process the actual catalyst —$HCo(CO)_3PR_3$—is formed from these compounds.

This catalyst is thermally more stable than cobalt hydrocarbonyl. Therefore the Oxo reaction needs only pressures around 100 atm compared with 200–300 atm in the other Oxo processes. However, the catalyst is much less reactive than cobalthydrocarbonyl. Thus at equal concentrations and even at 180°C which is far higher than normally used in the conventional Oxo process, it gives only reaction rates which are 1/5 to 1/6 of the rate of reaction which can be reached with cobalt hydro-carbonyl at such low temperatures as 145°C. This means that approximately a five- to six-fold increase in reactor volume is required for the same throughput compared with the classical processes. Since the hydrocarbonyl modified with phosphines is not only a hydroformylation but also a good hydrogenation catalyst the aldehydes formed are hydrogenated to alcohols in the Shell reactor at the tempera-ture of 180°C. This simplifies the Oxo process if alcohols are the desired reaction products. However, the Shell process is less suitable if aldehydes are the desired reaction products, e.g. in the manufacture of 2-ethylhexanol, the most important plasticiser alcohol, which is made by aldol condensation of n-butyraldehyde to 2-ethylhexenal followed by hydrogenation of the latter (see also Chapter 5.3).

The high thermal stability of the Shell catalyst allows the alcohols formed in the Oxo reaction to be separated from the catalyst and the heavy ends by a flash distilla-tion under pressure. The crude alcohols are purified by further distillations while the catalyst and the heavy ends are recycled to the Oxo reactor.[163] The fact that the hydrogenation can be achieved in the Oxo reactor itself partly makes up for the loss of capacity due to the lower activity of the catalyst in the Oxo step.

Compared with cobalt hydrocarbonyl the Shell catalyst gives also a higher n/iso ratio of the C_4-products formed (88:12 versus 80:20 in the Ruhrchemie process), which is an advantage since the straight chain compounds are generally of higher value than the branched ones.

In future this advantage may be less important since Ruhrchemie recently has developed a process to reconvert the undesired branched aldehydes into propylene and synthesis gas, the starting materials of the Oxo reaction[117, 118, 164, 165] (see also Chapter 3.7).

Severe disadvantages of the Shell process are the facts that about 10–15 per cent of the olefin feedstock is hydrogenated to propane against only 2–3 per cent in the classical processes (see Table IX.1) and that the catalyst is far more expensive than the one used in the other processes.

Shell operates two plants in the United States, one in England, and a very small one with a 8,000 tons per year capacity in India. In total about 6 per cent of the world capacity is produced via the Shell Oxo process.

9.3.4. Comparison of the Classical Oxo Processes with the Shell Process

In Table IX.1[125] a comparison is made between the Shell process and the conven-tional Oxo processes. The differences in the processes of Ruhrchemie, BASF, ICI, Kuhlmann, Mitsubishi and Union Carbide are relatively small compared to the difference of the Shell process towards the processes of the named companies.

On the basis of the data available so far, it looks as if the Shell process is well suited for the one-step production of n-butanol. In this respect it may compete with the classical processes, however, it is less suited for the manufacture of n-butyral-dehyde which is the starting material, e.g. for 2-ethylhexanol (via the aldol route) and for butyric acid (by oxidation of butyraldehyde). Moreover, it appears also less flexible for a switch in the production, e.g. from propylene to other olefin feedstock in a single reactor, since switching to another olefin with higher boiling point would require the use of another phosphine with a higher boiling point as catalyst modifier. Otherwise difficulties would arise in the pressure distillation.

TABLE IX.I

Hydroformylation of Propylene
(Comparison of the Classical Processes with the Shell Process)

	Classical processes (Ruhrchemie, BASF, ICI, Kuhlmann, Mitsubishi, UCC)	Shell process
Catalyst	$HCo(CO)_4$	$HCo(CO)_3P(n-C_4H_9)_3$
Reaction temp. (°C)	110–180	160–200
Pressure (atm)	200–350	50–100
Co-concentration (%)†	0·1–1	0·6
Throughput‡	0·5–1	0·1–0·2
Reaction products	aldehydes or alcohols	alcohols
Propylene hydrogenation to propane (%)	2–3	10–15
High boiling by-products (%)	3–7	1
Straight chain C_4-products (%)	76–81	88
Yield of straight chain C_4-products % of theory)§	67	67

* Compiled from patents and publications.
† Based on propylene.
‡ Volume of material per volume of reactor per hour.
§ Yield based on propylene converted.

9.3.5. Side and Consecutive Reactions

9.3.5.1. Side reactions   A number of side and consecutive reactions are observed in the hydroformylation. There are only two considerable side reactions with propylene as starting material in the Oxo reaction. One is the hydrogenation of propylene to propane. About 2–3 per cent of the propylene feed is hydrogenated to propane in the classical processes: in the Shell process the corresponding figures are 10–15 per cent depending on the reaction conditions.

$$H_3C{-}CH = CH_2 \xrightarrow[\text{Co-catalysts}]{\text{CO/H}_2} H_3C{-}CH_2{-}CH_3$$

The other is the formation of ketones according to the following equation:

$$2CH_2{-}CH = CH_2 + CO/H_2 \longrightarrow C_3H_6{-}\underset{\underset{O}{\|}}{C}{-}C_3H_6$$

Theoretically three ketones—dipropyl-, diisopropyl- and propyl-isopropyl ketone—can be formed. The percentage of ketones found among the reaction products is generally very low.

9.3.5.2. *Consecutive reactions* Of greater importance than the side reactions are the consecutive reactions. Parts of the aldehydes formed are hydrogenated in the Oxo reactor to the corresponding alcohols since the hydroformylation catalyst is to some extent also a homogenous hydrogenation catalyst.[4, 5] The amount of alcohol formed is different in the single processes. In the classical processes it is around 10 per cent, with phosphine modified cobalt catalysts of the Shell process alcohols are obtained as the main reaction products (see also Chapter 3, Sections 3.2 and 3.3).

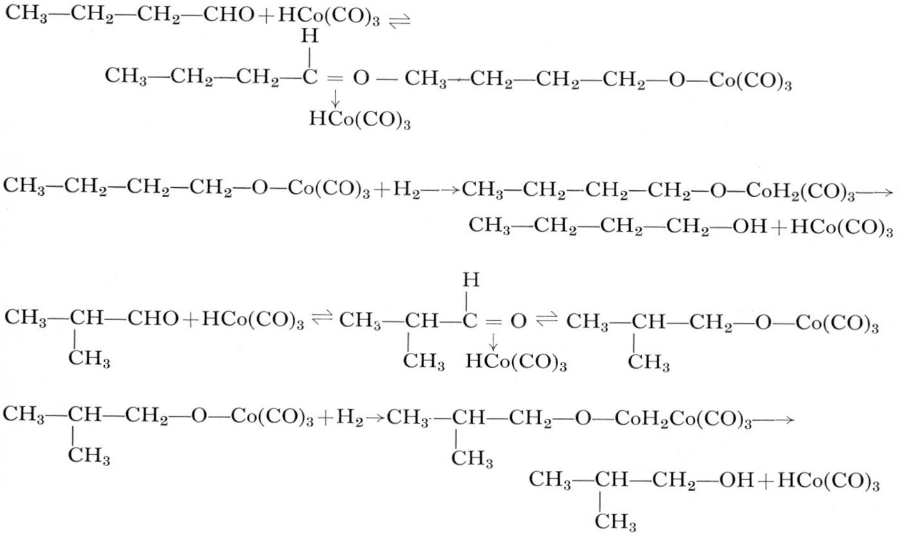

$$CH_3\!-\!CH_2\!-\!CH_2\!-\!CHO + HCo(CO)_3 \rightleftharpoons$$

$$\underset{\underset{HCo(CO)_3}{\downarrow}}{CH_3\!-\!CH_2\!-\!CH_2\!-\!\overset{\overset{H}{|}}{C}} = O - CH_3\!-\!CH_2\!-\!CH_2\!-\!CH_2\!-\!O\!-\!Co(CO)_3$$

$$CH_3\!-\!CH_2\!-\!CH_2\!-\!CH_2\!-\!O\!-\!Co(CO)_3 + H_2 \dashrightarrow CH_3\!-\!CH_2\!-\!CH_2\!-\!CH_2\!-\!O\!-\!CoH_2(CO)_3 \dashrightarrow$$

$$CH_3\!-\!CH_2\!-\!CH_2\!-\!CH_2\!-\!OH + HCo(CO)_3$$

$$\underset{\underset{CH_3}{|}}{CH_3\!-\!CH\!-\!CHO} + HCo(CO)_3 \rightleftharpoons \underset{\underset{CH_3}{|}\ \ \underset{HCo(CO)_3}{\downarrow}}{CH_3\!-\!CH\!-\!\overset{\overset{H}{|}}{C}} = O \rightleftharpoons \underset{\underset{CH_3}{|}}{CH_3\!-\!CH\!-\!CH_2\!-\!O\!-\!Co(CO)_3}$$

$$\underset{\underset{CH_3}{|}}{CH_3\!-\!CH\!-\!CH_2\!-\!O\!-\!Co(CO)_3} + H_2 \to \underset{\underset{CH_3}{|}}{CH_3\!-\!CH\!-\!CH_2\!-\!O\!-\!CoH_2Co(CO)_3} \dashrightarrow$$

$$\underset{\underset{CH_3}{|}}{CH_3\!-\!CH\!-\!CH_2\!-\!OH + HCo(CO)_3}$$

Another consecutive reaction is the formation of formates which can be understood by the following reaction sequence:

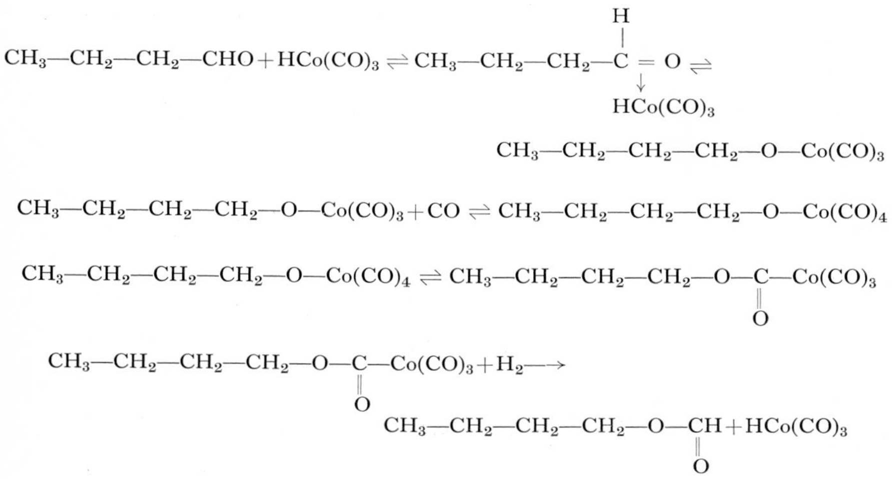

$$CH_3\!-\!CH_2\!-\!CH_2\!-\!CHO + HCo(CO)_3 \rightleftharpoons \underset{\underset{HCo(CO)_3}{\downarrow}}{CH_3\!-\!CH_2\!-\!CH_2\!-\!\overset{\overset{H}{|}}{C}} = O \rightleftharpoons$$

$$CH_3\!-\!CH_2\!-\!CH_2\!-\!CH_2\!-\!O\!-\!Co(CO)_3$$

$$CH_3\!-\!CH_2\!-\!CH_2\!-\!CH_2\!-\!O\!-\!Co(CO)_3 + CO \rightleftharpoons CH_3\!-\!CH_2\!-\!CH_2\!-\!CH_2\!-\!O\!-\!Co(CO)_4$$

$$CH_3\!-\!CH_2\!-\!CH_2\!-\!CH_2\!-\!O\!-\!Co(CO)_4 \rightleftharpoons \underset{\underset{O}{\|}}{CH_3\!-\!CH_2\!-\!CH_2\!-\!CH_2\!-\!O\!-\!C\!-\!Co(CO)_3}$$

$$\underset{\underset{O}{\|}}{CH_3\!-\!CH_2\!-\!CH_2\!-\!CH_2\!-\!O\!-\!C\!-\!Co(CO)_3} + H_2 \dashrightarrow$$

$$\underset{\underset{O}{\|}}{CH_3\!-\!CH_2\!-\!CH_2\!-\!CH_2\!-\!O\!-\!CH + HCo(CO)_3}$$

$$CH_3-\underset{\underset{CH_3}{|}}{CH}-CHO + HCo(CO)_3 \rightleftharpoons CH_3-\underset{\underset{CH_3}{|}}{CH}-\overset{\overset{H}{|}}{C}=O \rightleftharpoons CH_3-\underset{\underset{CH_3}{|}}{CH}-CH_2-O-Co(CO)_3$$
$$\underset{HCo(CO)_3}{}$$

$$CH_3-\underset{\underset{CH_3}{|}}{CH}-CH_2-O-Co(CO)_3 + CO \rightleftharpoons CH_3-\underset{\underset{CH_3}{|}}{CH}-CH_2-O-Co(CO)_4$$

$$CH_3-\underset{\underset{CH_3}{|}}{CH}-CH_2-O-Co(CO)_4 \rightleftharpoons CH_3-\underset{\underset{CH_3}{|}}{CH}-CH_2-O-\underset{\underset{O}{||}}{C}-Co(CO)_3$$

$$CH_3-\underset{\underset{CH_3}{|}}{CH}-CH_2-O-\underset{\underset{O}{||}}{C}-Co(CO)_3 + H_2 \longrightarrow CH_3-\underset{\underset{CH_3}{|}}{CH}-CH_2-O-\underset{\underset{O}{||}}{CH} + HCo(CO)_3$$

Acetals are formed from the aldehydes and the alcohols present in the reaction mixture.

$$CH_3-CH_2-CH_2-\underset{\underset{H}{}}{\overset{\overset{O}{||}}{C}} + 2HO-CH_2-CH_2-CH_2-CH_3 \rightarrow \qquad (1)$$

$$CH_3-CH_2-CH_2-\underset{\underset{O-(CH_2)_3-CH_3}{\diagdown}}{\overset{\overset{H}{\diagup}}{C}}-O-(CH_2)_3-CH_3$$

$$CH_3-CH_2-CH_2-\underset{\underset{H}{}}{\overset{\overset{O}{||}}{C}} + 2HO-CH_2-\underset{\underset{CH_3}{|}}{CH}-CH_3 \longrightarrow \qquad (2)$$

$$CH_3-CH_2-CH_2-\underset{\underset{O}{\diagdown}}{\overset{\overset{H}{\diagup}}{C}}-O-CH_2-\underset{\underset{CH_3}{|}}{CH}-CH_3$$
$$\underset{\underset{\underset{CH_3}{|}}{CH_2-CH-CH_3}}{\diagdown}$$

$$CH_3-\underset{\underset{CH_3}{|}}{CH}-\underset{\underset{H}{}}{\overset{\overset{O}{||}}{C}} + 2HO-CH_2-\underset{\underset{CH_3}{|}}{CH}-CH_3 \rightarrow CH_3-\underset{\underset{CH_3}{|}}{CH}-\underset{\underset{O}{\diagdown}}{\overset{\overset{H}{\diagup}}{C}}-OCH_2-\underset{\underset{CH_3}{|}}{CH}-CH_3 \qquad (3)$$
$$\underset{\underset{\underset{CH_3}{|}}{CH_2-CH-CH_3}}{\diagdown}$$

$$CH_3-\underset{\underset{CH_3}{|}}{CH}-\underset{\underset{H}{}}{\overset{\overset{O}{||}}{C}} + 2HOCH_2-CH_2-CH_2-CH_3 \longrightarrow CH_3-\underset{\underset{CH_3}{|}}{CH}-\underset{\underset{O-(CH_2)_3CH_3}{\diagdown}}{\overset{\overset{H}{\diagup}}{C}}-O-(CH_2)_3CH_3 \qquad (4)$$

$$CH_3-CH_2-CH_2-\overset{\displaystyle O}{\overset{\|}{C}}+HOCH_2-CH_2-CH_2-CH_3+HOCH_2-\underset{\underset{\displaystyle CH_3}{|}}{CH}-CH_3\longrightarrow$$

$$CH_3-CH_2-CH_2-\overset{\displaystyle H}{\underset{\displaystyle O-CH_2-\underset{\underset{\displaystyle CH_3}{|}}{CH}-CH_3}{\overset{\diagup}{\underset{\diagdown}{C}}-O-(CH_2)_3-CH_3}} \qquad (5)$$

$$CH_3-\underset{\underset{\displaystyle CH_3}{|}}{CH}-\overset{\displaystyle O}{\overset{\|}{C}}+HOCH_2-CH_2-CH_2-CH_3+HOCH_2-\underset{\underset{\displaystyle CH_3}{|}}{CH}-CH_3\longrightarrow$$

$$CH_3-\underset{\underset{\displaystyle CH_3}{|}}{CH}-\overset{\displaystyle H}{\underset{\displaystyle O-CH_2-\underset{\underset{\displaystyle CH_3}{|}}{CH}-CH_3}{\overset{\diagup}{\underset{\diagdown}{C}}-O-(CH_2)_3-CH_3}} \qquad (6)$$

Furthermore 2-ethylhexenal and 2-ethyl-4-methylpentenal are formed by aldol condensation of n-butyraldehyde and a mixed aldol condensation of n- and iso-butyraldehyde respectively.

$$2CH_3-CH_2-CH_2-CHO\longrightarrow\left[CH_3-CH_2-CH_2-\underset{\underset{\displaystyle OH}{|}}{CH}-\underset{\underset{\displaystyle C_2H_5}{|}}{CH}-CHO\right]\overset{\longrightarrow}{\underset{-H_2O}{}}$$

$$CH_3-CH_2-CH_2-CH=\underset{\underset{\displaystyle C_2H_5}{|}}{C}-CHO$$

$$CH_3-\underset{\underset{\displaystyle CH_3}{|}}{CH}-CHO+CH_3-CH_2-CH_2-CHO\longrightarrow\left[CH_3-\underset{\underset{\displaystyle CH_3}{|}}{CH}-\underset{\underset{\displaystyle OH}{|}}{CH}-\underset{\underset{\displaystyle C_2H_5}{|}}{CH}-CHO\right]\overset{\longrightarrow}{\underset{-H_2O}{}}$$

$$CH_3-\underset{\underset{\displaystyle CH_3}{|}}{CH}-CH=\underset{\underset{\displaystyle C_2H_5}{|}}{C}-CHO$$

Parts of them are hydrogenated to the saturated aldehydes or to 2-ethylhexanol or 4-methyl-2-ethylpentanol respectively.

Besides the products mentioned a great number of other by-products are formed each in very small amounts most of them having higher molecular weights. The sum of them is generally named 'thick oil' or 'heavy ends' in publications or patents.

9.3.6. Working-up and Utilisation of Side Products

Whereas the unconverted propylene and the propane formed by hydrogenation are vented with the off-gas the ketones and the products of the consecutive reactions remain in the bottoms of the aldehyde distillation. Special methods were developed by the various manufacturers for the working-up of these heavy ends. Some of them hydrogenate this residue under severe conditions in order to obtain additional alcohols,[123, 166, 167] others work-up for aldehyde[168] and some treat the bottoms with alkali, e.g. with sodium hydroxide at high temperatures (260–350°C) in order to produce carboxylic acids.[168-173] Even after these severe treatments some compounds remain unchanged. This part of the bottoms the so-called 'thick oil' is frequently used for preparation of catalyst slurries,[174] for flotation purposes[166] or as fuel oil.

9.3.7. Catalytic Cracking of Isobutyraldehyde into Propylene, CO and H_2 (see Fig. 9.3)

Isobutyraldehyde which is formed in a ratio of 1 part per every 3–4 parts of n-butyraldehyde must, from the commercial viewpoint, also be regarded a by-product since the production far exceeds the demand. The balance can only be credited with heating value.

Recently Ruhrchemie has developed a new process in which this unwanted by-product is catalytically cracked into propylene, carbon monoxide and hydrogen, the starting materials for the Oxo synthesis.[104, 165, 175]

$$CH_3-\underset{\underset{CH_3}{|}}{CH}-CHO \xrightarrow{\text{catalyst}} CH_3-CH=CH_2+CO+H_2$$

Recycling the cracked products to the Oxo synthesis results in complete reaction of propylene to n-C_4-products. The cracking reaction is carried out in the gas phase in a tube reactor at temperatures between 250–350°C at atmospheric or slightly higher pressure over a fixed bed catalyst containing a noble metal at a residence time of 2–4 sec and in presence of steam.

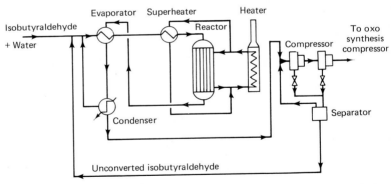

FIG. 9.3[168] *Ruhrchemie Process for the Catalytic Cracking of iso-Butyraldehyde*

About 80 per cent of the isobutyraldehyde feed is cracked. After separation of water and unconverted isobutyraldehyde the cracked gas is compressed and cycled to the Oxo plant. The unconverted isobutyraldehyde is recycled to the cracking reactor. The cracking catalyst is regenerable by burning the deposits with air in the reactor. A 300 kg per day pilot plant is at present operated by Ruhrchemie in Oberhausen-Holten. The process is available for licensing.

9.3.8. Oxo Capacities

TABLE IX.2

Oxo Plant Capacities 1970 (estimated)

Location	Company	Process	tons/year
Australia	C. S. R. Chemicals	Ruhrchemie	24,000[176]
Austria	Osterreichische Stickstoffwerke	BASF	45,000[177a]
Bulgaria	Technoimport		30,000[178b]
Canada	BASF	BASF	45,000[179c]
	Gulf Oil	Ruhrchemie	56,000[180]
Czechoslovakia	Chemapol	Mitsubishi	20,000[5b]
France	Kuhlmann	Kuhlmann	100,000[5]
	Oxochimie	Ruhrchemie	75,000[181]
Germany	BASF	BASF	370,000[182]
	Chemische Werke Hüls	BASF	90,000[183d]
	Farbwerke Hoechst	Ruhrchemie	56,000[185]
	Ruhrchemie	Ruhrchemie	320,000[5,184, 185]
Great Britain	ICI	ICI	200,000[186]
	Shell	Shell	110,000[187]
India	Nocil	Shell	8,000[5]
Italy	Montedison	Montecatini	30,000[188]
	Sincat	Union Carbide	50,000[188]
Japan			
	Chisso	Ruhrchemie	52,000[189]
	Kyowa Yuka	Ruhrchemie	71,000[190]
	Mitsubishi	Mitsubishi	68,000[5]
	Nissan Petrochem. Co.	Kuhlmann	80,000[191c]
	Tonen	Enjay	15,000[192]
Netherlands	Konam	Kuhlmann	50,000[5]
Puerto Rico	Grace/Corco	BASF	115,000[193]
Rumania	Riminicu Vilcea	Ruhrchemie	37,000[194]
Spain	BASF/Arrahona	BASF	28,000[195]
USA	Dow Badische	BASF	91,000[196]
	Enjay Chemical Co.	Esso	91,000[196]
	Getty Oil/Houdry	Kuhlmann	18,000[196]
	Gulf Oil	Gulf	23,000[197]
	Monsanto	Monsanto/Kuhlmann	68,000[198]
	Shell Chemical	Shell	159,000[196]
	Texas Eastman	Eastman	125,000[196]
	Union Carbide	Union Carbide	210,000[196]
	US Steel Chemicals	Standard Oil	32,000[196]
		Total	2,962,000

[a] 45,000 tons/annum enlargement planned.
[b] planned.
[c] under construction.
[d] 50,000 tons/annum enlargement planned.
[e] 24,000 tons/annum enlargement planned.

9.4. PROPERTIES OF N- AND ISO-BUTYRALDEHYDE

9.4.1. Physical Properties

9.4.1.1. Physical properties of n-butyraldehyde n-butyraldehyde is a mobile, colourless, highly flammable liquid with a characteristic pungent odour. It is readily soluble in all proportions in most of the organic solvents such as alcohols, ethers. benzene etc. but only slightly soluble in water (see Table IX.3).

TABLE IX.3

Solubility of n-Butyraldehyde in Water and of Water in n-Butyraldehyde[184, 185]

Solubility of n-butyraldehyde in water		*Solubility of Water in n-butyraldehyde*	
Temperature (°C)	*wt % n-butyraldehyde dissolved*	*Temperature (°C)*	*wt % water dissolved*
0	8·7	0	3·01
10	7·9	10	3·08
20	7·1	20	3·17
30	6·3	30	3·27
40	5·4	40	3·39

n-Butyraldehyde forms azeotropic mixtures with many solvents, see e.g. Table IX.4.

TABLE IX.4

Azeotropic Mixtures with n-Butyraldehyde[184, 185]

Added component	*Wt % in the vapour*	*Boiling point (°C at 760 mm)*
Binary mixtures		
ethanol	60·6	70·7
water	6·5	68·0
Ternary mixture		
ethanol/water	11/7	62·7

TABLE IX.5

Physical Properties of n-Butyraldehyde[184, 185]

Molecular weight	72·10	Specific gravity (d_4^{20})	0·809
Boiling point (°C, 760 mm)	75·7		
Freezing point (°C)	−97	Refractive index (n_D^{20})	1·3814
Ignition temp. (°C)	195		
Flash point (°C)	below −5	Surface tension (dyne/cm at room temp.)	24·6
		Viscosity at 20°C (cP)	0·447
		Dielectric constant at room temperature	14·9

9.4.1.2. Physical properties of iso-butyraldehyde Iso-butyraldehyde is a mobile, colourless, highly flammable liquid with a pungent, penetrating odour. Like n-butyraldehyde, iso-butyraldehyde is readily soluble in most of the organic solvents like alcohols, ethers, ketones, benzene etc. The solubility in water is somewhat higher than that of n-butyraldehyde, e.g. 10 per cent by volume at 20 °C.

Iso-butyraldehyde forms an azeotropic mixture with water, boiling at 60·5 °C. The mixture contains 6 wt per cent of water.

TABLE IX.6

Physical Properties of iso-Butyraldehyde[184, 185]

Molecular weight	72·10	Specific gravity (d_4^{20})	0·791
Boiling point (°C, 760 mm)	64·5		
Freezing point (°C)	−66	Refractive index (n_D^{20})	1·3742
Ignition temp. (°C)	225		
Flash point (°C) below	−15	Surface tension (dyne/cm at	
Boiling point of azeotrope		room temp.)	23·2
with H_2O at 760 mm °C/%		Viscosity at 20 °C (cP)	0·445
water	60·5/6·0	Dielectric constant at room	
		temperature	13·6

9.4.2. Chemical Properties and Economical Applications

9.4.2.1.   Chemical properties and economical applications of n-butyraldehyde n-Butyraldehyde undergoes the ordinary aldehyde reactions. Hydrogenation, in the presence of catalysts yields n-butanol, oxidation yields n-butyric acid.

The aldehyde group is one of the most reactive and versatile in its transformations in organic chemistry. n-Butyraldehyde is able to enter condensation or addition reactions with itself as well as with other aldehydes and ketones and with a great number of other organic compounds such as amines, alcohols, nitriles and others.

The main outlet of n-butyraldehyde is as intermediate for the manufacture of a great number of industrial products:

n-Butanol, the hydrogenation product of n-butyraldehyde (see 9.5.1) is mainly used as solvent[199] and for the production of plasticisers.

n-Butyric acid obtained by oxidation of n-butyraldehyde (see 9.5.4) has also found greater application.

Continuously increasing amounts of n-butyraldehyde are converted to 2-ethyl-hexanol by the aldol condensation and subsequent catalytic hydrogenation (see 9.5.3). 2-Ethylhexanol is required in large tonnage for the manufacture of plasticisers such as dioctylphthalate (DOP) and a number of other dicarboxylates. Smaller amounts are used as solvent.[199]

Trimethylolpropane, an intermediate in the production of alkyd resins, plasticisers and air-drying oils is another large outlet for n-butyraldehyde. It is formed by reaction of 1 mole of n-butyraldehyde with 2 moles of formaldehyde and subsequent hydrogenation (see 9.5.6).

n-Butyraldehyde condenses with phenol to give oil-soluble resins and with urea to give alcohol-soluble resins.

The condensation products of n-butyraldehyde with polyvinylalcohol (MOWITAL® of Farbwerke Hoechst AG)[184, 185] and with butylamine, thiourea, diphenylguanidine or methylthiocarbamate are used for the manufacture of laminated safety glass and as binders for lacquers.

The acetals of n-butyraldehyde with various alcohols are used as solvents for cellulose, resins and rubber and in the production of pharmaceuticals. Furthermore n-butyraldehyde finds use as an intermediate in the manufacture of pharmaceuticals and pesticides, they discourage 'game' birds from stripping the bark of trees.[215]

9.4.2.2. Chemical properties and economical application of iso-butyraldehyde
Iso-butyraldehyde undergoes like n-butyraldehyde the characteristic reactions of aldehydes. Hydrogenation yields iso-butanol and oxidation iso-butyric acid. It is also susceptible to a great number of condensation and addition reactions. Thereby, amongst others, compounds with neopentyl structure may be formed (see 9.5.7).

Iso-butyraldehyde is used for the manufacture of quite a number of industrial products. The most important one is iso-butanol (see 9.5.2). Iso-butanol can be used to a certain extent for many applications where n-butanol is also used.

Iso-butyric acid which is used for the production of reaction accelerators and as an intermediate for quite a number of flavourings and perfumes (esters) is formed by oxidation of iso-butyraldehyde (see 9.5.5).

The addition reaction of iso-butyraldehyde and formaldehyde leads to hydroxypivalaldehyde[200–207] which can be hydrogenated to neopentylglycol, a starting material for laquers, resins, fibres and lubricants (see 9.5.8). Fertilisers which release nitrogen slowly are manufactured by reacting iso-butyraldehyde with urea.[12] Iso-butyraldehyde is also used in the preparation of pharmaceuticals, amino acids and vitamins such as valine, leucine and pantothenic acid.[209–213, 243]

Besides the applications described there are a number of other uses.[184] However, it must be mentioned that at present there is an overcapacity of iso-butyraldehyde since the Oxo process yields both n- and iso-butyraldehyde in a rate around 4:1 and the production of iso-butyraldehyde exceeds the demand. Therefore large quantities of iso-butyraldehyde have to be used as fuel.

Since one can foresee a further rapid growth of the Oxo capacities the catalytic cracking process of iso-butyraldehyde, recently developed by Ruhrchemie may become of great importance. By this method the undesired iso-butyraldehyde is reconverted into propylene, hydrogen and carbon monoxide, the starting materials of the Oxo synthesis (see 9.3.7).

9.4.3. Analytical

n- and iso-butyraldehyde can be characterised *inter alia* as semi-carbazones, m.p. 106°C or 125–126°C respectively and as 2,4-dinitrophenylhydrazones, m.p. 122°C or 183–183·5°C respectively. Amongst the great number of analytical procedures described, reference must be made to gravimetric determinations using sodium bisulphite[214-217], hydroxylamine[218-226] or dimedone,[227-230] and to the determination of water by the Karl Fischer method.[231]

About half of the n-butanol output is used as solvent for oils, waxes, natural resins, plastics, and fats. The main customers are the paint and varnish industry. Although n-butanol is not a true solvent for nitrocellulose it is extensively employed in the manufacture of nitrocellulose lacquers as a valuable blending agent in combination with toluene and other specific solvents such as various esters. Addition of 5–10 per cent of n-butanol prevents the undesirable blushing of drying lacquer films which takes place when either excessive amounts of blending agents or too large proportions of highly volatile solvents are used. Besides this, n-butanol may be used to regulate the viscosity of lacquer solutions and improves the quality of lacquers in respect of good flow properties and clear drying.

n-Butanol is also utilised as a developing agent in thin layer or paper chromatography and as an extraction solvent in numerous chemical operations. Frequently it is used as the solvent component in floor cleaners or stain removers. Another important field of application is its use as an additive for solvent mixtures.

Some n-butanol is converted to esters of which the butylesters of acetic, butyric, valeric, acrylate, glycolic and lactic acid must be mentioned. They are excellent solvents, e.g. for chlorinated rubber or polystyrene. The esters in turn can be blended with up to 20 per cent of n-butanol without loss of solvent quality.

A considerable amount of n-butanol is used in the manufacture of plasticisers, the best-known of which being di-n-butylphthalate (DBP), besides this ester the butyl adipates, -sebacates, -oleates, -azelates and -stearates are used as plasticisers for many polymers and lacquer binders. The phosphoric esters of n-butanol are also widely used as plasticisers for PVC and cellulose derivatives. n-Butanol has found further application as flotation aid, humectant for guncotton in the explosive industry, and in paper production to prevent the foaming of dyestuff solutions. It is also used in the production of urea or melamine-formaldehyde resins where these are to be used in paints and need to be compatible with paint solvents.

9.5. SOME INDUSTRIAL DERIVATIVES

9.5.1. n-Butanol

n-Butanol is obtained by catalytic hydrogenation of n-butyraldehyde either in the gas phase or in the liquid phase. Normally nickel or copper catalysts are used.

$$H_3C—CH_2—CH_2—CHO + H_2 \xrightarrow{\text{catalyst}} H_3C—CH_2—CH_2—CH_2OH$$

n-Butanol is a colourless liquid of neutral reaction and characteristic odour. It is miscible with organic liquids in nearly all proportions and with water at 20°C in a proportion of 1:12. For special industrial purposes the formation of an 37 wt per cent water-containing azeotropic mixture with a boiling point of 92.3°C is of importance. The physical data are compiled in Table IX.7.

24

TABLE IX.7

Physical Properties of n-Butanol[184, 185]

Molecular weight	74·1	Specific gravity (d_4^{20})	0·8098
Boiling point (°C, 760 mm)	117·7		
Freezing point (°C) below	−80	Refractive index (n_{20}^D)	1·3991
Ignition temp. (°C)	340		
Flash point (°C)	34	Surface tension (dyne/cm at	
Boiling point of azeotrope		room temp.)	22·3
with H_2O at 760 mm °C/%		Viscosity at 20°C (cP)	3·0
water	92·3/37·0	Dielectric constant at room	
		temperature	17·8
		Evaporation number (ether	
		= 1)	33
		Lower ignition limit in air,	
		760 mm (% by vol)	1·4

9.5.2. Iso-butanol

Iso-butanol is obtained by catalytic hydrogenation of iso-butyraldehyde either in the gas or in the liquid phase. The catalysts used are the same as in the hydrogenation of n-butyraldehyde, namely fixed bed nickel or copper catalysts.

$$H_3C—CH—CHO + H_2 \xrightarrow{\text{catalyst}} H_3C—CH—CH_2OH$$
$$\quad\quad |\quad\quad\quad\quad\quad\quad\quad\quad\quad\quad\quad\quad |$$
$$\quad\quad CH_3 \quad\quad\quad\quad\quad\quad\quad\quad\quad CH_3$$

Iso-butanol is miscible with the common organic solvents in nearly all proportions. It takes up 19 wt per cent of water at 20°C while about 8 per cent of iso-butanol can dissolve in water. The azeotropic mixture with water, boiling at 89·9°C contains 66·8 per cent of iso-butanol.

TABLE IX.8

Physical Properties of Iso-butanol[184, 185]

Molecular weight	74·1	Specific gravity (d_4^{20})	0·803
Boiling point (°C, 760 mm)	107·9		
Freezing point (°C) below	−108	Refractive index (n_D^{20})	1·3959
Ignition temp. (°C)	430		
Flash point (°C)	28	Surface tension (dyne/cm at	
Boiling point of azeotrope		room temp.)	23·0
with H_2O at 760 mm °C/%		Viscosity at 20°C	4·0
water	89·9/33·2	Dielectric constant at room	
		temperature	18·8
		Evaporation number (ether	
		= 1)	24
		Lower ignition limit in air,	
		750 mm (% by vol)	1·7

The major part of iso-butanol is consumed by the solvent industry. It is suitable as solvent for fats, natural resins and polymers as well as for various condensation products. Since the solvent properties of isobutanol can be compared to that of n-butanol it can frequently replace this alcohol. Thus it also prevents the blushing of drying lacquer films and facilitates processing, flow and distribution of alkyd resins and oil lacquers by reducing their viscosity. The solubility of phenol/formaldehyde resins, however, is poorer in iso-butanol than in n-butanol. Like n-butanol, iso-butanol is also not a true solvent for nitrocellulose, but may serve as a blending agent or additive for nitrocellulose. It can also be used like n-butanol as a humectant for highly nitrated guncotton.

Owing to its low price it gained interest as a solvent component in industrial cleaning agents and printing inks. As an extraction solvent iso-butanol finds application in the same fields like n-butanol.

Diisobutylphthalate (DIBP) and triisobutylphosphate are frequently used as plasticisers. DIBP often replaces DBP in lacquer formulations, especially for PVC and nitrocellulose lacquers or chlorinated rubber. Diisobutylphthalate migrates more rapidly and its low temperature performance is slightly inferior to the corresponding normal esters.

The 2,4-dichloro- and 2,4,5-trichloroisobutyl phenyoxyacetate are known as efficient herbicides.

9.5.3. 2-Ethylhexanol

2-Ethylhexanol, also marketed under the name octanol, is produced by aldol condensation of n-butyraldehyde with subsequent hydrogenation of the aldolisation product 2-ethylhexenal. The hydrogenation may either be achieved in the gas or in the liquid phase in presence of fixed bed nickel or copper catalysts. As mentioned earlier many Oxo plants have adjacent 2-ethylhexanol facilities.

$$2H_3C-CH_2-CH_2-CHO \rightarrow \left[\begin{array}{c} CH_3-CH_2-CH_2-\underset{\underset{OH}{|}}{CH}-\underset{\underset{C_2H_5}{|}}{CH}-CHO \end{array} \right] \overset{\longrightarrow}{-H_2O}$$

$$H_3C-CH_2-CH_2-CH=\underset{\underset{C_2H_5}{|}}{C}-CHO \xrightarrow[\text{catalyst}]{H_2} H_3C-CH_2-CH_2-CH_2-\underset{\underset{C_2H_5}{|}}{CH}-CH_2OH$$

2-Ethylhexanol is a colourless liquid of characteristic odour, miscible in all proportions with ethanol, ether and various organic solvents. The solubility in water is 0·1 per cent at 20°C while 2-ethylhexanol itself takes up 2·6 wt per cent of water. The physical properties are collected in Table IX.9.

Esterification of 2-ethylhexanol with phthalic anhydride leads to di-2-ethylhexyl-phthalate (DOP) which is the most important plasticiser, in particular for PVC. Of special importance is its low volatility, excellent gelating, good resistance to high and low temperature and to water, its excellent electrical properties and physiological harmlessness.

Numerous 2-ethylhexyl esters of dibasic acids, especially of aliphatic dibasic

acids have also gained importance as plasticisers and as intermediates in the manufacture of synthetic lubricants or as additives for luboils. Their low pour points,
low volatility and favourable temperature/viscosity coefficient make the lubricants
produced on this basis particularly suitable for the use under extreme temperature
conditions. 2-Ethylhexanol may also serve as solvent.[199] It dissolves a number of
synthetic and natural resins and is used in combination with other solvents, in the
paint and lacquer industry. It may be used as a viscosity regulating agent to improve
the flow of alkyd resin lacquers and stove enamels based on phenol or melamine
formaldehyde resins. It also serves as a dispersing agent for the distribution of ink
and lacquer pigments as well as a slowly evaporating solvent in disinfectants or
insecticide sprays. In textile dying it is used mainly because of its antifoaming
effect.

TABLE IX.9

Physical Properties of 2-Ethylhexanol[184, 185]

Molecular weight	130·2	Specific gravity (d_4^{20})	0·8329
Boiling point (°C/ 760 mm)	184·7		
Freezing point (°C)	−75	Refractive index (n_D^{20})	1·4315
Ignition temp. (°C)	250		
Flash point (°C)	74	Surface tension (dyne/cm at	
Boiling point of azeotrope		room temp.)	25·0
with H_2O at 760 mm °C/%		Viscosity at 20°C (cP)	10·0
water	99·1/80·0	Dielectric constant at room	
		temperature	7·7
		Evaporation number (ether	
		= 1)	∼600
		Lower ignition limit in air,	
		760 mm (% by vol)	1·8

9.5.4. n-Butyric Acid

n-Butyric acid is produced by non-catalytic liquid phase oxidation of n-butyraldehyde
at lower temperatures.[232]

$$H_3C—CH_2—CH_2—CHO \xrightarrow{1/2\ O_2} H_3C—CH_2—CH_2—COOH$$

The acid is a colourless liquid with a pungent odour in concentrated form and a
sweet like odour when diluted. It is miscible with water and most organic solvents.
The physical data of n-butyric acid are collected in Table IX.10.

The major outlet for n-butyric acid — is for the manufacture of cellulose acetobutyrate, the preferred cellulose ester for molding powders, especially for automotive applications such as steering wheels. Due to its high resistance to moisture
heat and light cellulose acetobutyrate is particularly suitable for cable sheetings and
fishing equipments. The n-butyrates are used in the lacquer and plastics industry
as solvents or as perfumes because of their fragrant fruity odour and taste.

TABLE IX.10
Physical Properties of n-Butyric Acid[184, 185]

Molecular weight	88·1
Boiling point (°C, 760 mm)	164·0
Freezing point (°C)	−5·5
Flash point (°C)	84
Specific gravity (d_4^{20})	0·9577
Refractive index (n_D^{20})	1·3991
Viscosity at 20°C (cP)	1·6

9.5.5. Isobutyric Acid

Isobutyric acid is produced by liquid phase oxidation of isobutyraldehyde with oxygen or oxidising agents. The oxidation needs no catalyst.[233]

$$H_3C—CH—CHO \xrightarrow{1/2\,O_2} H_3C—CH—COOH$$
$$\underset{CH_3}{|} \qquad\qquad \underset{CH_3}{|}$$

Isobutyric acid is a colourless liquid of characteristic odour, miscible in all proportions with most of the organic solvents. It is partially miscible with water, e.g. at 20°C it forms a 55 per cent water containing mixture, while water takes up 22 per cent of isobutyric acid at this temperature. Above 26°C isobutyric acid and water are miscible in all proportions. The physical properties are collected in Table IX.11.

TABLE IX.11
Physical Properties of Isobutyric Acid[184, 185]

Molecular weight	88·1
Boiling point (°C, 760 mm)	154·7
Freezing point (°C)	−46·1
Flash point (°C)	74
Specific gravity (d_4^{20})	0·9504
Refractive index (n_D^{20})	1·3930
Viscosity at 20°C (cP)	1·1

The use of isobutyric acid parallels closely that of n-butyric acid. It is mainly applied for the production of the corresponding esters which are used as solvents or perfumes. It finds also use for the manufacture of lacquers and plasticisers. One of the largest outlet is as an intermediate in the production of isobutyronitrile which is converted to isobutylamidine hydrochloride, the raw material for the insecticide 'Diazinon'.

9.5.6. Trimethylolpropane

Trimethylolpropane is obtained by reaction of n-butyraldehyde with formaldehyde.[234]

$$H_3C-CH_2-CH_2-CHO+3HCHO \xrightarrow{NaOH} H_3C-CH_2-\underset{\underset{CH_2OH}{|}}{\overset{\overset{CH_2OH}{|}}{C}}-CH_2OH+HCOONa$$

It is a hygroscopic compound, which is soluble in water, alcohols, acetone and insoluble in aliphatic and aromatic hydrocarbons.

The physical properties of trimethylolpropane are given in Table IX.12.

TABLE IX.12

Physical Properties of Trimethylolpropane

Molecular weight	134·18
Boiling point (°C, 5 mm)	160
Melting point (°C)	58·8–59·0

The major use of trimethylolpropane is in alkyd resins and for the production of flexible polyurethane foams. It is also used as crosslinking agent, e.g. in acrylic resin systems. The esters of trimethylolpropane with fatty acids are applied to a certain extent as synthetic lubricants.

9.5.7. Neopentylglycol

Iso-butyraldehyde reacts with formaldehyde in the presence of alkali to give hydroxypivalaldehyde. The latter may be hydrogenated to neopentylglycol with hydrogen in presence of catalysts or by a crossed Cannizzaro reaction with formaldehyde.

$$\underset{\underset{CH_3}{|}}{\overset{\overset{CH_3}{|}}{CH}}-CHO+HCHO \xrightarrow{alkali} HOH_2C-\underset{\underset{CH_3}{|}}{\overset{\overset{CH_3}{|}}{CH}}-CHO$$

$$HOH_2C-\underset{\underset{CH_3}{|}}{\overset{\overset{CH_3}{|}}{C}}-CHO \xrightarrow[catalyst]{H_2} HOH_2C-\underset{\underset{CH_3}{|}}{\overset{\overset{CH_3}{|}}{C}}-CH_2OH$$

$$HOH_2C-\underset{\underset{CH_3}{|}}{\overset{\overset{CH_3}{|}}{CH}}-CHO \xrightarrow[NaOH]{+HCHO} HOH_2C-\underset{\underset{CH_3}{|}}{\overset{\overset{CH_3}{|}}{CH}}-CH_2OH+HCOONa$$

So far neopentylglycol has been produced by five companies on a small scale— Eastman Kodak, Mitsubishi, OSW, BASF and Ruhrchemic.

The physical properties of neopentylglycol are summarised in Table IX.13.

TABLE IX.13

Physical Properties of Neopentylglycol

Molecular weight	104·15
Boiling point (°C, 760 mm)	210
Melting point (°C)	128
Flash point (°C) (Cleveland Open Cup)	151·0
Fire point (°C) (Cleveland Open Cup)	151·6
Crystal density at 25°C (g/cm^3)	1·11

9.5.8. n-Butylamine

n-Butylamine is obtained either by alkylation of ammonia with n-butanol in the vapour phase over dehydrating catalysts such as aluminia[235-237] or by catalytic hydrogenation of an ammonia/n-butyraldehyde mixture.[238]

$$CH_3—CH_2—CH_2—CH_2OH + NH_3 \xrightarrow{\text{Al-catalyst}} CH_3—CH_2—CH_2—CH_2—NH_2 + H_2O$$

$$CH_3—CH_2—CH_2—CHO + NH_3 + H_2 \xrightarrow{\text{catalyst}} CH_3—CH_2—CH_2—CH_2—NH_2 + H_2O$$

n-Butylamine is a colourless liquid with odour resembling ammonia, completely miscible with water and with most organic solvents such as ether and alcohol. The physical data are shown in Table IX.14.

TABLE IX.14

Physical Properties of n-Butylamine

Molecular weight	73·14	Specific gravity (d_4^{20})	0·7401
Boiling point (°C, 760 mm)	77·8		
Melting point (°C)	−50·5	Refractive index (n_{20}^D)	0·4036
Ignition temp. (°C)	310		
Flash point (°C)	−2	Viscosity at 20°C (cP)	0·6
Boiling point of azeotrope with H$_2$O at 760 mm °C/% water	76·0/1·3	Surface tension (dyne/cm at 20°C)	23·81

n-Butylamine is an intermediate in the production of pharmaceuticals, especially of antidiabetics such as tolbutamid (e.g. Restinon® of Farbwerke Hoechst). It is a very active gum inhibitor for cracked gasoline. The viscosity of alkyd and urea enamels may be controlled by neutralisation of the acidity of the vehicle with butylamine. The fatty acid soaps of n-butylamine are flotation agents for nonferrous metals and are excellent emulsyfying agents for hydrocarbon oils. n-Butylamine has found special application in the petroleum industry, e.g. as selective solvent for dewaxing mid-continent petroleum and acts as a surface-tension depressant in breaking crude oil emulsions. It is also used in photographic developing compositions.

9.5.9. Iso-butylamine

Iso-butylamine is manufactured from iso-butanol and ammonia by passing the reactants in the vapour phase over a combined dehydration and hydrogenation catalyst especially alumina at temperatures above 300°C or by reaction of iso-butyraldehyde, ammonia and hydrogen.

$$\begin{array}{c} CH_3 \\ \diagdown \\ CH-CH_2OH+NH_3 \\ \diagup \\ CH_3 \end{array} \xrightarrow{catalyst} \begin{array}{c} CH_3 \\ \diagdown \\ CH-CH_2NH_2+H_2O \\ \diagup \\ CH_3 \end{array} \tag{1}$$

$$\begin{array}{c} CH_3 \\ \diagdown \\ CH-CHO+NH_3+H_2 \\ \diagup \\ CH_3 \end{array} \xrightarrow{catalyst} \begin{array}{c} CH_3 \\ \diagdown \\ CH-CH_2NH_2+H_2O \\ \diagup \\ CH_3 \end{array} \tag{2}$$

Iso-butylamine is a colourless liquid with the characteristic pungent odour of the lower aliphatic amines, miscible with water in all proportions and most of the organic solvents such as benzene, hydrocarbons, chloroform, carbon tetrachloride, ether or esters.

The physical properties of iso-butylamine are collected in Table IX.15.

TABLE IX.15

Physical Properties of Iso-butylamine

Molecular weight	73·14	Specific gravity (d_4^{20})	0·729
Boiling point (°C at 760 mm)	68·0		
Melting point (°C)	−85·5	Refractive index (n_{20}^D)	1·3977
Ignition temp. (°C)	395		
Flash point (Abel-Pensky)		Viscosity at 20°C (cP)	0·56
(°C)	−11	Vapour pressure at 20°C (mm)	125
		Surface tension (dyne/cm at 20°C)	22·25

The fatty acid soaps of iso-butylamine are flotation agents. Iso-butylamine itself may be used as anti-knock agent in gasoline. The hydrochloride of iso-butylamine serves as neutralisation agent in the production of washing and dispersing agents such as tridecylsulphobenzoate.

9.6. OTHER CARBONYLATION REACTIONS WITH PROPYLENE

9.6.1. Butanol via Reppe Synthesis

The Reppe alcohol synthesis[239, 240] is closely related to the hydroformylation. It is carried out in alkaline media with Fe(CO)$_5$-catalyst and differs from the Oxo reaction in that water is used instead of hydrogen. Alcohols are obtained directly at lower reaction temperatures. The active catalyst is believed to be a salt of iron hydrocarbonyl.

$$CH_3—CH = CH_2 + 3CO + 2H_2O \xrightarrow{\text{catalyst}} CH_3—CH_2—CH_2—CH_2OH + 2CO_2$$

The reaction was developed by BASF on pilot plant scale at Ludwigshafen.[240, 241] BASF itself never commercialised the process but preferred to make butanol via the Oxo synthesis. However, Japan Butanol Company Ltd, Yokkaichi, is operating a plant using the BASF technology

9.6.2. Reaction of Propylene with Carbon Monoxide and Amines

Propylene, carbon monoxide and amines react to give mixtures of butyric and iso-butyric acid amides.[5]

Thus, propylene reacts with aniline and CO to give butyric acid anilide plus iso-butyric acid anilide in a ratio of 4:1 in an overall yield of 87 per cent.[242] Up to now this reaction has not yet found technical application.

REFERENCES

[1] Roelen (Ruhrchemie AG), GP 849,548 (1938).

[2] Roelen, *Angew. Chem.*, 1948, **A60**, 213.

[3] Roelen, '*Naturforschung and Medizin in Deutschland*, Vol. 36, 'Präparative Organische Chemie Part I, ed. K. Ziegler (Wiesbaden, Dieterich'sche Verlagsbuchhandlung, 1948).

[4] Falbe, '*Synthesen mit Kohlenmonoxyd*' (Springer Verlag, Berlin, Heidelberg, New York, 1967).

[5] Falbe, '*Carbon Monoxide in Organic Synthesis*' (Springer Verlag, Berlin, Heidelberg, New York, 1970).

[6] '*Carbonylation of Unsaturated Hydrocarbons*', ed. Allunions Sci. Res. Inst. for Petrochem. Proc. Leningrad Chemistry Dept., 1968.

[7] Nienburg, Gemassmer and Eckard (Chem. Verwertungsgesellschaft Oberhausen), GP 888,687 (1942).

[8] Keulemans, Kwantes and van Bavel, *Rec. Trav. Chim.*, 1948, **67**, 298.

[9] Aldridge and Jonassen, *Nature*, 1960, **188**, 404.

[10] Aldridge, Fasce and Jonassen, *J. Phys. Chem.*, 1958, **62**, 869–70.

[11] Macho, Mistrick and Ciha, *Coll. Czech. Chem. Commun.*, 1964, **29**, 826.

[12] Hecht and Kröper, '*Naturforschung and Medizin in Deutschland 1939–46*, Vol. 36, 'Präparative Organische Chemie', Part I, p. 115, ed. K. Ziegler (Wiesbaden, Dieterich'sche Verlagsbuchhandlung, 1948).

[14] Kroper, '*Anlagerung von Kohlenmonoxid und Wasserstoff an Olefine*' (Hydroformylierung in: Houben–Weyl, Vol. IV/2, 367 (Georg Thieme, Verlag, Stuttgart, 1955).

[15] Adkins and Krsek, *J. Am. Chem. Soc.*, 1956, **78**, 5450.

[16] Orchin, '*Advances in Catalysis*', Vol. V, 383 (Academic Press, New York, 1950).

[17] Natta, Pino and Mantica, *Chim. e Ind.*, 1950, **32**, 201.

[18] Natta, Ercoli and Castellano, *Chim. e Ind.*, 1955, **37**, 6.

[19] Natta, *Brennstoff Chem.*, 1955, **36**, 176.

[20] Natta and Ercoli, *Chim. e Ind.*, 1952, **34**, 503.

[21] Wysokinskii, Gankin and Rudkowskii, '*Carbonylation of Unsaturated Hydrocarbons*, ed. Allunions Sci. Res. Inst. for Petrochem. Proc. Leningrad, Chemistry Department, 1968.

[22] Marko, *Ber. Ung. Mineralöl. u. Erdgasversuchsansanstalt*, 1961, **2**, 228.

[23] Orchin, Kirch and Goldfarb, *J. Am. Chem. Soc.*, 1956, **78**, 5450.

[24] Karapinka and Orchin, Abstracts 137th A.C.S. Meeting, Cleveland, Ohio, April 5–14, 1960, 92–100.

[25] Kirch and Orchin, *J. Am. Chem. Soc.*, 1958, **80**, 4428.

[26] Kirch and Orchin, *J. Am. Chem. Soc.*, 1959, **81**, 3597.

[27] Wender, Sternberg and Orchin, *J. Am. Chem. Soc.*, 1953, **75**, 3041.

[28] Pino, Pucci and Piacenti, *Chem. and Ind.*, 1963, 294.

[29] Almasy and Szabo, *Acad. Rep. Populare Romine, Studii Cercetari Chim.*, 1960, **8**, 531.

[30] Heck and Breslow, *J. Am. Chem. Soc.*, 1961, **83**, 4023.

[31] Heck and Breslow, *Chem. and Ind.*, 1960, 467.

[32] Nienburg, Helms and Pistor (Chem. Verwertungsgesellschaft Oberhausen), G P 914,375 (1954).

[33] Pino, Oxosynthese in, '*Ullmanns Encyklopadie der Technischen Chemie*', Vol. 13, 61 (München–Berlin, Urban and Schwarzenberg, 1962).

[34] Asinger, '*Chemie und Technologie der Monoolefine*', 656 (Berlin Akademie Verlag, 1957).

[35] Schuster and Eilbracht (Chem. Verwertungsgesellschaft Oberhausen), G P 892,287 (1953).

[36] Tramm, Kolling, Schnur, Büchner, Heger and Stiebling (Ruhrchemie A G), B P 736,875 (1955).

[37] Wilson (Standard Oil), U S P 2,695,315 (1954).

[38] Büchner and Kühne (Ruhrchemie A G), G P 854,216 (1952).

[39] Schiller (Chem. Verwertungsgesellschaft Oberhausen), G P 953,605 (1956).

[40] Esso, B P 801,734 (1956).

[41] Schuster and Eilbracht (Chem. Verwertungsgesellschaft Oberhausen), G P 877,300 (1953).

[42] Büchner (Ruhrchemie A G), G P 874,304 (1951).

[43] Gankin, Krinkin and Rudkowskii, '*Carbonylation of Unsaturated Hydrocarbons*', 45 ff., ed. Allunions Res. Inst. for Petrochem. Proc., Leningrad Chemistry Depart., 1968.

[44] Martin, *Chem. and Ind.*, 1954, 1536.

[45] Greenfield, Metlin and Wender, Abstract of Papers 126th Meeting of the A.C.S., New York, Sept. 1954.

[46] Chatt and Venanzi, *J. Chem. Soc.*, 1957, 4735.

[47] Sternberg and Wender, *J. Chem. Soc.*, Spec. Publ., 1959, **13**, 35.

[48] Heck, Private Communication to the author.

[49] Basolo, Brault and Poe, *J. Chem. Soc.*, 1964, 676.

[50] Heck, *J. Am. Chem. Soc.*, 1963, **85**, 657.

[51] Werner, *Angew. Chem.*, 1968, **24**, 1017.

[52] Osborn, *Endeavour*, 1967, **26**, 144.

[53] Angelici, *Organometal Chem. Rev.*, 1968, **3**, 173–226.

[54] Day, Basolo, Pearson, Kangas and Henry, *J. Am. Chem. Soc.*, 1968, **90**, 1925.

[55] Day, Basolo and Pearson, *J. Am. Chem. Soc.*, 1968, **90**, 6927.

[56] Tsutsui, Hancock, Ariyoshi and Levy, *Angew. Chem.*, 1969, **81**, 435.

[57] Mistrick and Durmis, *Chem. Zvesti.*, 1969, **23**, 286–94.

[58] Heck, *J. Am. Chem. Soc.*, 1965, **87**, 2572.

[59] Falbe, Lecture to session of the German Chemical Soc., Berlin, 12 Februarv 1968, Abstr. in *Angew. Chem.*, 1968, **80**, 568.

[60] Ungvary and Marko, *J. Organometal. Chem.*, 1969, **20**, 205–9.

[61] Heck, *Adv. Organometal Chem.*, 1966, **4**, 243.

[62] Nagy-Magos, Bor and Marko, *J. Organometal Chem.*, 1968, **14**, 205.

[63] Coffield, Kozikowski and Clossen, *J. Organometal Chem.*, 1957, **22**, 598.
[64] Noak and Calderazzo, *J. Organometal Chem.*, 1967, **10**, 101.
[65] Coffield, Kozikowski and Clossen, *J. Chem. Soc.*, Spec. Publ., 1959, **13**, 126.
[66] Mawby, Basolo and Pearson, *J. Am. Chem. Soc.*, 1964, **86**, 3994.
[67] Marko, Bor, Almasy and Szabo, *Brennstoff Chem.*, 1963, **44**, 184.
[68] Karapinka and Orchin, Abstract 137th A.C.S. Meeting, Cleveland, Ohio, April 5–14, 1960, pp. 92–100.
[69] Pino, Piacenti and Neggiani, *Chem. and Ind.*, 1961, 1400.
[70] Kniese (BASF), Private Communication.
[71] Kniese, Nienberg, Fischer, *J. Organometal Chem.*, 1969, **17**, 133.
[72] Heil and Marko, *Chem. Ber.*, 1969, **102**, 2238.
[73] Hughes and Kirshenbaum, *Ind. Eng. Chem.*, 1957, **49**, 1999.
[74] Falbe, Tummes and Weber, *Brennstoff Chem.*, 1969, **50**, 46.
[75] Sacco and Freni, *Ann. Chim. (Rome)*, 1958, **48**, 218.
[76] Hieber and Freyer, *Chem. Ber.*, 1960, **93**, 462.
[77] Falbe and Huppes, *Brennstoff Chem.*, 1967, 46.
[78] Bianchi, Benedetti and Piacenti, *Chim. Ind. (Milan)*, 1969, **51**, 613.
[79] Hershman, Robinson, Craddock and Roth, *Ind. Eng. Chem. Prod. Res. Develop.*, 1969, **8**, 372.
[80] Falbe, Tummes, Weber and Weisheit, *Tetrahedron*, 1971, **27**, 3603.
[81] Bergerhoff, Tihanyi and Hammes, *Tetrahedron*, 1971, **27**, 3593.
[82] Cannell, Slaugh and Mullineaux (Shell), GP 1,186,455 (1965).
[83] Mullineaux (Shell), Belg.P 603,820 (1961).
[84] Slaugh and Mullineaux (Shell), Belg.P 606,408 (1961); USP 3,239,569 (1966).
[85] Slaugh (Shell), Belg.P 619,344 (1962).
[86] Greene and Meeker (Shell), Belg.P 621,833 (1962); USP 3,274,263 (1966).
[87] Greene (Shell), Belg.P 623,213 (1962).
[88] Greene (Shell), Belg.P 627,365 (1963).
[89] Greene (Shell), Belg.P 627,371 (1963).
[90] Slaugh and Mullineaux (Shell), USP 3,239,566 (1966).
[91] Greene and Meeker (Shell), GP 1,212,953 (1962).
[92] Cannell, Slaugh and Mullineaux (Shell), GP 1,186,455 (1965).
[93] Shell, BP 995,459 (1965).
[94] Shell, BP 1,002,428 (1965).
[95] Van Winkle and Hasserodt (Shell), GP 1,282,633 (1968).
[96] Asinger, Fell and Rupilius, *Ind. Eng. Chem. Prod. Res. Develop.*, 1969, **8**, 214.
[97] Tucci, *Ind. Eng. Chem. Prod. Res. Develop.*, 1969, **V8**, No. 3, 286.
[98] Tucci, *Ind. Eng. Chem. Prod. Res. Develop.*, 1970, **V9**, No. 4, 516.
[99] Slaugh and Mullineaux, *J. Organometal Chem.*, 1968, **13**, 469.
[100] Evans, Osborn and Wilkinson, *J. Chem. Soc.*, A, 1968, 3133.
[101] Pruett and Smith, *J. Organometal Chem.*, 1969, **34**, 327.
[102] Evans, Yagupsky and Wilkinson, *J. Chem. Soc.*, A, 1968, 2660.
[103] Tucci, *Ind. Eng. Chem. Prod. Res. Develop.*, 1968, **7**, 32.
[104] Tucci, *Ind. Eng. Chem. Prod. Res. Develop.*, 1968, **7**, 125.
[105] Hershman and Graddock, *Ind. Eng. Chem. Prod. Res. Develop.*, 1968, **7**, 226.
[106] Tucci, *Ind. Eng. Chem. Prod. Res. Develop.*, 1968, **7**, 227.
[107] Piacenti, Bianchi and Benedetti, *Chim. Ind. (Milan)*, 1967, **49**, 245.
[108] Ibers, *J. Organometal Chem.*, 1968, **14**, 423.
[109] Simon, Nagy-Magos, Palagyi, Palyi, Bor and Marko, *J. Organometal Chem.*, 1968, **11**, 634.
[110] Fell, Rupilius and Asinger, *Tetrahedron Letters*, 1968, **29**, 3261–6.
[111] Orgel, '*Introduction to Transition—Metal Chemistry*', p. 132 (Wiley, New York, 1960).
[112] Hieber and Linder, *Chem. Ber.*, 1961, **94**, 1417.
[113] *V.W.D.—Chemie*, 1968, **226**, 6.
[114] *V.D.I.—Nachrichten*, 1969, **1**, 4.
[115] *G.I.T.—Fachz Lab.*, 1969, **13**, 560.

[116] *Chem. Ind. Düsseldorf*, 1968, 822.
[117] Weber and Falbe, Lecture given to A.C.S. 158th National Meeting, New York, 7–12 September 1969.
[118] Weber and Falbe, *Ind. Eng. Chem.*, 1970, **62**, 33.
[119] Falbe and Weber, Lecture given to S.K.R. Meeting, Stockholm, 12–14 October 1970.
[120] Hasche, BIOS (Mt. Vernon, Iowa), 1945, 27.
[121] Hall, BIOS (Mt. Vernon, Iowa), 1945, 447.
[122] Hohenschutz, *Eur. Chem. News*, large plant supplement, September 1966, 7.
[123] Dümbgen and Neubauer, *Chem. Ing. Techn.*, 1969, **41**, 974.
[124] Guccione, *Chem. Eng.*, 1965, 90–2.
[125] Falbe, *Ullmanns Encyklopädie der Technischen Chemie*, 1970, **13**, 90.
[126] Schiller (Chem. Verwertungsgesellschaft Oberhausen), GP 953,605 (1956).
[127] Wakamatsu, *Nippon Kagaku Zasshi*, 1964, **85**, 227–31.
[128] Imyanitov and Rudkowskii, *Petrol. Chem. (USSR)*, 1964, **3**, 91.
[129] Falbe, Huppes and Korte, *Brennstoff Chem.*, 1966, **47**, 207.
[130] Falbe and Huppes, *Brennstoff Chem.*, 1967, **48**, 24.
[131] Falbe and Huppes, Belg.P Appl. 33,538 (1966).
[132] Falbe and Huppes, Belg.P Appl. 33,539 (1966).
[133] Osborn, Wilkinson and Young, *Chem. Commun.*, 1965, **2**, 17.
[134] Millidge (Distillers), FP 1,411,602 (1965).
[135] Heil and Marko, *Chem. Ber.*, 1968, **101**, 2209.
[136] Imyanitov and Rudkowskii, *Neftehimiya*, 1963, **3**, 198.
[137] Gankin, Genender and Rudkowskii, *J. Appl. Chem. USSR*, 1967, **40**, 2029.
[138] Weigert (Degussa), Neth.P Appl. 6,516,193 (1966).
[139] Weigert (Degussa), GP 1,280,237 (1965).
[140] Bartlett and Hughes (Esso), USP 2,894,038 (1959).
[141] Yamaguchi, *Kogyo Kagaku Zasshi*, 1969, **72**, 671.
[142] Takegami, Watanabe and Massada, *Bull. Chem. Soc. Japan*, 1967, **40**, 1459.
[143] Imyanitov and Rudkowskii, *Kinetika i Kataliz*, 1967, **8**, 1051.
[144] Smith and Jaeger (ICI), BP 966,482 (1960).
[145] Smith and Jaeger (ICI), BP 1,159,926 (1963).
[146] ICI, Austr.P Appl. 30,352 (1963).
[147] Pichler, Firnhager and Koussis, *Brennstoff Chem.*, 1964, **44**, 337.
[148] Evans, Osborn, Jardine and Wilkinson, *Nature*, 1965, **208**, 1203.
[149] Lonza, FP 1,381,091 (1963).
[150] Lemke, *Supplement mensual a Chim. Ind. (Paris)*, 1963, **89**, 118.
[151] Green and Kirk-Othmer, '*Encyclopedia of Chemical Technology*', 2nd ed., vol. 4, 424 (Interscience Publishers J. Wiley & Sons, Inc., New York, London, Sydney, 1964).
[152] Van den Berg and Reinmuth, *Supp. Chem. and Proc. Eng.*, August 1970, 53.
[153] Büchner and Meis (Ruhrchemie AG), GP 1,024,499 (1958).
[154] Standard Oil, GP 944,728 (1965).
[155] Carter (Gulf Oil Corp.), BP 779,388 (1953).
[156] Nienburg, Eckert, Kölsch, Goilav and Pistor (Chemische Verwertungsgesellschaft Oberhausen), GP 953,606 (1956).
[157] Lemke (Kuhlmann), FP 1,089,983 (1953).
[154] Johnson and Cox (Union Carbide Corp.), USP 3,014,970 (1957).
[159] Lemke, *Hydrocarbon Process, Petrol Refiner*, 1966, **45**, 148.
[160] Lemke (Kuhlmann), GP 1,206,419 (1959).
[161] Meis and Tummes (Ruhrchemie AG), GP 1,235,285 (1967).
[162] Moell, Eckert, Kerber, App. Hohenschutz and Walz (BASF), GP 1,272,911 (1966).
[163] Greene (Shell), USP 3,369,050 (1968).
[164] Falbe, Tummes and Hahn, *Angew. Chem.*, 1970, **82**, 181; *Angew. Chem. Intern. Edition*, 1970, **IX**, 169.
[165] *Oil & Gas, J.*, 23 November 1970, 59.
[166] Mistrick and Rendko, *Chem. Techn. (Berlin)*, 1967, **19**, 154.
[167] Habeshaw and Rae (Anglo-Iranian Oil), GP 921,934 (1952).

[168] Tummes, Falbe and Cornils (Ruhrchemie AG), Belg.P 743,723 (1969).

[169] Tummes and Meis (Ruhrchemie AG), GP 1,258,855 (1965).

[170] Rehn and Theiling (Farbwerke Hoechst), GP 1,061,770 (1957).

[171] Van der Woude and Morris (Eastman Kodak), USP 2,763,693 (1951).

[172] Esso, FP 1,420,640 (1964).

[173] Bartlett, Kirshenbaum and Muessig (Esso), *Ind. Eng. Chem.*, 1959, **51**, 257.

[174] Häuber and Hagen (Chem. Verwertungsgesellschaft), GP 1,036,839 (1955).

[175] Falbe and Hahn (Ruhrchemie AG), BP 731,673 (1969).

[176] *Hydrocarbon Processing*, Vol. 49, October 1970, No. 10, C.R.-71.

[177] *Eur. Chem. News*, 1970, No. 450, 14.

[178] *Eur. Chem. News*, 1970, No. 452, 18.

[179] *Chem. Ind.*, XXI April 1969, 257; *Chem. & Eng. News*, 29 September 1969, 20.

[180] *Hydrocarbon Processing*, Vol. 49, No. 10, 1970, C.R.-24.

[181] *Eur. Chem. News.*, New Plants 1968, 28 February 1969, p.m.

[182] Private communication to the author.

[183] *Oil*, 2 February 1970, 44.

[184] '*Produkte aus der Oxosynthese*', special publication of Farbwerke Hoechst, AG and Ruhrchemie AG (1969).

[185] Oxo synthesis, 'Products from Ruhrchemie Aktiengesellschaft', Oberhausen-Holten. Aldehydes—Alcohols—Acids, 1971.

[186] *Oel—Zeitschrift für die Mineralölwirtschaft*, February 1970, 34.

[187] *Eur. Chem. News*, 1967, No. 275, 24.

[188] Private communication to the author.

[189] *Hydrocarbon Proc.*, June 1969, C.R.-59.

[190] *Hydrocarbon Proc.*, No. 10, October 1970, C.R.-63; *Chem. Eng.*, 22 September 1969, 174; *Eur. Chem. News*, 4 April 1969, 70.

[191] *Hydrocarbon Proc.*, 49, 1, No. 6, 1970, C.R.-64.

[192] *Chem. Ind.*, 20 October 1968, 784.

[193] *Eur. Chem. News*, No. 448, 4 September 1970, 28.

[194] *Hydrocarbon Proc.*, Vol. 10, October 1970, C.R.-43.

[195] *Chem. Ind.*, 21 October 1969, 665.

[196] *Oil Paint and Drug Reporter*, 12 July 1970.

[197] *Chem. Ind.*, 20 October 1968, 733.

[198] *C & EN*, 16 February 1970, 21.

[199] Falbe and Cornils, '*Oxoalkohole als Lösungsmittel in Fortschritte der Chem. Forschung*', Vol. 11/1 (Springer Verlag, 1968).

[200] Nienburg, Böhn and Elschnig (BASF), GP 1,014,089 (1956).

[201] Friedrichsen (BASF), GP 1,065,403 (1957).

[202] Haarer and Rühl (BASF), GP 1,057,083 (1957).

[203] Hagemeyer, jun. (Eastman Kodak), GP 1,140,563 (1957).

[204] Tummes, Schiewe, Cornils, Pluta and Falbe (Ruhrchemie), GP 2,045,669.

[205] Diekhaus, Hanisch and Falbe (Ruhrchemie), GP 2,045,668.

[206] Rottig, Tummes and Cornils (Ruhrchemie), GP 2,054,601.

[207] Farbwerke Hoechst AG, BP 1,205,899.

[208] Hamanto and Sakahi (Mitsubishi), GP 1,146,080 (1961).

[209] Farbwerke Hoechst AG, Belg.P 629,256 (1963).

[210] Feichtinger (Ruhrchemie), GP 1,468,591 (1963).

[211] *Ullmanns Enzyklopadie der Techn. Chemie*, Vol. XVIII, 205.

[212] Stiller, Harris, Finkelstein, Keresztesy and Folkers, *J. Am. Chem. Soc.*, 1940, **62**, 1785.

[213] Hoffman—La Roche & Co. AG, Swiss P 215,779 (1941).

[214] Houben-Weyl, Bd. 2. 463 Stuttgart (Georg Thieme Verlag, 1955).

[215] Parkinson and Wagner, *Ind. Eng. Chem. Anal.*, 1934, **6**, 433.

[216] Siggia and Maxey, *Ann. Chem.*, 1947, **19**, 1024.

[217] Rippler, M. *Monatshefte*, 1900, **21**, 1079.

[218] Houben-Weyl, v. 2, 459 (Georg Thieme Verlag, 1955).

[219] Hünig, *Ann.*, 1950, **569**, 224.

[220] Peret, *Helv.*, 1951, **34**, 1531.
[221] Stillmann and Reed, *Pertum. essent. Oil Rec.*, 1932, **23**, 278.
[222] Vadon and Anziani, Bl. 1937 [5], 2026.
[223] Trozzolo and Lieber, *Annal. Chem.*, 1950, **22**, 764.
[224] Trozzolo and Lieber, *Angew. Chem.*, 1951, **63**, 177.
[225] Mitchel, Smith and Bryant, *J. Am. Chem. Sec.*, 1941, **63**, 574.
[226] Schultes, *Angew. Chem.*, 1934, **47**, 258.
[227] Houben-Weyl, v. 2, 456 (Georg Thieme Verlag, Stuttgart, 1955).
[228] *Vorländer Fr.*, 1929, **77**, 321.
[229] Jonescu and Bodea, Bl. 1930 [4], 1408.
[230] Yoe and Reid, *Ind. Eng. Chem. Anal.*, 1941, **13**, 238.
[231] Bryant, Mitchell and Smith, *J. Am. Chem. Soc.*, 1940, **62**, 3504.
[232] Falbe, *Ullmanns Encyklopädie der Technischen Chemie*, Vol. 13, 1970, 126.
[233] Falbe, *Ullmanns Encyklopädie der Technischen Chemie*, Vol. 13, 1970, 130.
[234] Bauer, Danziger and Schulze (Farbenfabriken Bayer), GP 1,031,298 (1958).
[235] Deutsche Hydrierwerke, GP 611,924 (1931).
[236] I G Farbenindustrie, FP 747,905 (1932).
[237] Du Pont, USP 2,078,922 (1934).
[238] I G Farbenindustrie, GP 489,551 (1926).
[239] Reppe and Vetter, *Ann.*, 1953, **582**, 133.
[240] *Hydrocarbon Proc.*, 1967, **46**, 154.
[241] v. Kutepow and Kindler, *Angew. Chem.*, 1960, **72**, 802.
[242] Pino and Magri, *Chim. Ind. (Milan)*, 1952, **1**, 34.
[243] Rössher (BASF), GP Appl. 1,618,128 (1967).

CHAPTER 10

ACROLEIN, ACRYLIC ACID AND ITS ESTERS

By **D. J. Hadley and E. M. Evans**

10.1. ACROLEIN, CH$_2$ = CHCHO

by **D. J. Hadley**

10.1.1. Some Properties

Specific gravity	0·8402 (20 °C)
Freezing point	−87 °C
Boiling point (1 atm)	52·7 °C
Refractive index (n_D)	1·4017 (20 °C)
Specific heat of liquid	0·50 cal/g/°C (20 °C)
Vapour pressure at 20 °C	215 mm Hg
Heat formation (gas at 25 °C)	−17·79 kcal/mole
Heat combustion (liquid)	−389·6 kcal/mole
Solubility in water	20·6% wt/wt (20 °C)
Solubility of water in acrolein	6·8% wt/wt (20 °C)
Water azeotrope	
Boiling point (1 atm)	52·4 °C
Water content	2·6% wt/wt

Because it is the simplest member of the class of unsaturated aldehydes and therefore offers great promise as a highly reactive, versatile intermediate, acrolein has attracted many would-be producers and users. It appears as a product of several reactions, but pre-war text books usually referred to the dehydration of glycerol as the best method of prepatation.[1] This method has no commercial significance.

10.1.2. Production

10.1.2.1. Historical Acrolein was produced on an industrial scale, 10 tons/month, for the first time, in Germany, during World War II.[2, 3] The method employed was the catalysed vapour phase condensation of acetaldehyde and formaldehyde:

$$CH_3CHO + HCHO \rightarrow CH_2 = CHCHO + H_2O$$

The catalyst for this reaction is not highly specific, but silica gel, impregnated with sodium silicate, was preferred. Aqueous formaldehyde and a slight molar excess of acetaldehyde were vapourised and passed over this catalyst in a tubular reactor, heated by a gas-fired furnace, the temperature around the reactor being 300–320 °C. From the condensed product unreacted acetaldehyde and formaldehyde were easily separated by successive distillations and returned to the reactor, so that the overall yield approached the efficiency of conversion, which was about 65 per cent on formaldehyde and 75 per cent on acetaldehyde. A fresh catalyst gave 60 per cent conversion (per pass), declining over six days of operation to 40 per cent, because of accumulating carbonaceous material resulting from parasitic reactions. The carbon was then burned off by the passage of a mixture of steam and air through the catalyst while the exterior temperature was at 400°C. Since the catalyst is permanently impaired if it is heated above about 600°C, the combustion rate within it had to be carefully controlled by means of the air input.

It is reported that Degussa used their war-time production of acrolein for the manufacture of a material resembling Plexiglass (by condensation of acrolein and pentaerythritol), and for the manufacture of a crease-resisting agent for fabrics (by condensation of acrolein and thiourea). They were said also to be contemplating the production of methyl acrylate, β-picoline, allyl alcohol and various glycols, thus evincing their belief in the great potential of acrolein.

10.1.2.2. Development of propylene oxidation methods   In the early post-war period other companies shared this view of acrolein, but because olefin feed stocks were now becoming readily available, they were persuaded that propylene, which already has the required carbon skeleton, must be the most suitable raw material for acrolein manufacture if only a catalyst could be found which would promote the necessary partial oxidation of the methyl group by molecular oxygen:

$$CH_2 = CHCH_3 + O_2 \rightarrow CH_2 = CHCHO + H_2O$$

Some companies[4] attempted to utilise the reaction discovered by Deniges[5] in 1898; the oxidation of propylene to acrolein by mercuric sulphate suspended in dilute sulphuric acid. But the addition of air to the propylene feed did not prevent reduction of the mercuric ion to metal, and in fact reoxidation of the mercury could only be done with relatively expensive oxidising agents such as nitric acid.

In 1947 the Standard Oil Co. of New Jersey made the surprising discovery that cupric oxide deposited on silica (copper silicate) gives a significant yield of acrolein from a feed of propylene in air, whereas the yield is inappreciable if the cupric oxide is pure or if it is mounted on a non-siliceous support.[6] However, the yield was not enough to warrant further development.

The Shell Development Co. had for some time been making acrolein, on at least a pilot plant scale, by the decomposition of diallyl ether,

$$(CH_2 = CH—CH_2)_2O \rightarrow CH_2 = CHCHO + CH_2 = CHCH_3.[7]$$

In 1948 they filed the first of a series of patents relating to the oxidation of propylene to acrolein in the presence of a catalyst consisting essentially of cuprous oxide on an inert support.[8] Since cupric oxide did not function in the desired way it was essential to maintain the copper in the cuprous state by using a large excess of propylene over oxygen in the feed, which limited the conversion of propylene. For example a feed of 27·6 per cent propylene, 7·8 per cent oxygen, 35·3 per cent steam and 29·3 per cent nitrogen was used, and typically, the yield per pass on propylene was somewhat less than 10 per cent, but with an efficiency of 60 per cent or more it was possible to achieve an acceptable ultimate yield by recycling. The later patents are mainly concerned with the maintenance of catalyst activity by the removal of injurious material from the gas phase and with the addition of halogens and other promoters to the catalyst. From their discovery Shell developed the first process for the oxidation of propylene to acrolein to be operated on a commercial scale, and it is still in use. However, the considerable recycle of propylene that is required is a serious disadvantage; moreover, the cuprous oxide catalyst appears not to be very stable, being susceptible to either oxidation or reduction under reaction conditions.

Montecatini[9] have described what appears to be a particular modification of the Shell process in which very narrow copper or copper lined reactor tubes are used. Presumably the necessary cuprous oxide catalyst is formed on the internal walls, thus enabling the usual granular catalyst to be dispensed with.

The next development of note was initiated by a patent to the Battelle Memorial Institute[10] which described a catalyst for the oxidation of propylene that consisted of silver selenite, Ag_2SeO_3, supported on asbestos fibre. Cupric oxide was said to be a catalyst promoter. But the Distillers Co. Ltd. (DCL), by studying the rate of reaction as a function of time on stream[11] showed that silver selenite was not a catalyst, but simply a reagent, all the oxygen in which was available for the formation of acrolein from propylene, the inorganic product of the reaction being silver selenide, which had no catalytic activity. When cupric oxide was added to the supported Ag_2SeO_3 two concurrent reactions, both leading to acrolein, were detected: that already mentioned, the oxidation of propylene by Ag_2SeO_3, and a true catalytic reaction which was relatively very slow, but which persisted at a constant rate, when all the Ag_2SeO_3 had reacted. This same residual rate was also achieved by adding CuO to the Ag_2Se resulting from the reaction of the Ag_2SeO_3, and it was attributed to a small concentration of cupric selenite arising from the reaction of cupric oxide with oxygen, and selenium, resulting from the incipient dissociation of Ag_2Se. The cupric selenite catalytic function may be represented thus:

$$CuSeO_3 + C_3H_6 \rightarrow CuO + Se + C_3H_4O + H_2O$$
$$Se + O_2 \rightarrow SeO_2$$
$$CuO + SeO_2 \rightarrow CuSeO_3$$

High reaction rates and a yield of up to 80 per cent per pass with a feed of 2 per cent propylene in air, were achieved with a catalyst consisting of copper selenite distributed on a high area support such as silica or alumina, the contact time being about 2 sec with a bath temperature of 300°C.[12] The catalyst was most readily produced by passing a stream of air containing Se or SeO_2 over supported cupric oxide. It had one considerable drawback, arising from the fact that, under reaction conditions, the vapour pressure of Se (or SeO_2) in the system was not negligible, and although by calculation, each selenium atom went through the above reaction cycle many times, it was eventually, carried out of the reactor. Consequently, recovery of selenium at the reactor exit and a small transport through the reactor (about 0·5 mg/l gas) were necessary for continuous operation.[13] Selenium is not an abundant element, and in the early 1950's, when demand from the electronic industries increased, it became doubtful that a regular supply, such as would be required for make-up in a large plant, could be assurred, and DCL, although having successfully operated a large pilot plant, abandoned the project.

After an interval of a few years another and most important advance in the catalysis of propylene oxidation was disclosed by the Standard Oil Company of Ohio (Sohio). This was the discovery of a catalyst, essentially bismuth and molybdenum oxides, which did not suffer from the disadvantage of a volatile consistuent and which was capable of giving a sufficiently high yield per pass to render recycle of propylene unimportant.[14] In its preferred form the catalyst contains about 30 per cent silica, which somewhat enhances the performance, and phosphoric acid, whose function is obscure; it may be prepared simply by evaporating to dryness an aqueous mixture of bismuth nitrate, phosphoric and molybdic acids and silica sol. The best yield given in the Sohio patent is 41 per cent at 72 per cent efficiency, from a feed containing 10 per cent propylene, corresponding to a contact time of about 8 sec at 430°C. Although this yield is inferior to that obtained with the copper selenite catalyst, the concentration of acrolein in the reactor exit gas (an important factor in process economics) is about 4 per cent by volume, whereas from the copper selenite system it was only about 1·6 per cent and in the case of the Shell process it is possibly even less.

In 1959 DCL discovered a whole group of new catalysts, each member of which consisted of antimony oxide in combination with one of several other polyvalent metal oxides.[15] The preferred combination was antimony oxide/tin oxide, which has a catalytic performance similar to that of bismuth molybdate. A plant based on the new DCL process, owned and constructed by Ugilor in France, went on stream in 1965, to produce several thousand tons per year.

10.1.3. More recent Catalyst Developments

Following the disclosures of bismuth molybdate and antimony/tin oxides there has been a surge of research activity in the field of catalysis of propylene oxidation, which has, subsequently, been directed to the production of acrylic acid and acrylonitrile as well as acrolein, for there is a close relationship between these products and many instances of the same catalyst serving for the production of all three. Discussion of the relationship is deferred to the appropriate sections: here we are concerned only with the production of acrolein as the main reaction product.

The totally empirical character of the research is evidenced by the large number of relevant patents and by the diversity and complexity of the catalysts claimed in them. The patents cited below form no more than a good representative selection. For the sake of brevity the elemental composition of each catalyst is given, but each and every element is present in the form of its oxide or it has oxygen linkages.

10.1.3.1. Deniges method   Some interest in this method (which although not truly catalytic is conveniently dealt with here) has been revived by attempts to reduce the cost of reoxidation of the mercurous salt (or mercury metal). These include continuous electrolytic reoxidation.[16]

10.1.3.2. The Shell process   Research on catalysts for the Shell process seems to have been confined to additives (usually introduced by the vapour phase) for the improvement of the performance of the cuprous oxide.[17] No other oxide catalyst requiring an excess of propylene relative to oxygen, throughout the reactor, has been described.

10.1.3.3. Mixed oxide catalysts   The formation of acrolein in the absence of catalyst is virtually nil, at any temperature. Many of the polyvalent metal oxides (used singly) vigorously promote complete oxidation to CO_2, but it now appears that some have a slight selective activity, e.g. CuO, Sb_2O_4, SnO_2 and MoO_3 will each give traces or small yields of acrolein. However none of these, by itself, is of practical value. Upon mixing or combining two or more oxides there is often a synergistic effect, which in the better catalysts is very pronounced. For example, neither Sb_2O_4 nor SnO_2 will give an acrolein yield of more than about 2 per cent, but from the combination of the two the yield may rise to 50 per cent. Apart from the particular case of Cu_2O all useful catalysts contain two or more component oxides, and the gas fed to them contains at least as much oxygen as propylene. Amongst the numerous component oxides can be discerned, more or less clearly, certain principal or key components; selenium, tellurium, molybdenum and antimony. A key component will form useful catalysts by combination with each of several other oxides; and most, if not all, of the catalysts capable of giving yields of commercial interest contain at least one of these key components.

10.1.3.4. Selenium-containing catalysts A number of metal selenites have appreciable activity, but copper selenite appears to be the best. This system was abandoned by the original patentees, for the reasons already given, but continued to arouse limited interest elsewhere.[18]

10.1.3.5. Bismuth molybdate modifications A large number of patented catalysts appear to have been derived from the original Sohio catalyst, that is to say that they contain bismuth and molybdenum along with other components. The first, and possibly the most significant modification was the addition of iron. Various other single additives were suggested and then interest turned to the use of more than one. The culmination of this development appeared to be a composition comprising six elements other than oxygen: Bi, Mo, P, Fe, Ni, Co, for which a yield of more than 70 per cent was claimed. Such a satisfactory result was not demonstrated with a feed containing more than 1 per cent propylene. However, it is possible that this limitation may have been overcome by the inclusion of yet another element: Sm or Ta or As.

Catalyst elements	Patentee	Patent No.
Bi, Mo, P, Fe	Knapsack	BP 908,655
Bi, Mo, P, Te	Mitsubishi Rayon	Jap.P 5403/67
Bi, Mo, Al, Ta	Showa Electric	Jap.P 26283/67
Bi, Mo, As, one of (P, B, Si)	Shell	USP 3,432,558
Bi, Mo, P, Fe, B	Asahi Kasai Kogyo KK	BP 1,095,008
Bi, Mo, Sb, V, Al	Asahi Electrochem	Jap.P 23926/68
Bi, Mo, Fe, As	Daicel	BP 1,182,824
Bi, Mo, Fe, V	Toyo Soda	Jap.P 13609/68
Bi, Mo, Ba, Si	Sohio	USP 3,362,998
Bi, Mo, P, Ni, Co, Fe	Nippon Kagaku KK	BP 1,128,031
Bi, Mo, P, Ni, Co, Fe, As	Nippon Kagaku KK	FP 1,514,163
Bi, Mo, P, Ni, Co, Fe+Ta or Sm	Degussa	Dutch P 6913168
Bi, Mo, Te	DCL	BP 963,610
Bi, Mo, B	Sohio	USP 3,519,688

10.1.3.6. Catalysts containing molybdenum and tellurium There are few instances of the use of tellurium in the absence of molybdenum,[19] but many catalysts containing both these elements have been claimed and there is no doubt that tellurium in such combinations has a useful effect in improving catalyst efficiency. But while

Te is less volatile than Se (indeed a tellurium transport may not be necessary) it should be remembered that it is also less abundant, and if several large plants used Te containing catalysts there might be an embarrassing shortage of the element.

Catalyst elements	Patentee	Patent No.
Mo, Te, P, Si	ICI	Belg.P 671,437
Mo, Te, P, Ti	ICI	BP 1,146,870
Mo, Te, Cr	Toa Gosei	FP 1,579,839
Mo, Te, Ti	Mitsubishi Petrochem.	FP 1,530,334
Mo, Te, Cu	Röhm and Haas	USP 3,441,613
Mo, Te, Co, Fe	Union Carbide	USP 3,467,716
Mo, Te, Co, Fe	Union Carbide	USP 3,440,180
Mo, Te, Cr, Ni	Goodrich	USP 3,497,553
Mo, Te, Re, Ni	Goodrich	USP 3,492,247
Mo, Te + Group IIA or VIII metal phosphate	Goodrich	BP 1,146,637
Mo, Te, W and one or more of (Fe, Ni, Cu, Mn)	BASF	Dutch P 6908212
Mo, Te, Bi, Co, W Mo, Te, Co, V and Bi or Sb Mo, Te, V, Fe and one of (Sn, W, Bi) Mo, Te, V, W and one of (Bi, Sn, Sb)	Asahi Denka Kogyo	Jap.P 24645/68
One or more of (Mo, Cr, W) + Te	Shell	FP 1,342,903

10.1.3.7. Catalysts containing antimony

Sb and one of (Mo, W, Te, Cu, Ti, Sn, Co)	DCL	BP 864,666
Sb and one of (V, Co, Fe, Mn, Ni, Zn, Ga, Ge, Se, Rb, Y, Ir, Nb, Ru, Pa, Os, U)	DCL	BP 991,085
Sb, Sn, U	DCL	BP 1,026,477
Sb, Sn and one of (Fe, Cu, Co, Mn, Ti, Mo, V, W, Co, U, Ni)	DCL	USP 3,408,400
Sb, Mo and one of (Mg, Ca, Sr, Ba, Zn)	Mitsubishi Rayon	Jap.P 1,3963/68
Sb, Mo and one of (Mn, Ni, U)	Mitsubishi Rayon	Jap.P 27848/68
Sb + another metal	Nitto Chem. Ind.	Dutch P 6911897
Sb, Fe, Te + (P or B) + (V or Mo or W)	Nitto Chem. Ind.	Belg. P 730,696
Sb, Mo, Cr	Osaka Gas	Jap.P 26285/67
Sb, Fe	Sohio	GP 1,265,731
Sb, U and one of (Bi, Sn, Pt, B, Mg, Ag, Fe, Zr, Ca, As, W, P, Th, Zn, Co, Ni, Pb, Al, Cu, Sb, Cs)	Sohio	USP 3,328,315
Sb, U	Sohio	USP 3,428,674
Sb, Fe	Sohio	USP 3,468,958
Sb, Mn	Sohio	USP 3,445,520

10.1.3.8. Other catalysts

Relatively few patents relate to catalysts that cannot be included in the above four groups.

Mo and one of (Ca, Ce, Pr, Nd or Sm)	Asahi Denka Kogyo	Jap.P 23605/68

One of (B, P, Cr, Mo, W, V) with metals from the group of atomic nos. 25–30 and Ti, Ag, Cd, Sn, Ce, Pb, Bi, Th, U	Bayer	Can.P 588,908
Te, Rh and one of (Co molybdate Ni molybdate, Cu phosphate)	Goodrich	USP 338,571
V, Sn, P	Knapsack	GP 1,221,208
Mo, P, As	Soc. Nat. des Petroles d'Aquitaine	FP 1,426,166
Bi, V, P	Stamicarbon	BP 950,686
Group Vb and VIb metal, e.g. Mo, Cr or W and P	Hoechst	BP 998,465
Bi, As or As, Al, Mo	Koayo Gijutsuia	Jap.P 7765/68
Bi, Cr, V optional P	Stamicarbon	BP 1,015,180
Cu, Mo, Si optional As	I C I	BP 981,134

It is probable that plants for the production of acrolein from propylene that may go on stream during the next few years will utilise catalysts in which either antimony or bismuth/molybdenum is the principle component. A more detailed assessment of catalysts is not possible at this stage. High performance is claimed for many in the above lists; 50 per cent or more yield with a feed containing up to 10 per cent propylene is becoming commonplace. But patent examples are more often than not based on small-scale experiments which are no guarantee of commercial success. It may take years of costly development to ensure that, on the industrial scale, there will be no rapid deterioration of either the chemical or mechanical properties. Both the discovery and the development of catalysts have a highly empirical flavour that reflects the absence of useful penetrating theory.

10.1.4. The Mechanism of Propylene Oxidation

The earlier successes in evolving catalysts for the production of acrolein from propylene aroused great interest in academic as well as industrial circles. The kinetics of the reaction were found to be simple. Over cuprous oxide, where the propylene is in excess, the rate of conversion of propylene was first order with respect to oxygen and independent of propylene.[20] Over bismuth molybdate the rate was first order in propylene and zero in oxygen,[21] and over Sb/Sn catalyst the result was similar except that there was a slight dependence on oxygen.[22]

Adams et al.[21] considered that the reaction over bismuth molybdate was not retarded by the product acrolein, but Gel'bshtein et al.,[23] who used a reactor to which the product could be recycled, reported some inhibitory effect.

Typical kinetics, inclusive of the principal side reactions, for an Sb/Sn catalyst, are illustrated in Fig. 10.1, the curves for which were calculated on the basis of the following reaction scheme:

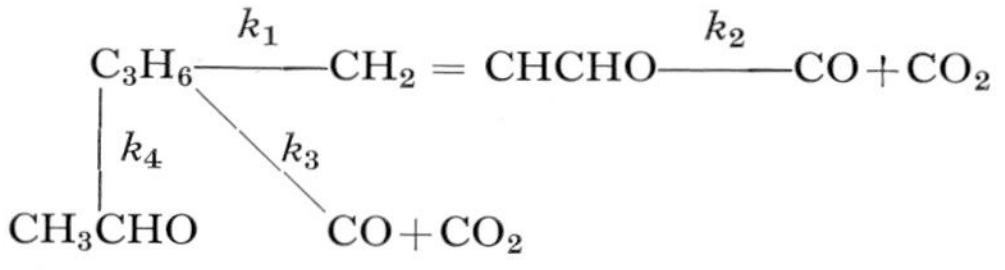

$$C_3H_6 \xrightarrow{\quad k_1 \quad} CH_2 = CHCHO \xrightarrow{\quad k_2 \quad} CO + CO_2$$
$$\Big\downarrow k_4 \qquad \searrow k_3$$
$$CH_3CHO \qquad CO + CO_2$$

using numerical integration and assuming first order dependence for propylene, and oxygen dependence based on the Langmuir adsorption isotherm. The differential equations were of the form:

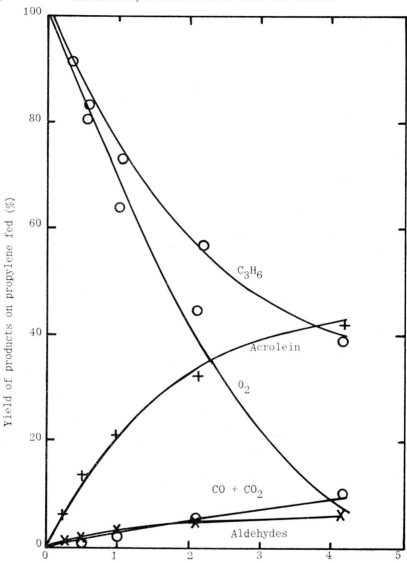

FIG. 10.1   *Variation of Yields of Products with Contact Time*

Catalyst $\dfrac{Sb}{Sn} = \dfrac{1}{3}$ Temperature 480°C

For calculated curves: $k_1 = 0.35/y$; $k_2 = 0.125/y$; $k_3 = 0.02/y$; $k_4 = 0.047/y$

$y = (1 + 24.3\ [O_2]\ \text{mm Hg})$

From paper No. 20 (Godin, McCain and Porter) 4th International Congress on Catalysis, Moscow, 1968.

$$\frac{d[C_3H_6]}{dt} = \frac{k[C_3H_6][O_2]}{C+[O_2]},$$ where k and C are constants. The predominating effect of reaction 2, the further oxidation of acrolein, on the efficiency of the overall reaction is evident. It can also be clearly seen from Fig. 10.1 that the efficiency decreases as the conversion increases, and the highest possible efficiency, according to this reaction scheme, is $\dfrac{k_1}{k_1+k_3+k_4}$, for this particular catalyst at this particular temperature.

The study of the reaction mechanism has been pursued in many parts of the world, with the aid of propylene in which one or more of the carbon atoms were labelled, either with deuterium or by the use of carbon isotopes. The first papers on the subject came from the Shell Development Co., where Voge and co-workers[24] passed a mixture of 1-propene-3-C^{13} ($CH_2 = CHC^{13}H_3$) and oxygen over a cuprous oxide catalyst and examined the acrolein product by mass spectrometry. The concentration in the product spectrum of mass 30 ions (attributable to $C^{13}HO$) was 1/5th of the concentration of mass 57 ions (attributable to $CH_2 = CHC^{13}HO$ and $C^{13}H_2 = CHCHO$) whereas in the normal acrolein spectrum the corresponding ratio of mass 29 ions (CHO) to mass 56 ions ($CH_2 = CHCHO$) was about 2/5ths, which indicated that 50 per cent of the tagged acrolein molecules had C^{13} in the carbonyl group and 50 per cent had C^{13} in the methylene group. Voge reasoned that the initial attack on the propylene molecule must, therefore, have been at the methyl group, giving an effectively symmetrical intermediate, which he suggested was the mesomeric allyl radical:

$$CH_2 = CHCH_3 \xrightarrow{\;-H\;} CH_2 = CH-\dot{C}H_2 \rightleftharpoons \dot{C}H_2-CH = CH_2$$

An interesting side inference from this work, based on the ratio $\dfrac{C^{13}O_2}{C^{12}O_2}$ in the reaction product, was that CO_2 comes from the total oxidation of a C_3 chain and not from selective oxidation of end carbon atoms.

Concurrently, and in the same laboratory, Adams and Jennings,[25] carried out similar studies with 1-propene-3d ($CH_2=CHCH_2D$), using both Cu_2O and Bi/Mo catalysts. They reached the same principal conclusion as a result of studying the distribution of deuterium in the product acrolein, and they wrote a reaction scheme for the abstraction of two hydrogen (deuterium) atoms from their monodeuterated propylene (assuming that the addition of oxygen is the final step in the formation of acrolein).

The numbers against the arrows show the relative probabilities of the alternative reactions and $Z = \dfrac{\text{probability of breaking a C—D bond}}{\text{probability of breaking a C—H bond}}$. Z is the isotopic discrimination effect $= \dfrac{k_D}{k_H}$. Assuming Z has the same value for both reaction stages, then its relationship with d_1, the fraction of monodeuterated acrolein to the total acrolein, is:

$$Z = -2{\cdot}5 + \sqrt{\dfrac{0{\cdot}25+6}{d_1}}$$

Fairly consistent values of Z (around 0·5) were derived from the experimental d_1 values, in reasonable agreement with that obtained by other and independent methods.

Subsequently[28] the monodeuterated acrolein derived from the oxidation of 1-propene-3d was analysed, and it was found that:

$$\dfrac{\text{moles } CH_2 = CHCDO}{\text{moles } CHD = CHCHO} = 0{\cdot}6$$

It is considered that this result is sufficiently close to the theoretical value of 0·5 to constitute a separate proof of the occurrence of a symmetrical intermediate.

McCain et al.,[26] working with $C^{13}H_2{=}CH{—}CH_3$, also arrived at the conclusion that a symmetrical intermediate was involved, but pointed out that an isopropyl radical was a possible alternative to the allyl radical:

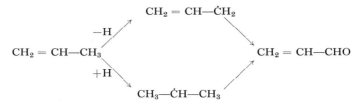

To decide between these they carried out the oxidation of normal C_3H_6 over a deuterated bismuth molybdate surface and since they obtained no deuterated acrolein concluded that hydrogen addition reactions are not involved in the formation of acrolein from propylene.

Further confirmation of the allylic intermediate was obtained by Sachtler[27] by oxidising $C^{14}H_2{=}CH{—}CH_3$, $CH_2{=}C^{14}H{—}CH_3$, and $CH_2{=}CH{—}C^{14}H_3$ over bismuth molybdate, and photochemically decomposing the resulting acroleins thus:

$$CH_2{=}CH{—}CHO \xrightarrow{\ h\nu\ } CH_2{=}CH_2 + CO.$$ The radioactivity of the CO relative to that of $CH_2 = CH_2$ was unity when the C^{14} was at either end of the starting propylene molecule and nil when the C^{14} was central. An alternative explanation of this result, i.e. the rapid isomerisation $C^{14}H_2{=}CH{—}CH_3 \rightleftharpoons CH_2{=}CH{—}C^{14}H_3$, was ruled out by experiments with n-butene which isomerised at a negligible rate.

In another paper Adams and Jennings[28] measured the velocity constants for the oxidation of various deuterated propylenes relative to that of C_3H_6 (over bismuth molybdate), and found them to be in good agreement with values calculated on the assumption that the first, and the rate controlling, step in the reaction is the abstraction of an allylic hydrogen atom, with a kinetic isotopic effect of 0·50 at 450°C, similar to that disclosed in the earlier paper.

| | *Relative rate constant at 450°C* | |
Compound	*Observed*	*Calculated*
C_3H_6	1·00	(1·00)
$CH_2 = CH—CH_2D$	0·85 ± 0·02	0·83
$CHD = CH—CH_3$	0·98 ± 0·02	1·00
C_3D_6	0·55	0·50

In Ref. 28 evidence is given in support of the previous assumption (that the addition of oxygen is the final step in the reaction). For the oxygen, if it attacked the allylic intermediate at all would be expected to attack both ends equally, whatever the deuterium content. In that case the proportion of deuterated acrolein in the product would be different from that actually observed. The distribution of deuterium in the acrolein derived from 1-propene-1-d is also reported in Ref. 28, and a reaction scheme for this reagent is shown, giving another relationship between Z and d_1, in good agreement with experimental results.

Questioning the validity of the assumption that the isotope effect has the same value in both of the successive hydrogen abstraction steps Godin and McCain[29] studied the oxidation of 1-propene-3d_3 over bismuth molybdate:

$$CD_3—CH = CH_2 \longrightarrow \overset{\displaystyle \centerdot}{\overbrace{CD_2 \cdots CH \cdots CH_2}}$$

$$\overbrace{CD_2 \cdots \overset{\displaystyle \centerdot}{CH} \cdots CH_2} \underset{\longrightarrow}{\overset{\longrightarrow}{}} \begin{array}{l} CH_2 = CH—CDO \\ CD_2 - CH—CHO \end{array}$$

The rate of oxidation of 1-propene-3d_3 divided by the rate of oxidation of ordinary propylene gave the kinetic isotope effect in the formation of the allylic intermediate $\left(\dfrac{k_D}{k_H} = 0·51 \right)$. Whilst the ratio of the amounts of the two deuterated acrolein products gave, unambiguously, the isotope effect in the second hydrogen abstraction step, $\dfrac{k_D}{k_H} = 0·55$. This work also confirmed that oxygen enters the molecule after both hydrogen abstraction steps have occurred, for if oxygen attacked this particular allylic intermediate the product acrolein would consist of equal amounts of the two types shown and there would be no measurable isotope effect.

The oxidation of propylene over Sb/Sn catalyst was also shown to proceed via an allylic intermediate.[22]

Sixma *et al.*[30] have also contributed to the impressive body of evidence in support of the view that the two end carbon atoms in the propylene molecule are equally likely to be part of the carbonyl group in the resulting acrolein, which implies not only intrinsic symmetry of the intermediate but also symmetry with respect to the catalyst surface. This could be achieved either by desorption of the radical before further reaction or by symmetrical terminal attachment to the surface.

Oxidation via the allylic intermediate is the usual but not the invariable result of the reaction of propylene and oxygen over oxide catalysts. If a large proportion of water is included in the feed, then over certain catalysts, at temperatures below 300°C, acetone is the main product.[31] In such cases propylene is attacked at the central carbon atom and water provides the oxygen in the product acetone. The intermediate may be the isopropyl radical.

10.1.5. Fundamental Catalyst Research

Oxidation catalysts are oxygen transfer agents, which, usually, will convert a limited amount of propylene to acrolein, even in the absence of gaseous oxygen. One may think of the oxygen as effecting continuous, concurrent regeneration of the catalyst. However, these simple principles, whose application to the $CuSeO_3$ catalyst has already been shown, are not of much use in the evolution of new or the improvement of known catalysts. A considerable effort has been devoted to the attempt to gain a better insight into catalytic action. For the most part it has been directed at Bi/Mo and Sb/Sn catalysts, and the method used has generally been that of attempting to correlate catalyst performance with bulk physical and chemical properties.

It has already been noted that Bi/Mo catalysts are made by precipitation from solutions of ammonium molybdate (or molybdic acid) and bismuth nitrate. Silica and phosphoric acid may be added before the precipitate is dried, but no heat treatment is necessary. Sb/Sn catalysts are usually produced by heating a mixture of any antimony oxide and SnO_2 of high surface area, in air, at a temperature of 600°C or more. The component oxides of the resulting solid are Sb_2O_4 and SnO_2.

Godin et al.[22] prepared a series of Sb/Sn catalysts by mixing the freshly precipitated oxides in various proportions and subjecting all to the same heat treatment at 850°C. Debye-Scherrer X-ray powder photographs of the finished catalysts showed that they consisted of SnO_2 and two phases of Sb_2O_4, but there was no compound formation. Proof of another structural possibility, i.e. a solid solution, by means of X-rays is difficult because it depends on relatively small changes in the unit cell dimensions of one of the constituents. The electrical resistance of these catalysts was then measured and plotted against antimony atom fraction $\left(\dfrac{Sb}{Sb+Sn}\right)$. Irrespective of the temperature of measurement there was a decided minimum resistance at about 0·07 antimony atomic fraction, and it was concluded that this composition was a solid solution saturated with respect to antimony. When the same series of catalysts was used in the oxidation of propylene the specific rate constant for the disappearance of propylene (rate constant per unit surface area) was found to be substantially constant over a wide range of composition—about 0·07–0·93 antimony atomic fraction. It was inferred that, within this range, the antimony-saturated solid solution is the essential catalyst, providing substantially all the sites for the desired reaction and substantially all the surface area, the catalyst action being ascribed to Sb^{5+} ions in octahedral co-ordination sites. However, the compositions with both the highest catalyst activity and surface area (per g) contained much more than 0·07 atomic fraction of antimony. In order to reconcile this fact with the theory it is necessary to consider that the role of the excess antimony is not merely that of diluent, but that it acts to restrict the reduction of surface area of the solid solution which occurs by sintering during the heat treatment.

Godin *et al.* point out that combinations of Fe, V, Cr, U, Cu, Ni, Co, Fe, Zn or Mn with Sb all have Sb^{5+} in octahedral co-ordination. Most of these compositions have been patented as catalysts for propylene oxidation.

After examining similar catalysts by X-ray and by infra-red spectroscopy Roginskaya *et al.*[32] reported solid solutions of Sb_2O_4 in SnO_2 (rutile structure) in catalysts (heated at 900°C), which had an antimony atomic fraction of up to 0·2, but with more antimony present a solid solution with an Sb_2O_4 structure was detected. They also reported the presence of Sb_2O_3, but this, assuming heating in air at 900°C had taken place, is difficult to understand, since Sb_2O_3 rapidly combines with oxygen at temperatures above 500°C and, in any case, has a high vapour pressure. Wakabayashi *et al.*[33] studied the conversion of propylene to acrolein as a function of antimony atomic fraction and temperature of heat treatment and found this complicated. They also obtained X-ray data which showed no compound formation and attributed the activity of the Sb/Sn catalyst to a solid solution. The same authors[34] carried out conductivity, reaction rate and surface area measurements similar to those of Godin *et al.* Their results were only in qualitative agreement, the maximum conductivity occurring at 3 per cent antimony atomic fraction. However, it should be noted that they used a different method of catalyst preparation. They mention the difficulty of determining the solubility of antimony oxide in tin oxide, because the former does not greatly change the cell dimension of the latter. Throughout their later paper, Wakabayashi *et al.* talk in terms of Sb_2O_5. This is considered to be an error; Sb_2O_4 is the stable oxide at the usual calcination temperatures.

Stroeva *et al.*[35] studied the systems Bi/Mo, Sb/Sn and Co/Mo, measuring specific surface, phase composition (by X-ray and thermographic methods), electron yield energies, and electro conductance as functions of composition. In no system did they find a simple connection between electrophysical parameters and catalyst performance in the oxidation of olefins when varying the composition. However, they reported some correlation between electroconductance and catalyst activity. Three different regions in the Sb/Sn system are claimed; the first (Sn/Sb < 0·5) is a solid solution of SnO_2 in Sb_2O_4 the second (Sn/Sb > 5/1) a solid solution of Sb_2O_4 in SnO_2 and the third a mixture of both these solid solutions. The promoting effect of Bi_2O_3 (on MoO_3) is considered to be due to the formation of bismuth molybdate, and the superior catalyst is that having Bi/Mo = 1. On the other hand cobalt molybdate in the Co/Mo catalyst is said to have no marked catalytic property and functions merely as a carrier for MoO_3. A twin mechanism involving two kinds of surface centres (S_1 and S_2) was suggested for the formation of the allyl radical:

$$H_2C = C-CH_3 \rightleftharpoons H_2C = CH-CH_2\ldots\ldots H-\rightarrow H_2C\cdots CH\cdots CH_2 + H$$
$$\downarrow \qquad\qquad \downarrow \qquad\qquad \downarrow \qquad\qquad \downarrow$$
$$S_1 \qquad\qquad S_2 \qquad\qquad S_1 \qquad\qquad S_2$$

As a result of studying the chemisorption of propylene/oxygen mixtures on Sb/Sn Derlyukova *et al.*[36] suggested the formation of pre-catalyst intermediates (hydrocarbon-oxygen complexes) on the surface, but with modern methods (ESR, IR) the same authors did not detect intermediate radicals.[37]

By thermal analyses and X-ray methods, Bleijenberg et al.[38] detected the following compounds in the Bi/Mo system.

	m.p. °C
$Bi_2O_3.3MoO_3 = Bi_2(MoO_4)_3$	676
$Bi_2O_3.MoO_3 = Bi_2MoO_6$	938
$3Bi_2O_3.MoO_3 = Bi_6MoO_{12}$	995

But in a later paper Batist et al.[39] using a new method of preparation of the catalyst, which consisted in slurrying a mixture of the two oxides (hydroxides), reported the compound $Bi_2O_3. 2MoO_3$ also, and found that the better catalysts were confined to the region $Bi_2O_3. MoO_3 \rightarrow Bi_2O_3 2MoO_3$. The bismuth has a dual function; firstly, it acts because it can be reduced (so affecting the re-oxidation of Mo to the MoO^{6+} state) and secondly it enforces a structural situation on Mo with respect to oxygen, that is favourable to the interaction of the Mo ion.

In a sequence of four papers[40] Peacock et al. describe ESR, electrical conductivity and kinetic studies with the Bi/Mo catalyst. When either bismuth molybdate or MoO_3 were exposed to propylene at 200–500°C, ESR signals were ascribed to MoO^{5+} ions formed by reduction, but there were no such signals in the presence of both propylene and oxygen. Bismuth appears to have an oxidising function. The drop in selectivity consequent on partial reduction of the catalyst is ascribed to a change in the relative strengths of C—C bonds (in reactant) and reactant to surface bonds. Bismuth molybdate and MoO_3, reduced with hydrogen or olefin, had increased electrical conductivity. This effect was reversible and the rates of reduction and re-oxidation were proportional to gas partial pressure. Conductivity measurements also showed that during the oxidation of propylene by oxygen over bismuth molybdate the latter was considerably reduced, whereas in the same circumstances MoO_3 remained in a well-oxidised state.

A possible correlation between the character of the metal-oxygen bond in a catalyst and its activity is discussed by Trifiro and Pasquin.[41] In general the best catalysts have a metal-oxgen linkage with a double bond character, e.g. bismuth and other molybdates. However, Sb/Sn, which has little double bond character, is also highly selective in the propylene to acrolein reaction.

At the last International Congress on Catalysis in Moscow, 1968, Margolis, in a paper (no. 19) that was largely a review of recent work on oxidation catalysis, concluded that selective catalysts must have elements capable of forming complexes with hydrocarbons, and oxygen linked with the surface by optimal energy and located at a suitable distance from the hydrocarbon centres. Knowledge of active centre structure and of surface bonding would make possible a rational choice of oxidation catalysts. Gel'bshtein et al. (paper 22) compared the activation energy E for olefin oxidation with the activation energy E_o for O_2 exchange for a series of oxides, E_o being considered to be a measure of the M—O surface bond strength. In the cases of Co, Mn, Ni, Cu, Fe, Sn and Cr which have high catalytic activity for complete oxidation and relatively low E_o values, E_o was directly proportional to E. On the other hand Bi, V, Sb and Mo oxides have low catalytic activity and high values of E_o, and E_o is not proportional to E. Gel'bshtein also measured the work function $\Delta\phi$ (work done in removing an electron) of oxides in O_2 at 400°C and found the following order: $Cr > V > Mo > Sn > Co > Cu > Fe > Mn > Sb > Bi$. The last two, with the lowest $\Delta\phi$, are therefore assumed to have the lowest capacity for co-ordinating olefins, electron donors. These authors believe that decrease in E_o will increase complete oxidation activity, whereas an increase will improve selectivity. Catalyst activity will also be dependent on $\Delta\phi$, and selective catalysts should have high values of both E_o and $\Delta\phi$. In a general way a correlation between E_o, $\Delta\phi$, activity and selectivity was exhibited by the series of binary combinations: Bi/Mo, Fe/Mo, Co/Mo, Fe/Sb, Co/Sb and Sn/Sb. However, these views were, in some details, in conflict with those of Alkhazov *et al.* (paper 17) who said that the binary oxide should include one member with high isomerisation selectivity and hence good olefin absorbing power (Mo, Sb, P, W, V) and one of moderate CO_2 selectivity to supply oxygen for the reaction (Cd, Co, In, Bi, Co, Sn, Mn).

The main conclusion from the considerable effort that has already been expended on the study of multi-oxide catalysts is that those which are effective are either compounds or solid solutions of the component oxides. This appears to be a necessary (but far from sufficient) condition and although it has not in all cases been demonstrated that combination of the oxides has taken place, in either a stoichiometric or random (solid solution) manner, it seems reasonable to assume that this is a general rule, simply because the pronounced synergism of the effective systems could hardly arise without molecular intimacy of the components.

Attempts to separate the functions of the components of binary oxide catalysts have had some success, and it seems probable that in some binaries at least (Sb/Sn, Bi/Mo) one component supplies oxygen to the hydrocarbon while the second serves to re-oxidise the first.

10.1.6. Manufacture of Acrolein by Oxidation of Propylene

An account of the oxidation of propylene to acrolein on a commercial scale must necessarily be generalised and lacking in definition since there are several versions of the process, differing considerably in design and operation detail and (most important) in catalyst performance, and there is no adequate published description of any.

10.1.6.1. Reactants The feed to the reactor will, for catalysts other than Cu_2O, contain between 2 and 10 per cent (by volume) of propylene (probably 5 to 10 per cent). It may contain steam, up to 40 per cent, for the purpose of rendering the gas mixture non-explosive, and for improving the selectivity of some catalysts. The remainder will be air, unless there is recycle of gas round the reactor, in which case it may be necessary to feed pure oxygen in order to maintain the concentration of that reactant in the gas entering the reactor. Recycle is essential if the Cu_2O catalyst is used, because there is more propylene and less oxygen in the feed, and the conversion of propylene per pass is, therefore, small.

10.1.6.2. Reactor system Fixed bed reactors of the shell and tube type appear to be generally favoured, the tube diameter being 0·5 to 2 in. Restriction of tube diameter, and the provision of a liquid heat transfer medium, commonly K/Na/ NO_2/NO_3 mixture, are necessitated by the considerable heat output of the principal reactions:

$$C_3H_6+O_2 \rightarrow CH=CHCHO+H_2O \qquad \Delta H_{298} = -83 \text{ kcal}$$
$$C_3H_6+4\tfrac{1}{2}C_2 \rightarrow 3CO_2+3H_2O \qquad \Delta H_{298} = -460 \text{ kcal}$$

Reactor conditions depend on the catalyst: usually the bath temperature will be in the range of 300–450°C and the contact time 1–10 sec. Pressures of more than a few atmospheres are not used.

Fluidised bed reactors are also feasible for this reaction, and are frequently mentioned, e.g. Ref. 42. But the susceptibility of acrolein to both homogeneous and catalysed oxidation militates against its production in a fluidised bed where the residence time of the reaction product may be increased by back-mixing. The rate of the catalysed further oxidation depends, of course, on the catalyst in use, but in general it is significant. The homogeneous gas phase reaction potential is such that it is desirable to cool the gas immediately after it leaves the catalyst zone, by direct injection of water or by indirect heat exchange.[43]

10.1.6.3. Reaction products   Apart from carbon oxides and water, the main by-products are: formaldehyde, acetaldehyde, propionaldehyde, acetone and acrylic acid. It is probable that all these componds (and several others) are produced whatever the catalyst in use, but the amounts relative to acrolein, vary greatly. Thus from the copper selenite catalyst the only significant by-products are acetaldehyde and formaldehyde, which together amount to some 2 per cent of the acrolein. Possibly the low reaction temperature (320°C) accounts for the small amount of by-product in this case. On the other hand Pierrote *et al.*[44] give the following composition for crude acrolein obtained, presumably, by oxidation of propylene over Cu_2O catalyst:

	% wt/wt
Acrolein	80–90
Acetaldehyde	3–10
Propionaldehyde	0·5–3
Acetone	1–5
High-boiling material	1–2
Water	2·4–6·0

Some catalysts, mentioned later in this chapter, give particularly large amounts of acrylic acid. This and other high-boiling substances may be removed from the exit gas stream by scrubbing it with water at a temperature above the boiling point of acrolein.

10.1.6.4. Recovery of acrolein   After being further cooled to ambient or lower temperature the product gas is passed up an absorption tower where soluble products are dissolved in a counter flowing stream of water. Gas leaving the absorber consists principally of nitrogen, oxygen, propylene, propane (a normal constituent of commercial propylene which escapes oxidation) and carbon oxides, together with a trace of acrolein. If the conversion of propylene is sufficiently great (say 50 per cent, depending on cost and availability of the hydrocarbon) there will be no recycle of the scrubbed gas, but before this passes to the atmosphere it will normally be necessary to remove the trace of acrolein, 1 or 2 p.p.m. of which is objectionable (because of the lachrymatory property) if not toxic. The acrolein may be destroyed by a reagent or by burning. The latter method also ensures recovery of the fuel value of the unreacted hydrocarbon. Unless at least part of the unreacted propylene is isolated from the gas stream, for instance by absorption in oil, it cannot be completely recycled to the reactor entrance, since part of the scrubbed gas must be bled off in order to limit the concentration of inert gas, principally N_2, CO_2, CO and parafins, although the bleed will be small if oxygen is fed in its pure form.

The molar concentration of acrolein in the gas entering the absorber depends simply on the yield per pass and the concentration of propylene in the feed gas (not taking account of the steam). Assuming efficient operation of the scrubber, the concentration in the aqueous effluent product is given approximately, at ambient temperature, by the following relationship:

$$\text{partial pressure (atm) acrolein in gas} = 0.011 \times \text{per cent in water.}$$

Aqueous acrolein is pumped to a stripper column, where it is subjected to distillation, the greater part of the acrolein coming off overhead in the form of its water azeotrope. Here concentrated acrolein is being handled for the first time and steps must be taken to prevent polymerisation. This is generally considered to be initiated by peroxides which are products of dissolved oxygen. Antioxidants, such as hydroquinone, are added to the top of the column to destroy peroxy radicals. According to Ref. 45 the requisite amount of antioxidant can be greatly reduced, if the aqueous acrolein is subjected to deaeration, i.e. is heated to near its boiling point at reduced pressure, before it is introduced into the stripper.

10.1.6.5. Purification The stripper product may already be sufficiently pure for some purposes. Further purification is effected by fractional distillation. Separation of acetaldehyde by this method is straight forward, but if propionaldehyde and acetone, with volatilities close to that of acrolein, have to be removed, extractive distillation must be used. Besides water[46] various polyhydroxy and other compounds have been disclosed as extractants.[47] According to Pierotte et al.[48] the addition of acid to the extractant water facilitates the separation of unsaturated from saturated carbonyl compounds by catalysing the formation of hydrates of the latter.

Acrolein is obtained from these various purification steps as the water azeotrope, which does not, upon condensation, separate into two layers. Complete dehydration may be effected by distillation in the presence of an entrainer such as a hydrocarbon[49] or ether.[50] Selective extraction of the water into glycerol has also been claimed.[51]

10.1.7. Uses of Acrolein

The discovery of practicable methods of manufacture of acrolein from propylene induced several companies, including Shell, Union Carbide and DCL, to study intensively its utilisation. Their efforts contributed much to the now extensive knowledge of the chemistry of acrolein, which is dealt with, comprehensively, by C. W. Smith.[52] Unfortunately, much of this knowledge has little relevance to established industrial applications, for these are few, and it must be admitted that the earlier beliefs in the development of acrolein as a major intermediate have not, as yet, been justified. There are at least two acrolein plants in the USA and others in France, Germany, Japan and, possibly, Eastern Europe. But, although precise data are not available it is clear that the present production capacity and demand for acrolein are small by comparison with the leading petrochemicals such as ethylene oxide. The most important contribution to the Petroleum Chemicals Industry from the research on acrolein production and its derivatives is probably the discovery of new routes to acrylic acid and acrylonitrile which were incidental to it. Acrolein is, probably, an intermediate in both these processes, although in the former, of which there are several variations (described later in this chapter) it does not necessarily have to be isolated, and in the latter it is no more than transient.

10.1.7.1.   Methionine $CH_3SCH_2CH_2CH(NH_2)COOH$ This amino acid, the most important industrial derivative of acrolein, is an essential growth-promoting constituent of an adequate diet. Its use as an additive to poultry and cattle feeds is increasing and there is a possibility that it may be included in certain foods for human consumption. It is believed that there is production from acrolein on a commercial scale in USA, Germany, France and Russia. The first step in one of the methods of synthesis[53] is the addition of methyl mercaptan to acrolein, catalysed by a base such as sodium ethoxide, to give 3-methylmercaptopropionaldehyde. This product is converted to the cyanhydrin by the addition of hydrogen cyanide in the presence of OH ions provided by KCN catalyst:

$$CH_3SCH_2CH_2CHO + HCN \xrightarrow{KCN} CH_3SCH_2CH_2CH(OH)CN$$

Upon heating the cyanhydrin with ammonia the OH group is replaced by an amino group,

$$CH_3SCH_2CH_2CH(OH)CN + NH_3 \longrightarrow CH_3SCH_2CH_2CH(NH_2)CN + H_2O$$

Finally, the CN group is hydrolysed by hot dilute acid,

$$CH_3SCH_2CH_2CH(NH_2)CN \xrightarrow{H_2O,\ HCl} CH_3SCH_2CH_2CH(NH_2)COOH + NH_4Cl$$

10.1.7.2.   Glycerol One of the several method of manufacture of glycerol from propylene proceeds by way of acrolein. This is hydrogenated to allyl alcohol by transfer of hydrogen from an alcohol such as isopropanol in a reversible reaction, which may proceed in the liquid phase with aluminium alkoxide catalyst,[54] or in the vapour phase over magnesium oxide:[55]

$$CH_2 = CHCHO + CH_3CH(OH)CH_3 - CH_2 = CHCH_2(OH) + CH_3COCH_3$$

Hydrogenation with molecular hydrogen is apparently not used because it is difficult to exclude addition to the carbon-carbon double bond.

Upon reaction with warm aqueous hydrogen peroxide solution, in the presence of tungstic acid, allyl alcohol yields glycerol, which is purified by vacuum distillation:

$$CH_2 = CHCH_2(OH) + H_2O_2 \longrightarrow CH_2(OH)CH(OH)CH_2OH$$

Shell began the commercial production of glycerol from acrolein in 1959.[56]

10.1.7.3. Other uses of acrolein Acrolein can participate in Diels-Alder reactions by filling either of two roles, that of diene:

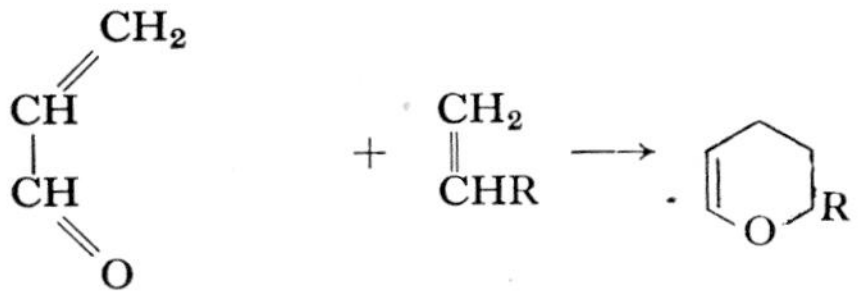

$$
\begin{array}{c}
CH_2 \\
\parallel \\
CH \\
\mid \\
CH \\
\diagdown \\
O
\end{array}
\quad + \quad
\begin{array}{c}
CH_2 \\
\parallel \\
CHR
\end{array}
\quad \longrightarrow \quad
\text{[pyran ring with O and R]}
$$

or that of dienophile:

$$
\begin{array}{c}
CHR_1 \\
\parallel \\
CH \\
\mid \\
CH \\
\mid \\
CHR_2
\end{array}
\quad + \quad
\begin{array}{c}
CH_2 \\
\parallel \\
CH\cdot CHO
\end{array}
\quad \longrightarrow \quad
\begin{array}{c}
CHR_1 \\
CH \quad CH_2 \\
\parallel \qquad \mid \\
CH \quad CHCHO \\
CHR_2
\end{array}
$$

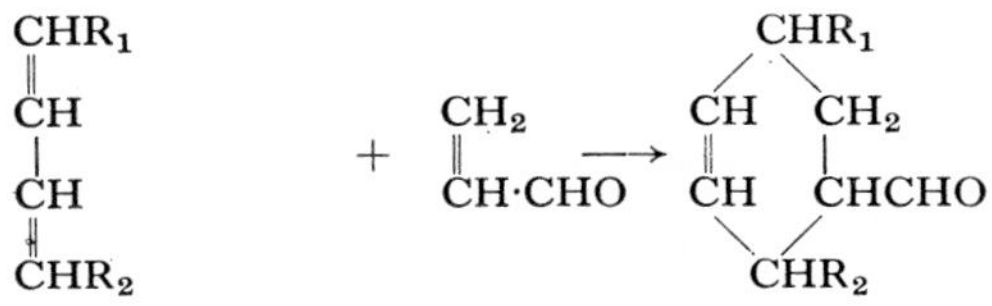

In the particular case of acrolein dimer formation[57] the monomer exhibits both functions:

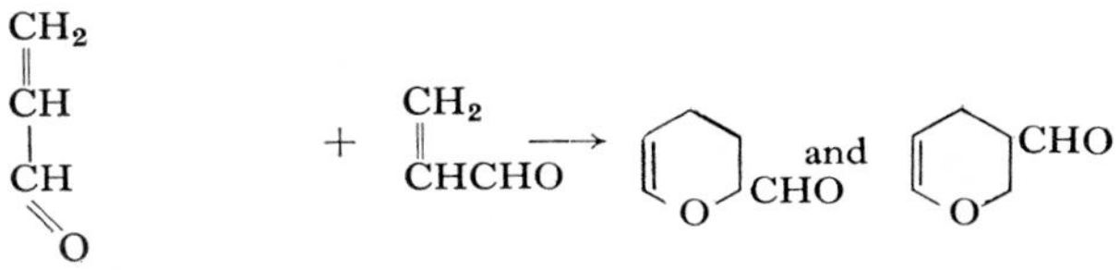

$$
\begin{array}{c}
CH_2 \\
\parallel \\
CH \\
\mid \\
CH \\
\diagdown \\
O
\end{array}
\quad + \quad
\begin{array}{c}
CH_2 \\
\parallel \\
CHCHO
\end{array}
\quad \longrightarrow \quad
\text{[ring with O, CHO]} \quad \text{and} \quad \text{[ring with O, CHO]}
$$

The major product is the first of these isomers, 2-formyl-3-4-dihydro-(2H)pyran.

Many Diels-Alder reaction products have been prepared from acrolein and many further derivatives of these have been made, but only a few are produced commercially or show immediate promise of exploitation.[58]

Acrolein dimer is hydrolysed in acidic medium to α-hydroxyadipaldehyde.

$$
\text{[pyran ring]}\ CHO + H_2O \longrightarrow OHC\ CH_2CH_2CH_2\overset{\displaystyle OH}{\underset{\displaystyle |}{CH}}\!\!-\!CHO
$$

And the addition product of acrolein and an alkyl vinyl ether, e.g. 2-ethoxydihydropyran, is similarly hydrolysed to glutaraldehyde:

$$
\text{[pyran ring]}\ OC_2H_5 + H_2O \longrightarrow OHCCH_2CH_2CH_2CHO + C_2H_5OH
$$

Both these dialdehydes are used as crosslinking or insolubilising agents for polyhydroxy compounds and proteins in the paper, textile, leather and photographic industries. Useful polyols are obtained by hydrogenation of their carbonyl groups.

26

Hydroformylation (Oxo reaction) of acrolein dimer leads to 2,6-dimethyloltetrahydropyran:

In a similar manner the Diels-Alder reaction product of acrolein and 1:3 butadiene (1,2,5,6-tetrahydrobenzaldehyde) yields a mixture of the isomeric di-(hydroxymethyl)cyclohexanes:

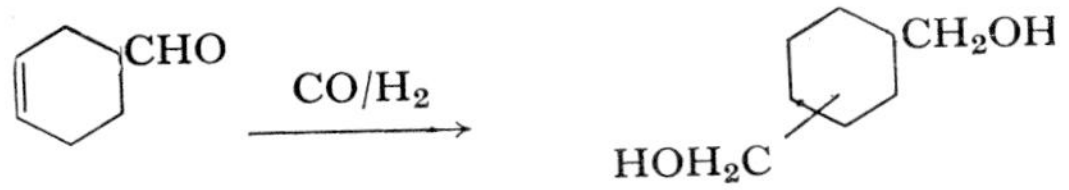

Both these diols may have applications in the manufacture of fibre-forming polymers by condensation with terephthalic acid.[59]

Tetrahydrobenzaldehyde is a cyclic olefin and by reaction with peracetic acid gives an epoxide the resins made from which have advantages over those made from glycidyl compounds where the epoxy group is in the chain.[60]

Canadian Industries Ltd. have developed a thermoplastic foam 'Kayfax Pyranyl Foam', by cationic polymerisation of acrolein dimer.[61]

10.1.7.4. Pyridine   The lower aliphatic aldehydes, or mixtures thereof, react with ammonia in the vapour phase over silica/alumina and similar acidic catalysts to yield pyridine and alkylpyridines. The product composition depends on catalyst, reaction conditions, presence or absence of oxygen in the feed, and on the aldehyde(s) employed. Particular advantages have been claimed for acrolein as the aldehyde feedstock for the production of pyridine. The acrolein does not have to be handled in liquid form, for the gaseous effluent from a propylene oxidation reactor may be passed, together with ammonia, over the pyridine-forming catalyst.[62]

10.2. ACRYLIC ACID, $CH_2 = CHCOOH$

10.2.1. Introduction

10.2.1.1. Some properties

Specific gravity	1·0511 (20 °C)
Freezing point	12·3 °C
Boiling point (1 atm)	141 °C
Refractive index (n_D)	1·4224 (20 °C)
Specific heat of liquid	0·45 cal/g/°C (20 °C)
Vapour pressure at 20 °C	3 mm Hg
Heat of formation (gas at 25 °C)	−79·80 kcal/mole
Heat of combustion (liquid)	−328 kcal/mole
Solubility in water	Completely miscible

10.2.1.2. Scale of manufacture The greater part of the acrylic acid made is esterified. According to data from BP Chemicals International Ltd. the production of the free acid in the USA was 5,400 tons in 1961 and about 37 000 tons in 1968, while the pattern of production and consumption of acrylic esters was:

Year	(Production) Ethyl acrylate	2-Ethylhexylacrylates	(Consumption) Total acrylates
1961	27,000		32,000
62	32,100		38,000
63	35,600	6,900	49,000
64	40,500	10,100	61,000
65	52,100	11,300	71,000
66	58,000		89,000
67	62,900		103,000
68	73,800	14,600	129,000
69			147,000
1970		Forecast	156,000
71		,,	174,000
72		,,	201,000
73		,,	234,000

Gaps in the table are due to lack of information. Consumption of acrylic esters in the UK rose from 1,200 tons in 1960 to 17,000 tons in 1969 and is likely to rise to 30,000 tons in 1975.

10.2.2. Earlier Methods of Manufacture

Although acrylic acid has been known for over a hundred years it was not until the period of World War II when Reppe discovered the synthesis from acetylene and carbon monoxide that it became of importance in the Chemical Industry, and even in the post-war years, which saw the rapid expansion of the Petroleum Chemicals Industry, acrylic acid production developed only slowly in comparison with that of other monomers such as vinyl chloride and styrene. But there is today a mounting interest in the acid and its esters which is, at least partly, attributable to the reduction of monomer price resulting from improved methods of production.

10.2.2.1. From acetylene Acetylene has been and, for the moment, remains the principal raw material for the manufacture of acrylic acid. It reacts, as Reppe discovered,[63] with water and CO in the presence of nickel compounds to give acrylic acid, or with alcohol and CO to give an acrylic ester. Whichever way the reaction is carried out—and there have been several variations on the theme—the overall result (in the case of acid production) is simply:

$$C_2H_2 + CO + H_2O \longrightarrow CH_2 = CHCOOH$$

According to the earliest version of the process there were two stages, in the first of which nickel carbonyl, aqueous HCl and acetylene reacted vigorously at 40–50°C and at atmospheric pressure, in the following way:

$$4C_2H_2 + Ni(CO)_4 + 2HCl + 4H_2O \longrightarrow 4CH_2 = CHCOOH + NiCl_2 + H_2$$

The hydrogen indicated in this equation does not leave the reaction zone; it reacts, at least partly, with acrylic acid, giving propionic acid, which is a particularly undesirable by-product because the physical properties of the two acids are so similar that they are difficult to separate.

In the second stage of this process the $NiCl_2$ was recovered and converted, by the addition of ammonia, to the hexamine dichloride from which carbonyl was regenerated by treatment with CO under pressure:

$$NiCl_2 \xrightarrow{\quad NH_3 \quad} Ni(NH_3)_6Cl_2 \xrightarrow{\quad CO \quad} Ni(CO)_4$$

This was called the stoichiometric process because the nickel carbonyl was a reagent and not a catalyst.

There are two improved process versions that are more nearly catalytic and have the incidental advantage of yielding much less propionic acid. In the first of these, which is operated at atmospheric pressure, a catalytic reaction is imposed on the stoichiometric reaction, that is to say after initiation of the stoichiometric reaction the reagents, C_2H_2, CO and H_2O, together with a relatively small amount of $Ni(CO)_4$, are passed continuously into the reactor. The stoichiometric reaction and a catalysed reaction then proceed concurrently, with the latter predominating. It is assumed that a concentration of reaction intermediate must be built up before the catalysed reaction will proceed at a significant rate.

The second improved process is entirely catalytic but requires more energetic conditions. The same reagents (CO, C_2H_2 and H_2O) together with a small propor- tion of nickel halide catalyst and a liquid reaction medium (tetrahydrofuran) are passed continuously to a tower reactor under a pressure of about 50 atm and at a temperature of about 200 °C. The mixture (gas and liquid) leaving the reactor passes to a degassing column for separation of unconverted C_2H_2 and CO. Tetrahydro- furan is then distilled from the aqueous product for recycle, leaving behind crude aqueous acrylic acid which is worked up by extraction and distillation methods.

There are numerous patents relating to additives that active or advantageously modify the action of the nickel catalyst, e.g. Ref. 64.

Acrylic esters may be made in a similar way by substituting alcohol for water as the third reagent, and it is the esters that are required usually in greater bulk than the acid. Whether a manufacturer will decide on this direct route to the esters depends on several factors. He will often wish to make a series of esters and he has to take into account that the rates of reaction and the by-products will vary with the alcohol employed. It may be more convenient to produce the esters from the acid in separate esterification processes. Alternatively he might make, e.g. ethyl acrylate directly and produce the other esters from it by transesterification.

10.2.2.2. From ethylene An alternative route, which has, however, largely fallen into disuse, starts from ethylene, which is converted to the epoxide either via ethylene chlorohydrin or by direct oxidation in the vapour phase, the latter, more modern method, being preferred. HCN is added to the epoxide in a liquid phase reactioncatalysed by a base such as diethylamine and the cyanhydrin so produced is heated with an alcohol and H_2SO_4 to give the required ester:

$$CH_2 = CH_2 \xrightarrow[\text{(Ag catalyst)}]{O_2} CH_2\text{—}CH_2 \xrightarrow[\text{(base catalyst)}]{HCN} CH_2OHCH_2CN$$

$$CH_2OHCH_2CN + ROH + H_2SO_4 \longrightarrow CH_2 = CHCOOR + NH_4SO_4$$

The Union Oil Co. of California have described a method for the preparation of acrylic acid from ethylene and carbon monoxide, by reaction in aqueous liquid phase with a palladium chloride catalyst system similar to that used in the Wacker process for the oxidation of ethylene to acetaldehyde.[65] The palladium, functioning as an oxidising agent, is reduced to the metallic state, and is separately re-oxidised, the overall result being:

$$CH_2 = CH_2 + CO + \tfrac{1}{2}O_2 \rightarrow CH_2 = CHCOOH$$

10.2.2.3. From ketene Another route uses ketene, obtained by the pyrolysis of acetone or acetic acid. This reacts with formaldehyde at ambient temperature in the presence of AlCl₃, yielding 2-propiolactone which in turn reacts with alcohol to give acrylic acid ester:

$$CH_2 = CO + CH_2O \xrightarrow[\substack{AlCl_3 \\ \text{catalyst}}]{} CH_2\text{—}CO \xrightarrow{+ROH} CH_2 = CHCOOR + H_2O$$
$$\qquad\qquad\qquad\qquad\quad CH_2\text{—}O$$

10.2.2.4. From acrylonitrile Acrylic acid is also readily made by hydrolysis of acrylonitrile, and a similar reaction may be carried out in the presence of alcohol, to give acrylic ester. Commercial interest in this route increased in the middle 1960's as a result of the fall in price of acrylonitrile consequent on its production by the ammoxidation of propylene. Nevertheless it will probably not be competitive with the newer method based on the oxidation of propylene, when the latter is operated on a substantial scale. Acrylonitrile hydrolysis is carried out with the aid of a relatively large amount of sulphuric acid, about 2 tons being required/ton of acrylic acid made. The excess of acid over the stoichiometric requirement is necessary for an adequate reaction rate, but this excess cannot be recycled to the process, because it is inextricably mixed with by-products. The equivalent amount of ammonium sulphate, which is necessarily made, has relatively little value, indeed its disposal may present problems.

10.2.3. Acrylic Acid by Oxidation of Propylene—Theoretical Considerations

The attraction of acrylic acid production from propylene lies largely in the relatively low cost of this feedstock, for it is generally a cheaper commodity than acetylene, ethylene or ketene. Certain companies with large existing capacities for acetylene production may be local exceptions to this rule, but interest in the propylene route is world-wide.

10.2.3.1. Liquid phase oxidation of acrolein Earlier attempts to oxidise propylene to acrylic acid were concerned with a two-stage route in which the acrolein potentially available from propylene oxidation was further oxidised in the liquid phase.

Several liquid phase processes for the oxidation of acrolein to acrylic acid have been developed; two that are typical will be briefly described.

In the first of these the carbon/carbon double bond in acrolein is protected by the addition of alcohol before the carbonyl group is oxidised:[66]

$$CH_2=CHCHO \xrightarrow[\text{(H}_2\text{SO}_4)]{ROH} \left[ROCH_2CH_2CH \underset{OR}{\overset{OR}{<}}\right] \xrightarrow[\text{oxalic acid}]{} ROCH_2CH_2CHO$$

$$\xrightarrow{O_2} ROCH_2CH_2COOH \xrightarrow[H_2SO_4]{} CH_2=CHCOOR$$

The alkoxy acetal (trialkoxypropane) is formed by the addition of 3 molecules of alcohol, in the presence of a little H_2SO_4. SO_4 ions are then removed from the reaction product by the addition of $Ba(OH)_2$ before the acetal is decomposed by heating at a suitable pH value (established by an organic acid) to yield 3-alkoxypropionaldehyde. This is oxidised to the alkoxyacid by the passage of air, with the aid of a catalyst, manganese acetate. Finally the vinyl group is regenerated by heating the alkoxyacid with H_2SO_4 and distilling off alcohol.

In the second liquid phase method sodium acrylate is produced by the oxidation of acrolein with oxygen in alkaline aqueous solution with a suspended silver oxide catalyst. Acrolein, oxygen and sodium hydroxide solution are passed to a reactor containing an aqueous suspension of silver oxide at a temperature of about 10°C. The flow of NaOH is such that the pH value in the reactor is about 12.[67]

Neither of these liquid phase processes is particulary attractive. The first is complicated and the second uses an expensive catalyst and consumes alkali and acid. Neither has been used commercially.

10.2.3.2. Vapour phase oxidation of propylene and acrolein   The earlier discoveries, during the late 1950's, of effective solid catalysts for the vapour phase oxidation of acrolein to acrylic acid were probably incidental to studies of catalysts intended for the oxidation of propylene to acrolein. That is to say that acrylic acid was observed as a significant reaction product in the presence of particular catalysts, and the natural assumption, readily confirmed by experiment, was that the acrylic acid was derived from acrolein.[68] Widespread interest was aroused, and many laboratories began working towards new and improved catalysts. A formidable number of patents have been issued on this subject. The number of patents and the unusual character of the reaction have led to a somewhat complicated situation which is best understood in terms of the following simple reaction model:

$$CH_2=CHCH_3 \xrightarrow{1} CH_2=CHCHO \xrightarrow{2} CH_2=CHCOOH \xrightarrow{3} \left.\begin{array}{l}\text{Further}\\ \text{oxidation}\\ \text{products}\end{array}\right.$$

which, it is assumed, is applicable to all oxide catalysts operating under conditions such that the allylic intermediate is formed. Catalysts are characterised by the values of the individual reaction velocity constants. The ideal catalyst would give a high yield (e.g. $\ll$70 per cent) of acrylic acid per pass (without the necessity for recycle of any kind), with a feed containing a substantial amount of propylene ($\ll$2 per cent and preferably > 5 per cent v/v) at a contact time of not more than a few seconds. This definition implies that over the ideal catalyst reaction 1 is fast, reaction 2 is at least as fast and reactions 3, 4 and 5 are negligible. The ideal catalyst does not seem to exist (as yet). If it did the following classification of catalysts would be of academic interest only.

10.2.3.3. Catalyst types of interest in the oxidation of propylene to acrylic acid
Whatever the catalyst system used it is clear that reaction 1 must be fast, to ensure a good output. It is equally clear that reaction 5, in parallel with the first main reaction step, will involve a fixed proportion of the propylene converted, whatever the extent of conversion. Therefore, overall efficient catalysts, reaction 5 must be slow.

Class A   There are catalysts which are or which could be used specifically for acrolein production, with only small co-production of acrylic acid. Both reactions 2 and 4 must be relatively slow, although k_3 could have any value. Notable members of this class are Cu_2O, $CuSeO_3$, Bi/Mo, Sb/Sn and Sb/U and other binary combinations including antimony. Of this group, $CuSeO_3$, giving the highest yield per pass of 80 per cent or more, would appear to have the highest value of $\dfrac{k_1}{k_2+k_4}$ (where k is the velocity constant). The others are, in this respect, inferior in varying degrees, but the total yield of acrolein from any may be increased by restricting the conversion per pass and recycling propylene. This follows simply from the fact that acrolein is the first intermediate in a series of reactions, competition to reaction 1 having already been ruled out.

Class B   In the oxidation of propylene these catalysts give acrolein and acrylic acid in comparable amounts. Therefore k_2 must be comparable with k_1, but since these catalysts are not ideal, i.e. they will not give high yields of acrylic acid from propylene in a single pass, reaction 3 and/or reaction 4 must be significant. If reaction 4 is unimportant, then, depending on the precise values of k_1, k_2 and k_3, it may be possible to obtain a high overall yield of acrylic acid by restricting the conversion per pass and recycling propylene and acrolein. However, further conversion of acrolein in a separate reactor may be preferred. Restricted per pass conversion will not reduce the destructive effect of reaction 4.

Catalysts are included in this class on the basis of patents claiming or illustrating the simultaneous production of acrolein and acrylic acid. Obviously, the demarcation between classes A and B is not always sharp and there are a few instances of the same catalyst appearing in both.

Among the class B catalysts are those that may be regarded as having been derived from bismuth molybdate:

Bi, Mo, Fe, P	Knapsack	BP 948,687
Bi, Mo, Fe, P, Ag	Knapsack	BP 1,184,402
Bi, Mo, Te, P and (Co or Sb) or (Mo+W+Ni) or (Mo+Ti) or (Mo+Sn+Fe)	Mitsubishi Rayon	Jap.P 12243/67
Bi, Mo, Te, P	Mitsubishi Rayon	Jap. P 5403/67
Bi, Mo, P, Ni, Co, Fe	Nippon Kayakin	Jap. P 16602/67
Bi, Mo and one of (V, Ni, Co)	Asahi Denka Kogyo	Jap.P 7522/67
Bi, Mo, P, Fe, B	Asahi KKKK	BP 1,095,008
Bi, Mo, Co	Grace	BP 1,115,116
Bi, Mo, P, Zi, Co, Fe+(Sm or Ta)	Degussa	Dutch P 6913168
Bi, Mo, Ni, Fe and one or more of (Ti, Ge, Zn, B, Cr)	Toa Gosei	GP 2,000,425

Most of the antimony-containing catalysts contain also vanadium:

Sb, V with or without one of (Ti, Co, Mn, Fe, Co, Ni, Cu, Zn, Mo, Sn, Cd, W, Th)	DCL	BP 1,034,914
Sb, V, Sn or Sb, V, Fe	DCL	BP 1,051,563
Sb, V	Hüls	GP 1,433,673
Sb, Fe	Sohio	USP 3,197,419

Other class B catalysts are:

Co, Mo, Te	DCL	BP 878,802
Co, Mo, Te	BASF	BP 1,193,489
Co, Mo, Te and B or Rh	Celanese	Dutch P 7000623
Co, Mo, Te, Sn	Nippon Shokubai KKKK	USP 3,475,488
Co, Mo, Te, Cu	Röhm and Haas	BP 1,118,497
Co or Ni, Mo and one of (B, P or V)	Du Pont	BP 893,077
Mo, Te, Ni, Re	Goodrich	USP 3,475,348
Mo, Te, P, Mn	Goodrich	BP 1,038,643
Mo, Te, P and a small amount of (Bi or Sb)	Goodrich	BP 1,146,637
Mo, Te, P, Mn, Fe	Goodrich	BP 1,131,162
Mo, Te and one of (Mn, Ni, Co)	Montecatini	Dutch P 6806577
Mo, Co, Te and optionally one of (Cd, Ca, Al, Co, Fe or Ni)	Toa Gosei	FP 1,579,839
H_3PO_4 and one of (Cu, Ag, Fe, Co, Ni, Sb, Bi, Mo, W or U)	Bayer	BP 967,241
Mo, As, P, Li	Soc. Nat. des Petroles d'Aquitaine	FP 1,433,572
2 or more of Mo, As, Sb, V and optionally one of 14 other metals	Sohio	Dutch P 6818481
Mo, P, As	Kodak	USP 3,379,652
W, V, Te, P, K	Mitsubishi Rayon	Jap.P 27401/68

Class C The third class comprises catalysts that are useful for the conversion of acrolein to acrylic acid. According to the reaction model reaction 2 over these catalysts is fast, and, since the efficiency is high, both reactions 3 and 4 are slow. Reaction 1 is also slow over class C catalysts, which means that they are unsuitable for the conversion of propylene to acrylic acid. In this class molybdenum is the predominating element:

Mo, P and one of (Bi, Sn, Sb)	Sohio	USP 2,881,212
Mo, P, B	Sohio	USP 3,326,817
Mo, Bi, Ni, Cd, Fe, P	Nippon Kagaku K K	USP 3,471,556
Mo, Bi, Ni, Co, Fe, As	Nippon Kagaku K K	Jap.P 2689/69
Co, Mo with or without Te	DCL	BP 878,802
Co, Mo	DCL	BP 915,799
Co, Mo and one of (Mn, Cr, Cd, Sn, Sb, V, Th, Al, Zr)	DCL	BP 924,532
Co, Mo	DCL	BP 915,800
Co, Mo and preferably B, P or V	Du Pont	USP 3,087,964
Mo, Co, Sn, Te	Nippon Shokubai K K	BP 1,035,147
Mo, Co and/or Ni, Te	Shell	BP 971,666
Mo, Co, Ni, As	Nippon Kayaku K K	Jap.P 2687/69
Mo, V, Te	Daiseru	Jap.P 14982/66
Mo, V, Sb	Daiseru	Jap.P 1662/67
Mo, Te	I C I	Belg.P 658,192
Mo and one of (Sn, V, Fe, Sb, Ce, Ti, Ni, W, Bi)	DCL	BP 903,034
Mo, Sb, V and one of (Ni, Cr, Ti or Sn)	DCL	BP 1,106,648
Mo, Sb, V, Fe	DCL	BP 1,107,478
Mo, Sb, V, Co	DCL	BP 1,111,440
Mo, Sb and a polyvalent metal	DCL	BP 1,131,132
Mo, Sb, Co	DCL	BP 1,127,677
Mo, Sb, V, Fe	U CB	BP 1,007,353
Mo, V, W	Nippon Kayaku K K	BP 1,185,109
Mo, V, Al	Rikagaku Kenkyujo	Jap.P 9045/68
Mo, V, W, Mn	Celanese	Dutch P. 7000894
Mo, V	Toyo Soda	Jap.P 1775/66
Mo, Sb, Sn, P, B	Asahi Kasei Kogyo K K	BP 1,094,328
Mo, W and one of (Fe, Ni, Mn, Cu)	BASF	Dutch P 6907079
Mo (Pd or Pt) and one of (Ag, Th, Sb, Bi, Cr, Se or Te) and one of (P, B, As or Si)	Japanese Geon	Dutch P 6911763
Mo, V and one of (Co, Cr, Mn. Bi, As)	Mitsubishi Rayon	Jap.P 6246/67
Mo, V and one at least of 17 metals	Rikagaku Kenkyujo	Dutch P 6907090
Mo, Pd and one of (As, P or B)	Toa Gosei	Jap. P 23405/67
Mo and/or Cr	Sohio	USP 2,881,213
Mo, P on silicon carbide	Sohio	USP 3,395,178

There is also a smaller group of patents relating to class C catalysts that do not contain molybdenum:

V, Cu or Ag and one of (Se, Te or As)	Asahi Kasei Kogyo	Jap.P 576/66
P, V optionally Sb	Degussa	BP 1,053,821
W, P optionally V or Mo	Hoechst	Belg. P 610,392
V, P, Sb	I C I	BP 991,836
V, P, Sb	Mitsubishi Yuka	Jap.P 17195/67
V, Sb	Mitsubishi Petrochemical	BP 1,088,099
V, P, Mn	Petro Tex	USP 3,238,253
W and one of (Sb, Sn, Bi, Cr, Co)	Mitsubishi Rayon	Jap.P 6244/67
V and/or Ag	Sohio	USP 2,881,214
W, P optionally Ag	Sohio	USP 3,021,266

10.2.3.4. Mixing catalysts   It might appear that the ideal catalyst system could be achieved by mixing a class C catalyst with one of either class A or class B. Knowing the kinetics over the component catalysts one can calculate the performance of a mixture of the two sorts of granules, assuming that the two velocity constants for each of the reaction steps are additive. Generally, it will be found (from such calculations) that the rate of destruction of acrylic acid is unduly large and that the great advantage of a good class C catalyst, i.e. the low value of the associated k_3, has been lost in the mixture. Even a class 1 catalyst, having a low value for k_2, may have a high value for k_3.

If there is more intimate mixing of the component catalysts, i.e. within the granules, then it is likely that new phases, hence new catalysts, will be formed. Oxide catalyst properties are not usually additive in this way. In several patents the elements of bismuth molybdate (a good class A catalyst) have been amalgamated with those of nickel and/or cobalt molybdate (class C catalyst), without, it seems resulting in the ideal catalyst.

10.2.4. Manufacture of Acrylic Acid by Propylene Oxidation

One plant is known to be in operation in the USA, that of Union Carbide which has a design capacity of 90,000 tons/annum and which utilises a catalyst developed by BP Chemicals International Ltd., who acquired the chemical interests of DCL in 1967. The Japan Catalytic Chemical Industry Co. Ltd., recently commenced production with a 10,000 tons/annum plant, in Japan. Several others are under active consideration by companies in USA, Europe and Japan.

For reasons already stated in the section on acrolein, it is not possible to give figures for commercial yields. According to patent examples, based presumably on small scale experiments, it is possible to obtain a yield of acrolein or of a mixture of acrolein and acrylic acid amounting to more than 70 per cent, from several catalysts. Class C catalysts give similar good yields, it appears, of acrylic acid from acrolein, the highest claimed being over 80 per cent.[69]

10.2.4.1. Reactor system   In addition to the exothermic reactions that occur in the production of acrolein from propylene the following is involved:

$$CH_2 = CHCHO(g) + \tfrac{1}{2}O_2 \rightarrow CH_2 = CHCOOH(g)$$
$$\Delta H_{298}°(g) = -62 \text{ kcal}$$

Heat removal is therefore a prime consideration, and as in the production of acrolein, fixed bed, shell and tube reactors, with tubes of not more than 2 in. diameter are likely to be preferred. There are several possible reactor arrangements.

10.2.4.2. Using one catalyst   The catalyst must obviously be class B. Unless acrolein is required as a final product it will have to be recycled to the reactor after it has been separated from the product stream. Probably, propylene also will have to be recycled, but this depends on the relative rates of reactions 1 and 2. Part of the reactor exit gas stream must be vented in order to limit the concentration of inert gases, and if the unreacted propylene in the vent is to be recycled it must first be isolated by, for instance, solvent extraction.

10.2.4.3. Using two catalysts

a. *In separate reactors*   With a catalyst of class A or B in the first reactor and one of class C in the second, in series, it becomes possible to achieve a high yield of acrylic acid from propylene in a single pass. Equipment will be simplest if the gas passes between reactors without any condensation occurring, but if class B catalyst is used in the first reactor it may be desirable to remove acrylic acid from the gas stream leaving it. Whatever the performance of each catalyst may be, restriction of the conversion per pass will improve the efficiency and hence the overall yield obtainable with recycle. But this must be weighed against the cost of recycling propylene and/or acrolein. The former must be recycled to the first reactor, and acrolein, perferably, to the second.

b. *In the same reactor*   The two catalysts are in separate zones, enclosed by the same heat transfer bath. In comparison with the previous system, loss of operating flexibility resulting from the fact that two catalysts have, necessarily, to be at the same bath temperature, has to be set against the saving in cost of reactors. It is important to achieve a good yield without recycle of acrolein, since this would have to be passed over both catalysts and the loss over the first may be unduly great, particularly if it is class B.

Grace, according to Ref. 70, use a bismuth promoted cobalt molybdate catalyst, apparently with the object of combining the two functions of Bi/Mo and Co/Mo (class A and class C catalysts), but in a subsequent patent[71] they describe a two reactor system, Bi/Mo at a temperature of 400–510° being followed by Co/Mo at 340–480°. In Ref. 72 propylene is oxidised over a class B catalyst, Sb/V/Sn, and the total product of this reaction is fed to a second reactor containing cobalt molybdate. Knapsack describe the operation of a single reactor system to which acrolein and propylene are recycled; in Ref. 73 a single catalyst, Bi, Mo, P, Fe, is used, and the efficiency is about 54 per cent. In Ref. 74 the catalyst is Bi, Mo, P, Fe, Ag, the efficiency going to about 63 per cent, and in Ref. 75, where the latter catalyst is again used but is followed by an additional layer of Co/Mo, the efficiency goes over 70 per cent. Output also rises with catalyst improvement to 161 g acrylic acid/l catalyst/hr. A patent to Shell International relates to the use of two catalysts in series, which may be within a single reactor.[76]

Having discussed the comparative merits of the one stage and the two stage reactor systems Toyo Soda[77] announce their intention of using the latter in the process they are developing. The reaction temperatures are 300–370°C first stage and 260–300°C second stage, and the yield is 65 per cent. Japan Catalytic Chemical Industry Co. Ltd. prefer a one-reactor system for their 10,000 tons/annum production unit.[78]

10.2.4.4. Recovery and purification The range of by-products is similar to that obtained when acrolein is the main product, but acetic acid now tends to predominate. Only a trace of propionic acid can be tolerated because its separation from acrylic acid is impracticable. According to Ref. 79 the proportion of by-products arising in the oxidation of acrolein to acrylic acid is reduced by restricting the conversion of acrolein to 65–90 per cent per pass.

The separation of acrolein and other volatile products may be effected by partial condensation of the reactor exit gas at 60–100°C. Gas escaping condensation has then to be scrubbed with water for acrolein recovery. Alternatively, all the acrolein may be extracted along with the acrylic acid from the gas by counter-current scrubbing or spraying with cold water.[80] Quenching the reactor exit stream with cold aqueous acrylic acid prevents polymer formation according to Ref. 81. Acrolein is then distilled from the extract. Another suggested method[82] consists in cooling the gas to an intermediate temperature, 90–200°C, so as to avoid condensation, and scrubbing it with a high boiling solvent such as tricresyl phosphate. From the extract so obtained acrylic acid is distilled off in the form of a concentrated aqueous solution.

All the various initial recovery steps that have been proposed result in a fairly dilute (about 20 per cent) solution of acrylic acid in water. In order to concentrate this Röhm and Haas[83] mix it with H_2SO_4, a sulphonic acid, or phosphoric acid, in an amount equal to between 0·1 and five times the weight of the acrylic acid, and then distil off most or all of the water at below 120°C. It is claimed that acetic and propionic acids distil with the water and only a little acrylic acid. Among the liquids claimed for the extraction of acrylic acid from water solution are butyl acrylate and 2-ethylhexyl alcohol[84] isophorone,[85] ethyl acetate,[86] ethyl acetate and acrylate.[87] According to BP 995,471 aqueous acrylic acid is contacted with a mixture of ethanol and ethyl acrylate to give a mixture of ethanol and water, and (in a separate phase) a mixture of ethyl acrylate and acrylic acid. From the latter acrylic acid is easily obtained by distilling off the ester. Instead of extraction, absorption of acrylic acid on to a material such as active carbon is proposed in Ref. 88. Acrylic acid may also be distilled from the water as one component of a ternary azeotrope; for this purpose a considerable variety of entraining agents has been claimed.[89]

Most of the methods for the preliminary concentration of acrylic acid give a wet mixture of acrylic and acetic acids with the extractant, acetic acid being an invariable by-product of the oxidation reaction. A suitable extractant, such as a low molecular weight ester, will form an azeotrope with the residual water, enabling this to be removed by distillation. Removal of excess solvent (extractant) by distillation then follows, leaving a mixture of, substantially, acrylic acid and a small proportion of acetic acid. The acids may be separated by fractional distillation, which must be at reduced pressure so that the kettle temperature does not exceed about 90°C, to avoid excessive polymerisation of acrylic acid.[90] Asahi KKKK[91] separate acrylic and acetic acids by forming an azeotrope of the latter, using a mixture of a hydrocarbon with water and/or an ester.

Finally, the acrylic acid may be flash distilled (at low pressure) in order to eliminate traces of polymer and other high boiling materials.

10.2.5. Acrylic Acid Esters

Some Acrylic Ester Properties

	Methyl	Ethyl	n-Butyl	2-Ethyl hexyl
Specific gravity (20°C)	0·9535	0·9234	0·8898	0·8852
Freezing point (°C)	−75	−75	−64	−90
Boiling point (1 atm °C)	80·0	99·5	147·4	216
Refractive index (20°C)	1·4040	1·4069	1·4185	1·4365
Specific heat (cal/g/°C) (20°)	0·47	0·47	0·47	0·45
Vapour pressure (20°C) (mm Hg)	68	30	4·0	0·1
Solubility in water (% wt/wt) (20°C)	9·3	2·0	0·2	<0·01
Solubility of water in ester (% wt/wt) (20°C)	2·6	1·5	0·7	0·14
Water azeotrope				
Boiling point (1 atm) (°C)	71	81	94·3	
Water content (% wt/wt)	7·2	15·1	38	

The greater part of the acrylic acid made by the oxidation of propylene will, according to present day requirements, be converted to esters. The methyl, ethyl, n-butyl and 2-ethylhexyl esters are the most important.

Esterification in the liquid phase is carried out according to classical principles, i.e. acid and alcohol are heated in the presence of an acid catalyst while one or both products of reaction, ester and water, are removed as quickly as possible in order to overcome equilibrium limitations. Modifications and complications of the standard esterification procedures are necessitated by the possibility of formation of polymers and the products of addition of water, alcohol, etc., across the double bond of acrylic acid. The most usual catalyst is sulphuric acid,[92] but others, including acid resins,[93] have been suggested.

The acetic acid by-product which always accompanies the acrylic acid may, as already indicated, be eliminated before the latter is esterified. Alternatively the mixture of acids may be esterified. Separation of the methyl and ethyl esters at least is easier than the separation of the acids because the esters are more volatile and have a greater volatility ratio. Knapsack[94] describe a method of preparation of methyl acrylate in which aqueous acrylic acid (4–15 per cent), containing 0·4–1·2 per cent acetic acid, which is obtained as condensate from a propylene oxidation reactor and has been freed of acrolein, is mixed with 0·1–0·5 per cent H_2SO_4 and fed to the upper part of a heated 46 plate esterification column, 5 mole methanol/mole acid being fed to the lower part. From the head of the column, operated at 1–20 atm (130–200°C), there is delivered in vapour form from an azeotropic mixture of methyl acrylate, methyl acetate and methanol, from which pure methyl acrylate is obtained by distillation methods. Water and H_2SO_4 leave the bottom of the column. The n-butyl ester of acrylic acid is prepared in similar equipment,[95] operating at 110–180°C with a residence time of 3/4–3 hr, the catalyst being a mixture of H_2SO_4 and ammonium sulphate. 5–20 moles n-butanol is fed per mole of acrylic acid.

In a method described by DCL,[96] an ethanol-acrylic acid mixture (molar ratio ≮1), which may contain water and acetic acid, is fed to a reactor fitted with a distillation column, and a concentration of 2 per cent H_2SO_4 is maintained in the reaction liquid, which is heated to a temperature of about 90°C at 1 atm pressure. Substantially all the ethyl acrylate and all the water of esterification, in addition to that introduced with the feed, distil from the reactor, but the acrylic acid is retained. The concentration of water in the reactor is maintained at 8–12 per cent: its removal may be facilitated by additional ester fed to the reactor to act as entrainer. Concentration of high-boiling substances is controlled by a continuous liquid bleed, which is heated, in separate equipment, for the recovery of monomers, which are mainly acrylic acid and ethyl ester. A modification of this system,[97] consists in passing the total distillate from the reactor to an intermediate point in a distillation column, the head product from which consists mainly of esters and ethanol. Water and unconverted acid leave at the base, and the acid is reconcentrated and returned to the reactor. Since acrylic acid is recycled its residence time in the reactor is reduced, with the object of reducing the amount of polymer and other non-volatile by-products.

According to Ref. 93 Knapsack collect the acrylic acid resulting from propylene oxidation by washing the hot reactor gas with a by-product mixture obtained during the reaction, which consists mainly of maleic ester and polyacrylic acid and ester. After low boiling materials have been distilled from the extract it is reacted with a C_1—C_8 alcohol in the presence of an acidic ion exchange resin. Acrylic ester is then distilled off and the residue is recycled to the reaction gas scrubber.

Production of acrylic esters in the liquid phase by reaction of the acid with an alkyl sulphate is described in Ref. 98:

$$2CH_2=CHCOOH+(RO)_2SO_4 \longrightarrow CH_2=CHCOOR+H_2SO_4$$

An improved preparation of ethyl acrylate consists, it is claimed,[99] in reacting acrylic acid with ethylene in the presence of at least 40 per cent by weight of H_2SO_4 at 95–105°C. In a similar manner t-butyl acrylate is prepared from isobutylene with phosphoric or phosphomolybdic acid catalyst.[100]

Hydroxyalkyl acrylates may be made by the reaction of acrylic acid with an epoxide:[101]

$$CH_2=CHCOOH+CH_2\!\!\overset{\diagdown}{\underset{O}{}}\!\!CHR \longrightarrow CH_2=CHCOOCH_2\!\!-\!\!CH(OH)R$$

There is considerable interest in vapour phase esterification. BASF[102] vaporise aqueous acrylic acid and pass it counter-current to a mixture of an alcohol with stabiliser (e.g. thiodiphenylamine) and H_2SO_4 catalyst (1 per cent molar) at 120–160°C. Product is recovered from the top of the esterification column in which the residence time is about 14 sec. Daicel[103] describes the preparation of methyl acrylate by passage of 2 mols methanol, 1 mol acrylic acid and 1–15 mols water over phosphoric acid on silica gel at 25–50 sec contact time. In Ref. 104 the lower alkyl acrylates are made by vapour phase esterification over silica/alumina catalyst in the presence of 4–13 moles water per mole acrylic acid. The gaseous product of a propylene oxidation reactor is, according to Ref. 105, without purification, reacted with a C_1—C_4 saturated or unsaturated alcohol at 200–350°C over a supported phosphoric acid catalyst. Sohio describe the preparation of alkyl acrylates, in the vapour phase, from acrylic acid[106] and from acrylonitrile.[107]

The higher alkyl acrylates may be made either by esterification or by transesterification, i.e. the interaction of a higher alcohol and methyl or ethyl acrylate:

$$CH_2{=}CHCOOC_2H_5 + ROH \underset{\longleftarrow}{\overset{\longrightarrow}{}} CH_2{=}CHCOOR + C_2H_5OH$$

Transesterification usually consists in heating the reactants to the boiling point, at atmospheric pressure, and continuously removing the lower alcohol that is liberated via a fractionating column, in the form of its azeotrope with either its acrylic ester or with an added entraining agent such as cyclohexane.[108] Various catalysts have been claimed for this reaction: alkyl titanates and zirconates,[109] neutral and basic catalysts,[110] alkali salts or mercaptobenzthiazole or phthalimide or Li(OH),[111] sodium methoxide,[112] H_2SO_4.[113] Röhm and Haas[114] describe a transesterification process in which the reactants are passed through a tube at 180–250°C, under a pressure of 9–20 atms, which is sufficient to maintain the liquid phase.

10.2.6. Inhibition of Polymerisation

Acrylic acid and its esters are particularly susceptible to polymerisation, which may be initiated by peroxides formed by the action of atmospheric oxygen on the monomer or impurities in it. They need protection from polymerisation during processing operations and in storage. Polymerisation inhibitors, which should, preferably, be present in both the liquid and the vapour phases function, according to theory, in more than one way.[115]

Vinyl polymers stem from free radicals (probably derived from peroxides by decomposition) which accumulate a succession of monomer units to produce a growing radical chain (propagation):

$$R{\cdot}+M{\rightarrow}RM{\cdot} \qquad RM{\cdot}+M{\rightarrow}RMM{\cdot} \quad \text{etc.} \tag{1}$$

where, in this case, M is —CH_2—CH—
                                     |
                                    COOH

Unless positive measures are taken to interrupt the chain at an early stage each original R· will, on the average, consume a large number of monomer units before the growing chain is terminated by a process such as mutual combination: $R(M){\cdot}_a + R(M){\cdot}_b \rightarrow R(M)_{a+b}R$.

Molecular oxygen inhibits vinyl polymerisation by attaching itself to the polymer radical:

$$\ldots\ldots M{\cdot}+O_2{\rightarrow}\ldots\ldots M{-}O{-}O{\cdot} \tag{2}$$

Reaction 2 being much the faster, largely eliminates the competing reaction 1. The product of reaction 2 can grow only by the addition of monomer and O_2 units alternately:

$$\ldots\ldots M{-}O{-}O{-}M{-}O{-}O{\cdot}+M{\rightarrow}\ldots\ldots M{-}O{-}O{-}M{-}O{-}O{-}M{\cdot} \tag{3}$$

But reaction 3 is much slower than reaction 1, at least in the case of acrylic acid and its esters,[116] and this determines the rate of polyperoxide formation.

The net effect of the oxygen is therefore to eliminate vinyl polymerisation substantially while creating a relatively small amount of polyperoxide. The molecular polyperoxide resulting from the termination of the oxidation chain, which may be represented by $R(M—O—O—)_nM$, where n is a small number, will itself decompose to give fresh radicals, $RMO\cdot$, that are potential new propagators of polymerisation, and a continuous supply of fresh oxygen is needed to restrain their effect.

Antioxidants, which are, primarily, phenols and aromatic secondary amines, destroy peroxy radicals, but not homopolymer radicals or molecular polyperoxide. They therefore reinforce the inhibiting effect of oxygen and reduce its consumption by shortening still further the polyperoxide chain. But it should be noted that, according to this theory, while oxygen alone has an inhibiting power, which is increased by the presence of antioxidant, the latter is, in the absence of oxygen, without effect. For an example of the use of oxygen alone see Ref. 117. In Ref. 118 the inhibiting action of oxygen with and without 'well known inhibitors such as methylene blue, phenothiazine or hydroquinone' is claimed. Some patents, however, refer to the use of antioxidants as inhibitors in the absence of oxygen.[119]

There are substances that inhibit in the absence of oxygen. These are of two kinds. The first have an unpaired electron, in fact they are themselves free radicals and they act simply by combining with the free radicals that would otherwise propagate the vinyl chain. Examples of these are nitric oxide and the N,N disubstituted nitroxides.[120] The second kind of true inhibitor comprises certain metal ions, in particular cupric and ferric ions, which terminate the homopolymer chain by oxidising the radical:

$$—\overset{|}{\underset{|}{C}}\cdot + Cu^{++} \longrightarrow \overset{|}{\underset{|}{C}}{}^{+} + Cu^{+}$$

Combinations of inhibitors may show synergistic effects. In practice two or more may be used together and the preferred inhibitor systems of particular processes are usually part of the guarded know-how.

Definitions

$$\text{Conversion (of reactant)} = 1 - \frac{\text{moles leaving reactor}}{\text{moles entering reactor}}$$

$$\text{Yield of product} = \frac{\text{weight carbon in product}}{\text{weight carbon in reactant entering reactor}}$$

$$\text{Efficiency} = \frac{\text{yield}}{\text{conversion}}$$

$$\text{Contact time} = \frac{\text{vol of catalyst zone}}{\text{vol gas (NTP) entering reactor per sec}}$$

10.3. ACRYLIC ACID AND ACRYLIC ESTERS—PROCESSING AND USES

By E. M. Evans

Although the polymerisation of acrylic acid to a polymer was first observed in 1872 and an investigation of the polymerisation of acrylic esters was made by Röhm in

1901, acrylic ester polymers in the form of their aqueous dispersions were first introduced commercially in Germany in 1927. The main outlets for these products prior to World War II were in leather finishing and textile treatment where their unique properties of softness and extensibility justified a relatively high priced product for a small tonnage market. The big upsurge in acrylate usage occurred in the post-war period when these still relatively expensive monomers were copolymerised with much cheaper monomers for applications such as the so-called emulsion paints and to a lesser extent the thermosetting acrylic surface coating resins and acrylic paper coating lattices, etc.

10.3.1. Polymerisation

Acrylic acid esters are very easily polymerised using free radical initiation. Moreover they copolymerise very easily with a wide range of other monomers.

The most commonly used process is that of emulsion polymerisation. In the simplest form of this process the acrylic ester is mixed with water (e.g. in equal weights) containing a few percentage parts by weight of a soap, e.g. sodium lauryl sulphate, in a vessel fitted with a reflux condenser. Emulsification of the monomer in the water is achieved by vigorous agitation with a high-speed stirrer and a water soluble catalyst which will generate free radicals is added (e.g. ammonium persulphate at, say, 0·05 per cent part by weight of monomer). The emulsion is then heated whilst continuing the agitation to, say, 80°C at which point polymerisation will commence with evolution of heat. The polymerisation of acrylic esters is highly exothermic with a heat of polymerisation of the order of 20 kcals/g mol of monomer. When polymerisation commences careful cooling must therefore be applied to avoid a run-away temperature rise and to maintain the reactants at the polymerisation temperature. Polymerisation will continue almost to completion (99 per cent) so long as there is a source of free radicals present. Any required adjustment of the pH value of the dispersion can then be made by the addition of an aqueous alkaline solution and the product is obtained as a low viscosity aqueous dispersion of the acrylic ester polymer.

In industrial practice the process is usually more sophisticated. It is common to add the monomer (or a major portion of it) and catalyst continually over a period of hours to the agitated soap solution at the polymerisation temperature. In this way better temperature control can be achieved and the risk of run-away polymerisations avoided. Moreover, by polymerising at a temperature at which the monomer/water mixture is refluxing, heat can be removed from the system as latent heat of evaporation. It is common to use a combination of surfactants, e.g. anionic together with non-ionic, and part of the aqueous solution of these may be added during the polymerisation process.

Other water soluble per-salts and hydrogen peroxide can be employed as catalysts and in one version of the emulsion polymerisation process, using a redox catalyst system, there is added to the reactants a reducing agent, e.g. sodium metabisulphite, as well as the per-salt. The redox system applied under an inert atmosphere will enable polymerisations to be carried out at much lower temperatures.

The molecular weight of the products is determined mainly by the concentration

27

of catalyst and polymerisation temperature employed and can be reduced by the addition of chain transfer agents such as mercaptans.

Most of the uses for acrylic ester polymers involve their application as aqueous dispersions and the emulsion polymerisation process is the natural choice.

However, there is one other large outlet for acrylic ester polymers and copolymers in conventional solvent based surface coatings and for these products a solution polymerisation process is employed. In the simplest form of this process an acrylic ester monomer is mixed with an approximately equal weight of a solvent such as toluene, or xylene, or a mixture of one of these with *n*- or *iso*-butanol, in a reaction vessel fitted with a reflux condenser, a small quantity (e.g. 1 part by weight to 100 of monomer) of an oil soluble peroxide catalyst such as benzoyl peroxide or cumene hydroperoxide is added, and the solution heated, preferably under an inert gas stream, with stirring at an elevated temperature (e.g. above 90 °C) for a number of hours until polymerisation is substantially (95 per cent) complete. For the same reasons that were given concerning the emulsion polymerisation process the industrial solution polymerisation processes are usually more sophisticated involving incremental addition of monomer, solvent and catalyst and the removal of heat of polymerisation by monomer and/or solvent reflux. Molecular weight of the product is controlled by similar variables, higher temperatures, higher catalyst concentrations, lower monomer concentrations, and the addition of chain transfer agents such as mercaptans leading to lower molecular weight. The products are obtained as almost colourless viscous solutions of the polymers in the solvent.

Acrylic acid, either on its own or as a partial salt, e.g. ammonium, sodium, can be homopolymerised in dilute aqueous solution, usually below 25 per cent concentration, using a water soluble per-salt as catalyst. Alternatively, acrylic acid can be polymerised by a peroxide or azobisisobutyronitrile in a solvent, e.g. benzene in which the monomer is soluble and the polymer insoluble, the product being filtered at the end of the process, dried and then often dissolved in aqueous alkali or ammonia. It is also sold as an unneutralised free flowing powder. Acrylic acid is, however, frequently employed as a minor co-monomer with other monomers when the emulsion and ordinary solution polymerisation processes can be employed.

10.3.2. Properties of Acrylic Ester Polymers and Copolymers

In order to appreciate the reasons why acrylic esters are now so widely used in industry, a consideration of the properties of the range of homopolymers is useful. The physical properties, such as hardness and flexibility, of all polymers are dependent on their molecular weight and the general comparisons which follow refer to polymers of reasonably high molecular weight such as are used industrially. The temperature dependence of these properties is also an important factor and reference will be made to the brittle point, which can be defined as the temperature at which the polymer will fracture on flexing, i.e. as it changes from the rubbery to the glass-like state. All other properties will be considered at room temperature.

The homopolymer of the lowest member of the *n*-alkyl acrylic ester series, polymethyl acrylate, is a tough, slightly rubbery material of low hardness and has a brittle point of 3 °C. Polyethyl acrylate is softer, tough, more rubbery with a very high extensibility, and has a brittle point of $-23\,°C$. Poly n-butyl acrylate is extremely soft and tacky with a brittle point of $-45\,°C$.

This progressive softening and lowering of brittle point as one proceeds through the homologous series of n-alkyl acrylate polymers holds true until it reaches a maximum with poly n-octyl acrylate, which has a brittle point of $-65\,°C$. Thereafter the polymers of still higher esters show rising brittle points.

Branching in the alkyl group of the esters leads to polymers of higher brittle point, for example poly 2-ethyl hexyl acrylate has a brittle point of $-55\,°C$ (cf. poly n-octyl acrylate above), and the softening points of the polymers derived from n-butyl, isobutyl, s-butyl and t-butyl acrylates are respectively, $-45\,°C$, $-24\,°C$, $-10\,°C$ and $40\,°C$.

Other properties common to the acrylic ester polymer series are excellent colour and colour retention (i.e. good resistance to ultra-violet initiated oxidative degradation). They are very much more resistant to hydrolysis than, for example, polyvinyl acetate, and this resistance improves with the higher alkyl ester polymers. Polymers of the lower acrylic esters show good resistance to oils and greases.

The properties given above explain the applications which acrylic esters find. Where a product of good flexibility, extensibility, colour or colour retention is required, an acrylic ester polymer, either in its pure form or modified with a very minor proportion of another monomer is employed, for example, in leather and textile treatment, non-woven fabrics and acrylic rubbers. In most applications, however, the acrylic ester monomer is copolymerised with a major proportion of a cheaper monomer, e.g. styrene, methyl methacrylate, which would normally give a hard homopolymer. In such copolymers the acrylic ester functions as an internal chemically combined plasticiser. The brittle point of the harder homopolymer (e.g. polystyrene, polymethyl methacrylate) is reduced in proportion to the amount of acrylic ester copolymerised and generally any compromise of hardness and flexibility intermediate between those of the extreme homopolymers can be achieved. Similarly, other properties such as light stability, chemical resistance, etc., will be intermediate between those of the homopolymers. From the comments made earlier on softness and brittle point it is clear that n-octyl acrylate will be a more efficient plasticising comonomer, weight for weight, than butyl acrylate which will in turn be more efficient than ethyl acrylate. The choice is usually based on economic considerations. Although n-octyl acrylate is more efficient than, say, 2-ethylhexyl acrylate, its price is so much higher that the use of a correspondingly greater concentration of 2-ethylhexyl acrylate will normally yield a cheaper product.

All the materials described above, polymers and copolymers, are thermoplastic. Acrylic polymers can also be made thermosetting by the incorporation of appropriate reactive groups. Acrylamide is typical of such modifying co-monomers. If it is copolymerised with ethyl acrylate in minor proportions, the pendant amide groups can be reacted with formaldehyde to give active methylol groups which can be reacted together by heating in the presence of an acid. Alternatively, the acid can be provided internally by incorporating a small proportion of a polymerisable acid such as acrylic acid into the terpolymer.

$$\sim\sim\sim CH\text{———}CH_2\text{—}CH\text{—}CH_2\sim\sim\sim\sim \text{Ethyl acrylate/acrylamide}$$

$$\underset{\displaystyle COOC_2H_5}{|} \qquad \underset{\displaystyle CO}{|} \qquad \text{copolymer}$$

$$\underset{\displaystyle NH_2}{|}$$

$$\downarrow HCHO$$

$$\sim\sim\sim CH\text{———}CH_2\text{—}CH\text{—}CH_2\sim\sim\sim$$

$$\underset{\displaystyle COOC_2H_5}{|} \qquad \underset{\displaystyle CO}{|}$$

$$\underset{\displaystyle NH}{|}$$

$$\underset{\displaystyle CH_2OH}{|}$$

CH₂
|
CHCONHCH₂OH   +   HOCH₂NHCOCH
| |
CH₂   CH₂
| |
CHCOOC₂H₅   CHCOOC₂H₅

$$\downarrow -H_2O$$

CH₂   CH₂
| |
CHCONHCH₂OCH₂NHCOCH
| |
CH₂   CH₂
| |
CHCOOC₂H₅   CHCOOC₂H₅

$$\downarrow -HCHO$$

CH₂   CH₂
| |
CHCONHCH₂NHCOCH
| |
CH₂   CH₂
| |
CHCOOC₂H₅   CHCOOC₂H₅

Other crosslinking groups can be introduced into acrylic acid ester polymers and will be described later. The degree of crosslinking introduced can obviously modify the acrylic polymer properties very considerably, increasing hardness and especially chemical resistance, e.g. resistance to dry cleaning solvents, detergents, etc.

10.3.3. Uses of Acrylic Ester Polymers

10.3.3.1. Surface coatings

a. *Emulsion paints* The biggest outlet for acrylic esters is in emulsion paints, products which came into vogue during and after World War II. They comprise an aqueous dispersion of a soft polymer to which pigments are added in paste form. When applied to a substrate these products dry by evaporation of water and the polymer particles must be soft enough to flow and form an integrated continuous film at room temperature (say, down to 5°C). The glass transition temperature of the polymer must therefore be of the order of 0°C. This can be achieved in three ways: (1) by adding a plasticiser, (2) by copolymerising a monomer such as vinyl acetate with a plasticising comonomer in sufficient quantity to give a glass transition temperature of 0°C, (3) by adding a small percentage of a so-called coalescing solvent which, when dissolved in a polymer, will lower the glass transition temperature by a few degrees, but which will subsequently evaporate out of the film. In practice all three methods are occasionally used together in one product.

The first polymer to be employed for emulsion paints was polyvinyl acetate, a development introduced in Germany during World War II. In order to lower the glass transition temperature of this polymer, an external plasticiser such as dibutyl phthalate or tricresyl phosphate was added in proportions of 15–25 per cent by weight. These products held sway in Europe for the next ten years. In the USA after the war styrene butadiene copolymers were the main products used in emulsion paints for about fifteen years although they suffered from embrittlement and discolouration on ageing resulting from oxidative attack at the unsaturation in the polymer.

The 1950's saw the emergence of two main developments the first of which was the introduction of so-called 'pure' acrylics in the USA for the high performance end of the emulsion paint market. These products were often used externally where their good colour retention and alkali resistance on cement surfaces and excellent adhesion and extensibility on wood, which is very widely used in the USA for exterior cladding, were considerably better than either externally plasticised PVA or styrene butadiene polymers. The products were basically terpolymers of ethyl acrylate, methyl methacrylate and a very minor amount of acrylic or methacrylic acid. The other development was the introduction in Europe of internally plasticised polyvinylacetate dispersions. In these products vinyl acetate was originally copolymerised with plasticising co-monomers such as dioctyl maleate, dibutyl fumarate, etc. Within a few years the maleate and fumarate co-monomers had been superseded in the UK by either butyl acrylate or 2-ethylhexyl acrylate which copolymerised with vinyl acetate more easily and completely and the prices of which were then economically competitive. They avoided embrittlement on ageing resulting from plasticiser volatility and diffusion into porous substrates, they had better binding capacity and they produced paints which were more water (scrub) resistant and better for exterior durability. Usually these products are composed of 80–85 per cent vinyl acetate and 15–20 per cent of acrylic ester. They are still widely used but are finding increasing competition from copolymers of vinyl acetate with olefins.

The 'pure' acrylics are still expanding, particularly in the USA, but they are finding increasing competition from some of the newer copolymer dispersions based on vinyl acetate copolymers containing, e.g. 40–50 per cent of the monomer Veo Va[10] (a Shell Chemicals product which is a vinyl ester of a branched C_{10} acid). These copolymers have good alkali resistance comparable with that of the acrylates and like them have good exterior durability on cement and wood. Products of good alkali resistance are also obtained from copolymerising styrene with an acrylic ester such as ethyl acrylate.

The future for acrylic esters both in high and medium performance emulsion paints is very difficult to predict with so many technical competitors potentially available. In the end their price relative to those of the competitors will be the deciding factor.

b. *Thermosetting acrylic resins*   The late 1950's saw a rapid growth of the so-called thermosetting acrylic resins which are basically copolymers of a monomer which contributes to hardness and chemical resistance, e.g. styrene or methyl methacrylate with a monomer contributing flexibility, e.g. an acrylic ester. A third monomer is copolymerised with these to introduce reactive groups, e.g. acrylamide, as mentioned earlier, the acrylamide groups being subsequently methylolated using formaldehyde. The crosslinked products obtained when films are stoved at elevated temperature have a high measure of chemical resistance and durability since they have a cross-linked carbon/carbon backbone. These useful properties resulted in the thermosetting acrylic resins, often used in admixture with minor amounts of epoxide resins, displacing alkyd and alkyd/amino finishes for many factory applied metal finishes where a high degree of resistance to alkaline detergents is required, e.g. in domestic appliances such as refrigerators and washing machines.

The monomers are copolymerised in solution in a mixed aromatic/butanol or isobutanol solvent. At the end of polymerisation the amide groups are post-reacted with formaldehyde. In the presence of butanol the methylol groups formed are etherified to N-butoxymethyl groups which are much more stable on storage and of better compatibility than unetherified methylol groups.

$$\sim\!\!\overset{O}{\underset{\|}{C}}NH_2 \xrightarrow{\text{HCHO}} \sim\!\!\overset{O}{\underset{\|}{C}}NH.CH_2OH \xrightarrow[\substack{\text{Mild acid}\\ \text{e.g. maleic}\\ \text{anhydride}}]{C_4H_9OH} \sim\!\!\overset{O}{\underset{\|}{C}}NH.CH_2OC_4H_9 \quad +H_2O$$

where $\sim\!\!\sim$ represents a segment of a copolymer chain. The water is removed by azeotropic distillation.

In the presence of added acids such as butyl phosphoric acid or by incorporating carboxyl groups into the polymer chain, e.g. by using a small proportion of acrylic acid in the original polymerisation recipe, the products can be cured in film form at 150°C in 30 min. Carboxyl groups also aid adhesion. Higher temperatures are required in the absence of acid catalysts. The crosslinking is analogous to that of the methylol acrylamide copolymers.

$$2 \sim\!\!\overset{O}{\underset{\|}{C}}.NHCH_2OC_4H_9 \xrightarrow[\substack{-C_4H_9OH\\ \text{and } -C_4H_9OC_4H_9}]{} \sim\!\!\overset{O}{\underset{\|}{C}}.NHCH_2OCH_2NH.\overset{O}{\underset{\|}{C}}\!\!\sim$$

$$\Big\downarrow -HCHO$$

$$\sim\!\!\overset{O}{\underset{\|}{C}}.NHCH_2NH\overset{O}{\underset{\|}{C}}\!\!\sim$$

A suitable crosslinking density to give excellent film properties is obtained by incorporating 10–15 per cent by weight of acrylamide in the original monomer mixture. Apart from using acrylamide and post-reacting with formaldehyde, the same product can be made by using a corresponding proportion of N-butoxymethyl acrylamide in the original polymerisation mixture. The acrylic esters employed not only contribute flexibility but also durability and stability to light. An advantage of using styrene as the monomer to contribute hardness and chemical resistance is its low cost.

The next most important thermosetting acrylic resins are those containing hydroxyl functionality. Here one again uses an appropriate blend of monomers conferring hardness and flexibility, a small proportion of acrylic acid to aid curing and 5–20 per cent by weight of a hydroxyalkyl acrylate or methacrylate such as β-hydroxyethyl acrylate. They are also polymerised in solution and are usually blended with etherified urea formaldehyde or melamine formaldehyde resins before pigmentation. A single stage in the crosslinking reaction is shown below:

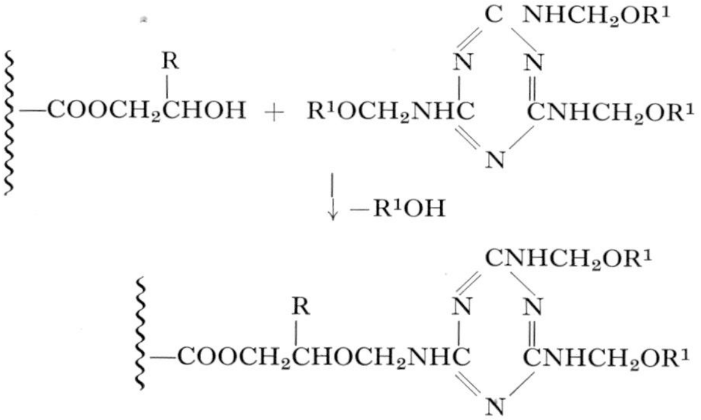

$$\begin{array}{c} \text{C NHCH}_2\text{OR}^1 \\ \text{R} \qquad\qquad \text{N} \quad \text{N} \\ \text{—COOCH}_2\text{CHOH} + \text{R}^1\text{OCH}_2\text{NHC} \quad \text{CNHCH}_2\text{OR}^1 \\ \text{N} \end{array}$$

$$\downarrow -\text{R}^1\text{OH}$$

$$\begin{array}{c} \text{CNHCH}_2\text{OR}^1 \\ \text{R} \qquad\qquad \text{N} \quad \text{N} \\ \text{—COOCH}_2\text{CHOCH}_2\text{NHC} \quad \text{CNHCH}_2\text{OR}^1 \\ \text{N} \end{array}$$

where R = H or CH$_3$ and R^1 = n- or iso-C$_4$H$_9$, leaving two butoxymethylamino groups to continue the crosslinking.

The hydroxyalkyl acrylates are relatively expensive and an alternative and sometimes cheaper route to these products is to substitute the hydroxy ester in the original recipe by a correspondingly increased quantity of acrylic acid. When polymerisation is complete a number of the carboxyl groups can be post-reacted with ethylene or propylene oxides in the presence of a catalyst such as triethylamine.

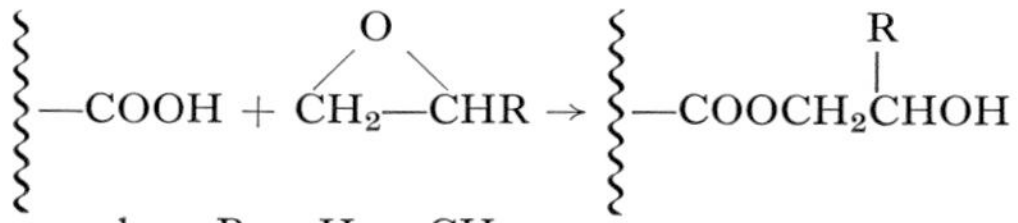

$$\begin{array}{c} \text{O} \qquad\qquad \text{R} \\ \text{—COOH} + \text{CH}_2\text{—CHR} \rightarrow \text{—COOCH}_2\text{CHOH} \end{array}$$

where R = H or CH$_3$

The thermosetting acrylic resins containing hydroxyl functionality and blended with amino resins can be stoved at lower temperatures than their acrylamide counterparts and produce excellent car finishes (including metallics) which have grown very rapidly in the last few years. They have contributed to saving car painting costs by enabling initial blemishes to be dealt with more easily.

A third type of thermosetting acrylic resin which has not made the spectacular growth of the earlier types described is that containing carboxyl functionality. Styrene or vinyl toluene is copolymerised with ethyl acrylate and with an acidic monomer such as acrylic or methacrylic acid. At least 6 per cent by weight of the acidic monomer must be incorporated to get adequate crosslinking. The terpolymer is cured by blending with epoxide resins and stoving at elevated temperature in the presence of basic catalysts such as amines.[121]

A fourth group of thermosetting acrylic resins is based on epoxide functionality. Epoxide groups are introduced into the copolymers by incorporating a monomer such as glycidyl methacrylate. These products can be blended with melamine resins and cured at elevated temperature in the presence of a base.[122]

where R = n- or iso-C_4H_9

Alternatively they can be made self-curing by copolymerising the monomers including glycidyl methacrylate with acrylic or methacrylic acid.

Thermosetting acrylic resins of all four types can obviously be formulated to cover a wide range of compromises in hardness and flexibility and apart from domestic appliance and car finishing they can be made more flexible for applications such as strip coating where their excellent adhesion and extensibility enable the coated metal sheet to be subjected to severe post-forming operations without rupture of the film which itself has excellent hardness, colour stability and exterior durability.

Water soluble analogues of the above resins can also be prepared by incorporating more acrylic acid in the copolymer and solubilising in a base such as aqueous ammonia. They have been suggested as a one-coat paint applied by electrodeposition.

c. *Miscellaneous coatings*   A large outlet for acrylic ester polymers and copolymers is in the coating of high-class printing paper where they are employed in aqueous dispersion blended with pigments such as china clay and titanium dioxide to produce a smooth glossy surface of good opacity, colour retention and water resistance with good printability. They also produce 'pick resistant' finishes so that a tacky printing ink applied from high-speed printing rollers will not tear out bound pigment.

Originally the binders for these coatings were based on natural polymers such as casein or starch. A desire to improve the water resistance of these finishes led to their reinforcement with synthetic polymer dispersions. Styrene-butadiene copolymers are effective for printing paper of short life but the unsaturation present leads to yellowing on ageing. Styrene/ethyl acrylate and vinyl acetate/butyl acrylate copolymers are obviously free from this disadvantage and are widely used for book and high-class magazine paper. Copolymers of ethyl acrylate with a minor proportion of methyl methacrylate are also excellent and other contenders which have found some application are butadiene/methyl methacrylate copolymers and vinyl chloride/ acrylic ester copolymers. Styrene-butadiene copolymers continue to dominate the tonnage of polymers used in the lower quality papers.

Copolymers of acrylic esters with vinylidene chloride of low film forming temperature have been used to provide flexible non-blocking paper coatings. Other acrylic ester copolymers have been employed for the coating of wallpaper and business machine paper.

There is an outlet for acrylic esters in acid-removable floor polishes. These are essentially acrylic ester copolymers into which a minor proportion of a basic comonomer such as an alkylaminoethyl acrylate has been incorporated. The coating produced is resistant to detergents but is easily removed by weak acid solutions. Acrylic esters are also used as flexibilising co-monomers in other polymer latices used in floor polishes.

Solution polymers of acrylic esters have been employed to provide fabrics with waterproof coats, e.g. in the manufacture of rainwear and shower curtains.

Finally there is the earliest coating application of acrylic ester copolymers already mentioned for the finishing of leather, e.g. in the finishing of shoes and gloves and car and furniture upholstery where the outstanding flexibility of copolymers containing a major proportion of ethyl acrylate make them the technically preferred materials. The finishes are applied from aqueous dispersion and can be air-dried, heated or plated against heated polished-steel platens.

Apart from the flexibility of such coatings, the copolymers improve the 'feel' and 'handle', scuff-resistance, water resistance and durability of the leather. They also bind pigments well where coloured finishes are involved. They are also used in conjunction with other resinous materials and form an excellent bond between, for example, leather and a subsequently applied nitrocellulose coat.

Although they represented the first commercial application of acrylic esters and are still the technically preferred materials, their rate of growth in recent years has been much lower than in many other applications for acrylic esters. With the current swing to leather substitutes it cannot be expected that the rate of growth will improve.

10.3.3.2. Adhesives and binders

a. *Bonding of miscellaneous materials* Acrylic ester copolymer latices, and to a lesser extent solutions, are widely used for the bonding of various types of sheet to decorative surfacing materials. Fabrics, polyvinyl chloride films and metal foil can be stuck to a wide variety of substrates such as plywood, hardboard and metal. The good colour and flexibility of the acrylic copolymers are particularly useful in such applications. In the shoe industry they have been used to bond canvas and leather and in the fixing of insoles, linings and decorative surfaces.

Copolymers of ethyl, butyl and 2-ethylhexyl acrylates with vinyl acetate and methyl acrylate are widely used. A 50/50 copolymer of vinyl acetate with butyl acrylate provides a copolymer of excellent flexibility. A permanently tacky adhesive which can be used as a contact adhesive is produced when vinyl acetate is copolymerised with a high proportion of 2-ethylhexyl acrylate.

b. *Bonding of textiles* Apart from the traditional use of acrylic polymers applied from aqueous dispersion to textile fabrics to give a soft full handle, ethyl acrylate copolymers are widely used in the production of non-woven fabrics. These are made by bonding mats of oriented or random dispersed fibres with an adhesive. The excellent flexibility and good colour retention of polyethyl acrylate provides the right combination of properties. A further requirement is that the bonding should be resistant to washing and dry cleaning, and these properties can be achieved by preparing an aqueous dispersion analogue of the thermosetting acrylic resins. Ethyl acrylate is therefore copolymerised in aqueous dispersion with minor amounts of a crosslinking agent such as methylol acrylamide and a catalyst such as acrylic acid so that when the non-woven fabric is heated crosslinking takes place resulting in insolubilisation of the polymer.

Other non-woven fabric binders include butadiene/styrene and butadiene/acrylonitrile copolymers but the acrylic esters score on colour retention and ageing owing to the absence of any unsaturation.

Acrylic ester copolymer dispersions have also been used for the bonding and spot welding of different types of textile fabrics, for the stabilisation of woollen and worsted fabrics against felting shrinkage, in the pigment printing of fabrics and, in conjunction with amino resins, to produce crease-resistant cotton fabrics of improved tear and tensile strength.

c. *Bonding of paper* Acrylic ester copolymers have been used in the binding of paper based disposable garments and by addition to the beater to provide paper of improved tensile strength and folding characteristics without a reduction in water permeability.

They have also been used to impregnate and saturate paper to provide a range of products such as artificial leather, pressure sensitive tapes and gaskets.

10.3.3.3. Acrylic rubbers These are copolymers of an acrylic ester such as ethyl acrylate with about 5 per cent of a chlorine containing monomer such as 2-chlorethyl acrylate ($ClCH_2CH_2OOCCH = CH_2$) or 2-chlorethyl vinyl ether ($ClCH_2CH_2OCH = CH_2$). They are prepared in aqueous dispersion and isolated by coagulation, washing and drying. Vulcanisation can be effected using an amine such as triethylene tetramine or using hexamethylene diamine carbamate in conjunction with dibasic lead phosphite. Sulphur and sulphur containing materials as well as reinforcing materials such as carbon black or white silica can be incorporated.

Polymers of lower acrylic esters display good oil resistance. Moreover, because they are saturated, the acrylic rubbers display good oxygen and ozone resistance and do not suffer from case hardening exhibited by many synthetic rubbers based on butadiene where the residual unsaturation can, at high temperature, react with sulphur containing materials used as stabilising additives for oils.

The acrylic rubbers therefore find application where a rubber is required to retain good physical properties when exposed to high temperatures and particularly in the presence of oils.[123]

The acrylic rubbers are expensive which limits their tonnage.

Even more expensive are the fluorine containing acrylic rubbers. These are prepared from 1,1-dihydroperfluoralkyl acrylates such as $R_FCH_2OOCCH = CH_2$ where R_F is a perfluoralkyl group.[124] The synthesis of the monomers is represented schematically below:

$$RCOOH \xrightarrow{HF} R_FCOF \xrightarrow{H_2O} R_FCOOH \xrightarrow{CH_3OH} R_FCOOCH_3 \xrightarrow{H_2} R_FCH_2OH$$

$$R_FCH_2OH \xrightarrow{CH_2 = CHCOOH} R_FCH_2OOCCH = CH_2$$

These monomers are also polymerised in aqueous dispersion and isolated by coagulation, washing and drying.

Typical commercial monomers have the structures:

$$CH_2 = CHCOOCH_2CF_2CF_2CF_3 \text{ and } CH_2 = CHCOOCH_2CF_2CF_2OCF_3$$

The polymers contain approximately 50 per cent by weight of fluorine and they are also saturated. The rubbers therefore display exceptional heat and oil resistance which is better than that of the acrylic rubbers. They are also resistant to synthetic diester lubricants at high temperatures. Their very high price and limited low temperature flexibility restrict their applications to highly specialised uses.

10.3.3.4. Acrylic fibres   Although these are based predominantly on acrylonitrile polymer, certain of the fibres do contain small percentages of copolymerised acrylic esters such as methyl acrylate which are believed to be introduced for the purpose of regulating dye diffusion into the fibre.[125]

10.3.3.5. Miscellaneous acrylic acid and acrylic ester uses   Acrylic acid polymers and copolymers with minor amounts of acrylic esters or acrylamide are widely used as thickeners in the form of aqueous solutions of their ammonium or alkali metal salts. Dependent upon the molecular weight and pH value, small concentrations of these polymers (e.g. 0·1 to 3 per cent by weight) in water can give a range of viscosities up to a gel-like structure. They have been used to thicken natural and synthetic rubber latices for dipping and coating applications and to control the viscosity characteristics of emulsion paints. They have been used as flocculating agents, as filler retention aids and dry strength agents in the paper industry, as additives during the drilling of oil wells to prevent mud caking, as warp sizes for nylon during weaving and in cosmetics.

The acrylic esters of long chain alcohols yield polymers which are soluble in oils so that copolymers of these esters have been used as the basis for stabilising viscosity/temperature relationships in lubricating oils, i.e. providing multigrade oils.

Small additions of acrylic ester copolymers to rigid PVC improve its processing characteristics and surface finish during calendering, vacuum forming and blow and injection moulding. They also improve the impact strength of rigid PVC in extruded pipe and profile extrusions.[126]

The cyanoacrylic esters which form the basis of very high priced, very small tonnage but nevertheless very efficient adhesives are mentioned only briefly here since they are not strictly propylene derivatives. They are produced by condensing formaldehyde with an alkyl 2-cyanoacetate to produce a polyalkyl 2-cyanoacrylate which is then depolymerised to the monomeric 2-cyanoacrylic ester.[127]

$$n-\text{NCCH}_2\text{COOR}+n\text{-HCHO} \rightarrow \left[-\text{CH}_2-\underset{\underset{\text{COOR}}{|}}{\overset{\overset{\text{CN}}{|}}{\text{C}}}- \right]_n$$

$$\downarrow \text{depolymerisation}$$

$$\text{CH}_2 = \underset{|}{\overset{\overset{\text{CN}}{|}}{\text{C}}}\text{COOR}$$

The 2-chloroacrylic ester polymers which have shown promise as hard, mar-resistant products of high softening point have not yet been commercialised probably owing to their very high cost.[128]

Other hard acrylic ester polymers which could be used as the basis for moulding materials similar to polymethyl methacrylate (such as Perspex, Lucite, Diakon, etc.) have been the subject of much, to date, commercially unsuccessful, research although high softening point polymers have been produced, e.g. pentachlorphenyl acrylate polymer (s.p. 140°C). Stereospecific polymerisation of some of the lower acrylic esters in the laboratory has also produced high melting point polymers. Moulding materials based on thermosetting acrylic resins could be a possibility for the future.

10.3.4. General

A profile review of acrylates[129] published during 1970 gave the USA capacity for the production of these esters as 645×10^6 lb. The demand for 1970 was forecast at 415×10^6 lb with a projected demand in 1974 of 725×10^6 lb and a growth rate of 15 per cent per annum through 1974. The breakdown of monomer uses was given as coatings (38 per cent), textiles (18 per cent), exports (13 per cent), miscellaneous acrylic acid uses (7 per cent), fibres (7 per cent), paper (6 per cent), polishes (6 per cent), leather (3 per cent) and miscellaneous (2 per cent).

The growth prospects of acrylic esters for the foreseeable future appear very good. Nevertheless it still remains to be seen how much of the very large use of acrylic esters as plasticising co-monomers, e.g. with vinyl acetate in emulsion paints and with styrene in thermosetting acrylic resins, will be replaced by olefins which also confer good plasticisation and alkali resistance.

At the same time newer uses could offset the inroads made from these other directions. An example of such developments is the range of AAS terpolymers recently described by Ugine-Kuhlmann.[130] The AAS resins are analogous to the well known ABS polymers which are produced by grafting styrene and acrylonitrile onto polybutadiene but the polybutadiene is replaced by an acrylic ester rubber containing a small percentage of copolymerised tetrahydrobenzyl acrylate to introduce reactivity into the rubber. The AAS resins which contain no polybutadiene unsaturation have much improved resistance to weathering and retention of impact strength compared with the ABS resins and could provide an outlet for acrylic esters in the moulding material field. Another example is that of the ethylene-ethyl acrylate copolymers recently announced by Union Carbide Corp. These maintain their rubbery properties over a wide temperature spectrum.[131]

REFERENCES

[1] Wohl and Mylo, *Ber.*, 1912, 2046.
[2] BIOS Final Report no. 783.
[3] FIAT Final Report no. 1157.
[4] Du Pont, USP 2,197,258. Acrolein Corpn. USP 2,270,705.
[5] Deniges, *Compt. rend.*, 1898, **126**, 1147.
[6] Standard Oil, USP 2,627,527.
[7] *Chem. Eng.*, 1947, **54**, 106–9.
[8] Shell, USP's 2,451,485; 2,486,842; 2,606,932 and 3; 2,608,585; 2,614,125; 2,620,358; 2,659,758; 2,690,457.
[9] Montecatini, BP 847,564.
[10] USP 2,383,711.
[11] DCL, Research and Development Dept., unpublished work.
[12] DCL, BP's 625,330; 658,179; 655,210; 694,362.
[13] DCL, BP's 661,882; 694,356; 875,160.
[14] Sohio, BP's 821,999.
[15] DCL, BP's 864,666; 906,328.
[16] Acrolein Corpn., USP 2,334,091. Asahi Denka Kogyo, Jap.P's 18668/63; 24447/65. T. Kahata, Jap.P 8575/61. NRDC, Dutch P. 68–05,024.
[17] Papers presented by Wise and Holbrook and by Mann and Yao at Symposium on Catalytic Oxidation, London, July 1970.
[18] Asahi Chem. Ind., Jap.P's 459/63; 18016/64.
[19] Mitsubishi Rayon, Jap.P 27401/68.
[20] Isaev and Margolis, *Kinetika Kataliz.*, 1960, **1**, 237.
[21] Adams *et al.*, *J. Catalysis*, 1964, **3**, 379–86.
[22] Godin *et al.*, Paper 20, 4th International Congress on Catalysis, Moscow 1968.
[23] Gelbshtein *et al.*, *Kinetika Kataliz*, 1965, **6**, 1025–32.
[24] Voge *et al.*, *J. Catalysis*, 1963, **2**, 58–62.
[25] Adams and Jennings, *J. Catalysis*, 1963, **2**, 63–8.
[26] McCain *et al.*, *Nature*, 1963, **198**, 989–90.
[27] Sachtler, *Rec. Trav. Chim.*, 1963, **82**, 243–5.
[28] Adams and Jennings, *J. Catalysis*, 1964, **3**, 549–58.

[29] Godin and McCain, *Symposium on Catalytic Oxidation*, London, July 1970.
[30] Sixma *et al.*, *Rec. Trav. Chim.*, 1963, **82**, 901.
[31] Yoshihiko Moro-oka *et al.*, *J. Catalysis*, 1968, **12**, 291–7, also *Symposium on Catalytic Oxidation*, London, July 1970.
[32] Roginskaya *et al.*, Kinetika Kataliz, 1968, **9**, 1143–7.
[33] Wakabayashi *et al.*, *Bull. Chem. Soc. Japan*, 1967, **40**, 2172–6.
[34] *Ibid.*, 1968, **41**, 2776–81.
[35] Stroeva *et al.*, *Neftekhimiya*, 1966, **6**, 412–22.
[36] Derlyukova *et al.*, *Izv. Akad. Nauk SSR, Ser. Khim*, 1968, **1**, 116–23.
[37] Derlyukova *et al.*, *Internat. Oxidn. Symposium*, San Fransisco, September 1967.
[38] Bleijenberg *et al.*, *J. Catalysis*, 1965, **4**, 581–5.
[39] Batist *et al.*, *J. Catalysis*, 1968, **12**, 45–60.
[40] Peacock *et al.*, *J. Catalysis*, 1969, **15**, 373–406.
[41] Trifiro and Pasquin, *J. Catalysis*, 1968, **12**, 412–16.
[42] Shell, Jap.P 19958/63. Sohio, BP 821,999.
[43] DCL, BP 975,686. Asahi Chemical, Jap.P 24976/65.
[44] Shell, USP 2,514,966.
[45] DCL, BP 693,187.
[46] Shell, USP 2,514,966; BP 664,414; USP 2,514,967.
[47] Union Carbide, USP 3,220,932. Celanese, USP 2,476,391. Shell, USP 2,767,216.
[48] Shell, USP 2,574,935.
[49] Shell, USP 2,542,752.
[50] Shell, USP 2,562,846.
[51] Shell, USP 2,767,216.
[52] *Acrolein*, edited by C. W. Smith (John Wiley & Sons, 1962).
[53] Catch *et al.*, *Nature*, 1947, **159**, 578.
[54] Shell, USP 2,779,801.
[55] Shell, USP 2,767,221.
[56] *Chem. Eng. News*, 25 October 1965, p. 40–42.
[57] Shell, USP 2,479,284.
[58] Isard *et al.*, 'Debouches Industriels de l'acroleine par l'Intermédiare des Syntheses Diéniques', Congres de Chimie Industrielle, Istambul 1969.
[59] *Brennstoff Chem.*, 1966, **47**, 207.
[60] *Chem. Week.*, 21/3/64, p. 39.
[61] Morrison, *J. Cellular Plastics*, August 1967, p. 364–8.
[62] ICI, BP 1,187,347.
[63] BIOS Reports nos. 266, 355, 358, 371.
[64] American Cyanamid, USP 3,268,579. Badische, GP 965,323. Union Carbide, FP 1,229,843.
[65] ECN, 24 October 1969, p. 54.
[66] Schulz and Wagner, *Angew. Chem.*, 1950, **62**, 5, 105–32.
[67] DCL, BP 740,005.
[68] Sohio, USP's 2,881,212; 2,881,214; 3,021,366. Du Pont, USP 3,087,964; BP 893,077; DCL, BP 878,802.
[69] Nippon Kayaku KK, BP 1,185,109. Kitahara and Tsuboyama, *Ind. Chem. Belge*, 1967, **32**, special issue. Gr VIII, S20 and 9, 698–701.
[70] Grace, BP 1,115,116.
[71] Grace, Dutch P. 67–14,549.
[72] DCL, BP 1,036,957.
[73] Knapsack, BP 948,687.
[74] Knapsack, 6,811,219.
[75] Knapsack, Dutch P. 69–13,173.
[76] Shell, FP 1,295,188.
[77] Nakatani, *Hydroc. Proc.*, May 1969, 152–4.
[78] Goro Kijima, *Japan Chemical Quarterly*, **V–III**, 12–16.
[79] DCL, BP 999,836.

[80] Knapsack, Dutch P. 67–10,799.
[81] DCL, BP 953,763.
[82] Knapsack, Dutch P. 67–16,963.
[83] Röhm and Haas, FP 1,573,704.
[84] Celanese, Dutch P 67–15,465.
[85] Daicel, GP 1,942,338.
[86] Knapsack, Dutch P 68–08,878
[87] DCL, BP's 995,471; 995,472; 997,324; 997,325.
[88] Dynamit Nobel, GP 1,518,723.
[89] Ugine, FP 1,359,885.
[90] DCL, BP 997,324. Daicel, GP 1,942,338. Kodak, FP 1,150,930.
[91] Asahi Chemicals, BP 1,120,284.
[92] Celanese, USP 2,916,512. DCL, BP 923,594. Knapsack, BP's 948,687 and 959,880.
[93] Knapsack, Dutch P 69–02,273.
[94] Knapsack, BP 948,687.
[95] Knapsack, FP 1,344,752.
[96] DCL, BP's 923,594; 923,595.
[97] DCL, BP 1,003,007.
[98] Toa Gosei, Jap.P 2689/66. Asahi Chemicals, GP 1,927,036. BASF, FP 1,332,186.
[99] BASF, FP 1,332,186.
[100] Sohio, USP 3,088,969.
[101] BASF, GP 1,248,036.
[102] BASF, GP 1,161,259.
[103] Daicel, Jap.P 20286/68.
[104] Dai Sellu, Jap.P 6324/67.
[105] Mitsubishi Rayon, Jap.P 5222/67.
[106] Sohio, USP 2,947,779.
[107] Sohio, USP 2,913,486.
[108] Esso, FP 1,474,114.
[109] DCL, BP 960,005.
[110] Lonza, FP 1,544,542.
[111] Du Viergier, BP 1,094,998.
[112] California Research, USP 2,891,991.
[113] Solvay, GP 1,067,805.
[114] Röhm & Haas, GP 1,067,806.
[115] Boguslavskaya, *Khim. Prom.*, 1967, no. 3, 177–83.
[116] Bailey, *Autoxidation of Chloroprene*, 'Advances in Chemistry Series', **75**, I, pp. 138–49.
[117] General Electric, BP 793,047.
[118] Knapsack, BP 920,353.
[119] Toa Gosei, Jap.P 2689/66.
[120] American Cyanamid, USP 3,253,015. BP Chemicals (UK) Ltd., BP 1,127,127.
[121] *Encyclopedia of Polymer Science and Technology, Plastics, Resins, Rubbers, Fibers* (Interscience Publishers, John Wiley & Sons Inc., 1964), Vol. 1, 274.
[122] *Ibid.*, 276.
[123] *Ibid.*, 226.
[124] *Ibid.*, 241.
[125] *Ibid.*, 342.
[126] *Ibid.*, 318.
[127] *Ibid.*, 337.
[128] *Ibid.*, 328.
[129] *Oil Paint and Drug Reporter*, 26 October 1970, 9.
[130] *Informations chimie*, no. 99, August/September 1971, 161.
[131] *Prod. Engng.*, 42, no. 10 (May 1970), 40.

CHAPTER 11

ACRYLONITRILE

By **D. J. Hadley**

11.1. INTRODUCTION

TABLE XI.1

Some Properties

Density	0·806 g/ml (20 °C)
Freezing point (1 atm)	−83·5 °C
Boiling point (1 atm)	77·3 °C
Refractive index (n_D)	1·3887 (25 °C)
Heat of formation (liquid) (25 °C)	+36·2 kcal/mole
Heat of combustion (liquid) (25 °C)	−420·8 kcal/mole
Solubility in water	7·4% wt/wt (2; °C)
Solubility of water in acrylonitrile	3·1% wt/wt (20 °C)
Vapour pressure	83 mm Hg (20 °C)
Water azeotrope	
Boiling point (1 atm)	70·6 °C
% wt/wt of water	14·3

Acrylonitrile was first prepared by Moureu in 1894,[1] by the dehydration of acrylamide and ethylene cyanhydrin with chemical dehydrating agents, but there was no technical application of this material until about 1930, when it was discovered, in Germany, that the introduction of polar cyano groups into synthetic rubber (by copolymerisation of butadiene and acrylonitrile) greatly increased its resistance to swelling in non-polar solvents. Production of the monomer was initially based on a route going via ethylene cyanhydrin but another method of manufacture, the addition of HCN to acetylene, was initiated in Germany during World War II[2] and this proved to be generally, more attractive. Buna N in Germany and equivalent nitrile rubbers in the USA were then produced on a modest commercial scale, thus establishing acrylonitrile as a monomer of significance. However, the subsequent great expansion in demand is attributable not to rubber but largely to acrylic fibres. About 60 per cent of the acrylonitrile consumed in the USA today and about 80 per cent of that consumed in W. Europe goes into acrylic fibre.

Du Pont commenced the first commercial production of fibres from acrylonitrile in 1948, under the brand name ORLON. As other companies, both in the USA and in the other technologically advanced countries, soon followed Du Pont's lead, the output of acrylic fibre (and hence of the monomer) increased rapidly, but the growth rate during the 1960's has been particularly notable. The following tables contain data furnished by BP Chemicals International Ltd.

TABLE XI.2

Acrylonitrile in the USA

Year	1,000 tons Production	Consumption
1960	102	89
1961	111	100
1962	161	121
1963	203	158
1964	265	182
1965	345	247
1966	320	266
1967	299	289
1968	456	361
1969	516	387
1970	500–525	413 Forecast
1971	498	463
1972	547	512
1973	598	563
1974	673	628
1975	749	694

TABLE XI.3

Acrylonitrile in Western Europe

Year	1,000 tons Production	Consumption
1961	42	47
1962	45	71
1963	46	90
1964	53	117
1965	71	157
1966	124	195
1967	165	234
1968	231	332
1969	328	427
1970	438	510 Forecast
1971	620	600
1972	740	699
1973	890	793
1974	940	891
1975	1,030	996

This quite spectacular increase in acrylonitrile monomer output, dating back some eight or nine years, is attributable to the fortunate conjunction of the consumer appeal of polyacrylonitrile fibre, with its wool-like characteristic, and the

28

improved methods of manufacture of the monomer which turned to propylene as the starting material. The improvement in manufacture is measured by the fall in price of the monomer from about £250–£300/ton, as it was in 1960, to less than £140/ton, as it was in 1967.[3]

Heyner and Seeboth[4] give the following American comparison of raw material costs, per pound of acrylonitrile, in the three principal methods of production:

Acetylene/HCN	12·5 cents
Ethylene (ethylene oxide) HCN	11·25 cents
Propylene/ammonia	4·0 cents

Cevidalli *et al.*[5] make a similar comparison:

Acetylene/HCN	188 lire⎱ per kg of
Propylene/ammonia	61 lire⎰ acrylonitrile

Heyner and Seeboth suggest that in 1970 more than 90 per cent of the plants producing acrylonitrile will use the propylene/ammonia method: this may well be an under estimate of the relative importance of the new route.

11.2. OLDER METHODS OF MANUFACTURE

11.2.1. From Acetylene

Until the early 1960's the most important method of production of acrylonitrile was the combination of acetylene and HCN:

$$HC \equiv CH + HCN \rightarrow CH_2 = CHCN$$

The process consists essentially in passing HCN and an excess of acetylene into a solution of cuprous chloride in dilute hydrochloric acid at about 80°C. Unreacted acetylene leaving the reactor entrains the product acrylonitrile and residual HCN, and these are dissolved out of the exit gas by scrubbing with water. Acrylonitrile is distilled from the water solution and is purified by further fractional distillation. The acetylene which leaves the water scrubber is freed of volatile by-products by washing with a selective solvent and is then recycled to the reactor.

The yield in the acetylene-based process is good, but there are some operational difficulties. The catalyst becomes fouled by by-products and loses activity, and some of the by-products, such as divinylacetylene and methyl vinyl ketone, which could have a deleterious effect on polymerisation, are difficult to separate from the monomer. However, the main disadvantage, in comparison with processes starting from propylene, is the high cost of the raw materials.

11.2.2. From Ethylene Oxide

The alternative older route starts from ethylene:

$$\begin{array}{ccccc}
CH_2 & O_2 & CH_2 & HCN & CH_2OH \\
\| & \xrightarrow{} & | \diagdown & \xrightarrow{} & | & \longrightarrow CH_2 = CHCN \\
CH_2 & \text{Ag catalyst} & | O & NHEt_2 & CH_2CN \\
& & CH_2 \diagup & \text{catalyst} &
\end{array}$$

Ethylene is converted to the epoxide either via ethylene chlorohydrin, or more probably (and as shown above) by direct oxidation in the vapour phase. HCN is added to the epoxide in a liquid phase reaction catalysed by a base such as diethylamine, and the cyanhydrin so produced is dehydrated, either in the liquid phase with a catalyst such as alumina, or in the vapour phase. Limitation on yield arises chiefly in the first reaction, the last two being both smooth and efficient. But ethylene oxide and HCN are, as we have seen, relatively expensive sources of the carbon and nitrogen required.

Both these processes although still in use, are obsolescent.

11.2.3. From Acetaldehyde

There is another method, attributable to Knapsack[6] and deserving of mention, although it was never used commercially, which is based on a series of reactions closely resembling the previous series:

$$\begin{array}{c} CH_2 \\ \| \\ CH_2 \end{array} \xrightarrow{O_2} CH_3CHO \xrightarrow[\text{(O\overline{H})}]{HCN} CH_3{-}\overset{\overset{\displaystyle H}{|}}{\underset{\underset{\displaystyle CN}{|}}{C}}{-}OH \longrightarrow CH_2 = CHCN$$

Acetaldehyde, another oxidation product of ethylene, is converted to lactonitrile by the addition of HCN, and the lactonitrile is dehydrated by being passed, together with phosphoric acid, in the vapour phase, through a tubular reactor at about 650°C. The essential and inventive feature of the dehydration step is that the acid and lactonitrile are sprayed directly into the hot reactor so that the nitrile spends little time at intermediate temperatures at which dissociation to HCN and aldehyde is favoured. This was a comparatively attractive process but it came too late to be exploited. Already it was becoming evident that acrylonitrile could be obtained in a relatively simple manner directly from propylene.

11.3. ACRYLONITRILE FROM PROPYLENE BY AMMOXIDATION

The first authenticated report of the preparation of acrylonitrile from propylene occurred in a patent to the Allied Chemical and Dye Corporation, for which application was made in 1947.[7] The principal claim of this patent is for the manufacture of aliphatic nitriles by reaction of an olefin with oxygen and ammonia at 400–600°C (preferably over a catalyst). What is principally illustrated is a degradative reaction of olefins, leading to nitriles with fewer carbon atoms. But, tucked away amongst the examples, is the statement that from propylene (in one experiment) there was obtained a 6 per cent yield of acrylonitrile, together with 10 per cent of acetonitrile and 10 per cent of HCN. Since the yield of acrylonitrile was small and the molar concentration of propylene in the feed was less than 1 per cent the concentration of acrylonitrile in the product gas was minute, which is probably the reason why the result had, apparently, no great significance for the patentee (or anyone else). In fact, this, the first real example of ammoxidation, as the reaction between propylene, ammonia and oxygen to yield acrylonitrile is now called, contributed nothing to the industrial development of the process, and the reaction was re-discovered, several years later, by a different process of thought and experimentation. All the examples in the Allied Chemical patent refer to a catalyst consisting of the oxides of vanadium, molybdenum and phosphorus on a support, but it is stated that the reaction will proceed effectively even in the absence of catalyst. It is now known that a catalyst is essential.

11.3.1. The Rise of Ammoxidation

In the early 1950's the Research Department of the Distillers Co. Ltd. (DCL) was studying the chemical reactions of acrolein, with the object of finding new uses for this material, for they had recently developed a process for its production by the vapour phase oxidation of propylene. This process, which comprised a supported copper selenite catalyst, requiring a transport of selenium, gave up to 80 per cent yield per pass of acrolein. A prime objective of the research on acrolein was acrylonitrile, which it was found could be prepared by the vapour phase reaction of certain derivatives of acrolein, viz. methyl acrylate and 3-methoxypropionic acid, with ammonia over dehydration catalysts:

$$CH_2 = CHCOOCH_3 + NH_3 \rightarrow CH_2 = CHCN + H_2O + CH_3OH$$
$$CH_2(OCH_3)CH_2COOH + NH_3 \rightarrow CH_2 = CHCN + 2H_2O + CH_3OH$$

The yields per pass, however, were small and the route from acrolein was protracted, so it was not pursued.

The catalysed vapour phase reaction of aldehydes in general with ammonia to yield nitriles had already been described or claimed,[8,9] and, in particular, there was a patent[10] relating to the production of acrylonitrile by reaction of acrolein and ammonia over a chromia/alumina catalyst at 950°F. But when the process therein described was attempted no acrylonitrile was obtained. However, the Shell Development Company had described the vapour phase oxidation of unsaturated amines to nitriles,[11] and the idea arose that if acrolein was passed together with ammonia and oxygen over a suitable catalyst the imine might be formed in situ and oxidised to the nitrile in a similar way:

$$CH_2 = CHCHO + NH_3 \longrightarrow CH_2 = CHCH = NH \overset{O_2}{\longrightarrow} CH_2 = CHCN$$

The desired overall result was achieved:[12]

$$CH_2 = CHCHO + NH_3 + \tfrac{1}{2}O_2 \rightarrow CH_2 = CHCN + 2H_2O$$

although the existence of the postulated intermediate imine has never been demonstrated.

The reaction above was catalysed by a number of metal oxides, the most effective being molybdenum oxide, with which yields of 80 per cent per pass were obtained. The molybdates of some polyvalent metals gave similar results. It is of particular interest that bismuth molybdate, although an excellent catalyst, showed no advantage over the simple molybdic oxide. Whereas, in all the related reactions, which have, subsequently, been intensively studied, viz. acrolein→acrylic acid, propylene →acrolein, propylene→acrylonitrile, catalysts comprising two or more oxides are necessary, and neither molybdic oxide nor any other oxide will by itself give a high yield (per pass) of product.

In the further development of a process for the manufacture of acrylonitrile from propylene via acrolein, DCL linked together the consecutive reactions by feeding a propylene/air mixture through a reactor containing a copper selenite catalyst, mixing the effluent gas with ammonia and passing the mixture to a second reactor charged with molybdic oxide catalyst:[13]

$$CH_2 = CHCH_3 + O_2 \rightarrow CH_2 = CHCHO + H_2O \qquad \text{followed by}$$
$$CH_2 = CHCHO + NH_3 + \tfrac{1}{2}O_2 \rightarrow CH_2 = CHCN + 2H_2O$$

The yield per pass of acrylonitrile from propylene was about 60 per cent, as expected. It was then realised that it might be possible to telescope the two reactions and thereby simplify the process by feeding a mixture of propylene, air and ammonia to a single reactor, provided that a catalyst capable of promoting both reactions could be found. The dual function catalyst that immediately sprang to mind was copper molybdate with selenium, this being virtually a combination of the catalysts used in the series reactors, and this indeed gave high yields of acrylonitrile from propylene. There was, however, no further development of this particular catalyst system because difficulty was foreseen in obtaining supplies of selenium (see chapter on acrolein and acrylic acid), and subsequently a research programme was initiated with the object of finding an alternative catalyst for the one-stage conversion of propylene to acrylonitrile.

Meanwhile, the Standard Oil Company of Ohio (Sohio) had been working on the oxidation of propylene, and in 1957 they filed a patent in which they revealed the important discovery of bismuth molybdate as a catalyst for the oxidation of propylene to acrolein.[14] Both bismuth molybdate and bismuth phosphomolybdate are claimed and these are preferably used with a silica support, the effect of which is illustrated by an example in which various mixtures of $Bi_9PMo_{12}O_{52}$ and SiO_2 were used as catalysts under given conditions (Table XI.4).

TABLE XI.4

% SiO_2	% Yield of acrolein	% Efficiency
90	8·4	64·5
55	29·6	54·7
30	40·8	71·8
0	34·9	69·2

A few months later Sohio filed a patent[15] in which they claimed the production of acrylonitrile by the reaction of propylene, ammonia and oxygen over a catalyst selected from the group: bismuth, tin and antimony salts of phosphomolybdic and molybdic acids and bismuth phosphotungstate. Silica support was again mentioned.

Patent examples and subsequent events indicated that bismuth molybdate and phosphomolybdate, which gave yields of over 50 per cent acrylonitrile with 10 per cent v/v propylene in the feed, were far superior to the other members of the group, and the preferred composition for industrial use appears to be bismuth phosphomolybdate/silica although the function of the phosphoric acid component is not clear. This was the first commercially viable catalyst for the manufacture of acrylonitrile from propylene, which is usually referred to as bismuth molybdate.

In view of the early work of D C L and Sohio in this field it appeared that molybdenum would be an indispensible component of any effective catalyst for acrylonitrile production from propylene. But in 1959 D C L made the surprising discovery that various binary oxide combinations that included antimony gave a catalyst performance similar to that of bismuth molybdate: yields of more than 60 per cent acrylonitrile were obtained with 5 per cent v/v propylene in the feed. Inititally D C L concentrated on the development of an antimony/tin oxide catalyst.[16] The new catalysts were effective also in the oxidation of propylene to acrolein and in the conversion of acrolein to acrylonitrile.

In 1960 D C L, who had been studying the preparation of nitriles in general, proposed the trivial but convenient name 'ammoxidation'[17] for the reaction

$$RCH_3 + NH_3 + 1\tfrac{1}{2}O_2 \rightarrow RCN + 3H_2O$$

It is now used to describe the reaction of various organic compounds with oxygen and ammonia.

II.3.I.I. Survey of propylene ammoxidation catalysts   Since the announcements of the first two important catalysts, Bi/Mo and Sb/Sn oxides, a remarkable world-wide activity has been displayed in the search for alternatives and for superior performance, and more than thirty companies have taken out patents in this field. Many of these patents confirm the close relationship between the oxidation and the ammoxidation of propylene by claiming the same catalysts for both purposes, but the economic incentive for the research is clearly acrylonitrile production. Almost always patents are concerned with increased yields on propylene, and more than 80 per cent yield per pass has been claimed, but it should be remembered that patent examples are almost always derived from small-scale experimentation, and that the catalyst performance therein described may not be reproducible on a full-scale plant. In the following list of selected patents only the qualitative composition of the catalyst is given, and, except in the few indicated cases, the elements mentioned are present as oxides or are linked to oxygen. For example if one or more patents claimed a mixture of copper, cobalt and molybdenum oxides, a mixture of copper oxide and cobalt molybdate, and a mixture of cobalt oxide and copper molybdate, these would all be indicated by Mo, Co, Cu.

II.3.I.2. Bismuth molybdate derived catalysts   A considerable number of catalysts contain bismuth and molybdenum together with other elements, and may fairly be regarded as modifications of the original bismuth molybdate.

Catalyst	Patentee	Patent number
Bi, Mo, P, Fe	Knapsack	BP 947,364
At least two of (Bi, Mo, P, Te, Cu, V)	Asahi Chem. Ind. Co. Ltd.	Jap.P 13290/67
Bi, As and two of (Mo, V, W)	ICI	BP 929,650
Bi, Mo, P and Ni or Co	Mitsui Petrochemicals	Jap.P 28906/66
Bi, Mo, P	OSW	BP 967,877
Bi, Mo, P	OSW	BP 1,021,328
Bi, Mo and Cu or Sr	OSW	BP 1,024,402
Bi, Mo, V	SNAM	FP 1,382,521
Bi, Mo, Ba, Si, optionally P	Sohio	USP 3,186,955
Bi, Mo, B	Sumitomo Chem. Ind. Co.	Jap.P 23393/64
Bi, Mo, Sb, V, As	Toyo Rayon	Jap.P 16778/66
Bi, Mo, Sb	Ube Industries Ltd.	Jap.P 1806/70
Bi, Sb and Mo or W	Ube Industries Ltd.	GP 1,919,840
Bi, Mo, As	UCB	BP 885,422
Bi, P, Al (Mo and/or V)	Zimmerverfahrenstechnik	Belg.P 623,578

The largest group is that of catalysts having antimony as the principal constituent. Three of those already included in the first group might, equally well, have been placed here.

Catalyst	Patentee	Patent number
Sb, Sn and one of (Mg, Sr, Zn, Cd, Bi, Ce)	Mitsubishi Rayon	Jap.P 29933/68
Sb, Sn and Co or Ni	Mitsubishi Rayon	Jap.P 62923/65
Sb, Te	Montecatini	BP 1,168,279
Sb, Cu	Nippon Shokubai	Jap.P 14093/66
Sb, V, Fe and P or B	Nitto Chem. Industry Co. Ltd.	Jap.P 75891/67
Sb, Fe, Te and one of (V, Mo, W) and P or B	Nitto Chem. Industry Co. Ltd.	Belg.P 730,696
Sb, Fe	Sohio	BP 973,338
Sb, U, Si and one or more of (Bi, Sn, Pt, Mg, Ag, Fe, Cu, Th, Co, Ni, Cs, Li, In)	Sohio	USP 3,431,292
Sb, Fe and one or more of (Re, Nb, Cu, Ag, Bi, Co, Sn, Mo)	Sohio	USP 3,338,952
Sb, U and one or more of (Bi, Sn, Pt, B, Mg, Ag, Fe, Zr, Cu, Th, Zn, Co, Ni, Pb, As, W, P, Al, Ca, Sb and Cs)	Sohio	USP 3,328,315
Sb, U	Sohio	USP 3,308,151
Sb, Mn	Sohio	USP 3,200,081
Sb, Fe	UCB	FP 1,333,040
Sb, Fe, As	UCB	BP 1,078,156
Sb and one of (Cr, Co, Cu, Mn, Ti, Ni)	DCL	BP 987,960
Sb, Sn, U	DCL	BP 1,026,477
Sb, U and an element of atomic number 22–41, 44–49, 73, 78–83, 90	DCL	BP 1,007,929
Sb, Sn and one of (Cu, Cr, Mn)	DCL	BP 1,033,829
Sb, V, Cr	DCL	BP 1,064,762

If the catalyst contains molybdenum but no bismuth it frequently contains tellurium, thus:

Catalyst	Patentee	Patent number
Mo, Te or Mo, P, Te	Asahi Chemicals	BP 948,014
Mo, Te, P, Mn	Goodrich	Belg.P 658,500
Mo, Te, P and a group IIA element	Goodrich	USP 3,417,125
Mo, Te, P, Cu and one of (Bi, Sb, Sn)	Goodrich	BP 1,045,112
Mo, Te, U, P	Goodrich	USP 3,412,135
Mo, Te, Cu, P	Goodrich	USP 3,417,128
Mo, Te, P and Bi or Sb	Goodrich	USP 3,426,060
Mo, Te, Fe, Cr, optionally P, B or V	Mitsubishi Rayon	Jap.P 7773/67
Mo, V, Th or Ce, Te or Bi	Sicedison	BP 1,054,895
Mo, Te or Bi and one of (Ti, Sr, Sn, Mn, Co, Ni, Cr, Al, Fe)	Sicedison	BP 1,055,307
Mo, Te, Ce	Sicedison	Ital.P 820,619
Mo, Te, Fe	Soc. Nat. des Pétroles D'Aquitaine	FP 1,563,988

However, catalysts containing molybdenum but no bismuth, tellurium or antimony have been claimed:

Catalyst	Patentee	Patent number
Mo, As, W and one or more of (Pb, alkali metal, elements of atomic number 22–30, Ag)	BASF	BP 1,136,744
Mo, Sn, B	BASF	BP 1,107,558
Mo, Cr, As, optionally one of (Mn, Fe, B)	Kodak	USP 3,293,280
Mo, P, As and one of (Mn, Cr, Fe)	Kodak	USP 3,287,394
Mo, P and a metal of the 1st period of periodic system	Knapsack	BP 947,365
Mo and/or V with Br	Nippon Carbide Ind.	Jap.P 18686/68
Mo, Cr	Ruhrchemie	GP 1,241,439
Mo, Fe	Toyo Rayon	Jap.P 23550/64
Mo, Co	DCL	BP 874,593

The list of propylene ammoxidation catalysts is nearly completed by a small group in which tellurium appears without either molybdenum or antimony.

Catalyst	Patentee	Patent number
Te, W	Asahi Chemicals	BP 909,907
Te, optionally P and an alkali metal	Asahi Chemicals	BP 937,380
Te, Ce	Montecatini	Dutch P 6,618,268
Te and Sn or Ti	Toyo Rayon	Jap.P 17803/67

It will now be evident, from the preceding lists of catalyst compositions, that there are three principal catalyst constituents, Mo, Sb and Te, a principal catalyst constituent being one that will produce a number of useful catalysts by its conjunction with other elements. In other words, according to our present knowledge, one at least of these principal constituents is essential for a catalyst performance of commercial significance. There are only a few patent claims that do not, apparently, conform to this statement, and most of these are so broad in their scope that they actually include a principal constituent amongst many other possibilities.

Catalyst	Patentee	Patent number
P and one or more of 18 metals	Bayer	Jap.P 14717/63
Ag, Cd alloy	Du Pont	USP 3,489,787
Cu, P	Esso	BP 930,773
Two or more of (U, As, Sb, Mo), optionally one or more of (Ce, Sb, W, Zn, Mn, Fe, Cr, Bi, Ni, Sn,) V, B	Sohio	Dutch P 6,818,481
Fe, P and one or more elements of groups V A and VI A	Shell	BP 943,374
Element of lanthanide or actinide series, W or Mo, one of (U, Sb, Bi, Sn, Cu, Te)	Sicedison	USP 3,226,421
V, one or more of (Bi, Sn, Pb, Fe) and one or more of (P, B, Mo, Se, Te)	Toyo Rayon	BP 1,025,676

11.3.1.3. The mechanism of the ammoxidation reaction Most of the publications containing information on the kinetics of the ammoxidation reaction are of Russian authorship. Dalin *et al.*[18] who experimented with a mixed oxide catalyst of unspecified composition (subsequent references indicate that it was bismuth molybdate), found that the rate of disappearance of propylene was proportional to propylene concentration and nearly independent of ammonia. Propylene conversion rate was also independent of oxygen concentration provided this did not fall below a certain level which, in the case of this catalyst system, corresponded to 1·5 mole oxygen per mole propylene in the reactor feed gas. The decrease in the rate of reaction of propylene caused by lack of oxygen, over catalysts in general, is associated with a reduction of the catalyst surface, which may rapidly become irreversible. In a second paper[19] Dalin and co-workers show the relationships between acrylonitrile and acrolein yields and the molar ratio $\frac{NH_3}{C_3H_6}$ in the feed, other conditions being constant (Fig. 11.1). According to these authors it is obvious from their results that acrolein is an intermediate in acrylonitrile production, even in a single stage process. The dependence of reaction product composition on contact time is given in this paper (Fig. 11.2), and it is deduced from these results that acrylonitrile, acetonitrile, hydrogen cyanide and carbon dioxide are formed in parallel reactions and are stable under reaction conditions. The stability of acrylonitrile was confirmed experimentally. Having also shown by experiment that formaldehyde, acetaldehyde and acetone will give rise, under ammoxidation conditions, to saturated nitriles such as are formed as by-products in propylene ammoxidation, Dalin proposes the following reaction scheme, in which, it must be pointed out, the cyclic peroxide intermediate is purely hypothetical:

$$CH_2 = CH—CN$$

$$CH_2 = CH—CHO$$

$$CH_2 = CH—CH_3$$

$$CH_3CH_2CN, CO_2, CO \leftarrow CH_3—CH_2—CHO$$

$$HCHO \rightarrow HCN, CO_2, CO$$

$$CH_2 — CH_2—CH_3$$

$$CH_3CN, HCN, CO_2, CO \leftarrow CH_3COCH_3 \quad O——O \quad CH_3CHO \rightarrow$$

$$CH_3CN, CO_2, CO$$

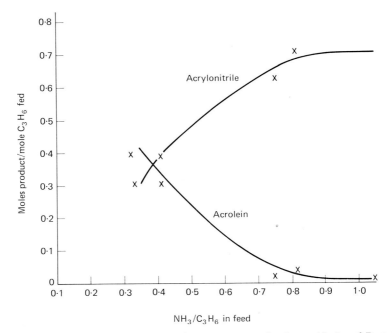

FIG. 11.1   *Effect of Ammonia Content of Reactants upon the Ammoxidation of Propylene According to Dalin* et al. *Azerb. Khim. Zhur. 1963 No. 4, 99–102*

To study the kinetics over a bismuth molybdate catalyst Gel'bshtein *et al.*[20, 21] used a 'circulating flow' reactor to which products were recycled. There was a complex interaction of products in the colder parts of the circulation system, which throws some doubt on the precision of the results. Nevertheless, the experimental findings of Dalin were generally confirmed. Gel'bshtein found that the reaction $C_3H_6 + 1\frac{1}{2}O_2 + NH_3 \rightarrow C_3H_3N + 3H_2O$ was not retarded by its products, but the reaction $C_3H_6 + O_2 \rightarrow C_3H_4O + H_2O$ was retarded by acrolein, and when the acrolein was frozen out of the cycle the rates of the two reactions (with similar propylene and oxygen concentrations) were almost identical, as were the energies of activation. In the following table the rate constants for three different batches of bismuth molybdate catalyst are given. The constants refer to 1 g of catalytically active mass, and k_1 and k_2 relate to ammoxidation and oxidation respectively.

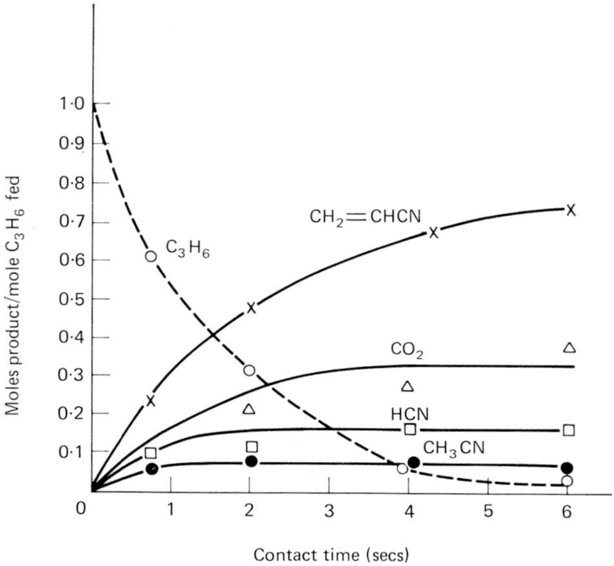

FIG. 11.2 *Dependence of Reaction Product Composition on Contact Time at 450°C.*
$C_3H_6:NH_3:O_2:H_2O = 1:1:1{\cdot}5\text{–}2{\cdot}5:1$
According to Dalin et al. Azerb. Khim. Zhur. 1963 No. 4, 99–102

TABLE XI.5

	$k, \frac{\text{P atm}^{-1}}{\text{g/hr}}$								
Catalyst	*410°*	*425°*	*430°*	*435°*	*450°*	*460°*	*475°*	*485°*	*500°*	*Temperature relation*
BM 1	1·04	—	1·52	—	2·20	—	—	—	—	$k_1 = 8{\cdot}0 \times 10^5 \exp.\left(-\frac{18500}{RT}\right)$
BM 2	—	—		1·50	—	2·33	3·00	3·50	—	$k_1 = 5{\cdot}8 \times 10^5 \exp.\left(-\frac{18000}{RT}\right)$
BM 3	—	1·30		1·53	2·10	—	2·72	—	4·03	$k_1 = 3{\cdot}8 \times 10^5 \exp.\left(-\frac{16000}{RT}\right)$
BM 3	—	—	1·45		1·92	—	2·81	—	—	$k_2 = 1{\cdot}0 \times 10^5 \exp.\left(-\frac{15500}{RT}\right)$

When acrolein was added to the feed to an ammoxidation reactor it was converted, nearly quantitatively, to acrylonitrile. In an analogous manner CH_3CHO was converted to CH_3CN and CH_2O to HCN.

Kinetic information, which is in accord with that of the Russian workers, has been published by Lichtenberger[22] and Blanc.[23]

Investigations of the mechanism of propylene oxidation by means of isotopically labelled propylene have been numerous. Ammoxidation has, in this respect, received much less attention. Adams and Jennings submitted 1-propene-3d to both oxidation and ammoxidation[24] and the results of their studies of the oxidation reaction have already been outlined in the chapter on acrolein. The distribution of deuterium in the acrylonitrile product of ammoxidation was consistent with the view that the same symmetrical allylic intermediate, $CH_2 = CH—\dot{C}H_2$ ($CH_2\text{------}CH\text{------}CH_2$), was involved in this reaction also. Assuming that the entry into the molecule of either oxygen or nitrogen would be the ultimate step in the reaction sequence, Adams and Jennings presented a reaction model consisting of series of hydrogen (deuterium) abstractions and including all possible routes to the corresponding acrylonitrile and acrolein precursors:

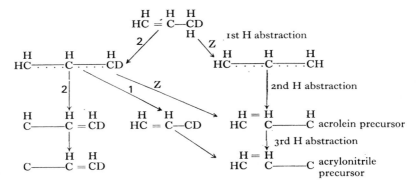

The numbers against the arrows are the relative probabilities of alternative reactions, and

$$Z = \frac{\text{probability of breaking a C—D bond}}{\text{probability of breaking a C—H bond}} = \frac{k_D}{k_H},$$

Z is called the isotope discrimination effect. Assuming (see chapter on acrolein for justification of the assumption) that Z has the same value in both of the first two dehydrogenation steps, then its relationship with d_1, the fraction of monodeuterated acrylonitrile in the total acrylonitrile product is:

$$Z = -2 \cdot 5 + \sqrt{0 \cdot 25 + \frac{4}{d_1}}$$

The value of d_1, corrected for the effects of isomerisation and exchange reactions, was about 0·46 for both copper oxide and bismuth molybdate catalysts, giving a calculated Z value of 0·49, in good agreement with that found in the plain oxidation experiments and with that reported by other authors who worked with different compounds and used different methods.

In a second paper[25] Adams and Jennings deal with the oxidation and ammoxidation of 1-propene-1d over the same two catalysts. The reaction scheme in this case becomes:

$$
\begin{array}{c}
\text{H} \qquad \text{H} \quad \text{H} \\
\text{HC}\!\!-\!\!\!-\!\!\!-\!\!\text{C} = \text{CD} \\
\text{H} \\
\quad\Big\downarrow \text{ 1st H abstraction} \\
\text{H} \qquad \text{H} \quad \text{H} \\
\text{HC}\cdots\text{C}\cdots\text{CD} \\
\end{array}
$$

H H H
HC - - - C - - - CD

  2/ |1   \Z   2nd H abstraction

H H H   H H   H H H
C—C = CD   HC = C———CD   HC = C—C   acrolein precursor

  | / 3rd H abstraction

H H   H H   H H
C—C = CD   HC = C———CD   HC = C—C   acrylonitrile precursor

and for acrylonitrile formation $Z = \dfrac{2}{d_1} - 3$.

The values of the Z factors obtained by calculation from the deuterium distributions in the products of the various systems, i.e. oxidation and ammoxidation of 1-propene-3d and 1-propene-1d over copper oxide and bismuth molybdate, were all in good agreement, but only after considerable corrections for isomerisation and exchange reactions had been made. These corrections were justified by separate experiments which showed that olefins isomerise over bismuth molybdate in the presence of oxygen but not in the presence of both oxygen and ammonia. On the other hand isomerisation does not occur over cuprous oxide during normal oxidation, but it does occur during ammoxidation and the reason probably is that the temperature has to be raised for ammoxidation in order to overcome the inhibitory effect of ammonia on the oxidation (the first step in the ammoxidation reaction) over this catalyst.

The data from the experiments with 1-propene-1d and 1-propene-3d were used to justify the assumption that hydrogen abstraction from the allyl radical precedes the addition of oxygen or nitrogen. For the oxygen or nitrogen would be expected to attack both ends of the molecule equally, and independently of the deuterium content, and if this attack occurred before the hydrogen abstractions the proportion of deuterated material in the product would be different from that actually observed.

Although Adams and Jennings were at pains to point out that their work threw no light on the way in which oxygen or nitrogen entered the molecule their reaction schemes might be taken to imply that acrolein and acrylonitrile, when produced together, as they are in the ammoxidation reaction, come from the same intermediate but by parallel routes, which would exclude acrolein as a precursor of acrylonitrile:

$$
CH_2 = CH - CH_3 \longrightarrow CH_2 = CH - CH_2 \longrightarrow CH_2 = CH - CH
\begin{array}{l}
\nearrow \quad CH_2 = CH - CHO \\
\searrow \quad CH_2 = CH - C \\
\qquad\qquad\quad \Big\downarrow \\
\qquad\quad CH_2 = CH - CN
\end{array}
$$

However, the distribution of deuterium in the acrylonitrile derived from the monodeuterated propylene would be the same whether the third hydrogen abstraction occurred from the molecular fragment before the addition of oxygen or from acrolein itself, since no alternative reactions are involved in either case. So that the work of Adams and Jennings neither confirms nor disproves that acrolein is a precursor of acrylonitrile, although it shows beyond reasonable doubt, that a symmetrical intermediate is common to the oxidation and the ammoxidation of propylene.

Publications reveal a conflict of views on the subject of acrolein intermediacy. It is important to realise that if acrolein were a precursor of acrylonitrile that did not leave the catalyst surface it would be difficult if not impossible, with present techniques, even to establish its existence in the reaction. But, in fact, it does appear in the gas phase, and indeed advocates of the acrolein intermediate theory consider that it is in or close to adsorption equilibrium. That is to say that most of the acrolein leaves the surface and has to be re-adsorbed before undergoing further reaction. The skeletonised picture of the reaction sequence accordingly becomes:

$$
CH_2 = CH-CH_3 \xrightarrow{-H} CH_2 = CH-\overset{\cdot}{C}H_2 \xrightarrow{-H} CH_2 = CH-CH \xrightarrow{O_2}
\begin{array}{c}
CH_2 = CHCHO \text{ (gas} \\
\text{phase)} \\
\updownarrow\ \uparrow \\
CH_2 = CH-CHO \\
\downarrow \\
NH_3,O_2 \\
\downarrow \quad \text{(Surface)} \\
CH_2 = CH-CN
\end{array}
$$

Still more abbreviated (for the purpose of discussion) it is:

$$
CH_2 = CH-CH_3 \xrightarrow{(1)} CH_2 = CH-CHO \xrightarrow{(2)} CH_2 = CH-CN
$$

Experimental facts in accord with the acrolein intermediate theory are as follows. Any catalyst suitable for the ammoxidation of propylene will catalyse each of the reactions 1 and 2 separately, and substances that will catalyse reaction 1 but not 2 (such as Cu_2O) or which will catalyse 2 but not 1 (such as MoO_3) are not effective in ammoxidation. If propylene and oxygen are fed at a constant rate to an ammoxidation catalyst while the feed of ammonia is varied from zero, both the conversion of propylene and the sum of the acrolein and acrylonitrile outputs remain substantially constant.[19, 20, 21] Reaction 1, considered separately, is first order with respect to propylene (except in the case of Cu_2O catalyst). Reaction 2 is first order with respect to both acrolein and ammonia not only over molybdic oxide[17] but also over all the effective ammoxidation catalysts.[26] The last important experimental fact is that acrolein added to the feed to an ammoxidation reactor leads to the formation of additional acrylonitrile.

However, another view of acrylonitrile formation has been put forward,[27] according to which there is an alternative reaction path.

$$
CH_2 = CH-CH_3 \xrightarrow{1} CH_2 = CH-CHO \xrightarrow{2} CH_2 = CH-CN
$$
$$
\underset{3}{\underbrace{\qquad\qquad\qquad\qquad}}
$$

Reaction 1 is considered to be largely suppressed under ammoxidation conditions (i.e. when there is a sufficiency of ammonia), while reaction 2 is slow compared with reaction 3.

In order to test this hypothesis Callahan et al.[27] disregard the effect of ammonia on reaction rates, putting rate of formation of acrylonitrile via acrolein $= r_2 = k_2$ [acrolein], and total rate of formation of acrylonitrile $= r = k$ [C_3H_6], where the k's are velocity constants. From their experiments in fluidised bed reactors they find that at $425°C$, over a bismuth molybdate catalyst, k_2 (measured by injection of iso-topically labelled acrolein in the reactor) is 0.38 sec^{-1}, and k is 0.21 sec^{-1}. Now, if acrolein is the main intermediate k_2 must obviously be much greater, in fact fifty to a hundred times greater, then k because [acrolein] is relatively small. But it is doubtful that a fluidised bed reactor is suited to the appreciation of a velocity con-stant of the order of 10 sec^{-1} for this would require the establishment of contact times of only a fraction of a second. These results are not, therefore, universally considered as a demonstration of the incorrectness of the acrolein intermediate theory.

As a result of studying the product spectrum in relation to reaction temperature Seeboth et al.[28] conclude that acrylonitrile results from the reaction of allyl radicals with NH or NH_2 fragments.

11.3.1.4. Catalyst studies In the chapter on acrolein, publications that were primarily concerned with relationships between the bulk properties of catalysts and their performance in the oxidation of propylene were reviewed. All these are relevant to the present subject. Studies of catalyst structure in relation to ammoxida-tion in particular are less numerous and are, mainly, confined to bismuth molybdate.

Kochin et al.[29] detected the following compounds in the Bi_2O_3/MoO_3 system, and determined their lattice structures:

α-phase $Bi_2O_3 . 3MoO_3$
β-phase $Bi_2O_3 . 2MoO_3$
γ-phase $Bi_2O_3 . MoO_3$

No solid solutions were found. Of the three compounds the β-phase (tetragonal lattice) had the greatest activity and selectivity in both the oxidation and ammoxida-tion of propylene, but the differences were more pronounced in oxidation. However, the activation energies for both acrolein and acrylonitrile production (19–21 kcals/ mole) were similar for all three phases and it is suggested that the differences in activity are attributable to structural and electronic factors. Kolchin et al.[30] studied the effect of phosphoric acid addition to the three bismuth molybdate phases previ-ously identified. They found that the kinetics of the oxidation and ammoxidation reactions were not altered. Much β-phase resulted from the addition of H_3PO_4 (equivalent to 0.2 per cent P) to a catalyst prepared under the previous conditions for α-phase production, but with 3.4 per cent and more of P the β-phase disappeared and a new phase, $2Bi_2O_3 . 3MoO_3 . P_2O_5$, was formed that was stable at high tem-peratures. P_2O_5 also entered the lattice of the γ-phase. With increasing P addition the activity (specific rate constant) of the initial β-phase steadily decreased, that of the γ-phase steadily increased, while that of the α-phase rose to a maximum at

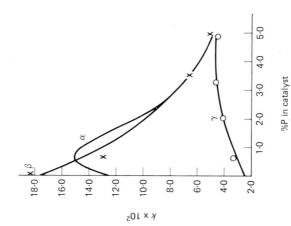

%P in catalyst

FIG. 11.4   *Acrylonitrile Production*

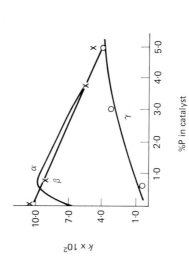

%P in catalyst

FIG. 11.3   *Acrolein Production*

Specific Rate Constants as Function of Phosphorus in Catalyst

$\alpha = Bi_2O_3 \cdot 3MoO_3$
$\beta = Bi_2O_3 \cdot 2MoO_3$
$\gamma = Bi_2O_3 \cdot MoO_3$

According to Kolchin et al., Kin. i. Kat., 1965, 6 No. 5, 878–883 which it approximately equalled the β-phase, and at 5 per cent P the activities of all three coincided and there was structural identity. The selectivity of the β-phase was scarcely affected by P but the other phases were improved. The effects of P on the catalyst performance in oxidation and ammoxidation were strikingly similar Figs. 11.3 and 11.4).

Gel'bshtein et al.[31] observed that MoO_3 had slight activity in oxidation and ammoxidation, while Bi_2O_3 had none, and agreed that the optimum catalyst composition corresponded to the β-phase of Kolchin, but they found that the Bi_2O_3/MoO_3 system was more complicated than is indicated in Ref. 29. The X-ray patterns of catalysts of optimum composition were not changed by the addition of 0·5 per cent wt/wt P_2O_5, but the activity and efficiency were reduced, while the stability of the catalyst, in prolonged use, was improved. The same authors in Refs. 32 and 33 consider that there are two kinds of centre in the catalyst for the chemisorption of propylene. On the first, shown by Z_1, there is dissociative adsorption leading to the allyl radical, and on the second, Z_2, there is adsorption with double bond fracture. Accordingly, the following reaction scheme is presented:

$$
\begin{array}{c}
\quad\quad\quad HCN \quad\quad\quad\quad CH_3CN \\
\quad\quad\quad \uparrow \quad\quad\quad\quad\quad \uparrow \\
Z_2(CH_2=CH-CH_3) \rightarrow Z_2(CH_2-CH-CH_3) \rightarrow HCHO \quad + \quad CH_3CHO \rightleftharpoons CH_3CHO \\
\uparrow\downarrow \\
C_3H_6 \quad\quad\quad\quad\quad\quad\quad\quad\quad\quad\quad\quad\quad\quad\quad\quad \rightarrow CO_2 \leftarrow \\
\downarrow\uparrow \\
Z_1(CH_2=CH-CH_3) \rightarrow Z_1(CH_2\cdots CH\cdots CH_2) \rightarrow Z_1\,(CH_2=CH-CHO) \rightleftharpoons CH_2=CH\cdot CHO \\
\downarrow \\
CH_2=CH-CN
\end{array}
$$

Since reaction efficiency depends on the preservation of the double bond, which in turn depends on the bond energy between catalyst and olefin, Gel'bshtein sought a relationship between electron energy yield (Q) and catalyst performance, both considered as functions of catalyst composition.[32] A single minimum value of Q, at $\dfrac{Bi}{Mo} = 4\cdot6$, is shown and there is no evident relationship between Q and the rate constant for ammoxidation or for the side reactions. This result differs markedly from that of Margolis et al.[34] who reported a maximum electron energy yield for $\dfrac{Bi}{Mo} = 1$. Gel'bshtein showed that the bismuth molybdate composition with the lowest electrical conductance had the maximum catalytic activity, and suggested that the required activity is attributable to the formation of mixed metal oxides with only slight semi-conducting properties. In contrast, Ioffe[35] has shown that selective oxidation of the double bond of unsaturated compounds is brought about by single component catalysts with distinct semi-conducting properties and large electron energy yields.

Studies of an antimony/uranium oxides system are reported by Callahan and Grasselli.[36] These catalysts were prepared by heating together, at about 930°C, Sb_2O_3 and a uranium oxide obtained by the decomposition of uranyl nitrate. X-ray diffraction data showed in addition to Sb_2O_4 and U_3O_8, two new phase which were identified, by means of infra-red absorption methods, as uranyl antimonate ($UO_2.Sb_3O_7$) and antimony uranate ($Sb_3U_3O_{14}$). Ammoxidation activity is particularly associated with the first of these since both catalytic activity and uranyl antimonate concentrations are at their maximum when the $\dfrac{Sb}{U}$ ratio is around 3.

11.4. MANUFACTURE OF ACRYLONITRILE BY AMMOXIDATION

Spread over the world there are now more than thirty plants producing acrylonitrile, more than 90 per cent of which is made by the ammoxidation method. Annual outputs from individual companies vary from over 100,000 tons (Monsanto, Ugine Kuhlmann, Erdölchemie) to 10,000 ton or less.

There are now more than twenty plants operating the process developed by Sohio and associates, including seven in the USA and at least seven in Japan. A process based on DCL research and developed by Ugilor of France and DCL jointly is utilised by two plants in France and one in the United Kingdom. A fourth plant based on the DCL/Ugilor process is being commissioned in Mexico. In 1967 the chemical interests of the DCL were acquired by BP Chemicals Ltd., and the UK plant is a joint venture of BP Chemicals Ltd. (now BP Chemicals International Ltd.) and ICI. Several other companies: SNAM and Montecatini (Italy), OSW (Austria), DDR (Sweden), Ube (Japan), UCB (Belgium) have developed their own processes, and have built, or announced their intention of building plants accordingly.

11.4.1. Catalyst and Reactors

The process is highly exothermic. By calculation $\triangle H°(g)$ for the reaction:

$$CH_2 = CHCH_3,(g) + NH_3(g) + 1\tfrac{1}{2}O_2(g) \rightarrow CH_2 = CHCN(g) + 3H_2O(g)$$

lies between -122 and -123 kcal/mole from ambient up to practical reaction temperatures. $\triangle H°(g)$ for the principal side reaction, the combustion of propylene, is about -460 kcal/mole.

It is believed that no recycle system is operated, and the yield per pass therefore is a most important factor. The yield is largely dependent on the catalyst, but information on the nature and performance of the catalysts used industrially is rather scanty and of variable reliability. Patent examples, usually derived from small scale experiments, must be considered as rough guides only to catalyst performance in industrial reactors.

Numerous references confirm that the original Sohio catalyst was bismuth phosphomolybdate with a minor proportion (*ca.* 30 per cent) of silica, but this has, to some extent, been displaced by antimony/uranium, the antimony being the major constituent.[36] Yield on propylene from the Bi/Mo catalyst is, according to reference 37, 50–55 per cent, and according to references 38 and 39 is 58 per cent. Grasselli and Callahan[36] reported > 80 per cent yield from the Sb/U catalyst in laboratory reactors: commercially this catalyst appears to give 63 per cent yield.[40] All Sohio type plants use fluidised bed reactors.

According to reference 38 the yield in the BP Chemicals/Ugine Kohlmann process, with fixed bed reactors and a catalyst containing antimony as the principal constituent, is 55–58 per cent.

OSW found that heat dispersal from a fixed bed of the usual Bi/Mo catalyst was difficult, and obtained better results by coating a layer of this material on granules of an inert support such as Alundum.[41] In a test reactor the yield was over 60 per cent.

The SNAM catalyst appears to be an improved version of the first Sohio catalyst; it consists of Bi, Mo, V. Pilot plant yields of 69 per cent in a fixed bed reactor are reported.[42]

Montecatini Edison use a tellurium containing catalyst in a fluidised bed reactor. In reference 43 the composition Te,Mo,Ce supported on silica is mentioned. No tellurium volatilises and the yield (pilot plant) was 70–75 per cent.

A 2 tons/month pilot plant with a catalyst having antimony as the main ingredient, in a fixed bed, has been operated by Ube Industries Ltd. The proportion of by-products and the consumption of feed materials were 'small'.[44]

Pilot plant fixed bed reactors have also been operated by UCB[45] and the Soc. Nat. des Pétroles D'Aquitaine,[46] both of whom mention yields of about 60 per cent. The UCB catalyst was Sb/Fe.

According to reference 45 fluidised beds are more suitable for larger plants but the yields from fixed beds are generally greater. Whereas in reference 46 it is stated that a higher selectivity was obtained from a fluidised bed because of better temperature control. Sohio have described a fluidised bed divided into a number of compartments[4–12] by horizontal, perforated plates, which are intended to reduce back mixing and increase the plug flow character of the reactor.[47] The same company has also claimed a reactor with at least five connected compartments, oxygen being fed in the first, and propylene and ammonia being fed subsequently.[48]

Some carry-over of the relatively finely divided catalyst that is used in a fluidised bed into the recovery system is inevitable, whereas mechanical loss of catalyst from a fixed bed is negligible, and if the catalyst does not deteriorate, because of reduction or other chemical change, its useful life will be very long.[22]

Fixed bed reactors are of the usual shell and tube construction, with circulated molten salt as the heat transfer medium. Tubes of about 1 in. diameter and at least 10 ft. in length seem to be generally preferred, and there may be upwards of 10,000 in a single shell. To control the rate of reaction and so to prevent undue temperature rise in the catalyst OSW deposit the active material on an inert support. DCL have described the dilution of the catalyst granules with granules of inert material.[49] Heat is removed from fluidised catalyst beds, which because of solid and gas mixing tend to be at a uniform temperature, by the insertion of cooling coils or other heat disposal devices. In reference 50 a fluidised bed (with communicating compartments) of high surface area catalyst is used, which enables unreacted ammonia to be recycled. The catalyst in the last compartment, being maintained at a lower temperature than that in the foregoing compartments, absorbs ammonia from the gas stream about to leave the reactor and is continuously returned to the first compartment.

A typical contact time in a fixed bed reactor is 2–4 secs, but in a fluidised bed it may be considerably greater (25 secs is mentioned in reference 37), partly for mechanical reasons and partly to provide a reserve of activity to compensate for reduction or other catalyst deterioration. Temperature within the fluidised bed depends on the catalyst: the range 750–925°F is indicated for the Bi/Mo catalyst.[39]

11.4.2. Reactor Feed

Several writers[22, 45] have quoted the following range of composition (per cent molar) for the reactor feed: propylene 5–8, ammonia 5–10, steam 10–30, remainder air. Propylene derived from a cracker unit, will usually contain about 90 per cent C_3H_6. Of the impurities, paraffins for the most part pass through the reactor unchanged, while the unsaturated compounds, other than ethylene which is not very reactive in the ammoxidation environment, are mainly converted to substances that are already produced from propylene itself as by-products, i.e. carbon oxides and the lower aldehydes and nitriles. Large amounts of butenes in the feed are undesirable since the C_4 nitriles to which they give rise increase the load on the purification system.

11.4.2.1. Ammonia The typical relationship between yield of acrylonitrile, related to propylene, and the $\dfrac{NH_3}{C_3H_6}$ ratio in the feed, which is shown in Fig. 11.1, is readily understood in terms of the acrolein intermediate theory. For simplicity's sake assume that ammoxidation is taking place in a fully stirred reactor and is completely described by:

$$C_3H_6 \xrightarrow{k_1} C_3H_4O \xrightarrow{k_2} C_3H_3N$$

The ammonia concentration in the feed is varied while all other conditions are fixed. Let A, B, C and N be the concentrations of propylene, acrolein, acrylonitrile and ammonia respectively in the gas within and leaving the reactor, t = contact time, then

$$B+C = k_1 t A \text{ and } C = k_2 t BN$$

$$\text{and } C = \frac{A k_1 k_2 t^2 N}{1+k_2 tN} \quad \text{giving } \frac{C}{B} = k_2 tN$$

As N increases so does $\dfrac{C}{B}$ and C approaches the value $A k_1 t$ asymptotically. If, therefore, C and B are plotted against N the curves obtained will obviously be similar to those of Fig. 11.1. And N is a linear function of r, the $\dfrac{NH_3}{C_3H_6}$ feed ratio:

$N = rA^* - (B+C)$, where A^* = concentration of C_3H_6 in the feed. The precise shape of the curve relating C to r will, of course, depend on the values of k_1 and k_2 and the efficiency of the reaction, possibly also on the oxygen feed concentration, since at the end of a plug flow reactor, or throughout a fully stirred reactor, the oxygen may fall below the level at which the reaction rate is independent of oxygen concentration. For all industrial catalysts the maximum value to which r may be usefully increased, in the sense that further increase will give only a small increment (<1 per cent) in yield, lies within the range of 0·9–1·2 approximately.

11.4.2.2. Steam Steam, an optional component, is firstly a useful diluent with which to adjust the feed composition in order to avoid the explosive region. The use of additional steam during the start-up of a reactor to prevent a dangerous condition arising, even temporarily, before a steady state is established is described in references 22 and 51. The more subtle function of steam is to improve the performance (of some catalysts) by increasing the rate of reoxidation and/or by retarding the combustion of ammonia. Quantitatively, these effects vary greatly from catalyst to catalyst some of which would be quite impracticable in the absence of steam because of the rapid destruction of ammonia. One of the earlier catalysts, MoO_3/alumina[52] gave a maximum yield of only 3 per cent acrylonitrile with a propylene feed concentration of up to 15 per cent vol/vol, when there was no steam in the feed. The addition of steam increased the yield threefold. By contrast Sb/Sn catalyst does not appreciably promote the oxidation of ammonia even in the absence of steam. If there is considerable back mixing in the reactor, as there may be in a fluidised bed, the water produced by the desired reaction may be sufficient to serve as a moderator.

11.4.2.3. Oxygen The feed composition is thus a compromise between several requirements, and the amount of oxygen (usually introduced as air) may be less than would be employed if reaction rate were the only consideration. Below a certain limit, however, oxygen deficiency may lead to a disastrous irreversible reduction of the catalyst. Monitoring of the oxygen in the reactor exit gas with regulation of the amount introduced is described by Sohio.[53]

11.4.3. Recovery and Purification

Veatch *et al.*[39] disclose the following plant yields from a bismuth molybdate catalyst (Table XI.6).

TABLE XI.6

	% Yield on C_3H_6 feed	Moles/mole acrylonitrile
Acrylonitrile	58	1
Acetonitrile	7·5	0·20
HCN	7	0·34

TABLE XI.7

	% Yield on C_3H_6 feed
Acrylonitrile	50–60
Acetonitrile	1–2
HCN	5–7
Acrolein	1–7
$CO+CO_2$	16–20
Other by-products	1–2
Unreacted propylene	15–18

Veatch does not mention acrolein, but numerous patents relating to its elimination from acrylonitrile show that it causes general concern. The yield of acrolein usually lies within 0·5–2 per cent. Lichtenberger[22] quotes the following yields, but without relating them to any particular catalyst. (Table XI.7.)

Product compositions from various catalysts containing tellurium or selenium are indicated in reference 54, including the following from an Fe, Si, Li, Se catalyst. (Table XI.8.)

TABLE XI.8

	% Yield on C_3H_6 consumed
Acrylonitrile	74·3
Acetonitrile	1·4
Acrolein	1·2
CO_2	23·1

It is probably true to say that whatever catalyst is used there are more than fifty by-products, although most of them occur in traces only. Among these, nitriles figure largely, but there are also hydrocarbons, acids, alcohols, esters, carbonyl compounds and others. Conspicuous by their absence, however, are those substances, vinylacetylene and unsaturated ketones, which were the most troublesome by-products in the manufacture of acrylonitrile from acetylene.

11.4.3.1. Removal of unreacted ammonia   In most recovery systems the first step is the elimination of ammonia from the reactor exit gas, because, in the condensed crude product, various and complex polymerisation, addition, and condensation reactions are liable to occur, which would be catalysed by hydroxyl ions. The polymerisation of acrolein, the addition of HCN to acrolein and acrylonitrile, and the polymerisation of HCN are examples of parasitic reactions that could proceed rapidly in the presence of ammonia. It is common practice, therefore, to quench the partly cooled (*ca.* 200°C) reactor exit gas in dilute sulphuric acid which is cycled through a tower fitted with plates or trays. The temperature in the tower may be sufficiently high, about 90°C, to ensure negligible absorption of acrylonitrile. Sulphuric acid (0·5–2 per cent wt/wt) and ammonium sulphate (20–40 per cent wt/wt) concentrations are maintained by fresh water and sulphuric acid additions and a bleed of liquor from the column.[55] The water supply may be in the form of steam in the entering gas. In a patent relating to the acid wash tower[56] Erdölchemie mention that insoluble catalyst particles are trapped in the wash liquid.

This method largely, but not entirely, prevents polymerisation in the liquid phase, as is indicated in reference 57, where the removal of a tarry product from the surface of the ammonium sulphate liquor is claimed. There has been no disclosure of the composition of the polymers that may be formed in the acid wash column. In relation to the acrylonitrile product they are small in amount, but they can have a disproportionate effect on operational efficiency and on the quality of the ammonium sulphate which is crystallised from the liquor to be sold as fertiliser. Other patents relating to the recovery of ammonium sulphate are references 58, 59, 60 and 61. Sohio have patented the concentration of the ammonium sulphate along with the polymeric by-products followed by incineration.[62] Alternatives to sulphuric acid have been suggested, including acetic and citric acids,[63] CO_2,[64] and formic acid,[65] which have the advantage that free ammonia is more readily recoverable from their ammonium salts than from ammonium sulphate, but it is doubtful if any but the last of these is sufficiently strong for the primary purpose of retarding polymerisation and addition reactions.

According to Erdölchemie[66] polymer deposition in the acid wash column is avoided by a preliminary scrubbing of the reactor exit gas with water containing no acid, in which polymer and other high boiling materials, as well as catalyst fines collect, as a solution or slurry. In an example a reactor exit gas, at 230°C, containing (per cent vol/vol); ammonia 1, oxygen 2, acrylonitrile 5·2, HCN 1·2, steam 30, 'as well as organic polymers' enters a jet washer in which water is circulated at 70–77°C. It is claimed that only about 0·3 per cent of the total acrylonitrile and HCN is lost in the water washer.

Sohio[67] describe a system in which ammonia and Sb/U catalyst fines are selectively absorbed from the reactor gas stream in a molten sulphur salt or acid melt, preferably in a mixture of sodium and potassium hydrogen sulphates. Both ammonia and the elements of the catalyst are recoverable from the melt. Other recovery methods utilising non-aqueous absorption media have been described,[68] according to which a gaseous mixture containing acrylonitrile and ammonia is scrubbed with diphenyl-diphenyl oxide mixture (preferably) or toluene, xylene, etc., ammonia then being expelled from the wash liquor by the passage of inert gas and the acrylonitrile then being distilled out. No such method has yet found commercial application.

11.4.3.2. The absorber   The gas leaving the acid wash column is scrubbed with cold water in a column known as the absorber. Scrubbed gas from the top of the absorber, consisting principally of nitrogen together with carbon oxides, propylene and oxygen is normally vented to atmosphere, provided that the content of acrolein and hydrogen cyanide is sufficiently small. In the liquid effluent there is substantially all of the normally liquid products, including acrylonitrile at a concentration of 3–5 per cent wt/wt.

11.4.3.3. Purification   There are two principal schemes for the production of pure acrylonitrile, starting from the absorber liquor.

Scheme 1   The essentials of this scheme are outlined in reference 41 and in various patents.

Absorber liquor is passed to a stripping tower where it undergoes fractional distillation, with decantation of the condensed, two-phase distillate, the aqueous phase being returned to the tower. Most of the water leaving the stripper base, which contains only traces of organic matter, is recycled to the absorber, but the excess represented by the water brought into the stripper/absorber cycle by the gas entering it, is bled off. It is difficult if not impossible to avoid traces of nitrogen compounds in this kettle effluent, which may, therefore, have to receive decontamination treatment before it is passed to waste. The overhead condensate from the stripper may contain 70–80 per cent wt/wt of acrylonitrile which is still allied with the bulk of the liquid by-products. Lichtenberger[22] gives the following typical composition range of crude acrylonitrile, in per cent wt/wt:

Acrylonitrile	70–75
HCN	12–15
Acrolein	2–4
Acetonitrile	1–2·5
Higher nitriles	0·05–0·2
Acetaldehyde	0·1–0·3
Water	5–7

In the next operation acrolein (along with the other aldehydes, of which the chief is acetaldehyde) is largely eliminated from the crude acrylonitrile by combining it with HCN, of which a large excess over the stoichiometric requirement is usually present. The cyanhydrin produced has small volatility and so is easily separated from the remainder of the mixture. The method of operation follows from the chemistry of the cyanhydrin formation:

$$CH_2 = CH—CHO + HCN \underset{\longleftarrow}{\overset{\longrightarrow}{}} CH_2 = CH—CH(OH)CN$$

The equilibrium constant of this reaction (exothermic from left to right) $k = \dfrac{[\text{cyanhydrin}]}{[\text{acrolein}][\text{HCN}]}$ has a value of over 300 litre mole^{-1} at 25°C,[69] but the rate of formation and dissociation of the cyanhydrin is very dependent on OH ion concentration, and if the pH of the crude acrylonitrile (i.e. the pH of an aqueous layer in equilibrium with it) is less than about 4 the reaction rate, in either direction, is infinitely slow at room temperature, whatever the displacement from equilibrium.[70] But with increasing pH the approach to equilibrium can be made very fast. Treatment of the crude acrylonitrile with a suitable amount of alkali will, therefore, ensure that the acrolein (and other aldehyde), in the presence of excess HCN, is substantially converted to cyanhydrin. The pH of the liquid is now reduced by the addition of (preferably) a strong organic acid,[70, 71] and the liquid is subjected to distillation at low temperature and pressure to separate the volatile material (mainly acrylonitrile, HCN and water) from the residual cyanhydrin, with little loss of nitrile at the column base. This is the first column handling concentrated acrylonitrile and it is much smaller than the preceding columns which handled twenty or more times as much liquid.

Acrolein cyanhydrin may be returned to the ammoxidation reaction where it will be converted to acrylonitrile and HCN.[72]

A simplified version of the acrolein elimination step is feasible, according to which cyanhydrin is formed in the aqueous solution of crude acrylonitrile leaving the absorber. Upon distillation of the solution the cyanhydrin remains with the water in the kettle effluent,[69, 73] but the separation of acrolein is less complete in this method, if it depends solely on cyanhydrin formation, because in dilute solution more dissociation of the latter occurs at equilibrium and more acrolein passes overhead during the distillation.

Specific reagents, e.g. polyhydric alcohols, hydrazine, nitrogen oxide, activated alumina, have been claimed for the elimination of carbonyl compounds in crude acrylonitrile,[74] but it is doubtful if any of these is used commercially.

The product from the cyanhydrin separation column, consisting of acrylonitrile, hydrogen cyanide and acetonitrile with traces of acrolein and other by-products, passes to the next distillation column, where HCN is taken off overhead in a very pure form. This is a straightforward operation that has been the subject of little patenting activity. At the top section of the column, where concentrated HCN is present, the usual acidic polymerisation inhibitor is introduced.

The next column is used for the separation of acetonitrile. This cannot be done practically by simple fractionation since the two nitriles have similar volatilities. Good separation is achievable by various methods such as double extraction of the nitrile mixture (at room temperature) with countercurrent streams of water and a water-immiscible solvent,[75] extractive distillation with a polyhydroxy compound,[76] or with high boiling carbonyl compounds and water.[77] But none of these can compete on economic grounds with the extractive distillation with water that is everywhere employed. This operation consists in feeding the mixture of nitriles to an intermediate point in a distillation column to the top of which a stream of water is supplied, which has the effect of increasing the volatility ratio of the nitriles. With a sufficient number of plates (Du Pont[78] suggest fifty), and suitable temperature control, and with an adequate amount of extractive water (DCL[79] indicate $\lessdot 4 \cdot 5$ volumes water per volume of nitrile) most of the acrylonitrile is obtained overhead substantially free of acetonitrile in a distillate which separates into two phases, the lower, aqueous, phase being returned as reflux. The water effluent from the column contains most of the acetonitrile, a small fraction of the acrylonitrile, and most of the minor reactor products such as the higher nitriles. The first patent on this subject came from American Celanese,[80] who said that the concentration of water in the liquid phase in the column should be 93–98 per cent. Ugine[81] says that with eleven plates in the column 7·3 kg water per kg of nitriles is preferably used, but with forty-two plates the same degree of separation is achieved with only 2·8 kg water. From an introduced mixture of the following composition,

Acrylonitrile	91·84 per cent
Water	4·1 ,, ,,
Acetonitrile	3·8 ,, ,,
Propionitrile	0·18 ,, ,,
Allyl alcohol	0·06 ,, ,,
Methanol	0·02 ,, ,,

there was obtained acrylonitrile containing,

Acetonitrile	0·024 per cent
Propionitrile	< 0·001 ,, ,,
Allyl alcohol	< 0·001 ,, ,,
Methanol	< 0·0005 ,, ,,

The kettle effluent contained 0·8 per cent acetonitrile and practically all the organic impurities.

In reference 82 the use of water extractant at pH 5–8, resulting in a base residue of pH 9–11, ensures that small quantities of HCN fed to the hydroextractive column are destroyed. Sohio[83] remove trace compounds having nitrate or peroxide characteristics from the azeotropic distillate by the addition of sodium carbonate. According to reference 84 minor impurities such as propionitrile and butyronitriles may be withdrawn as a side stream from an intermediate plate of the column, so that the acetonitrile solution at the base is practically free of them.

Acetonitrile is distilled out of the base effluent from the hydroextractive column in the acetonitrile stripping column. It may be burned,[85] but there are various patents relating to its purification.[86]

Water and residual impurities, which may include acrolein and HCN, are reduced to the necessary level in the acrylonitrile leaving the hydroextractive column by a final fractionation step, usually a topping and tailing operation.

Scheme 2 Various publications indicate that Ugine, Border, S N A M and O S W at least follow scheme 1. Vekemans *et al.*[45] briefly describe another procedure, which apparently,[39] is used by Sohio. According to this the dilute aqueous acrylonitrile leaving the absorber passes directly to a hydroextractive column to the top of which extra water is fed. Disregarding the HCN the products are similar to those from the hydroextractive column of scheme 1. In scheme 2, therefore, one column (which must be of increased dimensions), does the duty of both the stripping and the hydroextractive columns of scheme 1. Also in this column, it appears, acrolein is destroyed, although the mechanism of this destruction is not clear.[87]

The subsequent stages of scheme 2, removal of HCN and final fractionation of acrylonitrile, are similar to those of scheme 1, but there may be a modification to deal with residual acrolein, although according to Societa Edison[87] this disappears spontaneously.

11.4.4. Product Quality

Because the product is of high purity there is no adequate method of assay. According to impurity estimations the $CH_2 = CHCN$ content is over 99·9 per cent, disregarding water and added inhibitors. Impurities may be placed in two classes, the first comprising unsaturated compounds such as acrolein, which could affect polymerisation, and which are kept at a very low level. Belonging to the other class are such compounds as acetone and acetonitrile, which appear to be mere diluents; the objection to these is that they may build up in a polymerisation system involving only partial polymerisation and recycle of monomer. Accordingly they are limited to a few hundred p.p.m. Lichtenberger[22] gives the following composition per cent wt/wt for pure acrylonitrile, 'on a dry basis':

Acrylonitrile	99·98 minimum	
Hydrogen cyanide	0·0005 maximum	
Acrolein	0·005	,,
Acetonitrile	0·02	,,
Higher nitriles	0·005	,,
Acetaldehyde	0	

In 1966 it was found that acrylonitrile made by ammoxidation of propylene and subjected to the usual purification procedures was liable to give rise to coloured fibres. The impurity responsible for the colour was identified as oxazole

$$
\begin{array}{ccc}
HC & = & CH \\
| & & | \\
N & & O \\
& \diagdown \diagup & \\
& CH &
\end{array}
$$ [79, 88]

and found to be peculiar to the ammoxidation route,[89] the yield on propylene being < 0·1 per cent. The volatility of oxazole is such that it cannot be readily eliminated by fractionation. At least the greater part of it may be removed along with the acetonitrile in the hydroextractive distillation step,[79] but reduction of the oxazole level below about 20 p.p.m. seems to be best ensured by a final extraction with an acid. Cation exchange resins are particularly suitable for this purpose.[90]

11.4.5. Inhibition of Polymerisation

Although acrylonitrile polymerises less readily than acrylic acid, inhibitors are usually added during distillation procedures and to the stored material. Derivatives of hydroquinone are most commonly employed,[91] in conjunction with, but some-times (intentionally) without oxygen. See chapter on Acrylic Acid for further remarks on this subject.

Water is considered to have some inhibiting effect and a small amount ($\not> 0·5$ per cent) is included in all stored acrylonitrile. Its action has not been explained. Alone, it does not give adequate protection.

11.4.6. Main By-Products

11.4.6.1. Hydrogen cyanide Reported outputs of HCN from various catalyst systems vary considerably, but it is to be expected that there will be 0·1 ton or more of HCN per ton acrylonitrile, which represents a remarkable turning of the tables. For, whereas previously it was an essential and expensive raw material for the production of acrylonitrile (from acetylene), HCN now becomes a valuable by-product. Its chief application is probably in the manufacture of methacrylic ester via acetone cyanohydrin. Whether supply will eventually exceed demand as the output of acrylonitrile expands is not certain. At present there is no difficulty in making use of it. On the contrary, it is said that the availability of HCN from acrylonitrile plants in Japan has stimulated the expansion of methacrylate capacity. Until recently it seemed that nitrilotriacetic acid (for detergent manufacture) would soon become another important outlet, but there have been adverse reports on the toxicity of this material.

The Asahi Chemical Co. has developed a method of manufacture of acrylonitrile from HCN and ethylene, depending on a gas phase, catalysed oxidation reaction:

$$CH_2 = CH_2 + HCN + \tfrac{1}{2}O_2 \longrightarrow CH_2 = CH-CN + H_2O$$

The catalyst is, typically, Pd/V/Ce on alumina, moderated by HCl in the vapour phase.[92]

11.4.6.2. Acetonitrile   Some of the earlier patents indicated high yields of acetonitrile, but with improved catalyst systems there has been a drastic reduction to as
little as $\frac{1}{40}$ ton per ton of acrylonitrile, which is perhaps fortunate since acetonitrile
finds only limited use; chiefly as a selective solvent in the separation of dienes and
olefins.

11.4.6.3. Ammonium sulphate   It has become increasingly difficult to dispose of
ammonium sulphate, the only use of which is as a fertiliser, although the problem
this presents to acrylonitrile manufacturers is small compared to that facing the
producers of caprolactam by the older methods who have to cope with several tons
of ammonium sulphate per ton of product. Complete consumption of the ammonia
in a single bed ammoxidation reactor is not practicable. Selective destruction of
unreacted ammonia in a second bed of catalyst is feasible, but a suitable catalyst for
this purpose remains to be evolved, and none of the claimed selective absorption
methods, such as have already been mentioned, seem to have been applied
industrially.

11.5. ANOTHER ROUTE TO ACRYLONITRILE FROM PROPYLENE

Du Pont discovered that propylene and nitric oxide will react over certain catalysts
at elevated temperatures thus:

$$4C_3H_6 + 6NO \longrightarrow 4CH_2 = CH-CN + N_2 + 6H_2O$$

Catalysts containing silver seem to be preferred.[93, 94] Although good yields may
be obtained in this process it is not used industrially, and according to reference 94
the reason is that excess propylene must be used to achieve a good efficiency and
this cannot be directly recycled because it is diluted with nitrogen.

It is unlikely that ammoxidation is simply a variant of the Du Pont process in
which the preliminary step is the oxidation of NH_3 to NO because:
1. Little if any nitrogen is produced in the ammoxidation reaction when this is
   proceeding efficiently.
2. No nitrogen oxide is detected in the product from the ammoxidation of propylene.
3. Catalysts for the reaction of NO with C_3H_6 do not effectively promote ammoxidation, and the converse is true. Moreover, ammoxidation catalysts do not promote
   the oxidation of NH_3 to nitrogen oxides.

11.6. ACRYLONITRILE FROM PROPANE

ICI[95] have described a process for the manufacture of acrylonitrile in which
propane, ammonia and oxygen are passed over a mixed oxide catalyst containing
antimony (e.g. Sb/U, Sb/Sn or Sb/Ce), in the presence of a halogen at the catalyst
surface. According to the patent examples the reactor temperature (485–520°C) is
somewhat higher than it is for ammoxidation of propylene, and it is possible that

preliminary dehydrogenation of propane to propylene occurs. Efficiencies up to 99 per cent are recorded, but yields are less than 30 per cent. Monsanto have also shown an interest in this reaction.[96] The Lummus Co. have described the production of acrylonitrile by passage of a mixture of ammonia, oxygen and either propane or propylene through a molten salt mixture such as KCl, CuCl and $CuCl_2$.[97]

For a process based on propane to become comparable with or superior to the ammoxidation of propylene either propane would have to become much cheaper than propylene or else a considerable operating advantage would have to be associated with its use.

ACRYLIC FIBRES
By R. E. Sanders

11.7.1. Acrylic Fibres—Their Nature and History

Acrylic fibres are defined by the US Federal Trade Commission as fibres made of a fibre forming polymeric material containing not less than 85 per cent acrylonitrile units by weight.

They are the most wool-like of the truly synthetic fibres, those fibres made from synthetic, man-made, polymers. The 'handle' of acrylic fibres is very soft and warm. A list of the important end-uses of acrylic fibre indicates its replacement of wool in many products.

Acrylic fibres were first made as a result of the work of Rein[100] of I. G. Farbenindustrie of Germany in 1931. The obstacle to the use of polyacrylonitrile as a fibre-forming polymer had been that of finding a solvent. Polyacrylonitrile is insoluble in most common organic and inorganic solvents. As the then commonly known plasticisers were ineffective,[98] and the polymer decomposes before melting, it was obvious that solution spinning would have to be employed. The nature of the solvents discovered by Rein coupled with the theories of Marvel et al.[101] indicated that it was intermolecular forces between adjacent chains causing the insolubility and not cross-linking. It has been postulated that these inter-chain forces are due to hydrogen

bonds between adjacent polymer chains (HC CN . . . . HC CN . . . .

etc., where chains run top-to-bottom of the page) and that dissolution is only achieved after these bonds have been broken and replaced by a similar bond with the solvent.[99] Those solvents most likely to achieve this are strongly polar.

World War II produced a new incentive for the development of acrylic fibres. The raw material, acrylonitrile, was available in bulk, due to its use in the production of synthetic rubber.

Acrylic fibres were first produced commercially by Du Pont of the US but research was concurrently being carried out by Chemstrand in the US and in Germany Rein had joined Cassella and taken up where he had left off before the war. Finally, however, the Cassella process was bought by Bayer.[98]

Acrylic fibres might not have been successful, and certainly would not have been so quickly developed without the work of W. H. Carothers, who can rightly be called the founder of the truly synthetic fibre industry. Staudinger had put forward the concept of the macromolecule in 1922[102] and Carothers (who joined Du Pont in 1928, was the inventor of Nylon 66 in 1934 and mainly responsible for the melt extrusion spinning process) confirmed and expanded the macromolecular concept by experiment, his work being published in 1940.[103] It was also Carothers who first investigated the potential of esters as the monomers of fibre-forming polymers, but he did not discover a polymer of high enough melting point. This was left to Whinfield, of Calico Printers' Association, in the UK, and the final result was ICI's 'Terylene'.[98]

It was not surprising therefore, that Du Pont also turned their attention to acrylic fibres. An experimental group set up in 1943 finally selected dimethyl formamide as solvent for polyacrylonitrile, and succeeded in developing a saleable product in 1948, first called 'Fibre A'. This fibre renamed 'Orlon' was in commercial production in July 1950. Following, as it did Carother's work on the melt spinning processes, Orlon was dry-spun and the first products were in continuous filament form. Next came 'Acrilan' made by Chemstrand (now Monsanto Textile Division) in 1952, also in the US and which was at first plagued by striation problems. In West Germany, 'Dolan' and 'Redon' were first produced in 1952, and Bayer's 'Dralon' followed in 1954.

Du Pont's 'Fibre A' was a 'straight' polymer of 100 per cent acrylonitrile. This fibre, without a chain opening copolymer or one providing active dyeing sites, proved extremely difficult to dye and its elasticity was inadequate for the subsequent spinning processes. Du Pont first thought that their Fibre A would be best suited to specialist outdoor and industrial applications because of its excellent acid and sunlight resistance. However, the problem of dyeing and elasticity were satisfactorily overcome by the addition of co-monomers and Du Pont then saw a very wide range of uses for their new fibre in the apparel field.

In the first years of production they had evaluated the fibre as one best suited to woollen-type goods in staple form, and they successfully developed an acrylic staple fibre in 1952. In 1953, the copolymerised acrylic fibres superseded the straight polymers. The manufacture of continuous filament product was discontinued, but was taken up some time after 1959 when their bicomponent, 'Orlon Sayelle' fibre, came on the market.[104]

The late fifties and then the sixties saw developments in many parts of the world, many developing countries even taking licences to erect small plants.

There are sixteen different processes in use today employing eight polymer solvents: dimethyl formamide, dimethyl acetamide, dimethyl sulphoxide, ethylene carbonate and solutions of $NaSCN$, HNO_3, $ZnCl_2$, $NaZnCl_3$. The number of operating commercial plants in the world is about seventy-five, which are currently producing a total of a million tons per year of acrylic fibres.[104–106]

It should be noted that since acrylic fibre process know-how is jealously guarded there is inevitably some disagreement between various references covering the same subject. The author has endeavoured to deduce the correct information wherever possible, and has indicated where doubt exists.

11.7.2. Production of Polymer

11.7.2.1. The reaction—major features That acrylonitrile polymerises appears to have been known for a considerable time; perhaps since soon after its discovery in Germany in 1893.[98] The polymerisation may be represented most simply by the addition polymerisation equation.

$$CH_2 = CH-CN + CH_2 = CH-CN + \quad \ldots \ldots \ldots \ldots \ldots \ldots \ldots$$

$$\rightarrow CH_2-CH-CH_2-CH-CH_2-CH- \ldots \ldots \ldots \ldots \ldots \ldots \ldots \ldots$$
$$\qquad\quad | \qquad\quad | \qquad\quad |$$
$$\qquad\quad CN \qquad\ CN \qquad\ CN$$

Generally the pure polymer is unbranched, or very little branched, except in the case of high molecular weight polymer made in aqueous systems above 50°C (particularly with active initiators, such as persulphates).[107]

Pure monomer is stable to heat up to 200–300°C at which temperature a cyclic dimer is formed.[108]

$$CH_2-CH-CN$$
$$|\qquad\quad |$$
$$CH_2-CH-CN$$

However, as polymerisation is initiated, in general, by any source of free radicals and by visible light, acrylonitrile is normally stored with the addition of a few parts per million of an inhibitor such as ammonia or hydroquinone monomethyl ether.

Although a considerable amount of work has been carried out to investigate the kinetics of the polymerisation reactions of acrylonitrile, the reaction is still only imperfectly understood in many areas. A comprehensive review of the literature dealing with the homopolymerisation in various media and with various initiation systems up to the beginning of 1960 has been prepared by W. M. Thomas.[109]

The kinetics of copolymerisation are discussed in 'Copolymerisation', by Mayo and Walling.[110]

Acrylonitrile polymerisation is also discussed by L. B. Morgan,[111] with particular reference to redox systems.

The reaction has the following stages:

(i) *Initiation*
Initiation by free radicals In this stage a monomer molecule is converted into the beginning of a growing polymer chain normally denoted M°. It is often preceded by an induction period when impurities and inhibitors react and are thereby effectively removed. The reactive intermediate M° may be achieved by an acrylonitrile molecule reacting with a free radical, resulting from the decomposition of an initiator, decomposition being brought about thermally or by irradiation.

It may also be achieved by free radicals, produced by the redox systems of catalysts (with or without metal ion promoters). Ionising radiation may also provide free radicals.

Anionic initiation Reactive intermediates with negatively charged growing chain ends can be produced by ionising radiations, by strong bases or by reagents providing an anion with which the monomer molecules combine to form the reactive intermediate onto which other monomer molecules may add.

(ii) *Propagation*

This is the stage of chain growth normally represented by the equation $M^\circ + M \to M^\circ$ (where M is monomer).

(iii) *Termination*

The growing polymer chain is finally rendered inactive, and 'dead' polymer results. There are a number of ways in which the growing chain is terminated. The majority of growing chains are terminated by combination (with another growing polymer chain) where monomer is present in relatively high concentration. $M^\circ + M^\circ \to X$, where X is dead polymer. They are also terminated by reaction with free radicals, $M^\circ + I^\circ \to X$ (where I° is a free radical).

In certain cases the termination reaction produces a new free radical (e.g. by extraction of a hydrogen atom from dimethyl formamide which is a solvent for monomer and polymer), $M^\circ + S \to S^\circ + X$ (where S = solvent). They can be terminated by electron transfer $M^\circ + Fe^{+++} \to X + Fe^{++}$, and by reaction with other compounds present $M^\circ + U \to U^\circ + X$. (In reactions employed commercially these compounds are often added in accurately measured quantities to control chain length.) In ionic reactions the termination may be by reaction with water or CO_2.

The principal polymerisation techniques are discussed below.

11.7.2.2. Polymerisation in homogeneous solution (solution polymerisation)   Thomas[109] has defined the stages of free radical homogeneous solution polymerisation thus:

Radical formation

$$I \xrightarrow{\hspace{2cm}} 2I^\circ \qquad ----(1) \quad \text{(where I = initiator)}$$

Initiation

$$M + I^\circ \xrightarrow{\hspace{2cm}} M^\circ \qquad ----(1') \quad \text{(where M = monomer}$$
$$M^\circ = \text{growing polymer)}$$

Propagation

$$M^\circ + M \xrightarrow{\hspace{2cm}} M^\circ \qquad ----(2)$$

Termination

By transfer with the solvent

$$M^\circ + S \xrightarrow{\hspace{2cm}} X + S^\circ \qquad ----(3) \quad \text{(where X = dead polymer}$$
$$S = \text{solvent)}$$

By transfer with another compound present

$$M^\circ + U \xrightarrow{\hspace{2cm}} U^\circ + X \qquad ----(3')$$

By combination with another growing chain

$$M^\circ + M^\circ \xrightarrow{\hspace{2cm}} X \qquad ----(4)$$

By metal ion electron transfer

$$M^\circ + Fe^{+++} \xrightarrow{\hspace{2cm}} X + Fe^{++} \qquad ----(5)$$

By combination with a primary radical

$$M^\circ + I^\circ \xrightarrow{\hspace{2cm}} X \qquad ----(6)$$

If the rate coefficients (or 'constants') of the above reactions are represented by $k_1 k_2$ etc., then a typical steady state free radical polymerisation rate equation is:

$$\frac{-d[M]}{dt} = k_2 \left(\frac{k_1 f}{k_4}\right)^{0.5} [I]^{0.5} [M]$$

where [ ] indicated molar concentrations, f is 'catalyst efficiency' and is introduced since if all the free radicals formed initiated a polymer chain then the rate of formation of activated monomer would be the same as that of free radicals (i.e. for given conditions normally proportional to [I]). However, a considerable proportion of free radicals is used up in side reactions not leading to polymer. (In the case of polymerisation initiated by azobisisobutyronitrile in N.N'. dimethyl formamide the catalyst efficiency is often about 0·5).

The rate of formation of initiating free radicals is not always independent of monomer concentration (e.g. low monomer concentrations). The so-called rate constants vary in fact with temperature, with the nature of the solvent and with other parameters.

The rate equations for the polymerisation of acrylonitrile in homogeneous solution, initiated by free radicals are more predictable than the rate equations for the other reaction schemes. The exponent of [M] is constant over a wide range of monomer concentrations, and that of [I] is also fairly constant for many initiators in different solvents.

Post polymerisation effects after the source of free radicals have been removed have been observed in solution polymerisation.

11.7.2.3. Heterogeneous polymerisation in a dispersing medium (suspension polymerisation) Many reactions have been studied by many workers. An examination of the literature leads one to the view that reaction kinetics in heterogeneous polymerisations are influenced by many variables of state which are difficult to measure. The initial state of the monomer (whether it is all in solution or in solution and emulsion or suspension) the intermediate and final states of polymer precipitate, and hence the rates of change of state, and the degree of agitation, can each have an effect.

Several alternative schemes of heterogeneous polymerisation in a dispersant are possible. These schemes are discussed below.

Alternative 1 Solubility or miscibility of monomer in dispersing medium negligible; solubility of polymer also nil.

Alternative 2 Solubility (miscibility) of the monomer in the dispersing medium relatively small; solubility of polymer, nil (as in the case of water). The two reaction schemes which are possible are discussed below (alternatives 2a and 2b).

Alternative 2a Catalyst can be added to an aqueous monomer solution causing the precipitation of polymer. An example is found in Smeltz and Dyer's experiments[115] in which they produced radicals in an aqueous solution of acrylonitrile. This particular reaction is 'Precipitation polymerisation'.

30

The Orlon polymerisation reaction in (aqueous) 'suspension' in which the monomer concentration is reported to be 7·5 per cent[104] is probably an example of this type of reaction.

The solubility of acrylonitrile in water is as follows (Table XI.9).[116]

TABLE XI.9

Temperature °C	wt % AN in H₂O	wt % H₂O in AN
0	7·2	2·1
20	7·35	3·1
40	7·9	4·8
60	9·1	7·6
80	10·8	11·0

In descriptions of processes from patents the monomer is sometimes added to the aqueous catalyst solution.[99]

Alternative 2b   Alternative 2 also means that the complex scheme is possible, in which some monomer is in solution and some dispersed. The dispersed monomer would polymerise leading to a dispersion of polymer, while the polymer from monomer in solution would precipitate in the solution phase.

From a reaction and polymer product control point of view this scheme would appear to present quite difficult problems.

The kinetics are complicated by the preferential wetting and adsorption of monomer on particles of polymer where a relatively rapid reaction is thought to take place.[117] The rate equations are, of course, different for the reactions occurring simultaneously in the aqueous solution phase, in the dispersed monomer, and at the surface of, and possibly within, the particles of polymer.

Dispersion of monomer may either be a suspension of droplets or an emulsion.[118, 119] The form of polymer precipitate depends on reaction conditions and concentrations of initiator, electrolyte, and dispersant, and surface active agents. A very wide range of particle sizes appears to be possible. Palit and Guha[120] showed that, as the concentration of redox catalyst was increased, the polymer precipitate form changed through fine sol and milky dispersion to coarse precipitate, indicating an electrolyte concentration dependence, while the polymerisation rate first increased, then decreased and finally increased again.

Dispersion stabilisers[121] are used to stabilise suspensions or emulsions of monomer and to control polymer precipitation.

An example of aqueous suspension polymerisation is reported by E. Cernia.[118] Citing work of Hohenstein and Mark, and of Fryling, carried out between 1940 and 1944, he states that the suspension is achieved by violent stirring in the presence of stabilisers. The droplets containing initiator, co-polymers etc. polymerise to beads of polymer, and the reaction may be likened to a very large number of bulk polymerisations in miniature. The initiator is, in this case, monomer soluble.

In emulsion polymerisation it is normal to use water-soluble initiators, such as the redox systems.[118]

Emulsion co-polymerisation of acrylonitrile in water, with co-monomers of varying solubility, has been studied.[122] It is worthwhile recording here the solubilities in water of these co-monomers. (Table XI.10).

TABLE XI.10

	Solubility (%)	Temperature (°C)
Styrene	0·28	25[123]
Methyl methacrylate	1·5	30[123]
Vinyl acetate	2·5	20[124]
Methyl acrylate	5·2	—[124]

Alternative 3   Monomer soluble to only a limited extent in the dispersion medium, polymer soluble (e.g. inorganic salt solutions).

This would mean an emulsion or suspension polymerisation with the polymer going into solution after it was formed.

A line of enquiry to throw more light on this reaction scheme, could begin with an examination of solubilities of acrylonitrile in the various aqueous inorganic polymer solvents used in direct spinning processes, such as NaSCN and $ZnCl_2$.

Note   Heterogeneous polymerisation reactions have been studied in benzene, toluene, carbon tetrachloride, benzene/dimethyl formamide, toluene/dimethyl formamide, and methyl cellulose solution.

The monomer is miscible with benzene and toluene in all proportions, and soluble in methyl cellulose solution.

Quantitative miscibility data has not been located in the literature for the other liquids. The benzene and toluene systems are further examples of the 'alternative 2a'.

However, there is no evidence that any of these reactions are used commercially, and for that reason the original published data has not been cited. The results are summarised and references are given by Thomas.[109]

11.7.2.4. Bulk polymerisation   Bulk polymerisation is not used commercially because of the difficulty of control. The important characteristics of the reaction will be given here for completeness.

It is reported that light, azo and peroxy compounds, and ionising radiation initiate the polymerisation of pure acrylonitrile. The polymerisation of pure acrylonitrile is under many conditions uncontrollable and can be explosive.

The polymer is insoluble in the monomer and precipitates. The heat of reaction is 17·3 kcal/mole.[112] This heat must be removed in commercial reactions.

The degree of polymerisation can be very high indeed. Molecular weights as high as one million have been reported.[113]

The intrinsic viscosity (and hence degree of polymerisation) of polymer produced by bulk polymerisation shows a maximum in the range of preparation temperature,

50–60°C for many initiators. (Intrinsic viscosity which is a function of degree of polymerisation, is a standard test measurement of viscosity of polymer in dimethyl formamide solution.)

There is evidence that the polymer traps active radicals thus removing them from the reaction. The effects of heating and of adding swelling agents, both of which release these occluded radicals, is more complex than would be expected. An interesting theory to account for the observed results has been given by Bamford and Jenkins.[114] Radical occlusion may, in fact, at certain stages of polymerisation be a reason for increased polymerisation rate due to its prevention of radicals taking part in the radical termination reaction.

The rate equations for bulk polymerisation with different initiators under various conditions show a wide variation. The rate of the reaction itself is characterised by a chort period of acceleration, followed by a constant rate to about 50 per cent sonversion, followed by a diminishing rate. It has been demonstrated that 'post polymerisation' is exhibited when the source of initiating radicals is removed, i.e. the reaction continues but at a greatly reduced rate. (See reference 109, page 412).

11.7.2.5. The reactions employed commercially   A wide choice of polymerisation reactions exist, although relatively few are used commercially. Most anionic polymerisations do not appear to give a polymer suitable for spinning to fibres and for this reason are unlikely to be used.

There is no *direct* evidence that the 'monomer emulsion or suspension' methods of polymerisation are used commercially. The author has suggested the use of suspension polymerisation in certain processes which are followed by direct spinning and which employ solutions of inorganic salts. On the other hand, in nearly all such cases, a spinning, and, therefore, presumably a polymerisation dope, concentration of polymer of 10 per cent is employed[104] (discounting subsequent dliution and dissolving). No aqueous redox initiated reaction commercially employed is reported to have a polymer concentration above 10 per cent.[104] It is very possible, therefore, that these polymerisations are in homogeneous solution leading to polymer precipitate. (Monomer 'solubilisers' are mentioned in the literature, e.g. a solution of aryl sodium sulphonate.)[128]

A U S Patent to American Cyanamid[126] describes a continuous aqueous solution polymerisation. Another U S Patent to American Cyanamid[127] describes an aqueous process leading to relatively concentrated slurries of polymer in water, in excess of 40 per cent. In fact, Cyanamid's process is a continuous and, probably, homogeneous solution polymerisation in NaCNS solution, leading to a polymer concentration of 10 per cent.

The low polymer concentration in aqueous systems may, in fact, indicate continuous reaction processes. Perhaps the difficulties of control of a reaction, in which not only polymer, but monomer is present in more than one phase, have outweighed its advantages of lower volume throughput, and cheaper reactants recovery.

Batch processes are reported, and one can envisage a multi-stage continuous process, in which monomer is added and reacted until a relatively high polymer concentration is attained, but again, reported polymer concentrations for commercial aqueous reactions lend no support to the view that these processes are employed.

a. *Aqueous redox systems* The polymerisation rate in aqueous redox initiated systems is a fast one, probably including the fastest of those employed commercially. An example of 86 per cent conversion to polymer in one hour has been reported.[125] This was a batch reaction in aqueous emulsion with potassium persulphate catalyst. A similar system has been described where, no doubt, a Du Pont patent is being quoted.[99] In this case an ammonium persulphate/sodium bisulphite activated redox catalyst was used. The monomer/co-monomer mixture was added over two hours.

The majority of polymerisation processes use a redox catalyst system in aqueous solution. Redox catalyst is reported to be used in the processes of U C B,[129] Bayer,[129] Du Pont,[99] Monsanto, Asahi Kasei and U. F. S. Savinesti.[129] All such aqueous systems require the production of dry polymer and subsequent solution in a suitable solvent to produce the spinning dope. This type of process is called an indirect spinning process.

b. *Aqueous salt solution systems* $ZnCl_2$ solution is employed in the process of the Dow Chemical Company and sodium thiocyanate solution is employed by both Courtaulds and Cyanamid.

These three processes result in a spinnable solution of polyacrylonitrile in the aqueous salt solution. Dry polymer is not obtained. Such processes are continuous and called 'direct spinning processes'.

The solubility of the monomer in the salt solutions has not been published. That redox, or other inorganic systems, of initiation have not been mentioned in connection with these processes lends a little support to their employing organic free radical initiators, and this, in turn, may suggest solution polymerisation.

A relatively new process is that of Dr J. Löbering K.G. Munich. The monomer is polymerised in homogeneous solution with other reactants in 55 per cent $NaZnCl_3$/HCl solution and an extrudable solution of polymer is produced. The initiation is achieved by ultra-violet light.[130] Although the Löbering process is the only one of its kind in commercial use, ultra-violet initiation has received attention in the past. Onyon has investigated it[131] and at least one patent has been taken out.[132] That redox catalysts are not used is claimed to be an advantage in that impurities resulting from redox decomposition are avoided.

High molecular weight fractions produced in aqueous systems tend to be branched. This is due to transfer on to dead polymer chains, whose concentration is of course relatively high at the end of the reaction. The greater the activity of the catalyst and the higher the temperature, the more marked this effect becomes.[109] This is one possible disadvantage of the high reactivity of the aqueous redox systems. Percentage monomer conversion may be limited to achieve the desired molecular weight distribution. Conversions as low as 60 per cent[130] and 67 per cent[133] are reported. Low conversions are troublesome in direct spinning processes. Unreacted monomer has a tendency to polymerise to undesirable low molecular weight polymer in the extruded filaments. For this reason, below certain limits of conversion, monomer is normally removed from the reaction mixture by evaporation.

At some stage of the reaction, chain transfer agents are added to control the degree of polymerisation. The following chain transfer agents are mentioned in the literature: alkyl and aryl mercaptans, carbon tetrachloride, chloroform, dithioglycidol and alcohols,[125] Fe^{+++}, and also aliphatic hydrocarbons, ketones, halogenated compounds and amines.[134]

The way in which the co-monomer or co-monomers are incorporated is normally by simple addition with one or more reaction feed streams (usually including the monomer stream), but 'graft' co-polymerisation is also reported to be used.

c. *Organic solvent systems*   There are two processes employing homogeneous solution polymerisation, those of Snia Viscosa[129] and of Toray Industries Inc.[104] The former employs dimethyl formamide and the second dimethyl sulphoxide as the solvent. Both probably use organic free radical initiators. Solution polymerisation in DMF initiated by azobisisobutyronitrile and by benzoyl peroxide have been well reported in the literature.

11.7.2.6. Polymerisation processes   Though there are a number of polymerisation reactions, the processes which are employed have many features in common. The basic requirements of all polymerisation processes are the same.
a. to produce a polymer of predictable molecular weight normally in the range 32,000–110,000[135] (i.e. a degree of polymerisation of 600–2,000).

It has been reported that the lower half of this range 35,000–50,000 is best suited to dry spinning and the upper half best suited to the wet spinning process (e.g. Bayer's dry spinning process employs molecular weights from 40–60,000).[139]
b. to control the reaction temperature.
c. to incorporate in the polymer chains co-monomers to increase dyeability and to improve physical properties, particularly elasticity and thermoplasticity. Comomoners are normally reacted with the acrylonitrile but there are exceptions.

Dow blend a graft co-polymer of polyvinyl pyrrolidone with their polyacrylonitrile/methyl acrylate co-polymer[104] and Monsanto are reported to add a 50 per cent acrylonitrile and vinyl acetate/methyl vinyl pryidine co-polymer to the polyacrylonitrile before solution.[129]

The process of Superfosfat, Sweden, which is no longer in commercial production[136] was of interest in that it deliberately set out to produce branched polymer chains by the addition of triacrylohydrothiazine to the polymerisation reaction.[129]

Co-monomers used in particular processes are summarised by the author.[104] Methyl acrylate is the most common. Classes of chemicals which are polymerised with acrylonitrile are given in reference 135, page 320.

The temperatures at which polymerisation reactions are carried out are between ambient and 60°C. The pressure of reaction is normally atmospheric or a little above atmospheric and is carried out under a blanket of inert gas, either nitrogen or carbon dioxide.

Some exceptions have been noted. One patent (to Chemische Werke Huls—not a commercial producer of acrylic fibre) describes a process under 10 atm absolute of CO_2,[137] and a pressure of 5 atm of CO_2 has been suggested to serve the dual purpose of pH control and an inert gas.[125]

11.7.2.7. Process flowsheet for aqueous systems   Although batch preparations are the subject of patents it is likely that the processes now used, particularly by the

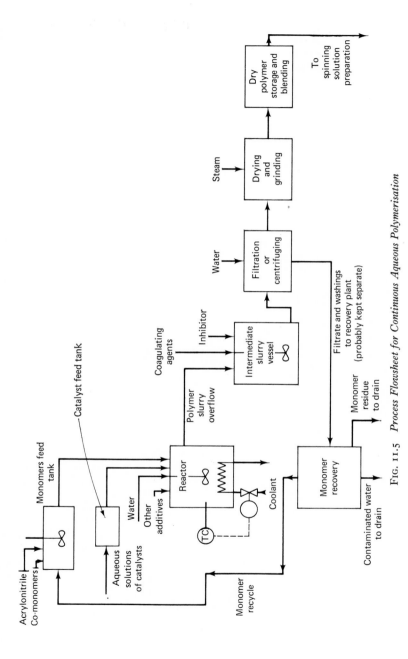

FIG. 11.5 *Process Flowsheet for Continuous Aqueous Polymerisation*

established acrylic fibre producers, are continuous. The advantages of continuous processes are obvious. The product is more uniform, blending requirements are minimised, labour and capital cost are reduced. A typical process scheme is illustrated. (Fig. 11.5.)

The reaction additives include: acid or other means for pH control, chain length controllers, reaction accelerators, whitening agents, polymer stabilisers, and polymer precipitation controllers. The coolant may be ordinary cooling water from a conventional circulating system or chilled water, or refrigerated brine in the case of ambient temperature polymerisation.

The redox system promoted by cupric ions is discussed and a mechanism and rate equation is given.[111] It is suggested that the Cu^{++} promoted persulphate/thiosulphate redox system will provide a wide range of reaction rates proportional to $\sqrt{[Cu^{++}]}$ without the need to increase actual initiator concentration. It is further suggested that $CuSO_4$ is added in p.p.m. quantities in commercial aqueous redox systems to swamp the effect of other trace ions.

In emulsion polymerisation reactions, an emulsifying device would be used, into which would be fed monomers, water and emulsifying agent.

11.7.2.8. Equipment design and process control

a. *The reactor*   The reactor may be a straight-forward jacketed vessel, but other types are used, e.g. an elongated vertical vessel (a reactor column) with top inlet and bottom outlet. In this case the stirrer may be of the scraped wall design, and as the reaction rate per unit volume is likely to be high, the heat transfer surface area per unit volume must also be high. One patent mentions the use of an external heat exchanger through which the reactant mixture is circulated.

Various automatic controls are feasible, pH, temperature and monomer feed concentrations, based on automatic analysis, refractive index or other measurements. The degree of polymerisation will be regularly checked by intrinsic viscosity measurement of the standard solution of dried polymer in DMF. Catalysts and monomer analyses and pH are critical. Catalysts and certain other liquid additives may be added by metering pumps. Solids must be accurately weighed. Monomers and water will be accurately flow ratio controlled.

Temperature and/or pH could be controlling factors for the degree of polymerisation but it is more likely to be the concentration of chain transfer agent. The reaction is carried out under a nitrogen blanket. In addition to its inhibiting property, oxygen can cause polymer discolouration.

b. *Slurry storage, coagulation and chain stopping*   Redox systems may be rendered inactive by increasing the pH to just above normal. Additions of chemicals at this stage to coagulate the polymer and to short-stop the chains are also possible.

c. *Monomer recovery plant*   A boiler and rectifier column at reduced pressure are normally used to remove monomers including other lights from the polymer filtrate. The water washings may be discarded. The recovered monomers must be inhibited to prevent polymerisation.

The condensed water/monomer top product from the rectifier will be cooled, separated into layers, and the monomer-rich layer will be further distilled at low pressure to obtain monomers pure enough to be recycled to the polymeriser. Analysis of the recovered monomer and the desired polymer product will decide the quantity of fresh monomer and co-monomers to be added to the monomer preparation vessel. Residues from recovered monomer distillation will be sent to drain.

The bottoms product from the rectifier boiler which is mostly water, with dissolved monomer (a trace), higher boiling water-soluble by-products of the polymerisation reaction and the dissolved catalyst residue, is probably sent to drain.

d. *Filtration or centrifuging*   Rotary vacuum filters are often represented in process flowsheets of aqueous polymerisation processes. Centrifuges may also be used.[127] The cake is washed with water which may be followed by an acetone wash to remove low molecular weight and branched or modified polymer.[138]

In the process of Du Pont, the wet polymer is extruded in the form of 'noodles'.[144]

e. *Drying, grinding, storage, blending*   The dryers are probably either the moving conveyor type or rotary drum (vacuum) type. Hot, steam-heated air at, say, 80°C (higher temperatures could cause discolouration) or hotter inert gas are likely drying media.

Polymer samples will be tested for intrinsic viscosity and the data fed back to the reaction section. The degree of polymerisation will also assist in determining the blending requirements. Some specially designed blenders may be employed if large quantities of dry polymer must be blended to uniform specification. Otherwise blending may conveniently be carried out in the spinning solution preparation vessels.

11.7.2.9. Spinning solution preparation   Blended polymer from storage will be chosen depending on end use. There are a wide range of solvents used in the indirect processes: N.N'. dimethyl formamide, N.N'. dimethyl acetamide, nitric acid solution, and ethylene carbonate. In some cases a slurry of polymer in solvent is made at ambient or reduced temperature, then heated in a dissolving tank to effect solution. At this stage a delustrant (finely divided titanium dioxide), whitening agents and pigments may be added. Finally the solution is passed through a screw mixer/homogeniser and through filters to the spinning machine.

The spinning solution is heated and evacuated to degas it before storage, in order to reduce as much as possible undesirable voids in the filaments.[139]

The process is sometimes a batch one, partially or fully-automated, or may be continuous.

The simplest process is one in which one or more mixer/homogenisers are employed, supplied continuously with blended polymer, solvents and additives.[140]

11.7.2.10. Typical mass balance   The following mass balance has been derived from what little published data is available. A solid polymer slurry content of 10 per cent has been assumed and the redox catalyst consumption is based on that given in reference 133, i.e. 0·023 lbs/lb acrylonitrile, 2·5 : 1 ammonium persulphate: sodium metabisulphite (but for other formulae see references 99 and 138). On this basis, catalyst is a minor cost item amounting to $0·076 per kg polymer.[141] A conversion of 65 per cent has been assumed.

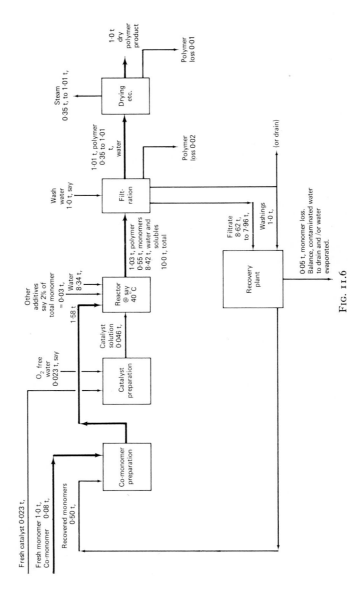

Fig. 11.6

The polymer is assumed to be about 8–10 per cent co-monomer. The filtered solid polymer water content is taken to be 25–50 per cent water on total cake weight. Losses and other throughputs are guestimates (similar mass balances are possible for the other systems).

11.7.2.11 Process flowsheet for aqueous salt solution systems The following solution compositions are reported to be in use.[104] (Table XI.11.)

TABLE XI.11

Company	Solution composition	% polymer in solution
Dow (and Licensees)	$ZnCl_2$ 54% to 60% (with $MgCl_2$, $CaCl_2$ and with 4% NaCl also)	10
Cyanamid (and Licensees)	NaSCN 50%	10
Courtaulds (and Licensees)	NaSCN 51% and 35 to 45% NaSCN with 30 to 50% methanol	10
Lobering (and Licensees)	$NaZnCl_3$/HCl 55%	not known

60 per cent $ZnCl_2$ solution is reported to crystallise at $77°F$,[130] 55 per cent $NaZnCl_3$ at $50°F$[130] and 50 per cent NaSCN between 10 and $15°C$.[142] Hot water jackets or steam tracing will therefore be necessary wherever the solution/ambient temperature may fall below these temperatures. In any case, the viscosities of the dope, particularly that sent to the spinning area may necessitate heated pipelines, and equipment.

Premixing of solvent and reactants may be employed as in the Löbering process[130] or reactants and solvent may be charged to the reactor as in the Courtaulds process.[129]

Salt solution polymerisations are characterised by low temperature. Courtaulds is carried out at ambient temperature. Löbering's uses refrigerated water coolant indicating ambient or near ambient temperatures. Dow's process will be at temperatures above $77°F$ ($25°C$).

The order of the equipment may differ from that shown on the process flowsheet, but the essential operations will be common to all processes. (Fig. 11.7.)

All salt solution processes at present employed are followed by direct wet spinning. (The process is continuous apart from buffer storage through the polymer production/dope extrusion stages.) Hence, to complete the flowsheet the spinning and washing machines are included.

11.7.3. Equipment Design

11.7.3.1. The reactors The number and design of reactors varies, of course, from process to process. Two reactors in series are used in the Löbering process and these appear to be identical in design although they operate at different temperatures. High monomer conversion to polymer is claimed for the Löbering process; so high, in fact, that monomer evaporation from the dope and recovery from the spinning coagulating solution are not necessary. One assumes that the small amount of monomer in the vapour from the degasser is passed to atmosphere or to drain.

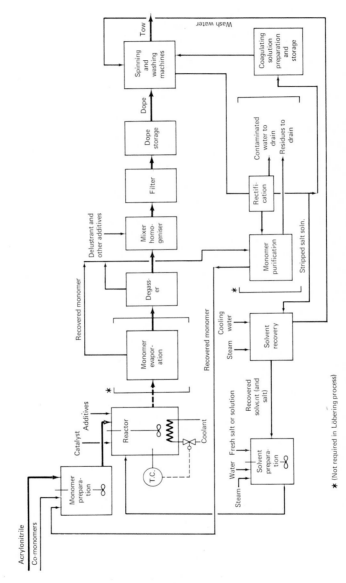

Tow

Wash water

Spinning
and
washing
machines

Coagulating
solution
preparation
and
storage

Dope

Dope
storage

Contaminated
water to
drain

Residues to
drain

Filter

Mixer
homo-
geniser

Rectifi-
cation

Monomer
purification

Stripped salt soln.

Delustrant and
other additives

Degass-
er

Recovered monomer

Monomer
evapor-
ation

Cooling
water

Solvent
recovery

Steam

Recovered monomer

✶

Catalyst

Additives

Reactor

Coolant

T.C.

Recovered
solvent (and
salt)

Monomer
prepara-
tion

Water   Fresh salt or solution

Solvent
prepara-
tion

Steam

Acrylonitrile

Co-monomers

✶ (Not required in Löbering process)

Fig. 11.7 *Process Flowsheet for Continuous Aqueous Salt Solution Polymerisation*

In addition to the already discussed postulation of a monomer conversion limitation in aqueous systems, an optimisation of the process is feasible between monomer concentration, monomer conversion, and reaction residence time (and hence capital cost). The design of reactors to increase conversion in the later stages of a high conversion process is likely to be expensive as heat transfer must be adequate and the residence time, when the rate of reaction is diminishing, will be relatively high. No doubt the use of catalysts of higher and higher reactivity has necessitated such an optimisation.

The Löbering process is initiated by ultra-violet light using a 'sensitiser', which is probably a source of free radicals, leading in turn to a growing chain.[138]

As homogeneous solution is normally being reacted, the reactors may be fitted with internal cooling devices for which there is little evidence in heterogeneous reactors, thus improving heat transfer.

11.7.3.2. Monomer evaporation and degassing These two stages could be combined. An evacuated thin film evaporator may be employed. Condensed monomer is passed to the recovery area for purification in the monomer still.

11.7.3.3. Mixer/homogeniser, filter and dope storage This stage is similar to that required for the spinning solution in the aqueous polymerisation process.

11.7.3.4. Monomer and solvent recovery The temperature and salt concentration of the liquid in the spinning bath is such that the polymer is insoluble. As the dope is extruded through the spinnerets, which are immersed in this solution, the polymer coagulates into filaments. There follows a multi-stage stretching and countercurrent water washing process with the result that solvent, residual monomer, excess delustrant and other reagents are washed from the filaments.

The monomer is removed from the recovered salt solution by rectification at reduced pressure. After layer separation purified monomers are produced by further distillation at reduced pressure.

Solvent recovery is normally carried out in two stages, which are characteristic of solution polymerisation/direct wet spinning processes. The solution from the coagulating bath (which includes water used for washing the fibres, plus the polymer solvent salt solution, monomers and other excess reactants and additives) is the feed for the rectifying column. Part of the rectifier boiler bottoms product are recycled to the coagulating bath.

The remaining stripped solution passes to the second stage of solvent recovery. This essentially involves the removal of water from the solution to provide the solvent recycle for the polymerisation process.

The concentrations required are indicated in Table XI.12.

The evaporated water may be condensed and used for fibre washing, but this will depend on whether any by-products in the evaporated water cause contamination of the fibre product.

With what is virtually total solvent recycle, impurities gradually build up in the system. Titanium dioxide may be filtered from the spinning solution while other impurities (by-products of the polymerisation reaction) are either removed by chemical/carbon adsorption means, or by evaporation/distillation processes of

TABLE XI.12

Process	Polymer solvent concentrations	Recovered salt solution concentration
Courtaulds	51% NaSCN	11% NaSCN
Cyanamid	50% NaSCN	10% NaSCN
Dow	54–60% ZnCl$_2$	40% ZnCl$_2$ average
Löbering	55% NaZnCl$_3$/HCl	5% NaZnCl$_3$/HCl

separation. One of the most troublesome of solution impurities are those which are found to impair the whiteness of the fibres by reaction in the polymerisation and coagulation stages. In the Löbering process the concentrated salt solution is treated with zinc dust, peroxide and activated carbon to effect their removal.[130]

In all salt solution systems, of course, these impurities could be removed in the mother liquor after the crystallisation of at least part of the recycled stream.

11.7.3.5. Process flowsheet for organic solvent systems   Solution polymerisation in organic solvents is similar to salt solution polymerisation. The solvent recovery stage is a distillation unit with a water top product and a pure DMF or DMSO bottom product for recycle to polymerisation section. Fresh make-up solvent may not be pure enough for the polymerisation process and so may be distilled before use.

Apart from these differences the equipment employed will be similar to that used for aqueous salt solution polymerisation.

The polymer content of the final spinning dope is around 20 per cent; this higher concentration being evidence, perhaps, of the faster reaction rates for aqueous systems reported by several workers. Monomer conversions are relatively high also.[104]

11.7.4. Production of the Primary Fibre Products, Staple, Tow and Filament Yarn

11.7.4.1. Extrusion, coagulation, stretching and washing processes   There are two types of spinning processes employed in the manufacture of acrylic fibres; solution dry spinning and wet spinning.

(There is no evidence that a third method, melt spinning using plasticisers (non-solvent plasticising impregnants or perhaps solvents in low concentration) is used commercially but it has been the subject of patents.)[125]

Dry spinning processes in commercial use are indirect—(Du Pont's, Bayer's Suddeutsche Chemiefaser's and perhaps Ph. Welker's) and the polymer solvent is always DMF.[104]

Wet spinning may be either indirect or direct. The polymer solvents for the indirect processes are DMF, DMA, 70 per cent HNO$_3$, ethylene carbonate (with some added water), and 74 per cent H$_2$SO$_4$.[104] Direct polymerisation/spinning solvents have already been discussed.

Normally just before the extrusion process the temperature of the dope is raised to 80–150 °C. An examination of individual filaments of synthetic fibres in an article of clothing will give some idea of the very small holes in the spinnerets (down to one or two thousandths of an inch diameter). Because of this, and to ensure uninterrupted operation, it is normal to incorporate a filter in each dope line to the spinnerets. It is essential to ensure constant dope throughput per hole, which leads in turn to uniform filament cross section, and weight per unit length. Each spinneret may be served by a metering gear pump driven by a common drive shaft. This is true of most wet processes where the throughput per spinneret is relatively high. Dry processes may employ spinning heads in which one pump serves two or more spinnerets, but in this case the dope flow pattern is critical.

The machine drive may also be a common one for the stretching (or drawing) devices which stretch the collected filaments from a number of spinnerets, called a 'tow' (pronounced 'toe'). These techniques to produce a uniform final product, coupled to the need for close speed and tension control throughout the multi-stage extrusion stretching and after treatment processes, require, in turn, a reliable electricity supply of reasonably constant voltage and frequency. However, electricity specification is not so critical as in the case of melt spinning and particularly filament yarn processes.

The dry and wet extrusion process are shown diagrammatically in Fig. 11.8 on the following page.

11.7.4.2. Dry spinning

a. *Extrusion and coagulation*   The dry spinning process is essentially a simple one. The concentration of polymer in dope is 20–30. The temperature of the air or inert gas in the filament drying chamber is 230–260°C.[118] The DMF, boiling at 153°C at atmospheric pressure, evaporates leaving a dry solid filament which is wound up or passed to further processing stages at the bottom of the drying chamber. Extrusion is always downwards and vertical for what would appear to be obvious reasons, but upwards extrusion has been attempted. A closed circuit extrusion device with continuous condensation and recovery of solvent has been reported in the literature.[143] Air is circulated by the movement of the filaments, themselves. The solvent is condensed and the air recirculated. In-circuit fans or blowers would obviously increase the throughput and efficiency of such devices. Concentrations of dimethyl formamide in the air below the lower explosive limit of about 50–55 g/m³ have been suggested by E. Cernia[118] for safety reasons. If this is so, the use of inert gas or a modification of the air composition with inert gas are possibilities. Du Pont has been reported to use inert gas dry spinning for their Orlon manufacture.[144] Temperatures as high as 400°C are reported.[104]

The gas inlet and exit in Fig. 11.8 indicate co-current flow with the fibres but countercurrent flow can be employed with numbers of inlets and/or exits.

The number of holes in a dry-spinning spinneret is relatively low at 200–600. This is to achieve uniformity of product. In fact, to ensure that coagulation and drying conditions throughout the drying chamber are the same for all the filaments, the holes may be positioned in a relatively narrow circular band towards the edge of the spinneret face. The limit of winding up speed is presently of the order of 500–700 m/min.

The advantages of the dry-spinning system are those of maximum flexibility of operation with small compact units producing a dry fibre, a slightly simplified processing flowsheet as contaminants are minimised (this is true of course of all indirect

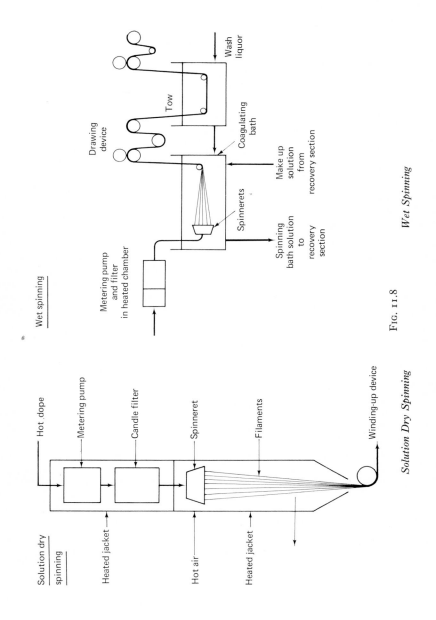

Wet Spinning

Fig. 11.8

Solution Dry Spinning

processes in which pure polymer is dissolved to form the spinning solution), and the ready ability to produce a wide variety of products including filament yarns, in relatively small quantities.

The market generally reports dry spinning to be particularly suited to the production of good quality products of the heavier deniers. (The success of Dralon in furnishing fabrics and carpets is evidence of this.)

The disadvantages are that fine denier products may not be so good as those obtained from wet spinning processes and the large number of spinning units leads to high capital cost for this section, which will be particularly unfavourable in the case of large plants.

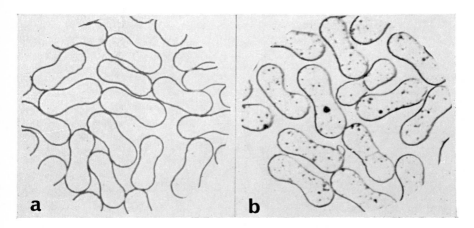

FIG. 11.9 *Typical Dog-Bone Cross-Sections of Fibres Dry Spun with Dimethyl Formamide as Solvent*

b. *Stretching and washing* The collected filaments from a spinneret may either be wound up directly on bobbins or a number can be collected to form a tow. Direct bobbin take-up would normally be used when the final product is to be filament yarn. The early treatment of filament yarn is similar to that of the collected tow, but the total denier of the collection of filaments to be treated is relatively low and hence the equipment may well be comparatively simple in design. A design of washing bath which features in the patents is an advancing immersed roller type.

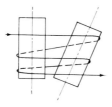

The principle is illustrated above. This design is unsuitable for wide high denier tows as it imparts a differential degree of stretching to the fibres according to which side of the tow they happen to be. It would be suitable, however, for filament yarn product. For those not familiar with the term, filament yarn is that product which is converted directly to a yarn, in which the individual filaments are continuous.

The washing and stretching process for a collection of filaments in the form of a tow is indicated diagrammatically in Fig. 11.10.

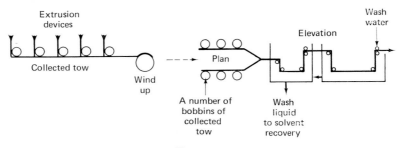

FIG. 11.10

The number of stages in the stretching process is variable, but the final filaments will have been stretched four to ten times their original length. A hot stretching zone in which the tow passes through a steaming bath, or hot water bath is sometimes incorporated. The stretching is achieved simply by increasing the speed of subsequent drawing devices. The design of these devices is dependent on total tow denier, speed, tension and draw-ratio. (The simplest design is indicated.) The 'mangle' action of the double pressure (or nip) rollers minimises liquor carry-over by the tow from one stage to the next.

The stretching process gradually aligns the long chain polymer molecules parallel to the longitudinal axis of the filament. Thus strength is increased. The degree of orientation in acrylic fibres is relatively high.

11.7.4.3. *Wet spinning*

a. *Extrusion, coagulation and washing*   The extrusion/coagulation process is shown in Fig. 11.8. Coagulation is normally carried out in a bath of polymer solvent/water mixture with perhaps additives to improve certain characteristics of the coagulated fibre. The temperature of the bath is normally low, at 0–30°C. Superfosfat's (Sweden) process, not now commercially employed, was an exception using heavy petroleum at 210–250°C or kerosene at 130°C.[104] Methanol in the coagulating solution is probably used to dissolve out low molecular weight polymers.[104]

The coagulating bath compositions and temperatures are those which produce an optimum filament, i.e. one that is uniform, free from voids or bubbles, and not of so low a density as to affect adversely fibre properties. The dope polymer concentration is also a critical factor and it is not unlikely that coagulation difficulties have limited polymer concentration in some cases. Coagulation must not be too rapid, thus forming a solid skin which inhibits diffusion from within the filament or adversely affects fibre structure.

Wet spinning has been reported to have the disadvantage of fibrillation of the filaments (i.e. longitudinal splitting) leading to a 'frosting' of the finished product.

Salt solution spun filaments are produced in an intermediate 'aquagel' (water-swollen or hydrated) form,[125] and a US patent to Dow Chemical Co.[145] for a chemical treatment to prevent fibrillation suggests that this defect is met with particularly in salt solution spinning processes. The swollen aquagel filament is collapsed irreversibly by drying. The high quality and the success of salt solution spun products in the market witness to the fact that fibrillation is a soluble problem. Disperse dyeing may be used to advantage in the coagulating stage.

The advantages of wet spinning are those of a high production rate continuous process, and achievement (more easily than in the case of dry spinning processes) of very uniform end products. The wide variety of products on the market made by the wet spinning process indicates the suitability of these processes to the production of products for the whole range of end uses for acrylic fibres.

The number of holes per spinneret in the wet spinning process for tow and staple production is high at 20,000 up to possibly 50,000 in some cases. Normally, because of the wide variety of products to be made, a number of parallel lines of production for the spinning and tow treatment stages are employed. (This is true of all processes, of course, dry as well as wet spinning.) The capacity of these lines depends (a) on the stage of development of the particular process (b) on the total capacity of the plant, (c) on the product mix to be made, and (d) to some extent, on the demand pattern of the market.

The speed of extrusion in wet spinning processes has been given at 140–150 yd/min, and the final take-up (after stretching), 30–120 yd/min.[146] These figures are seen to be reasonable for a process in which polymer is 15–20 per cent of the spinning dope solution and the resulting filament is stretched say four or five times its original length.

b. *Stretching*   As in the process for collected tow following dry spinning the tow from the wet spinning bath (i.e. the first coagulating bath) is passed through a varying number of additional stretching baths where the coagulation of the filaments is completed and the first stages of extraction take place.

A hot stretching zone may be incorporated.

c. *Washing*   The washing process is similar to that described for tow formed from dry-spun filaments. The solvent/reactant/by-products extraction process is largely controlled by diffusion from within the filaments to the surface and is commonly carried out in a series of separate washing troughs with intermediate drawing devices. The wash water is fed into the last stage and flows counter-current to the tow often passing from the first stage into the last coagulating bath. However this is by no means a universal scheme as economics or fibre characteristics may favour one of a wide variety of possible schemes e.g. separate recovery of wash liquor.

The wash water must be fairly pure as certain ions take up the reactive dyeing sites supplied by the incorporation of co-monomers particularly of course those giving an affinity for ionic dyestuffs. Apart from the virtual waste of the co-monomer intended to increase dye-ability there is inevitably a lack of uniformity in the distribution of the remaining reactive dyeing sites and this leads to patchy or streaky dyed products.

To increase the rate of diffusion of contaminants from within the fibres wash water is normally fed at elevated temperature.

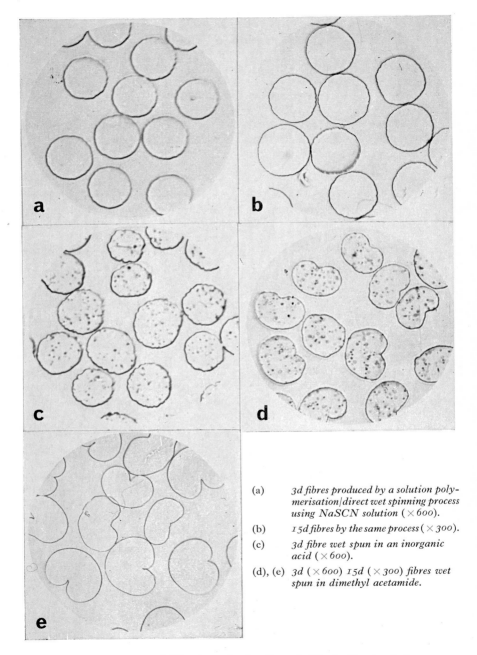

(a)   3d fibres produced by a solution poly-
       merisation/direct wet spinning process
       using NaSCN solution (×600).

(b)   15d fibres by the same process (×300).

(c)   3d fibre wet spun in an inorganic
       acid (×600).

(d), (e) 3d (×600) 15d (×300) fibres wet
       spun in dimethyl acetamide.

FIG. 11.11   *Photos of Fibre Cross-Sections Recently Obtained by the Author*

FIG. 11.12(a) *A Selection of Engelhard Precious Metal Spinnerets*

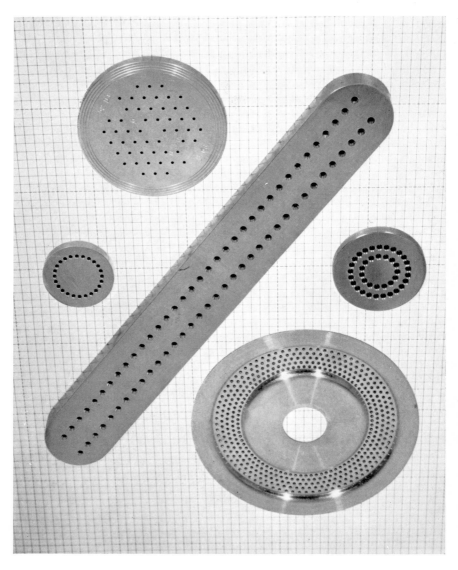

FIG. 11.12(b)   *Base Metal Spinnerets*[147]

FIG. 11.12(c)

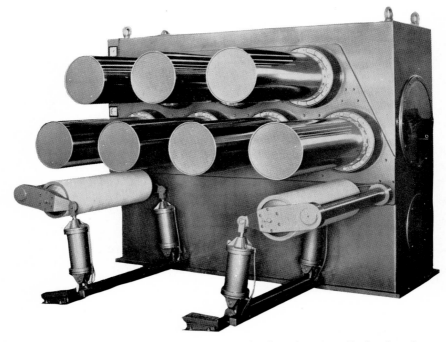

FIG. 11.13   *A Large Drawing Septet (drawing force 6 tonnes, roller length 1 m)*

d. *Recovery section*   The recovery of solvent and unconverted monomers from the extrusion bath liquor in a direct wet process has already been described. In an indirect wet process a similar two-stage recovery is common. First, if necessary, solids are removed followed possibly by purification treatment. Coagulating solution is recycled to the spinning machine. Secondly, polymer solvent is recovered by evaporation/distillation. This purified solvent is recycled to the polymer solution section.

In the dry process the recovered condensed DMF is distilled for recycle to the dope preparation section. If economic DMF will be recovered from the washing liquor and also recycled.

11.7.4.4. Treatment of spun tow   The tow of filaments emerging from the washing machines is thereafter submitted to a series of treatment processes, most of them common to all processes, the order of which, however, may vary slightly from process to process.

The fibres leaving the washing machine have adequate chemical and surface purity and orientation giving the desired strength, but too much residual shrinkage. They are also very wet (perhaps 100 per cent water adhering on dry weight basis) so they are dried under conditions of very low tension to allow shrinkage and to prepare for the subsequent processes.

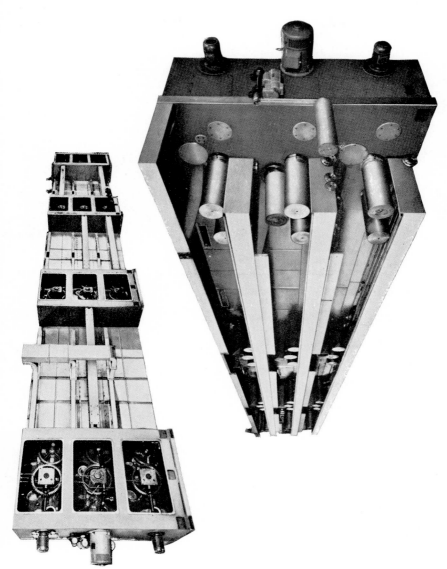

FIG. 11.14   *Three-Tier Washing Machine for Acrylic Tows Showing Front and Back of Machine*

3

FIG. 11.15(a)   *Drive Side of 4-Drum Dryer with Insulating Panels Removed*

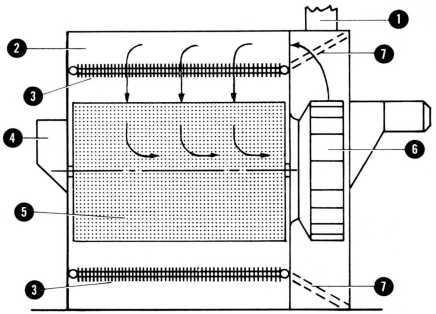

FIG. 11.15(b)   *The principle of the Perforated Drum Dryer with Insulating Panels Removed*

A widely used tow drying technique is that of the suction drum drier. The tow enters the drier and is 'sucked' onto a rotating perforated drum. Air is circulated through the drum via tubular heater banks. By a system of internal baffles the tow is released from one drum and passes to the next. 'Overfeeding' is possible (i.e. the tow is fed faster than the peripheral drum speed and is laid by an oscillating feeder onto a conveyor). This is called 'plaiting' usually pronounced 'plate-ing', and is met with in other types of fibre drier, e.g. the perforated conveyor type). However, care must be taken particularly with tow product to maintain the parallel nature of the filaments. The reason for this will become clear later.

At some stage of the process softening and antistatic oils are applied. These applications are often sufficient for later processing, but lubricating finishes may also be applied at a later stage. Subsequent heat treatments like that in the Turbo-Stapler, and in dyeing and scouring, may remove these oils necessitating a further application.

Crimping is carried out to impart a crinkle to the individual filaments so that they will adhere together in the subsequent processing to staple yarns (called spun yarns) and will possess additional elasticity in the yarn relative to each other. A degree of bulkiness is also achieved by the crimping process, which is normally done at elevated temperature using steam to ensure permanence.

Steam is also used to relax under controlled conditions residual shrinkage, particularly any non-uniform residual shrinkage previously imparted to the fibres. The water thus introduced is finally removed in a drier, often of the conveyor type. Alternatively final relaxing and drying may be combined. Steam setting is sometimes carried out under vacuum, as air in contact with fibre at elevated temperature for any length of time will cause yellowing.

FIG. 11.16(a)

Fig. 11.16(b)   *Typical Fibre Setting Machines by Firma Ph. Welker, K.G., West Germany*

11.7.4.5. Process variables and control   In the coagulating, stretching, washing and after-treatment processes, the parameters which may be varied, and which must be controlled, are those of temperature, draw ratio, tension and speed, and concentration of treatment fluids. All these may be chosen at various stages of the process to suit particular end uses, and where necessary particular filament deniers. Speeds of drawing devices and hence draw ratios are controlled by mechanical means. Tension is often controlled by a Dancer roller unit whose principle is illustrated thus:

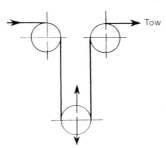

The position of the lower roller depends on tension and a signal is used to control upstream relative to downstream speed.

Processing rate (i.e. tow speed) after extrusion is normally limited by two factors. For low deniers, the diffusion and heat treatment processes are fastest, but the strength of individual filaments is low, so breakage and wrapping around rollers can be a significant problem. For heavy denier filaments the reverse is the case. (We have already seen the possible variables of co-polymer composition and degree of polymerisation which can be adjusted in the polymerisation process.)

11.7.4.6. Staple production Wet tow, before final drying, or dry crimped tow, may be cut to staple in a cutting machine. This process may be said to produce fibres similar to ginned graded cotton from the cotton plant, or clean wool. An acrylic staple of variable cut length may be made, giving a staple diagram suited to particular end uses. By this means the many varieties of cotton and wool may be simulated.

Cutting machines are now available to receive tows of high total denier (probably up to 10^6 denier).

Because of their very high volume (S.G. down to 0·025) staple fibres are normally pressed into bales for transport.

11.7.4.7. Tow and tops production Tow products are made for processing to 'sliver' on convertor machines. There are basically two types of conversion machines. Both are specifically designed to maintain the filaments in the parallel state in which they were produced in the tow. 'Tops' is the name given to a blended and homogeneous continuous sliver or 'rope', of fibres in which the staple lengths are randomly distributed but in which all the fibres are parallel along the longitudinal axis.

The first type of machine (a cutting machine—for example, the Pacific Converter) can blend other synthetic fibres with the acrylic tow. The tow is passed through a cutting section where the staple diagram may be selected, and then through a zone in which the cut lengths become arranged in the random manner characteristic of spun yarns and which is necessary to give strength to the final yarn. This is achieved by a diagonal 'rolling up' or 'shuffling' of the cut tow. The cut tow, now called a sliver, is finally re-crimped.

The second type (a 'stretch breaking' machine, e.g. the Turbo Stapler) utilises the fibre's thermoplastic properties. The fibres are first heated and stretched, set by cooling in the stretched state, and then passed through a random breaking zone, before re-crimping. By taking a part of this product, relaxing it in a vacuum steam-setter, and recombining this relaxed portion with the remainder, high bulk yarns (or yarns with high 'loft') may be made. When the two-component yarn is subjected to heat at a later processing stage (e.g. dyeing), that part of the yarn set in a stretched state by the convertor relaxes (or shrinks), and the relaxed fibres crinkle up. As a result the yarn volume increases considerably. This gives a very soft, warm and light handle to the final product.

The tow for conversion in this way must be virtually free of kinks and non-parallel fibres, as these would cause breakages in the convertors, leading to unacceptable down-time. The convertors are designed for single tow deniers of about half a million and hence a tow of this denier is normally processed in the latter stages of the tow treatment section.

11.7.4.8. Filament yarn production In 1969, only about 0·5 per cent of all acrylic (and modacrylic) fibre production was in continuous filament form,[105] and this has been the case since 1963. Acrylic fibre is largely processed on the machinery already available for wool and cotton.

Filament yarn is spun as previously described, stretched, washed and treated and taken up on bobbins. The number of combined filaments depends on the end use of the particular denier being produced. The yarn, after twisting, is ready for use, mainly in the knitting field and in 100 per cent acrylic carpets.

11.7.4.9. Special and 'premium' products Man can control synthetic fibres—their form and properties—but he cannot, as yet, exercise the same control over the climate or nature which control directly or indirectly the natural fibre products from animal or vegetable origins. To improve upon the natural products, often to the extent of changing established practices to much cheaper ones or to copy nature more closely, man has invented new forms of synthetic fibre (second and third generation synthetic fibre technology is often the convenient term used to describe these, sometimes surprisingly recent, developments). Some of these special products are called 'premium' products since, as they have an especially desirable characteristic, they can command higher prices. This characteristic may result in the short-cutting or replacement of established processing methods or may give the final consumer a new or more attractive product. Often both these advantages are achieved in the same product. Some 'special' products are discussed briefly below. (Although one or two may now almost be considered first generation products, they are grouped together for convenience.)

a. *Bicomponent fibres (sometimes called conjugate fibres)* These are produced by several companies by extruding a filament consisting of two co-polymers with different physical and thermal characteristics through a special spinneret. This gives an inherent and permanent crimp to the fibre. Thereby aesthetic qualities of appearance and handling, and dimensional stability, particularly after repeated washings, are sought to be improved. (These fibres were first produced by Du Pont as Orlon Sayelle in 1959.)

b. *Additives* Additives to improve light reflectivity, anti-static properties and flame-resistance are examples.

c. *Dope dyed products* Pigment may also be added to the dope giving permanent colours, particularly desirable in carpets, curtains, furnishings and outdoor fabrics which are subject to long periods of sunlight.

d. *Special filament cross-sections* Physical and light reflective properites may be enhanced in the finished product by using different orifice shapes in the spinneret.

e. *Spin-bath or gel dyeing* Dye may be applied in the early stages of coagulation.

f. *Differential dyeing*   A technique receiving considerable attention at the present time is the manufacture of fibre products with different polymer combinations and hence different dye uptake rates, or affinities for different dyestuffs. The spun yarns, woven cloth or carpet, may thus be subsequently dyed to give special colour effects and designs.

g. *High shrinkage staple*   Heat stretching, followed by setting of the tow in the stretched state before cutting, and blending this stretched staple with relaxed staple product is an alternative and cheaper route to high-bulk spun yarns.

11.7.5. Product Characteristics and Properties

Acrylic fibre properties are shown in the table which, for comparison purposes, includes the common natural and synthetic fibres. The range of properties are intended to include products for all the various applications but new products appear so frequently that certain of these products may well fall outside these ranges. References 99, 104, 149 and 150 have been used for physical properties and end-uses. Prices are from reference 151.

It is clear that acrylic fibre has adequate strength and elasticity for fibre processing through to weaving, knitting and other machinery.

Wet processes using circular spinneret orifices produce round to kidney or bean-shaped filament cross sections while dry spinning processes produce characteristic dog-bone or dumb-bell shapes.[104]

All the acrylics show excellent resistance to sunlight, are non-allergenic, are not attacked by insects and mould and other micro-organisms, and the physical properties are unaffected after long storage. Acrylics are insoluble in most organic solvents, including, of course, those used for dry cleaning. They are attacked by hot concentrated alkalis; weak acids, weak alkalis, and oxidising agents, however, having little effect. They will dissolve in concentrated nitric and sulphuric acids. This relative inertness of acrylic fibres to most chemical reagents is important, as it results in little or no restriction on chemical processes, such as scouring, bleaching and the like used in the textile industry.

The specific gravity of acrylic fibres is seen to be low among fibres. In addition, it is extremely bulky and soft. This bulkiness leads to warm but light knitwear and fabrics with obvious economies on a weight for weight basis over wool.

The abrasion and crush resistance of acrylics are good, making them highly suited to use in carpets.

With a glass transition temperature of about 60°C, above which cool-set deformation is permanent unless the fibres are reheated to a higher temperature, garments with permanent creases can readily be made. This thermal property is also utilised of course in the bulking processes.

The dyeability of most of the acrylic fibres is now extremely good, most well-known brands offering a choice of dye affinities and techniques. The two functions of co-monomers previously mentioned are to improve dyeing and elastic properties. Reference 135, pages 320 and 327 give the type of dyestuffs used and a comprehensive list of co-monomers.

TABLE XI.13

Properties and Uses of the Common Natural and Man-Made Fibres

	Cotton	Wool	Acrylic	Polyamides	
				Nylon 66	Nylon 6
Ultimate tenacity, g/denier					
dry	3·0–4·9	1·0–1·7	2·5–4·5 (up to 7·0 g/d for filament yarn)	4·3–8·8	4·0–8·0
wet	3·3–6·4	0·76–1·63	2·0–4·5	3·5–7·5	3·9–7·0
knot	—	—	1·8–3·5	3·7–7·5	—
Elongation at break, %					
dry	5–10	25–35	27–48	18–45	18–45
wet	—	25–50	27–58	—	40–53
Initial modulus, g/d	68–93	11–25	25–62		8–30
Elastic recovery (% elongation), %	75 (2%)	99 (2%)	95–99 (2%)	95–100	similar
	45 (4%)	63 (20%)	70 approx. (10%)	(4–8%)	Nylon 66
Specific gravity	1·54	1·32	1·12–1·19	1·14	1·14
Moisture regain (20°C 65% RH), %	7 (25–27 @ 100% RH)	16–18	1·0–2·5	4·2	3·5–5·0 (4.0 average)
Melting point, °C	decomposes at 150	decomposes at 130 chars at 300	decomposes before melting (at 230 approx.)	250–263	215
Uses of filaments	Shirts, dresses, underwear, sheets, etc. etc.	Similar to acrylic (or vice versa)	Suiting fabrics, dresses, pile fabrics and artificial fur, sweaters and knitwear of all kinds, underwear, blankets, furnishings, carpets, awnings, limited industrial use particularly where acid resistance is important	Hosiery, clothing of all kinds, sheets, fur, carpets, nets, drive belts, bristles, etc., parachutes, cords, harnesses, ropes, satin, threads, pneumatic tyres, 'tarpaulins', canvas, filter fabrics, etc.	
Typical quoted price for 1·7 to 3d regular staple (new English pence/kg)			65·5–65·5	88·67–77·67	

Polyester	Polyolefins		Cellulosics		
	Polyethylene	Polypropylene (isotactic)	Viscose rayon (high strength in brackets)	Cellulose acetate	Cuprammonium rayon
4·5–7·5	1–8	2–9·0	2·0 –2·6 (3·3–3·8)	1·1–1·7	1·7–2·3
4·5–7·5	1–8	2–9·0	0·95–1·15 (2·1–3·0)	0·65–0·9	—
4·0–5·0	—	—	—	—	—
15–30	10–50	15–25	17–25 (19–24)	23–30	10–17
30–50	—	—	23–32 (24–29)	35–45	17–33
30–130	—	25–90	approx. 50		
90–95 (?)	good	88 (5%)	60 (2%) (70–100 (2%))	89 (5%) 59 (10%)	poor
1·38	0·92–0·96	0·90–0·94	1·50–1·52	1·32	1·52
0·4–0·5	Very low virtually nil	Very low virtually nil	12–13	6·5	12·5
260	85–132	149–155	decomposes at 185–205 without melting	232 with some decomposition	chars at 250 without melting
Suitings, tropical suits, carpets, ties, underwear, fillings, bags, felt, nets, conveyor belts, electrical insulation, hoses, tyres, awnings, sails etc.	Twines, nets, ropes, filter fabrics, outdoor fabrics	Ropes, fishing gear, carpets, blankets, knitwear, paper, conveyors	Textiles, hosiery, underwear, dresses, linings, carpets, furnishings, sportswear (high tenacity rayon used for tyres)	Pyjamas, shirts, underwear, dresses, satins, ribbons (stiffened material can be made— Trubenising), wire insulation	Apparel, hosiery, underwear, substitute for silk, sheer fabrics
60–57		55·0–55·0	25–30 approx.	probably about 40	

11.7.6. Statistical Section

Acrylic fibre production in the world has risen rapidly since the introduction of the fibre in 1950. The production and capacity figures for acrylics and modacrylics are given below in ton per annum.[105]

TABLE XI.14

	Production	Available capacity
1950	3,200	—
1960	110,000	—
1965	400,000	—
1967	540,000	—
1969	850,000	950,000 (March)
1970	—	1,160,000 (March)
1971	—	1,430,000 (Dec.)

According to Textile Organon filament yarn production capacity has risen from 5,800 ton/annum in March, 1969 (the same capacity is reported for March, 1970) to more than 10,000 ton/annum predicted for December, 1971. However, as this includes modacrylics—(>35 per cent <85 per cent acrylonitrile units by weight)— and products made from these fibres are frequently found in filament yarn form it is difficult to draw any conclusions. The world production of modacrylics is relatively small.[104-106] The author estimates that about half this continuous filament production is of true acrylic fibres.

World wool production has been constant for about ten years at roughly 1·5 million tons/annum. World cotton production has risen very slightly over the same period and is currently about 11 million tons/annum. Rayons have risen by about 30 per cent over the last ten years but show little increase over the last five years. (Current world production about 3·5 million tons/annum.) Present indications of consumption and profitability suggest that both rayon and wool have found it difficult to maintain their positions in the last year or so. The general climate of restricted industrial and economic growth in many developed countries has had a greater effect on wool and rayon sales than on the true synthetics.

TABLE XI.15

Fibre	Actual production (tons)		% average increase	Available capacity (tons/annum)
	1963	1969	1963–9	Dec. 1971
Acrylics (and modacrylics)	210,000	850,000	400	1,430,000
Nylons	740,000	1,800,000	240	2,700,000
Polyester	260,000	1,360,000	520	2,500,000
Polyolefines	36,000	230,000	640	480,000
Totals	1,246,000	4,240,000	340	7,110,000

Taking the last six years as a guide to the growth of the four major synthetics the figures are as follows in Table XI.15. Also included is the available capacity predicted for December, 1971 from announcements of planned new capacity and capacity extensions.[147]

Acrylic fibre production has been rising in recent years at rates between 20 per cent and more than 30 per cent per annum. America has a rate of increase at the bottom of the range but in both Japan and Europe the production of the fibre has shown a growth rate at the top of the range.

11.7.7. End Uses

11.7.7.1. Uses in Textiles The main outlets for acrylic fibres are in knitted goods (probably nearly half is sold in this form) in carpets and furnishing fabrics (about 20 per cent), and in woven fabrics (about 10 per cent). The remaining outlets are in handknitting yarns, linings and fillings and miscellaneous industrial applications, pile fabrics.

These proportions are what might be expected of a developed European country but particular countries use different proportions depending on consumer demands and to some extent on the particular characteristics of the available indigenous supply. For example there is a substantial and increasing demand for carpets in the developed countries, particularly those with cold to moderate climates, but less demand in, say, Italy and Spain and in Japan.

11.7.7.2. Carbon fibres There is one secondary product made from acrylic fibres which should not go unmentioned—carbon fibres. By a special heat treatment process developed at the Royal Aircraft Establishment, Farnborough, acrylic fibre filament yarns of high purity and strength are converted to carbon fibres of very high strength and stiffness. The fibre has found special applications in the US in their aerospace programme, and is also used in the textile industry itself for the heddle frames of high speed looms[147] and has been developed (as yet not with complete success) by Rolls Royce for the blades of their RB 211 jet-engine. The fibre is used moulded with a polymer such as epoxy resin. There is thought to be potential for this new structural material in airframes and other uses where high stiffness and strength combined with lightness and/or small size will result in worthwhile economies. However, the material is still very expensive. The latest news is of its use in tubes from which bicycles are being made for a British cycle team.[148]

NITRILE RUBBERS
By **P. T. Mapp**

11.8.1. Historical Development

Nitrile rubbers were discovered in 1930 during co-polymerisation studies of butadiene and mono-olefins by Konrad and coworkers[152] at I. G. Farbenindustrie AG Germany who found that the co-polymerisation product of butadiene with acrylonitrile yielded an elastomeric material showing excellent oil and solvent resistance; the product was named Buna N. Following this discovery, subsequent work led to the pilot plant production during 1934 and full scale manufacture of Buna N in 1937. Exports of Buna N to the UK and USA commenced almost immediately but for commercial reasons the name was changed to Perbunan nitrile rubber.

The outbreak of World War II in 1939 interrupted the exports of Perbunan to America at a time when USA consumption was running at around 150 tons per annum. This prompted various USA rubber companies to initiate nitrile rubber (NBR) manufacture.[153] Standard Oil Company of New Jersey, who controlled Jasco Inc. to whom American rights had been assigned by I. G. Farbenindustrie, first started local nitrile manufacture in 1939 under the product name 'Paracril'. They were closely followed by Firestone Tyre and Rubber ('Butaprene'), Goodyear Tyre and Rubber ('Chemigum') and also B. F. Goodrich (Hycar). International Latex, now Standard Brands Chemical Industries Inc. started manufacture of their 'Tylac' nitrile rubbers early in the 1960's. Outside USA, Polymer Corp. ('Krynac') commenced nitrile rubber production in Canada during 1948 and British Geon ('Breon') and ICI ('Butakon') started up in UK in 1957; the ICI 'Butakon' nitrile polymer interests were sold to Revertex Ltd. during 1969. In Europe, apart from Germany, Ugine SA produced 'Butacril' in France during 1960 while Montecatini ('Elaprim') and ANIC (Europrene N') commenced manufacture in Italy early in 1961.

Production spread to USSR during the 1950's and to Japan in 1962.

Currently the total reported world capacity for nitrile rubber both latex and dry polymer, is around 285,000 tons being broken down into North America 149,000 tons, Europe 83,000 tons, Japan 23,000 tons and Eastern Europe 30,000 tons.[154]

11.8.2. Manufacture

Nitrile rubber manufacture is carried out by the co-polymerisation of butadiene with acrylonitrile using an emulsion polymerisation technique generally by a batch process although continuous manufacture is also practised. Processing details are basically similar to conventional emulsion polymerisation procedures. A portion of the water phase is added to the stainless steel or glass-lined reaction vessel fitted with temperature control and stirring facilities. The emulsifier system is then added followed by the acrylonitrile and modifier. Butadiene and activater then follow, the mixture is brought to the correct temperature and polymerisation started under pressure and with gentle agitation. For certain activator systems it is necessary to exclude oxygen from the vessel before commencing.

Not all the recipe ingredients are necessarily added initially, certain materials which readily decompose or are rapidly used up during polymerisation are usually added in incremental amounts during the course of polymerisation.

Polymerisation is carried out generally at a temperature between 5°C and 40°C and proceeds at a conversion rate of 5–14 per cent per hour depending on conditions and recipe until a conversion of 70–85 per cent is achieved. Having achieved the desired and predetermined conversion the latex is immediately pumped into a stripping vessel and the reaction killed by the addition of short-stop. Free monomers are then stripped from the latex under reduced pressure and regenerated for subsequent re-use. The monomer-free latex is then compounded with stabiliser and passed on to the coagulation process.

Efficient coagulation and washing out of excess polymerisation materials such as surfactants is extremely important. Coagulation is normally achieved by feeding a fine stream of latex through a series of tanks containing coagulant and precipitating the polymer by agitation into a fine crumb. Coagulants are usually electrolytes such as chlorides or sulphates but cationic amine types are also used.

The fine crumb is then filtered and washed thoroughly to remove all excess coagulant and other materials by passing over vibrating screens through a bank of water spray treatments. After washing, the crumb is dried by either band feeding through a drying oven or alternatively passing the crumb through an expeller-extruder drier. The dried polymer is then weighed, pressure baled and wrapped ready for despatch.

Recipe and temperature of polymerisation are important considerations which influence polymer properties and polymerisation techniques. Following on from the basic principles of nitrile rubber manufacture detailed above an attempt is made below to identify more closely the choice and parts played by individual recipe ingredients.

11.8.2.1. Monomer ratio   The ratio of butadiene to acrylonitrile is important in nitrile rubber manufacture. Besides influencing the final polymer properties the ratio has a tremendous effect on polymerisation. Because of the different reactivities of the two monomers (acrylonitrile polymerises faster than butadiene), the ratio of monomers in the polymer chain do not correspond to the ratio of the monomers in the reaction mixture except under azeotropic conditions. For nitrile rubber polymerised at 25°C the azeotropic mixture ratio is 37 per cent ACN: 63 per cent butadiene; at this ratio and temperature the monomeric mixture is similar to the polymer ratio and remains constant during co-polymerisation. In the manufacture of acrylonitrile polymers, containing less than 37 per cent acrylonitrile, the amount of acrylonitrile required in the recipe is less than that obtained in the final polymer. Conversely, to achieve bound polymer acrylonitrile polymers higher than 37 per cent acrylonitrile the amount of acrylonitrile required in the recipe is substantially higher. A typical relationship between monomer and polymer ratio for acrylonitrile/ butadiene polymerisations at 5°C and also 25°C and taken to 70 per cent conversion is illustrated in Fig. 11.17.

The azeotropic ratio changes slightly with polymerisation temperature and at 5°C the ratio is 43 ACN/57 Bd.

In addition to having a marked effect on polymer composition the level of acrylonitrile also effects the rate of polymerisation. Being the more reactive monomer acrylonitrile greatly accelerates rate of polymerisation producing similar polymer yields in a shorter time or at lower temperatures.[155]

11.8.2.2. Emulsifiers   The main function of the emulsifier is to form micelles into which the monomer droplets are solubilised. Polymerisation proceeds within the monomer micelles and as polymer is formed the emulsifier serves its secondary function of stabilising the polymer against agglomeration. Occasionally, mixed emulsifiers are used, one being responsible for micelle formation the other acting as

a protecting agent for polymer particles. The choice of emulsifier depends on monomer ratio, catalyst system and polymerisation temperature.

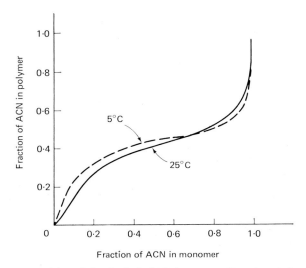

FIG. 11.17   *Composition of Acrylonitrile in Polymer as a Function of Acrylonitrile in
Monomer Mixture*

Anionic materials represent the most important class used in nitrile rubber production. Typical examples being alkyl salts of fatty acids, rosin acids or long chain
aryl sulphonates and alkyl sulphates. The anionic emulsifier provides a negative
charge to the latex particles. Of the various anionic types available most effective
systems are usually based on alkyl salts of the alkyl and alkylaryl sulphonic acids.
These materials have the attraction of being very soluble and effective in acidic
conditions and also they can be easily washed out and show little sensitivity to pH
changes.

Cationic and non-ionic emulsifiers find little use in nitrile rubber manufacture.

11.8.2.3. Activators/initiators   Polymerisation is initiated by a chemical reaction
which forms free radicals. Two types of activators are commonly used; peroxidic
activators and redox activators. Redox systems are preferred because of improved
processibility and physical performance of the subsequent polymer and also because of their versatility, especially at low polymerisation temperatures.

Peroxidic activators are typified by persulphates, peroxides and hydroperoxides;
alkali persulphates are safe to handle and break down into free hydroxyl and sulphate
radicals. Typical peroxide activators would be benzoyl peroxide and cumene hydroperoxide. Unfortunately the ease with which the above materials form radicals is
dependent on temperature and when used as the sole initiator a satisfactory polymerisation rate is only achieved at temperatures above 30 °C. The choice of activator

combination is very much dependent on polymerisation temperature and for low temperature reactions redox systems are more commonly used.

Typical redox systems are those based on organic hydroperoxides catalysed by metal reducing agents; complexing agents such as pyrophosphates or hexaphosphates are also included, especially under alkaline polymerisation conditions. Other systems using persulphate compounds activated by long chain aliphatic mercaptans also find interest in nitrile rubber manufacture.

11.8.2.4. Modifiers Modifiers affect the chain length of the polymer and act as chain transfer agents modifying the molecular weight and controlling the degree of branching and crosslinking. During the course of polymerisation the ratio of modifier to monomer concentration decreases with conversion. Highly reactive modifiers are rapidly used up during polymerisation and in order to prevent subsequent branching it is often desirable to add modifiers in portions during the reaction.

Typical modifiers currently being used are long chain aliphatic mercaptans with between eight to fourteen carbon atoms. Aliphatic mercaptans having less than six carbon chains markedly retard the polymerisation rate[156] as do mercaptans with greater than sixteen carbon chains.

Besides acting as modifiers, mercaptans also act as secondary activators in conjunction with peroxy compounds.

Other low valency sulphur compounds such as di- and polysulphides and sulphides of xanthogenic acid have also successfully been used as nitrile rubber modifiers.

11.8.2.5. Short-stops Uncontrolled high conversion of nitrile rubber leads to an insoluble crosslinked polymer which is difficult to process. Short-stops are required to stop polymerisation at a predetermined conversion. This is achieved by neutralisation of polymerisation radicals and activators. Dithiocarbonates, hydroquinone, hydroxylamine and phenyl napthylamine are all useful short-stops and selection is generally dependent on solubility, discolouration, staining and toxicity considerations.

11.8.2.6. Stabilisers Nitrile rubbers are easily oxidised by air because of remaining unsaturation in the polymer. To prevent oxidation antioxidant stabilisers are added. In addition to preventing oxidation during polymer storage stabilisers are also generally adequate to protect the polymer during subsequent processing.

Almost all types of conventional antioxidants are suitable but again the choice is dependent on discolouration, volatility, toxicity, and odour. Stabilisers are not included in the polymerisation recipe but are added to the latex before coagulation.

11.8.3. Processing and Compounding

11.8.3.1. Processing Nitrile rubber compositions can be readily prepared on conventional rubber machinery. Processibility depends primarily on the inherent make-up of the polymer which has been fixed during manufacture. Unlike general purpose rubbers, nitrile polymers show minimum degradation during processing and hence raw polymer viscosity is extremely important in deciding compound viscosity. With correct selection of base polymer, polymer premastication is not necessary for most applications.

Compounding of nitrile stocks is normally carried out by either open mill or internal mixing. Milling should be undertaken on a cold mill using a tight nip and it is preferable to select a low speed and friction ratio thereby keeping heat generation to a minimum. Low viscosity polymers form a smooth sheet immediately whilst the higher viscosity grades need a few passes before compounding can start. The batch size is usually kept somewhat smaller than for other polymers to ensure effective mixing and the sequence of ingredient addition is important. Good dispersion of ingredients is extremely important if optimum physical properties are to be obtained. Because of its low solubility in nitrile rubber, it is advisable to add sulphur, either as powder or master-batch, as early as possible during open mill mixing and therefore sulphur is generally the first ingredient added followed by fillers and other compounding ingredients. Carbon blacks must also be added carefully, good initial dispersion of part of the black is essential and mixing temperatures should be kept as low as practicable to achieve this; a dull surface appearance indicates poor or inadequate dispersion, while a smooth glossy surface denotes good black dispersion. Blacks are generally added in at least two portions mixed with softeners. Accelerators are added at the end of the mixing process and in certain cases, where compound is likely to be stored, they are omitted until the subsequent compound warm-up process.

Where internal mixers are used it is again essential to keep temperatures as low as possible and to follow a similar addition sequence to that detailed for mill mixing. For high viscosity 'hot' polymers and compounds where heat build-up is high and temperature control more difficult, sulphur addition is preferably delayed in Banbury compounding until the end of the mixing cycle to avoid prevulcanisation. In such cases the accelerator is added on a warm-up mill after cooling off the batch.

Calendering conditions vary with the type of compound being processed. Temperatures are somewhat lower than for general purpose rubbers and range from 80°C for compounds high in rubber content down to 30°C for compounds rich in filler. Semi active blacks, factice and pre-crosslinked nitrile rubber are useful ingredients in compounding to give smooth calendered sheets showing minimum shrinkage.

Properly compounded, nitrile rubbers are readily extruded; processing aids as detailed for calendering are also applicable for extrusion compounds. Screw and jacket temperatures should be kept as cool as possible and the die head heated sufficiently to produce the required surface texture and appearance.

Curing of nitrile rubbers may be effected over a wide range of temperatures by the use of hot air, open steam, hydraulic presses with steam, superheated water or electric heating for compression, transfer and injection moulding. There is little danger of reversion with nitrile compounds; consequently nitrile rubbers may be safely processed at temperatures higher than those normally used for natural rubber. Curing times and temperatures required will depend upon the vulcanising system used and the size of the article to be cured; any temperature between 125°C and 200°C may be considered suitable.

11.8.3.2. Compounding Because of lower unsaturation nitrile rubbers require lower sulphur levels and higher accelerator levels than natural rubber. Between 1·0 and 2·0 phr sulphur is usual for the curing of nitrile compounds; alternatively sulphur donors such as TMTD, at levels of up to 3·0 phr can be used. These latter systems are of special importance where low compression set and good heat resistance are required. Peroxide curing systems find only occasional use in nitrile rubber technology while polyvalent metal salts systems are used solely for carboxylated polymers.

Almost all conventional rubber accelerators can be used with nitrile rubber, in general, sulphenamides, thiazole and thiuram types are preferred as primary acce-lerators. The rapid rate of vulcanisation shown by nitrile polymers means that secondary accelerators are kept to a minimum but in practice it is usual to activate thiazole accelerators by thiuram or guanidine boosters. For normal sulphur curing systems the total accelerator level varies between 0·75 and 1·5 phr; using thiuram systems it is normal to use a sulphenamide accelerator up to a level of 3·0 phr.

Zinc oxide at a level up to 5 phr activates vulcanisation of nitrile rubbers whilst for carboxylated nitrile polymers the use of a masterbatch of zinc peroxide in nitrile rubber is recommended for this purpose.[157]

The majority of nitrile polymers contain sufficient non-staining, non-discolouring antioxidant to protect them for a considerable period of time under normal storage conditions. Nevertheless it is generally necessary to include additional antioxidants during compounding especially where good thermal and oxidation resistance is required in application. Stabilisers of the phenylene diamine, naphthylamine or diphenylamine types are particularly useful.

Without the addition of fillers the properties of gum nitrile compounds are poor; reinforcing fillers are therefore of extreme importance in developing the best proper-ties. Carbon black is by far the most outstanding ingredient in imparting high strength.

Practically all fillers used in rubber compounding can be used with nitrile rubber. Reinforcing behaviour is similar to that in other rubbers; highly active fillers give the greatest reinforcement but produce hard, boardy and difficult-to-process stock whilst inactive materials are easy to process but achieve low reinforcement. Optimum reinforcement by active blacks, such as channel and high abrasion furnace blacks, is achieved at a filler loading of between 40 and 60 phr similar to natural rubber; less reinforcing blacks like GPF, SRF, thermal and lamp black are frequently used at high loadings thereby producing cheaper compounds. Typical non-black reinforcing fillers are hydrated silica and aluminium silicate, while inert fillers of the whiting and barytes type are often used to lower costs. China clay, talc and mica are used in much the same way as lamp black.

Correct selection of softener is essential in nitrile rubber technology. Since the polymer itself is oil resistant special types of softeners are required. The addition of softeners to nitrile rubber affects the viscosity, tack and processing characteristics in addition to the elastic, tensile, hardness, cold flex and swelling properties. Softener effectiveness is largely dependent on the compatibility of the materials. Because nitrile rubbers have strongly polar properties, only high polarity softeners are compatible. In general, softeners can be divided into two classes; fats, resins, waxes

33

and polymeric materials on one hand and ester types on the other. The former types are especially useful for improving tack and processing while the latter are employed where good low temperature performance is needed. For use in applications where oil and solvent resistance are required softener volatility and extractibility are important considerations and here the polymeric types are of particular usefulness. The amount of softener must be determined with care, too high a level can adversely influence tack and cure rate and although up to 50 phr is used for some applications generally 15–25 phr is a satisfactory compromise.

11.8.4. Properties and Applications

11.8.4.1. General properties In many industrial and engineering applications rubber components are required to convey fluids, act as seals and to carry out a multitude of other functions where the well known elasticity of rubber is particularly useful. Environmental conditions involving oils, greases and other chemicals, preclude the use of either natural rubber or styrene-butadiene rubber and it is in these applications where nitrile rubbers provide the best balance of properties. In general the properties of nitrile rubbers can be summarised as:

a. Excellent oil and solvent resistance in contact with aliphatic hydrocarbons, such as gasoline, mineral oil, animal and vegetable fats and oils, and non-polar solvents.
b. High abrasion resistance and good mechanical properties.
c. Good heat resistance up to 120°C.
d. Low air and gas permeability.

On the debit side, nitrile polymers exhibit poor ozone resistance and electrical properties, limited resilience and poor gum strength properties. Their poor electrical properties are, however, useful in antistatic applications.

The oil resistance of nitrile rubber is dependent on the amount of acrylonitrile present in the polymer; the higher the acrylonitrile content the greater will be its oil and solvent resistance (Fig. 11.18).

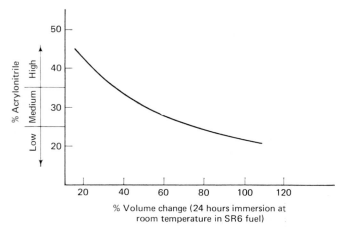

FIG. 11.18   *Effect of Acrylonitrile Content on Volume Swell in SR6 Fuel*

Acrylonitrile content also influences the rubbery properties of the co-polymer and this fact is of special importance with regard to low temperature performance where high levels of acrylonitrile impair low temperature flexibility. (Fig. 11.19.)

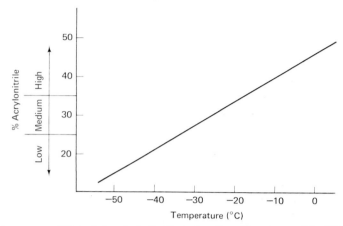

FIG. 11.19 *Effect of Acrylonitrile Content on Low Temperature Flexibility*

In addition the acrylonitrile content also has some influence on other physical properties such as resilience, permeability, hysterisis, compression set and softener compatibility, all of which are adversely affected by increased acrylonitrile. On the other hand, high nitrile polymers generally show improved tensile, modulus, heat resistance and also greater compatibility with polar polymers. Thus nitrile rubber technology demands a compromise between oil and solvent resistance on one hand and low temperature and physical performance on the other.

11.8.4.2. Processing properties The butadiene/acrylonitrile ratio in the polymer also influences polymer processibility; high acrylonitrile generally being associated with processing difficulties. This effect is noticeable in 'hot' polymers more so than for the more generally preferred 'cold' polymers and can be minimised by the correct choice of modifiers during polymerisation and careful control of the molecular weight and gel characteristics of the polymer.

Molecular weight is normally identified by reference to Mooney viscosity and nitrile rubbers are generally characterised by Mooney viscosity in addition to acrylonitrile content. Viscosity provides a measure of molecular weight which in turn influences processing behaviour; easy processing being generally associated with low molecular weight. Easy processability is illustrated in terms of lower power consumption at lower mixing temperatures which in turn leads to freedom from scorch during processing. In addition low viscosity also benefits mould flow, easier extrusion and better calendering together with reduced shrinkage. Unfortunately too low a Mooney viscosity results in lower tensile and tear strength, higher compression set and greater difficulty in achieving good filler dispersion.

Gel content, which is defined as the polymer portion which does not dissolve in recognised nitrile rubber solvents, is a further important variable which influences polymer processability. Three types of gel can be present in nitrile polymers: (a) gel produced by crosslinking and branching caused by uncontrolled polymerisation, (b) gel formed during drying and pos-tpolymerisation processes and (c) deliberate and controlled gel introduced by the polymer manufacturer. Permanent gel, especially predominant in 'hot' polymers, as identified in types (a) and (b) adversely affects processing properties. Type (c) gel on the other hand is deliberately introduced into certain polymers which are marketed as processing aids for addition to normal conventional nitrile polymers.

11.8.4.3. Applications   Nitrile rubbers because of their excellent resistance to oils and hydrocarbons coupled with good heat stability, high abrasion and low permanent set properties, make them especially suited to applications where these properties can be exploited to the full. Typical products for which nitrile polymers find their major interest are oil seals, diaphragms, gaskets, printing roller coverings, fuel hose, oil resistant belting, flexible fuel tanks, ebonites, sponge, rubber shoes and textile aprons. In addition nitrile rubber finds application in the adhesive field and also as a modifier in PVC, phenolic resin and bitumen technology.

FIG. 11.20

| Natural rubber | Neoprene | Nitrile rubber | Butakon AC6040 |
| 15 sec exposure | 9 min exposure | 3 min exposure | 20 min exposure |

Resistance to 1 per cent Ozone

Although the above applications represent quite a comprehensive number of outlets for nitrile polymers the relatively poor ozone resistance has limited their wider use especially in outdoor applications. The technique of either dry blending or latex blending nitrile polymers with PVC as a means of up grading ozone resistance has widened considerably the usefulness of nitrile in other fields. Not only is excellent ozone resistance obtained for NBR/PVC blends but flame and tear resistance is significantly improved whilst at the same time maintaining other important properties such as oil, fuel and abrasion resistance.[158] These polymer blends are finding increasing interest in applications like cable jackets, hose and belting covers and other areas where good ozone resistance is required. The excellent ozone resistance of NBR/PVC blends is evident by reference to Fig. 11.20.

Another recent development in nitrile rubber technology aimed at extending the field of application is the introduction of carboxylated nitrile polymers.[159] Carboxylation, achieved by the inclusion of a suitable acid component in the polymerisation recipe, enhances the tensile, modulus, hardness, tear and abrasion properties of nitrile rubbers thereby making the polymer particularly suitable for roller coverings, oil well packers, shoe soling and adhesives. Carboxylated nitrile rubbers also exhibit a marked improvement in compound green strength over conventional non-carboxylated nitrile polymers and this could generate further interest.

TABLE XI.16

Compound	A	B	C
Polymer 100			
Zinc oxide, 5			
Stearic acid, 1			
Sulphur, 1·5			
MBTS, 1·0			
Elastomer	NBR	NBR	CNBR
Acrylonitrile content rating	Med low A/N	Med high A/N	Med high
Tensile strength, kg/cm^2	20	45	148
100% modulus, kg/cm^2	4	5	21
300% modulus, kg/cm^2	10	14	70
Elongation at break, %	530	650	410
Hardness, IRHD	41	46	58
Volume change			
in ASTM no. 3 oil, %	23·6	9·6	8·7
in ASTM fuel B, %	51·8	24·6	23·3
Low temperature flex, °C	−33	−21	−20

The higher inherent hardness shown by carboxylated nitrile polymers should permit the attainment of high hardness compounds at lower filler loadings resulting in compounds showing the minimum drop in resilience properties and processing difficulties associated with high filler loadings.

Acknowledgements

The writer wishes to acknowledge the help and suggestions given by his colleauges in writing the above and also Revertex Limited for their permission to publish.

REFERENCES

[1] Moureu, *Ann. Chim. Phys.*, 1894, **2**, 187, 191.
[2] FIAT Final Report no. 836.
[3] *Chem. Pro. Eng.*, June 1967, 85–7.
[4] Heyner and Seeboth, *Chem. Technik*, 21 Jg., Heft 5, May 1969, 301–6.
[5] Cevidalli *et al.*, *Chim. e Ind.*, 1967, **49**, 809–13.
[6] Knapsack, BP 792,572, 880,952, 901,559.
[7] Allied Chemical, USP 2,481,826.
[8] Maihle and de Godon, *Comptes Rend.*, 1918, *166*, 215.
[9] Du Pont, USP 2,443,420, 2,452,187.
[10] Phillips Petroleum Co., USP 2,412,437.
[11] Shell, *Ind. Eng. Chem.*, 1948, **40**, 2046.
[12] DCL, BP 709,337, 744,011.
[13] DCL, BP 723,003.
[14] Sohio, USP 2,941,007.
[15] Sohio, USP 2,904,580.
[16] DCL, BP 864,666, 876,446, 897,226.
[17] Hadley, *Chem. and Ind.*, 1961, 238–43.
[18] Dalin *et al.*, *Dokl. Akad. Nauk SSSR*, 1962, **145**, 1058–60.
[19] Dalin *et al.*, *Azerb Khim. Zhur.*, 1963, No. 4, 99–102.
[20] Gel'bshtein *et al.*, *Kinetika Kataliz.*, 1965, **6**, 1025–32.
[21] Gel'bshtein *et al.*, *Neftekhimiya*, 1964, **4**, 906–15.
[22] Lichtenberger, *Revue AFTP*, 1967, No. 183, 29–35.
[23] Blanc, *Chim. et Ind.-Genie Chimique*, 1970, **103**, 1359–63.
[24] Adams and Jennings, *J. Catalysis*, 1963, **2**, 63–8.
[25] Adams and Jennings, *J. Catalysis*, 1964, **3**, 549–58.
[26] BP Chemicals International Ltd., unpublished work.
[27] Callahan *et al.*, Division of Petroleum Chemistry Inc., ACS, September 1969, C13–C27. Preprints Vol. 14, No. 4.
[28] Seeboth *et al.*, *Brennstoff Chem.*, 1969, **50**, 268–72.
[29] Kolchin *et al.*, *Neftekhimiya*, 1964, **4**, 301–7.
[30] Kolchin *et al.*, *Kinetika Kataliz*, 1965, **6**, 878–83.
[31] Gel'bshtein *et al.*, *Neftekhimiya*, 1964, **4**, 906–15.
[32] *Ibid.*, 1965, **5**, 118–25.
[33] *Ibid.*, 1969, **9**, 400–8.
[34] Margolis *et al.*, *Kinetika Kataliz*, 1962, **3**, 181.
[35] Ioffe *et al.*, *Kinetika Kataliz*, 1962, **3**, 194.
[36] Grasselli and Callahan, *J. Catalysis*, 1969, **14**, 93–103.
[37] *Ind. Eng. Chem.*, 1963, **55**, p. 39.
[38] *Hydroc. Proc.*, 1967, **46**, 141–2.
[39] *Ibid.*, 1962, **41**, 187–90.
[40] *Ibid.*, 1969, **48**, 146–7.
[41] Schoubeck *et al.*, *Chem. Ind. Technik*, 1966, **7**, 701–5.
[42] *Chem. Eng. News*, 14 December 1964, 46–7.
[43] *Chim. e Ind.*, 1967, **49**, 809–13.
[44] *Japan Chem. Week*, 1 January 1970, p. 2.
[45] Vekemans *et al.*, *Ind. Chim. Belge*, 1967, **32**, special issue, Grp. VIII, S20 and 9, 694–7.
[46] Blanc, *Chim. et Ind.—Genie Chimique*, 1970, **103**, 1359–63.
[47] Sohio, USP 3,230,246.
[48] Sohio, BP 1,126,617.
[49] DCL, BP 978,520.
[50] Sohio, USP 3,472,892.
[51] Ugine, FP 1,355,619.
[52] DCL, BP 848,924.
[53] Sohio, USP 3,200,141.

[54] Ashahi K K K K, BP 913,832.
[55] Ugine, BP 1,006,815. Nitto Chem. Ind. Co. Ltd., BP 1,199,697.
[56] Erdölchemie, G P 1,808,306.
[57] Erdölchemie, Dutch P 68–12,659.
[58] O S W, Austria P 278,046.
[59] Nitto, Dutch P 1,954,450.
[60] Sohio, U S P 3,468,624.
[61] Erdölchemie, Dutch P 69–05,798.
[62] Sohio, U S P 3,404,947.
[63] Ugine, FP 1,336,342.
[64] Bayer, BP 995,007.
[65] D C L, BP 1,063,403.
[66] Erdölchemie, BP 1,051,080.
[67] Sohio, U S P 3,489,788.
[68] Knapsack, BP 928,777.
[69] D C L, BP 1,012,013.
[70] Ugine, Dutch P 65–00,866; BP 990,013.
[71] Sohio, BP 965,351.
[72] Knapsack, BP 1,038,069.
[73] Sohio, U S P 3,185,636.
[74] Knapsack, U S P 2,987,451. Mitsubishi, Jap.P 4865/63 and 14660/64. Monsanto, U S P 3,262,966.
[75] D C L, BP 901,555. Bayer, Jap.P 8213/64.
[76] D C L, BP 970,629.
[77] S N A M, BP 987,347.
[78] Du Pont, U S P 2,681,306.
[79] D C L, BP 1,131,134.
[80] Celanese, BP 719,911.
[81] Ugine, BP 984,725.
[82] Ugine, BP 998,639.
[83] Sohio, U S P 3,442,771.
[84] Ugine, BP 1,058,379.
[85] Erdölchemie, Dutch P 68–08,562.
[86] Ugine, FP 1,431,919 and 1,473,072; BP 984,725.
[87] Sohio, U S P 3,352,764. Societa Edison, BP 1,138,143.
[88] Ugine, FP 1,493,519.
[89] B P Chemicals, BP 1,130,846.
[90] Ugine, FP 1,493,519. Courtaulds, BP 1,180,556.
[91] Monsanto, BP 708,108 and U S P 2,683,163. American Cyanamid, U S P 2,783,269.
[92] Asahi Chemical Co., BP 1,084,599, 1,127,355, 1,139,398, 1,156,620. Kominami et al. 'Symposium on Catalytic Oxidation' (London, July 1970).
[93] Du Pont, BP 2,736,739, 3,489,787.
[94] Ind. Chem., May 1963, 242–5.
[95] I C I, FP 1,556,127.
[96] Monsanto, G P 1,964,786.
[97] Lummus, Dutch P 69–18,396.
[98] Carress and Lyne, Chem. and Ind., 23 September 1967, 1579.
[99] Moncrieff, 'Man-Made Fibres', Heywood Books, 4th ed., revised 1966.
[100] U S P's 2,117,210 and 2,140,921.
[101] Marvel et al., J. Am. Chem. Soc., 1940, 62, 3109.
[102] Staudinger, H., Helv. Chim. Acta., 1922, 785.
[103] Carothers, W. H., 'High Polymers', Vol. I (Interscience Publishers, New York, 1940).
[104] Sanders, Acrylic Fibres—Process Survey, C P E, Sept. 1968.
[105] Textile Organon, June 1970.
[106] Sanders, Acrylic Fibres—Process Survey, 2, C P E, Sept. 1970.
[107] Several references support this, among them: Miller et al., Paper presented at meeting

in miniature, N Y Section of *J. Am. Chem. Soc.*, 16 March 1956; Peebles, *J. Am. Chem. Soc.*, 1958, **80**, 5603–7.

[108] Coyner, E. C. and Hillman, W. S., *J. Am. Chem. Soc.*, 1949, **71**, 324–6.

[109] Thomas, W. M., *Adv. Polymer Sci.*, 1961, **2**, 401–41.

[110] Mayo and Walling, *Chem. Rev.*, 1950, **46**, 191–287.

[111] Morgan, L. B., Polyacrylonitrile in *'Fibres from Synthetic Polymers'*, ed. R. Hill (Elsevier Publishing Co., 1953).

[112] Tong, L. K. H. and Kenyon, W. O. J., *J. Am. Chem. Soc.*, 1947, **69**, 2245.

[113] Forney, *Chem. Eng. Prog.*, 1966, **62**, 91.

[114] Bamford and Jenkins, *Proc. Roy. Soc.*, 1955, A**228**, 220–37.

[115] Smeltz and Dyer, *J. Am. Chem. Soc.*, 1952, **74**, 623–8.

[116] *'Fibres from Synthetic Polymers'*, ed. R. Hill (Elsevier Publishing Co., 1953).

[117] See ref. 109, 426.

[118] Cernia, Chapter on Acrylic Fibres, *'Man-Made Fibres—Science and Technology'*, Vol. III (Interscience Publishers, 1968).

[119] Hohenstein and Mark, *'High Molecular Weight Organic Compounds'*, Chapter 1, 'Polymerisation in Suspension and Emulsion' (Interscience Publishers, 1949), 1–74.

[120] Palit and Guha, *J. Polymer Sci.*, 1959, **24**, 243–50.

[121] See Thomas, Gleason and Mino, *J. Polymer Sci.*, 1957, **24**, 43–56. This paper entitled *Acrylonitrile Polymerisation in Aqueous Suspension* studies the chlorate-sulphite redox initiated polymerisation of homogeneous aqueous acrylonitrile solutions leading to polymer precipitation. In addition to showing an increase in polymerisation rate due to the presence of emulsifiers (such as sodium lauryl sulphate) it also shows that ferrous and cupric ions in p.p.m. concentrations greatly increase rate. The effect of temperature is also indicated. Neutral salts reduced reaction rate. Agitation had the effect of reducing rate and increasing aggregation of the polymer.

[122] Uchida and Nagao, *Bull. Chem. Soc. Japan.*, 1957, **30**, 311–14.

[123] Bovey *et al.*, *'Emulsion Polymerisation'*, 1955, 157.

[124] Blout *et al.*, *'Monomer'* (Interscience Publishers Inc. N V, 1949, 23, in chapters on 'Vinyl Acetate and Esters of Acrylic Acid.'

[125] U S P 2,639,279.

[126] U S P 2,847,405 (1958).

[127] U S P 2,777,832 (1957).

[128] Marshall Sittig, *'Synthetic Fibres from Petroleum'*, Noyes Development Corporation (n.d.c.), 1967.

[129] Sandor, *Materiale Plastice*, 1965, **2**, 14.

[130] Ellwood, P., *Chem. Eng.*, 25 August 1969, 90.

[131] Onyon, *Trans. Faraday Soc.*, 1956, **52**, 80–88.

[132] U S P 292,394 to Courtaulds.

[133] Alexander, Arthur, *'Man-Made Fibre Processing'*, n.d.c., 1966.

[134] See Das, Chatterjee and Palit, *Proc. Roy. Soc.*, 1955, **227**A, 252–8.

[135] Othmer, Kirk, *'Encyclopaedia of Chemical Technology'*, Vol. I (Interscience Publishers), 2nd ed.

[136] Textile Organon, June 1969 and June 1970.

[137] U S P 2,642,418 (1953).

[138] Schildknecht, C. E., *'High Polymers'*, Vol. X, 'Polymer Processes' (Interscience Publishers, 1956), 196.

[139] U S P 2,779,746.

[140] U S P 3,107,971 to Asahi Kasei.

[141] *Oil, Paint and Drug Reporter*, 18 January 1971.

[142] Weaste, R. C. (ed. in chief), *'Handbook of Chemistry and Physics'* (Chemical Rubber Co.), 47th edn., 1966–7.

[143] U S P 2,472,842.

[144] Hayes, *Production of Synthetic Fibres*, American Dyestuff Reporter, 16 May 1960, 79.

[145] USP 3,129,273.

[146] Siclari, F., Fundamental aspects of wet spinning solutions in *Man-Made Fibres—Science and Technology*', Vol. I (Interscience Publishers, 1967).

[147] *Machinery and Production Engineering*, 17 June 1971.

[148] *Daily Express*, 21 April 1971.

[149] Cook, Gordon, '*Handbook of Textile Fibres*' (Merrow Publishing Co., 1964), 3rd ed.

[150] '*Man-Made Fibres—Science and Technology*', Vols. I and II (Interscience Publishers).

[151] *Textile Month*, April 1971.

[152] BP 360,821.

[153] *Rubber Chem. and Tech.*, 1963, **36**.

[154] *Rubber Journal*, September 1970—Ruebensaal.

[155] Hofmann, W., *Rubber Chem. and Tech.*, 1963, **36**, 70.

[156] Smith, W. V., *J. Am. Chem. Soc.*, 1946, **68**, 2046.

[157] Bryant, C. L., *JIRI*, 1970, **4**, No. 5.

[158] Idem, *Kautschuk u. Gummi*, 1962, **15**, No. 2, 34 WT.

[159] Brown, H. P., *Rubber Chem. and Tech.*, 1957, **30**, 1347.

INDEX OF NAMES

The number in front of the parenthesis indicates the page of the book, the number in parenthesis gives the literature reference.

SUBJECT INDEX

34

INDEX OF COMPANIES

The number in front of the parenthesis indicates the page of the book, the number in parenthesis gives the literature reference.

Type set by Gloucester Typesetting Co. Ltd.
Printed in Great Britain at
The Pitman Press · Bath